Biological Microcalorimetry

Biological Microcalorimetry

edited by

A. E. BEEZER

Chelsea College
University of London

1980

ACADEMIC PRESS
A Subsidiary of Harcourt Brace Jovanovich, Publishers
LONDON   NEW YORK   TORONTO   SYDNEY   SAN FRANCISCO

Academic Press Inc. (London) Ltd
24–28 Oval Road
London NW1

US edition published by
Academic Press Inc.
111 Fifth Avenue,
New York, New York 10003

Copyright © 1980 by Academic Press Inc. (London) Ltd

All Rights Reserved

No part of this book may be reproduced in any form,
by photostat, microfilm or any other means,
without written permission from the publishers

British Library Cataloguing in Publication Data

Biological microcalorimetry.
 1. Calorimeters and calorimetry
 2. Biology – Technique
 I. Beezer, A. E.
 574. 1'9285   QH324.9 C/   79–41236

 ISBN 0-12-083550-9

Printed in Great Britain by
John Wright and Sons Ltd,
at the Stonebridge Press, Bristol BS4 5NU

Contributors

D. BACH
    Department of Membrane Research, Weizmann Institute of Science, Rehovot, Israel

A. E. BEEZER
    Department of Chemistry, Chelsea College, University of London, Manresa Road, London SW3 6LX, England

J. P. BELAICH
    Laboratoire de Chimie Bacterienne du CNRS, 31 Chemin Joseph Aiguier, Marseille, France

K. A. BETTELHEIM
    Department of Health, National Health Institute, 52–62 Riddiford Street, Newtown, Wellington South, New Zealand

R. L. BILTONEN
    Departments of Biochemistry and Pharmacology, University of Virginia, School of Medicine, Charlottesville, Virginia 22903, USA

D. CHAPMAN
    Department of Biochemistry and Chemistry, Royal Free Hospital School of Medicine, University of London, 8 Hunter Street, London WC1, England

B. Z. CHOWDHRY
    Department of Chemistry, Chelsea College, University of London, Manresa Road, London SW3 6LX, England

D. J. EATOUGH
    Thermochemical Institute and Department of Chemistry, Brigham Young University, Provo, Utah 84062, USA

M. EFTINK
    Departments of Biochemistry and Pharmacology, University of Virginia School of Medicine, Charlottesville, Virginia 22903, USA

L. D. HANSEN
    Thermochemical Institute and Department of Chemistry, Brigham Young University, Provo, Utah 84062, USA

T. E. JENSEN
    Thermochemical Institute and Department of Chemistry, Brigham Young University, Provo, Utah 84062, USA

R. B. KEMP
    Department of Zoology, University College of Wales, Penglais, Aberystwyth SY23 3DA, Wales

I. LAMPRECHT
    Freie Universität Berlin, FB23 WE1, Habelschwerdter Allee 30, 1000 Berlin 33, German Federal Republic

K. LEVIN
    Västmanslands Lans Landsting, Cetrallasarettet, 721 89 Västerås, Sweden

R. D. NEWELL*
    Department of Oral Medicine and Pathology, Guy's Hospital, St Thomas Street, London SE1 9RT, England

P. L. PRIVALOV
    Institute of Protein Research, Academy of Science of the USSR, Poustchino, Moscow Region, USSR

S. L. RANDZIO
    Institute of Physical Chemistry, Polish Academy of Science, PL-01-224, Warsaw 49, Poland

E. J. SHAW
    Department of Medical Microbiology, St. Bartholomew's Hospital Medical College, London EC1 7BE, England

J. SUURKUUSK
    Thermochemistry Laboratory, Chemical Centre, University of Lund, S-220 07 Lund, Sweden

I. WADSÖ
    Thermochemistry Laboratory, Chemical Centre, University of Lund, S-220 07 Lund, Sweden

R. C. WOLEDGE
    Department of Physiology, University College, University of London, Gower Street, London WC1E 6BT, England

---

* Present address: British Gas Corporation, Research and Development Division, London Research Station, Michael Road, London SW6 2AD, England.

Preface

This book is a natural successor to the splendid book edited by Harry Brown ("Biochemical Calorimetry", Academic Press, 1969). However in this book the scope is widened to include a greater concentration upon microbiological systems, tissue cells and drug interactions. There is, too, a greater concentration upon model studies, e.g. membrane, polymer and heat capacity studies. This reflects, in part, the way in which research activities have changed since the publication of Brown's book and also the extent to which practising biologists have accepted calorimetry.

There is nowhere in this book an attempt to survey the whole field of microcalorimetric research activities. Some of these omitted fields are excellently dealt with elsewhere (e.g. instrumentation and analysis in Wadsö and Spink's contribution to "Methods in Biochemical Analysis" (D. Glick, ed.), vol. 23, Wiley, 1976, and more generally the volume "Calorimetry in the Life Sciences" (I. Lamprecht and B. Schaarschmidt, eds), de Gruyter, 1977).

The selection of topics is entirely the responsibility of the editor and has undoubtedly been influenced by the many friends and colleagues met through mutual interests in Biological Microcalorimetry. This book is therefore the product of the influences of these friends (many of whom have contributed chapters), my graduate students who have moulded my thoughts and attitudes, and of my own graduate studies supervisor, Professor C. T. Mortimer. To all these I express my thanks.

February 1980   A. E. BEEZER

Contents

Contributors	v
Preface	vii
Growth and Metabolism in Bacteria *J. P. Belaich*	1
Growth and Metabolism in Yeasts *I. Lamprecht*	43
Microcalorimetric Studies of Tissue Cells *in vitro* *R. B. Kemp*	113
Studies on Blood Cells *K. Levin*	131
The Heat Production of Intact Organs and Tissues *R. C. Woledge*	145
The Identification and Characterization of Microorganisms by Microcalorimetry *R. D. Newell*	163
Microcalorimetry in Diagnostic Medical Microbiology *K. A. Bettelheim and E. J. Shaw*	187
Microcalorimetric Investigations of Drugs *A. E. Beezer and B. Z. Chowdhry*	195
Some Problems in Calorimetric Measurements on Cellular Systems *I. Wadsö*	247
Calorimetric Studies of Biomembranes and their Molecular Components *D. Bach and D. Chapman*	275
Interpretation of Calorimetric Thermograms and their Dynamic Corrections *S. L. Randzio and J. Suurkuusk*	311

Thermodynamics of Interacting Biological Systems 343
M. Eftink and R. Biltonen

Heat Capacity Studies in Biology 413
P. L. Privalov

Applications of Continuous Titration Isoperibol and Isothermal Calorimetry to Biological Problems 453
L. D. Hansen, T. E. Jensen and D. J. Eatough

Subject Index 477

Growth and Metabolism in Bacteria

J. P. BELAICH

A. Introduction and history

Dubrunfaut (1856) was the first author who reported a quantitative study on heat production by a microbial fermentation. His experiment was performed on a culture of 21 400 litres containing about 3·5 tons of sugar and cannot be considered microcalorimetry. The first usage of a calorimeter was in 1895 when Bouffard applied Berthelot's instrument to measure heat evolved by 1 litre of culture. This first calorimetric experiment was surprisingly good and the heat quantity corresponding to the fermentation of glucose was found to be 26·5 kcal mol^{-1}, a value very close to the one obtained sixty-five years later by Battley (1960c). At the beginning of the century Rubner (1903, 1904, 1908) reported many observations on the thermogenesis of yeast and bacteria, using a primitive calorimeter. The first attempts to measure the enthalpy incorporated in microorganisms were carried out by Tangl (1903) and Terroine and Wurmser (1922), using a calorimetric bomb. The energetic studies of bacterial growth were developed to a great extent by Meyerhof (1924), Hill (1912), Shearer (1921) and Bayne-Jones (1929). Combining respirometric measurements with measurements of heat production, Meyerhof determined the caloric quotient of some phases of bacterial growth, expressed as the ratio of heat produced per gram of oxygen consumed. Hill had developed a differential calorimeter which was later perfected by Bayne-Jones (1929). With this instrument it was found that the order of magnitude of the heat produced by a growing bacterium was 10^{-9} cal per cell. This can be considered as the first determination of the rate of cellular catabolic activity. In addition, it was shown that the thermogenic activity was dependent on the growth phase. The earlier workers were handicapped by the technical limitations of their instruments and by the fact that the kinetic laws of the bacterial growth were not sufficiently known.

The prototype of the modern heat fluxmeter was devised by Tian (1923) and perfected and used for microbiological studies by Calvet (1948) and Prat et al. (1946). The high sensitivity of Calvet's calorimeters permitted observation of the detail in the thermal events which occur when bacteria grow in a complex medium.

With the development of modern equipment as well as the increase in knowledge concerning bacterial growth kinetics, it became possible to obtain valuable information on the significance of heat quantities evolved by growing bacteria (Battley, 1960a, b, c; Forrest et al., 1961; Belaich, 1962, 1963; Boivinet, 1964). More recently Monk and Wadsö (1968) have devised a flow microcalorimeter which allows continuous heat measurements. This apparatus has been widely used and its applications were reviewed recently by Spink and Wadsö (1976).

B. Thermochemistry of microbial growth

The main feature of a living cell is self-replication. This means that a bacterium, for example, is able to generate an absolutely identical bacterium in a finite period of time. When a microbial population is put into a good culture medium containing all the nutrients which are necessary for its growth, the relationship which defines the increase of biomass with time is

$$\frac{dm}{dt} = \mu m \quad (1)$$

with

$$\mu = \frac{0 \cdot 69}{t_{\frac{1}{2}}} \quad (2)$$

In these relationships, m, t, μ and $t_{\frac{1}{2}}$ are the biomass, the time, the Naperian growth rate and t the generation time, respectively. The first work on the energetics of the bacterial growth was carried out by Monod (1942) who defined the growth yield and the generation time. From a quantitative point of view relationships (1) and (2) show that for each $t_{\frac{1}{2}}$ the biomass doubles, i.e. the mass of proteins, nucleic acids, polysaccharides and lipids is increased by a factor of two.

The biosynthetic reactions are endergonic and must be driven by the energy which is supplied by the exergonic reactions of the catabolism (see Fig. 1). We must now pay attention to the definition of the thermodynamic system which we are going to consider. So, if we consider the isolated living cell which exchanges the nutrients necessary for its growth with its surroundings, we must apply the thermodynamic laws of open systems. In contrast, when considering the whole cell culture, which is surrounded by a wall and which does not exchange matter with its exterior, we have a closed system to which we can apply the laws of classical thermodynamics. In the following we shall consider such a

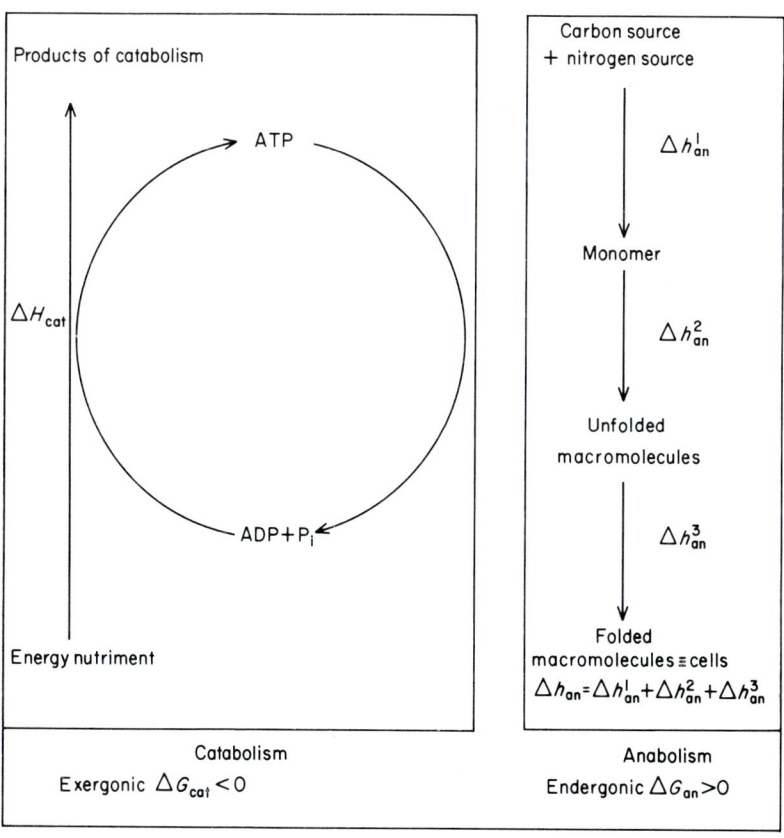

FIG. 1. General metabolic scheme.

closed system constituted by the entire cell culture, i.e. by the living cells plus the culture medium. Moreover, we shall consider that all the biochemical reactions occurring in the culture are at constant pressure and temperature. The system, so defined, can exchange heat only with its surroundings. With these conditions we can write:

$$\Delta G_{met} = \Delta H_{met} - T\Delta S_{met} \tag{3}$$

In this classical formulation of the second law of thermodynamics ΔG_{met}, ΔH_{met}, ΔS_{met} are the variations of free energy, enthalpy and entropy of our system (cell culture + medium) respectively, the subscript "met" means metabolism and T is the absolute temperature.

As the increase in biomass is a spontaneous phenomenon, ΔG_{met} is high and negative, ΔS_{met} is high and positive, and we shall see that ΔH_{met} is negative.

At constant temperature and pressure the heat quantity Q evolved by the culture during the growth is equal to ΔH_{met}. Let us now consider, separately, the exergonic catabolic and the endergonic anabolic reactions, identified in the following relationships by the subscripts "cat" and "an" respectively. We have

$$Q = \Delta H_{met} = Q_{cat} + Q_{an} \tag{4}$$

with

$$Q_{cat} = \sum m_i \, \Delta H_{cat,j} \tag{5}$$

and

$$Q_{an} = \sum n_j \, \Delta H_{an,j} \tag{6}$$

In the two last relationships m_i and n_j are the numbers of moles of reaction for components i and j of catabolism and anabolism respectively which occurred during the growth; and $\Delta H_{cat,j}$ and $\Delta H_{an,j}$ are the corresponding enthalpies of reaction per mole.

The catabolic reactions can be easily described by the following relationship:

$$\text{Energy source} \longrightarrow \text{products of catabolism} \quad \Delta H = \Delta H_{cat} \tag{7}$$

When the energy source and the products of catabolism are simple and well-known molecules such as glucose, organic acids, alcohol, CO_2, the ΔH_{cat} can be simply calculated from the enthalpies of formation of these compounds, given in tabulated thermodynamic data (see Index of values, p. 39). The same applies for the calculation of ΔG_{cat} and ΔS_{cat}. Examples of such calculations are given in Table 1. The estimation of the heat corresponding to the ionization of protons can also be made from the literature data. Thus it is possible in the most simple case to take into account all the chemical events corresponding to the catabolism. When the catabolic reactions are complex, or even unknown, none of these calculations are possible. Such is the case for growth occurring on complex medium, the composition of which is ill defined. There is therefore insufficient knowledge of the nature of the energy source and of the end-products of catabolism to permit calculation.

TABLE 1
Enthalpy variations corresponding to some important catabolic reactions

Catabolic reaction	ΔH_{cat} (kcal mol^{-1})
Lactic fermentation	
1 Glucose →2 lactic acid[a]	− 27·97
Alcoholic fermentation (Entner and Doudoroff)	
1 Glucose →1·7 ethanol + 1·7 CO$_2$ + 0·2 lactic acid[b]	− 32·22
Alcoholic fermentation (Glycolysis)	
1 Glucose →2 ethanol + 2 CO$_2$[c]	− 33·05
Respiration	
1 Glucose + 6 O$_2$ →6 CO$_2$ + 6 H$_2$O[d]	− 670·00

[a] Data from Boivinet (1964).
[b] Data from Belaich (1967).
[c] Data from Battley (1960b).
[d] Theoretical.

Anabolism is more difficult to investigate than catabolism. Two methods can be used for the estimation of the enthalpy variation corresponding to biosynthesis (ΔH_{an}). The first one is based on the heat of combustion of the entire cells. It is well known that the heat of combustion of entire cells is practically identical, or very close to, the heat of combustion of glucose. So it is generally assumed that the cell material is equivalent, from a thermochemical point of view, to glucose. This assumption must now be discussed in detail. The anabolic processes can be summarized by

$$C(H_2O) + NH_4OH \longrightarrow C_xH_yO_zN + nH_2O, \quad \Delta H = \Delta H_{an} \quad (8)$$

From the elemental analysis of microbial cells different stoichiometries can be established for relationship (8). This is shown in Table 2 which was compiled from the data of Battley (1960a), McCarty (1965), Luria (1960) and Burkhead and McKinney (1969). Table 2 is not exhaustive and we can assume that many other stoichiometries with minor differences can be written. The heat of formation of the cell material can be calculated from the elementary analysis and from the heat of combustion of microorganisms. Such heats of formation corresponding to one gram of cell dry weight are also shown in Table 2 from the data of Sedlaczek (1964) and Prochazka *et al.* (1970). It can be seen that the enthalpy of formation of microbial cells varies from 4·07 to 0·29 kcal g^{-1} dry weight.[1] This great difference between the two

[1] From this point all references to grams dry weight cellular material are abbreviated to g or multiples of g.

TABLE 2

Heats of formation of bacterial cells calculated from heats of combustion and enthalpy variations corresponding to the synthesis of 1 g dry cells from glucose + NH_4OH

ΔH combustion = $-3·7$ kcal g^{-1} dry weight (Sedlaczek, 1964)
ΔH combustion = $-5·4$ kcal g^{-1} dry weight (Prochazka et al., 1970)

Bacterial formulation	MW	% C	ΔH^S_f kcal g^{-1}	ΔH^P_f kcal g^{-1}	Reference
$CH_{2·4}O_{0·3}N_{0·25}$	22·7	52	−4·07	−2·37	(a)
$CH_{1·75}O_{0·45}N_{0·20}$	23·75	50	−2·79	−1·09	(b)
$CH_{1·4}O_{0·4}N_{0·2}$	22·6	53	−2·58	−0·89	(c)
$CH_{1·8}O_{0·6}N_{0·2}$	26·2	46	−2·25	−0·55	(c')
$CH_{2·2}O_{0·8}N_{0·2}$	29·8	40	−1·99	−0·29	(d)

Δh_{an} formulation	Δh^S_{an} kcal g^{-1}	Δh^P_{an} kcal g^{-1}	Reference
$CH_2O + 0·25\ NH_4OH \rightarrow CH_{2·4}O_{0·3}N_{0·25} + 0·425\ H_2O + 0·26\ O_2$	−2·18	−0·49	(a)
$CH_2O + 0·2\ NH_4OH \rightarrow CH_{1·75}O_{0·45}N_{0·20} + 0·62\ H_2O + 0·06\ O_2$	−1·75	−0·05	(b)
$CH_2O + 0·2\ NH_4OH \rightarrow CH_{1·4}O_{0·4}N_{0·2} + 0·8\ H_2O + 0·4\ O_2$	−2·02	−0·33	(c)
$CH_2O + 0·2\ NH_4OH \rightarrow CH_{1·8}O_{0·6}N_{0·2} + 0·6\ H_2O$	−1·24	−0·45	(c')
$CH_2O + 0·2\ NH_4OH \rightarrow CH_{2·2}O_{0·8}N_{0·2} + 0·4\ H_2O$	+1·05	−0·65	(d)

Data from: (a) Gunzalus and Schuster (1961); (b) Battley (1960a); (c) and (c') McCarty (1965); (d) Burkhead and McKinney (1969).

ΔH^S_f, ΔH^P_f: Enthalpy of formation in kcal g^{-1} of cells dry weight calculated from the enthalpy of combustion of bacteria dry weight reported by Sedlaczek (1964) and Prochazka et al. (1970) respectively.

Δh^S_{an}, Δh^P_{an}: Enthalpy variation corresponding to the synthesis of 1 g of bacteria dry cells from CH_2O and NH_4OH calculation done with ΔH^S_f and ΔH^P_f respectively.

extreme values comes from the uncertainties in the bacterial formulation and especially from the different heats of combustion reported for whole dry cells. The enthalpy variations Δh_{an} corresponding to the synthesis of 1 g of cellular material from glucose and ammonium hydroxide are also reported in Table 2. The discrepancy between the heats of combustion strongly influences the Δh_{an} value. The extreme values of Δh_{an} are $-2 \cdot 18$ and $+1 \cdot 05$ kcal g^{-1} of cell dry weight synthesized. These simple considerations show that it is very hazardous to predict the heat evolved during the microbial growth (ΔH_{met}) from thermodynamic data in the literature available at the present time even in a simple case such as when the carbon source and the nitrogen source are glucose and ammonium hydroxide respectively. This can be illustrated by the following example.

Consider an organism growing anaerobically on glucose as the sole energy and carbon source and ammonium hydroxide as the nitrogen source. We shall suppose that the fermentation occurs according to the glycolysis pathway. The theoretical growth yield (Y_g) will be 20 g of cell dry weight per mole of glucose fermented. Taking the percentage of cellular carbon as 50 per cent, it is possible to estimate that $10 \cdot 30/12 \cdot 180 = 0 \cdot 14$ mol of glucose is transformed into protoplasm and that, consequently, $0 \cdot 86$ mol of glucose is converted into equal quantities of ethanol and CO_2 each time that 1 mol of glucose is metabolized. So, for each mole of glucose metabolized the Q_{cat} will be: $0 \cdot 86 \times -33 \cdot 05 = -28 \cdot 42$ kcal and the Q_{an}: $20\Delta h_{an}$. The two extreme values of Δh_{an}, $-2 \cdot 18$ and $+1 \cdot 05$ (see Table 2) give Q_{an} as $-43 \cdot 6$ and $+21$ kcal, respectively. Thus the heat quantity evolved during the growth, i.e. $\Delta H_{met} = Q_{cat} + Q_{an}$ will be $-72 \cdot 02$ or $-7 \cdot 42$ kcal mol^{-1} of glucose metabolized, depending on whether the highest or the lowest value of Q_{an} is used for the calculation of ΔH_{met}. This very simple example shows how the uncertainties in the heat of combustion of entire cells can disturb the calculation of the heat quantities evolved during the microbial growth. It is commonly assumed that the cellular carbon is in the same oxidation state as the carbon of carbohydrates; in other words, that ΔH_{an} is negligible provided that the energy source is a sugar. I think, personally, that this assumption is good or, rather, that it is not bad. However, no data are presently available to support it.

The second method which can be used for the estimation of Δh_{an} involves the scrutinization of anabolism step by step. As is shown in Fig. 1, we can consider that anabolism occurs in three main steps. The first is the conversion of the carbon and nitrogen sources into monomers, i.e. mainly of amino acids, purine and pyrimidine bases and fatty acids. The second step is the polymerization of monomers giving unfolded

TABLE 3
Enthalpy incorporated in the cell

Nature of the polymer	% of dry weight[a]	Mean of molecular weight	Quantity of compound (μM g^{-1} of dry cell)	$\Delta h^1{}_{an}$[b] cal g^{-1}	$\Delta H^2{}_{an}$ (contribution kcal per monomer)	$\Delta h^2{}_{an}$ (cal g^{-1} dry weight of cells)	$\Delta H^3{}_{an}$ (contribution kcal per monomer)	$\Delta h^3{}_{an}$ (cal g^{-1} dry weight of cells)
Protein	60	110	5454	−107	+ 2[c]	+10.9	−0.261[g]	−1.42
Nucleic acid	20	300	666	+ 17	+ 4[d]	+ 2.6	−3.43[g]	−2.28
Polysaccharides	10	166	602	0	+ 3[e]	+ 1.8	0	
Lipids (monomer, dipalmitoylphosphatidylethanolamine)	10	675	148	+172	+10[f]	+ 1.48	−8.65	−1.28
TOTAL				+ 82		+16.78		−4.96

$\Delta h^1{}_{an}$ are calculated from Table 4.
$\Delta h^2{}_{an}$ are calculated from the enthalpies of hydrolysis of the main biological bonds.
$\Delta h^3{}_{an}$ are calculated from the data of Table 5.
[a] From Gunsalus and Schuster (1961).
[b] See Table 4.
[c] Taking −2 kcal as a mean value of the enthalpy of hydrolysis of the peptide bond (from Sturtevant, 1953; Ravitscher et al., 1961).
[d] Taking −4 kcal as a mean value of the enthalpy of hydrolysis of the ester phosphoric bond (from Oesper, 1951; Smullach et al., 1959; Lehninger, 1965).
[e] From Lazniewski (1959) and Sozaburo (1963).
[f] This is the sum of the enthalpy of hydrolysis of two acyl ester bonds (from Calvet and Prat, 1956) and two ester phosphoric bonds (Lehninger, 1965).
[g] See Table 5.

macromolecules (protein, nucleic acids). The last event consists of the spatial organization of macromolecules and each step is accompanied by an enthalpy variation so that

$$\Delta h_{an} = \Delta h^1_{an} + \Delta h^2_{an} + \Delta h^3_{an} \quad \text{(see Fig. 1)}$$

Let us consider now how it is possible to quantify Δh^1_{an}, Δh^2_{an} and Δh^3_{an}.

Δh^1_{an} is the enthalpy variation corresponding to the synthesis of monomers from carbon and nitrogen sources. Table 3 shows the average content of polymers and monomers in bacterial cells. From the heat of formation of each monomer it is possible to estimate the Δh^1_{an} contribution corresponding to the synthesis of each monomer according to the relationship:

$$\text{glucose} + NH_4OH \longrightarrow \text{monomer} \quad (9)$$

Table 4 shows the stoichiometric relationship corresponding to the reaction of synthesis of amino acids, purine and pyrimidine bases, ethanolamine and fatty acids, from glucose and ammonium hydroxide. The contribution $\Delta h^1_{an, protein}$ corresponding to the synthesis of amino acids contained in 1 g of cell dry weight is obtained from the sum of all the $\Delta h^1_{an, j}$ according to

$$\Delta h^1_{an, protein} = \sum_j \Delta h^1_{an, j}$$

The synthesis of all the amino acids corresponding to 1 g of cell dry weight is accompanied by an enthalpy decrease of -0.107 kcal. This means that the process is slightly exothermic. The $\Delta h^1_{an, na}$ corresponding to the synthesis of monomers of nucleic acids (na) is impossible to calculate with precision because the enthalpies of formation of pyrimidine bases are not available in the literature. So we have estimated $\Delta H^1_{an, na}$ taking into account only the purine bases and we have done the calculation assuming that the $\Delta H^1_{an, j}$ for the synthesis of one base is 26 kcal mol^{-1}. The $\Delta h^1_{an, na}$ with this approximation is $+0.017$ kcal. Last but not least is the $\Delta h^1_{an, lipid}$ corresponding to the synthesis of lipids. The monomers chosen for the calculation of this Δh value are the components of the dipalmitoic phosphatidyl ethanolamine which is the main component of the membrane of *Escherichia coli*. The $\Delta h^1_{an, lipid}$ is positive and is $+0.172$ kcal. The Δh^1_{an} contribution resulting from the synthesis of polysaccharides from glucose is evidently negligible and is not considered in this contribution. So, the Δh^1_{an} corresponding to the synthesis of all the monomers contained in 1 g of cell dry weight is $+0.082$ kcal (see Table 3). This value is probably erroneous as not all

TABLE 4

Monomer	Synthetic reaction from glucose and ammonium hydroxide	$\Delta H^1_{an,f}$ kcal mol^{-1} monomer synthesized	Quantity of monomer g^{-1} dry weight (μM)^b	$\Delta h^1_{an,f}$ kcal g^{-1} dry weight
PROTEIN				
Aspartic acid	0.66 C$_6$H$_{12}$O$_6$ + NH$_4$OH + O$_2$ → C$_4$H$_7$O$_4$N + 3 H$_2$O	−150.00	512	−0.076
Lysine	C$_6$H$_{12}$O$_6$ + 2 NH$_4$OH → C$_6$H$_{14}$O$_2$N$_2$ + 4 H$_2$O + O$_2$	+39.98	362	+0.014
Methionine	0.83 C$_6$H$_{12}$O$_6$ + NH$_4$OH + H$_2$SO$_4$ → C$_5$H$_{11}$O$_2$NS + 3 H$_2$O + 2.5 O$_2$	+159.35	175	+0.027
Threonine	0.66 C$_6$H$_{12}$O$_6$ + NH$_4$OH → C$_4$H$_9$O$_3$N + 2 H$_2$O	−41.46	242	−0.010
Isoleucine + leucine	C$_6$H$_{12}$O$_6$ + NH$_4$OH → C$_6$H$_{13}$O$_2$N + 2 H$_2$O + 1.5 O$_2$	+99.7	650	+0.064
Glutamate	0.83 C$_6$H$_{12}$O$_6$ + NH$_4$OH → C$_5$H$_9$O$_4$N + 3 H$_2$O	−107.84	545	−0.058
Proline	0.83 C$_6$H$_{12}$O$_6$ + NH$_4$OH → C$_5$H$_9$O$_2$N + 3 H$_2$O + 0.5 O$_2$	+8.18	240	+0.0019
Arginine	C$_6$H$_{12}$O$_6$ + NH$_4$OH → C$_6$H$_{14}$O$_2$N$_4$ + 9 H$_2$O + 0.5 O$_2$	−114.9	280	−0.032
Serine	0.5 C$_6$H$_{12}$O$_6$ + NH$_4$OH + 0.5 O$_2$ → C$_3$H$_7$O$_3$N + 2 H$_2$O	−72.5	317	−0.022
Glycine	0.33 C$_6$H$_{12}$O$_6$ + NH$_4$OH + 0.5 O$_2$ → C$_2$H$_5$O$_2$N + 2 H$_2$O	−77.67	415	−0.032
Cysteine	0.5 C$_6$H$_{12}$O$_6$ + NH$_4$OH + H$_2$SO$_4$ → C$_3$H$_7$O$_2$NS + 3 H$_2$O + 1.5 O$_2$	+113.56	87	+0.0098
Alanine	0.5 C$_6$H$_{12}$O$_6$ + NH$_4$OH → C$_3$H$_7$O$_2$N + 2 H$_2$O	−33.4	658	−0.022
Valine	0.83 C$_6$H$_{12}$O$_6$ + NH$_4$OH → C$_5$H$_{11}$O$_3$N + 6 H$_2$O + 0.5 O$_2$	+114.41	285	+0.033
Tyrosine	1.5 C$_6$H$_{12}$O$_6$ + NH$_4$OH → C$_9$H$_{11}$O$_2$N + 6 H$_2$O + 0.5 O$_2$	−30.61	110	−0.003
Phenylahenine	1.5 C$_6$H$_{12}$O$_6$ + NH$_4$OH → C$_9$H$_{11}$O$_2$N + 6 H$_2$O + O$_2$	+18.28	170	+0.003
Tryptophane	1.83 C$_6$H$_{12}$O$_6$ + 2 NH$_4$OH → C$_{11}$H$_{11}$O$_2$N$_2$ + 10.5 H$_2$O + 0.5 O$_2$	−89.32	54	−0.005
Histidine	C$_6$H$_{12}$O$_6$ + 3 NH$_4$OH → C$_6$H$_9$O$_2$N$_3$ + 7 H$_2$O + 2 H$_2$	Not known	50	

$$\Delta h^1_{an,\,protein} = \sum \Delta h^1_{an,\,f} = -0.107$$

NUCLEIC ACID

Adenine	$0.83\ C_6H_{12}O_6 + 5\ NH_4OH \rightarrow C_5H_5N_5 + 10\ H_2O + 5\ H_2$	$+28.88$	$+0.017$
Guanine	$0.83\ C_6H_{12}O_6 + 5\ NH_4OH \rightarrow C_5H_5ON_5 + 9\ H_2O + 6\ H_2$	$+25.31$	
	No values are available for cytosine and thymine		
	Mean for 1 base =	$+26$	666^a
			$\Delta h^1_{an,\ na} = +0.017$

LIPID

Palmitic acid	$2.66\ C_6H_{12}O_6 \rightarrow C_{16}H_{32}O_2 + 7\ O_2$	$+593$	296^a $+0.175$
Glycerol	$0.5\ C_6H_{12}O_6 + H_2 \rightarrow C_3H_8O_3$	-1.23	148^a -0.0002
Ethanolamine	$0.33\ C_6H_{12}O_6 + NH_4OH + H_2 \rightarrow C_2H_7ON + 2\ H_2O$	-15.85	148 -0.0023
			$\Delta h^1_{an,\ lipid} = +0.172$

[a] Data from Table 5.
[b] From Roberts *et al.* (1955).

the data necessary for its estimation are yet available in the literature. Nevertheless, this value gives a good approximation of the enthalpy variation accompanying the synthesis of all the monomers from glucose and ammonium hydroxide as energy and nitrogen sources. The same type of calculation can be done when the carbon and energy sources are oxidized to a different extent than glucose and ammonium hydroxide. Such is the case, for example, for growth on organic acids or on paraffin as sources of carbon and energy, and on nitrate as a source of nitrogen.

The Δh^2_{an} represents the enthalpy variation which occurs during the polymerization of monomers. From the data on the enthalpy of hydrolysis of the main bonds involved in the biological macromolecules (peptide, glycoside, acylester and phosphodiester bonds) it can be estimated that Δh^2_{an} equals about 0.017 kcal g^{-1} of cell dry weight (see Table 3).

The enthalpy accompanying the spatial organization of linear macromolecules and the formation of membrane bilayer structure can be estimated from the results of scanning calorimetry. It is shown in Table 5 that an average value of -0.261 kcal per amino acid can be taken for the heat of folding of protein. The averages of -3.43 kcal per base of nucleic acid and -8.65 kcal per phospholipid can be used to account for the spatial organization of nucleic acids and membrane bilayers respectively. With these values Δh^3_{an} has been estimated at -0.005 kcal g^{-1} of cell dry weight.

From the estimation of Δh^1_{an}, Δh^2_{an} and Δh^3_{an} it is possible to predict that the order of magnitude of Δh_{an} is $+0.094$ kcal g^{-1} of cell dry weight. The value of Δh_{an} so calculated can be compared with the values calculated from the heat of combustion of entire cells reported in Tables 2 and 3. It can be seen that this value is close to those calculated from the heat of combustion reported by Prochazka et al. (1970). Using $+0.094$ as Δh_{an} the Q_{an} corresponding to the example cited above is $20 \times 0.094 = 1.88$ kcal so the ΔH_{met} becomes:

$$-28.42 + 1.88 = -26.54 \text{ kcal mol}^{-1} \text{ of glucose metabolized}$$

The low value of Δh_{an} estimated according to the second method of calculation is in good agreement with the assumption that Q_{an} is negligible and that the cellular material is equivalent, from a thermochemical point of view, to carbohydrate.

It is now possible to formulate a relationship which allows the estimation of ΔH_{met} from the following three factors: ΔH_{cat}, the molecular growth yield (Y) and the quantity of carbon incorporated in the cellular

TABLE 5
Enthalpies of denaturation of biological macromolecules

Macromolecule	MW g	T °C	ΔH of denaturation kcal mol^{-1}	ΔH of denaturation cal g^{-1}	References
PROTEIN					
Lyzozyme	1.43×10^4	45	+55	+3.84	O'Reilly et al. (unpublished)
Horse serum albumin	6.9×10^4	55	+90	+1.30	Privalov and Monaselidze (1963)
Pepsin	3.5×10^5	35	+69	+0.197	Buzzell and Sturtevant (1952); Sturtevant (1954)
Trypsin	2×10^4	25	+8	+0.400	Gutfreund and Sturtevant (1953)
Ribonuclease A	1.37×10^4	43	+70	+5.109	Beck et al. (1965)
				Average +2.17	
NUCLEIC ACID					
P.S. fluorescens DNA		25	Denaturation	+7.83 kcal per base pair	Bunville et al. (1965)
Poly A poly U		25	Heat of mixing	−5.9 kcal	Steiner and Kitzinger (1962)
			Mean for heat of mixing per base pair	−6.86 kcal	
LIPIDS					
Dipalmitoylphasphatidylcholine		41	Heat of transition +8.65 kcal mol^{-1}		Chapman (1973)

ΔH^3_{an} corresponding to one amino acid (MW 100). The contribution is taken to be equal to $-2.17 \times 100 = -217$ cal per monomer (see Table 3).
ΔH^3_{an} corresponding to one base will be taken equal to -3.43 kcal.
ΔH^3_{an} accompanying the formation of lipid bilayer membrane structure will be taken equal to -8.65 kcal mol^{-1} phospholipid.

material (x). Let us consider an organism growing on glucose as the sole energy source and using a simple or complex nitrogen source. The simplest would be ammonium hydroxide while a more sophisticated source would be a mixture of all the amino acids, purine and pyrimidine bases, i.e. all the nitrogen monomers necessary for the synthesis of macromolecules. To simplify, we shall call "minimal medium" the medium containing only glucose and ammonium hydroxide and "complex medium" the one containing glucose and all the nitrogen compounds. In both cases the media will be supplemented with all the classical oligo elements and with convenient buffering molecules. Whatever the medium used for growth, the organism considered will incorporate some of the glucose into its cellular material. In the case of growth on minimal medium, 100 per cent of the cellular carbon will come from glucose, while in the case of complex medium only a percentage, x, the quantity depending on the richness of the medium, will be derived from glucose. When 1 mol glucose is metabolized, the part used for synthesis, i.e. which is anabolized, is $\frac{1}{2} Y_g (x/100)(30/12)(1/180)$. In this relationship, Y_g is the molecular glucose growth yield, $\frac{1}{2}$ means that cellular carbon represents 50 per cent of dry weight; 30, 12 and 180 are respectively the molecular weight of (CH_2O), the atomic weight of C, and the molecular weight of glucose. So the quantity of glucose catabolized is

$$\left(1 - \tfrac{1}{2} \times Y_g \times \frac{x}{100} \times \frac{30}{12} \times \frac{1}{180}\right) \quad \text{or} \quad (1 - 6{\cdot}94 \times 10^{-5} \, x Y_g)$$

The heat quantity associated with the metabolism of 1 mol glucose is

$$\Delta H_{met} = (1 - 6{\cdot}94 \times 10^{-5} \, x Y_g) \Delta H_{cat} + Y_g \Delta h_{an} \tag{10}$$

According to relationship (10) each time that 1 mol glucose is metabolized, the heat quantity evolved, the Q_{cat} and the Q_{an} are: $(1 - 6{\cdot}94 \times 10^{-5} \, x Y_g) \Delta H_{cat}$ and $Y_g \Delta h_{an}$, respectively.

Notice that x and Δh_{an} are strongly dependent on the nature of the culture medium. In the minimal medium these two quantities are 100 and 0·094 respectively. In the complex medium x and Δh_{an} are always lower than 100 and 0·094. The lowest values for these two quantities would be attained in a complex medium supplying all the monomers, including lipids. In such a case x and Δh_{an} would be theoretically zero and 0·011 (see Table 4) respectively. Unfortunately, it seems practically impossible to obtain experiments where $x = 0$. Belaich and Senez (1965b) reported a well-defined growth of *Zymonomas mobilis* on a complex medium containing high concentrations of a yeast extract and a bactopeptone as nitrogen and carbon sources and glucose as the sole energy source. Even in such a rich medium x was measured as 48 per cent. This relatively high x value is probably due to the incorporation of some of the glucose into polysaccharides, into the sugar-moiety of nucleic acids and also into the glycerol and the fatty acids of lipid. Table 6 shows the ΔH_{met}, Q_{cat} and Q_{an}, calculated in some examples of experimental and theoretical growth. The results used for these calculations were reported by Battley (1960c), Boivinet (1964), Belaich (1967), Belaich et al. (1968), Murgier and Belaich (1971) and Forrest (1969). The main feature of Table 6 is that the absolute value of Q_{an} always represents a small percentage of Q_{cat} whatever the nature of the medium, the aerobic or anaerobic nature of the incubation, and the organism studied. The average is about 2 per cent. So we can, as a first approximation, neglect Q_{an} and write:

$$\Delta H_{met} = Q_{cat} \qquad (11)$$

In other words, when bacteria grow on glucose as the energy source the heat quantity evolved is equal to the heat associated with catabolism. It is evident that this is true only when a sugar is the energy source. The same type of calculations can be done when a substrate oxidized to a different extent than glucose is used as the energy source.

The experimental values of ΔH_{met} which have been called the ΔHs of conservative reactions by Battley (1960c) and K_{exp} by Belaich et al. (1968) are in very good agreement with the ΔH_{met} calculated using relationship (10) and the Q_{an} of Table 3. The differences observed between the experimental and calculated values are both positive and negative and, except in the case of maltose, the mean which is given by

$$[\Delta H_{met}(calc) - \Delta H_{met}(exp)]/[\Delta H_{met}(calc)] \times 100$$

is less than ± 5 per cent. These slight differences are probably due to experimental errors. Thus as the errors on values of ΔH_{met} are of the same order of magnitude as $(Q_{an}/Q_{cat}) \times 100$ it seems, at first sight, impossible to measure Q_{an} by direct calorimetric experiments. This last fact demonstrates the need for very precise determinations of heats of combustion of bacteria which alone would permit the estimation of Δh_{an}.

C. Microcalorimetric growth studies

When a small quantity of bacteria is inoculated into a microcalorimetric vessel containing a convenient culture medium the growth is always

TA**
Enthalpies corresponding to the metabolism

Organism and medium	Catabolic reaction	ΔH_{cat} (calculated, kcal mol^{-1})	Y_g	x (%)	Minim medium Δh_{an} $= 0.094$
Yeast, minimal	1 Glucose → 2 ethanol + 2 CO$_2$	−33.05	22[a]	100[a]	+2.0?
Yeast, minimal	1 Glucose → 2 ethanol + 2 CO$_2$	−33.05	22.75[b]	100[a]	2.2?
Yeast, complex ideal	1 Glucose → 2 ethanol + 2 CO$_2$	−33.05	22[a]	0[a]	
Yeast, complex experimental	1 Glucose → 2 ethanol + 2 CO$_2$	−33.05	22.7[c]	50[c]	
Yeast, complex experimental	1 Fructose → 2 ethanol + 2 CO$_2$	−34.6	26[d]	50[d]	
Yeast, complex experimental	1 Galactose → 2 ethanol + 2 CO$_2$	−31.0	26[d]	50[d]	
Yeast, complex experimental	1 Maltose → 4 ethanol + 4 CO$_2$	−85.48	40[d]	50[d]	
Yeast, minimal	1 Glucose + 6 O$_2$ → 6 H$_2$O + 6 CO$_2$	−670	44.57[b]	100[a]	4.28
Z. mobilis, complex	1 Glucose → 1.8 ethanol + 1.8 CO$_2$ + O$_2$ lactate	−32.22	6.5	48	
S. faecalis, complex	1 Glucose → 2 lactate	−27.97	15[e]	0[a]	
S. lactis, complex	1 Glucose → 2 lactate	−27.97	19.6[f]	0[a]	
E. coli, minimal	1 Glucose + 6 O$_2$ → 6 CO$_2$ + 6 H$_2$O	−670[a]	100[a]	100[a]	9.4
E. coli, complex	1 Glucose + 6 O$_2$ → 6 CO$_2$ + 6 H$_2$O	−670[a]	380[a]	0	

* $z = \left[\dfrac{\Delta H_{met}(\text{calc}) - \Delta H_{met}(\text{exp})}{\Delta H_{met}} \right]$

[a] Theoretical values.
[d] Data from Murgier and Belaich (1971).

e microbes growing on different culture media

ol of abolized Complex medium Δh_{an} 0·011 kcal	Q_{cat} when 1 mol of substrate is metabolized	ΔH_{met} (calculated, kcal mol^{-1})	ΔH_{met} (experimental, if available)	(Q_{an}/Q_{cat}) × 100	z × 100*
—	− 28·00	− 25·93	—	7·5	—
—	− 27·60	− 25·37	− 23[b]	8·2	+ 8
0·24	− 33·05	− 32·81	—	0·72	—
0·25	− 30·44	− 30·19	− 28·5[c]	0·82	+ 5·5
0·28	− 31·47	− 31·19	− 29·67[d]	0·88	+ 4·8
0·28	− 28·20	− 27·92	− 29·75[d]	0·59	− 6·2
0·44	− 73·61	− 73·17	− 60[d]	0·59	+18·6
—	−462·75	−458·47	−479[b]	0·1	− 4·5
0·071	− 31·52	− 31·45	− 32·00	0·2	− 1·5
0·21	− 27·97	− 27·76	− 27·22[e]	—	+ 1·9
0·21	− 27·97	− 27·76	− 28·9[f]	0·7	− 4·1
—	−205·02	−195·8	—	4·6	—
4·18	−670	−665·82	—	4·6	—

ata from Battley (1960c).   [c] Data from Belaich et al. (1968).
ata from Forrest et al. (1961).   [f] Data from Boivinet (1964).

accompanied by a production of heat which is easily detected by modern instruments. The curves showing such heat quantities versus time are called growth thermograms. The study of bacterial growth by calorimetry can be done with batch adiabatic, isothermal calorimeters or flow isothermal instruments. A voluminous literature (Calvet and Prat, 1956; Forrest, 1961; Calvet, 1962; Benzinger and Kitzinger, 1963; Monk and Wadsö, 1968) exists on the different calorimeters commercially available which are now all suitable for studies of bacterial growth. I shall not develop here the theoretical description of materials and the procedures of utilization which are described at length in specialized literature. I shall describe only some of the results obtained by microcalorimetry.

1. ORGANISMS, CULTURE MEDIUM AND PHYSIOLOGICAL GROWTH CONDITIONS

Theoretically, all microbial organisms can be studied by microcalorimetry. The only limitations are due to technical difficulties in stirring, bubbling a gas and illuminating the calorimetric vessel. The simplest experimental conditions are those required for anaerobic studies. For this reason the largest quantity of results of bacterial calorimetry is available from studies of fermentations.

The earlier thermograms were derived from studies of bacteria, whose metabolism was unknown and which grew on complex media. The experimental growth conditions were not described precisely. Thus, the first thermograms obtained with modern microcalorimeters were often complex, revealing many peaks as shown in Fig. 2 (Boivinet, 1964). It became evident that growth conditions must be clearly defined for quantitative interpretations of heat evolution accompanying the bacterial

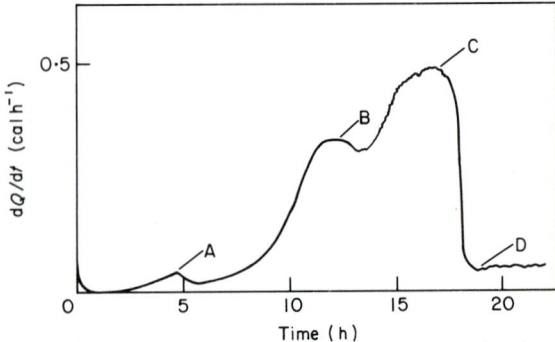

FIG. 2. Growth thermograms of *Aerobacter aerogenes*. The growth was successively limited by air, acid production and glucose. The four peaks A, B, C, D correspond to an aerobic growth phase, the saturation of the buffer, the limitation by the glucose and a residual unknown final metabolism. (Boivinet, 1964.)

growth. The main conditions which must be controlled can be summarized as follows.

a. *Anaerobiosis and aerobiosis.* For anaerobiosis the culture medium must be carefully deaerated. Air can be replaced by an inert gas such as nitrogen or argon. For aerobic experiments it is necessary to be sure, by means of control experiments, that aerobiosis is complete from the beginning to the end of the growth.

b. *pH.* The medium must be well buffered and it is necessary to know the different heats of ionization which are necessary to correct the experimental heat quantities.

c. *Physiological conditions.* The limiting growth factor and the energy source must be known. Different limiting factors can be used: energy or nitrogen sources, phosphorus sulphur, vitamins, etc. In all cases it is necessary to perform some classical growth studies for the determination of the different growth yields. Inhibition by end-products must be taken into account.

Some of these conditions cannot always be fulfilled as, for example, in growth studies on natural complex media or studies on mixed microbial populations. The thermogram data are therefore difficult to interpret. However, as can be seen below, microcalorimetry can in such cases give valuable qualitative information.

2. GROWTH LIMITED BY THE ENERGY SOURCE

The first quantitative study of the energetics of bacterial growth limited by the energy source was reported by Forrest *et al.* (1961) who showed that the kinetics of the heat evolution by a growing culture of *Streptococcus faecalis* is identical to the kinetics of the increase in biomass (Fig. 3). The experimental determination of the ΔH_{met} was in excellent agreement with the pioneer work of Battley (1960a, b, c) and showed, as demonstrated in the preceding section, that the experimental heat quantity is very close to the enthalpy change accompanying the catabolic reaction. The work of Forrest *et al.* (1961) was performed with an adiabatic differential microcalorimeter recording heat quantities versus time. The second series of quantitative investigations was carried out by Belaich (1962, 1963, 1967) using Calvet's microcalorimeter. This instrument is not exactly a calorimeter but a fluxmeter. The theory and practice of such instruments were extensively described by their developers (Calvet, 1962; Benzinger and Kitzinger, 1963; Wadsö, 1970). The principal and useful property of these new types of microcalorimeters is to record directly the heat flux (dQ/dt) evolved by a bacterial culture. Figure 4 shows a typical thermogram for a bacterial growth

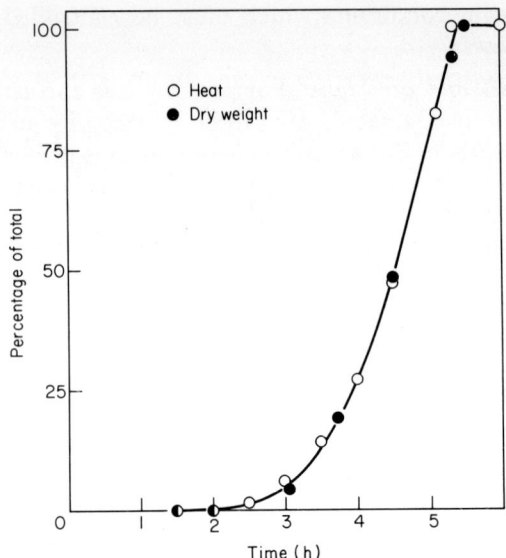

FIG. 3. Heat production in a substrate-limited culture of *S. faecalis* with glucose as energy source (Forrest *et al.*, 1961).

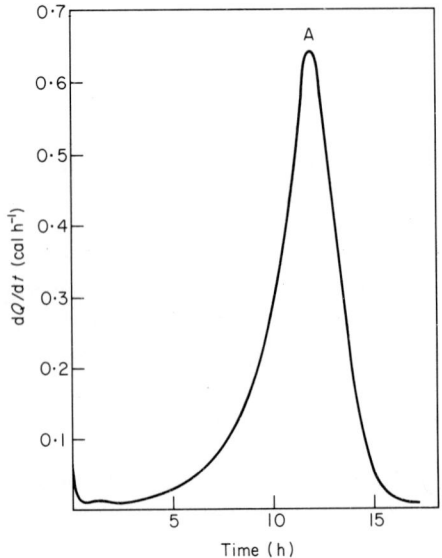

FIG. 4. Heat production in a substrate-limited culture of *Z. mobilis* with glucose as energy source (Belaich, 1963).

limited by the energy source. This Figure was obtained with the bacterium *Zymomonas mobilis* growing anaerobically in a complex medium containing glucose as the sole energy source. After the short initial break (time 0) due to the inoculation, the microcalorimetric curve shows a lag phase, the length of which depends on the age of the cells and the inoculum density. The lag phase is followed by an exponential phase identical to the log phase obtained by conventional turbidimetric measurement. When the concentration of the energy source becomes limiting (about 1 h before the point A in Fig. 4) the thermogram comes back to the base line, which is reached when the glucose is completely exhausted. The interpretation of such thermograms shows, as reported in Fig. 5, that the quantity of cells and the heat evolved are linearly proportional to the quantity of glucose metabolized. The ΔH_{met} (see Table 6) obtained in anaerobiosis with studies performed firstly with *Z. mobilis* and later with *Saccharomyces cerevisiae* (Belaich, 1962, 1963; Belaich *et al.*, 1968; Murgier and Belaich, 1971) confirm the previous results of Battley (1960a, b, c) and Forrest (1969). The same type of results were also obtained by Boivinet (1964) with *Streptococcus lactis*. Belaich (1967) and Belaich *et al.* (1968, 1971; Belaich and Belaich, 1976b) used the properties of Calvet's fluxmeters to calculate the affinity of the entire microbial cells for their energy substrate. The cellular rate of catabolic activity A_c, i.e. the quantity of energy substrate metabolized per gram dry weight and per unit of time is given by

$$A_c = \frac{dS_{met}}{dt} \frac{I}{m} \qquad (12)$$

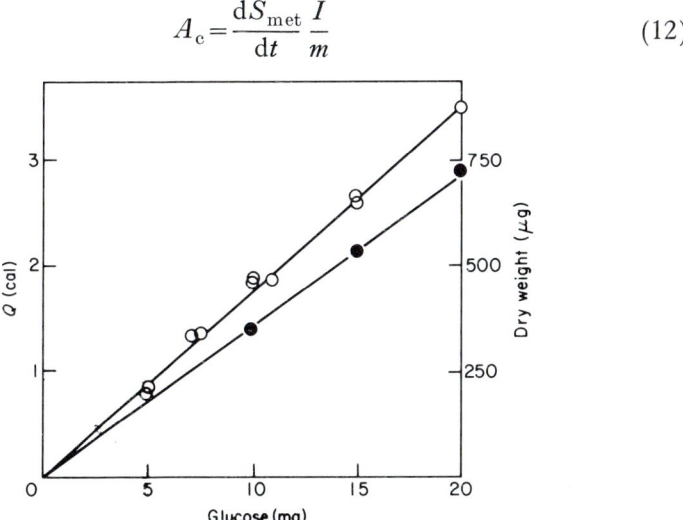

FIG. 5. Heat evolved and dry weight of cells produced as a function of glucose metabolized by growing cultures of *Z. mobilis* (Belaich *et al.*, 1968).

with
$$m = Y_s S_{met} \qquad (13)$$

In these relationships S_{met}, m and Y_s are the quantity of energy substrate metabolized by the culture, the biomass and the molecular growth yield of the organism studied growing on the energy substrate S, respectively.

The heat quantity Q_t at any time t of the thermogram is proportional to S_{met}:
$$Q_t = K_{exp} S_{met} \qquad (14)$$

the constant K_{exp} of proportionality being the ΔH_{met}.

From equations (12), (13) and (14) A_c can be expressed as
$$A_c = \frac{dQ_t}{dt} \frac{I}{Q_t} \frac{I}{Y_s} \qquad (15)$$

The concentration of the energy substrate of the culture medium at any time is also given from the thermogram by
$$[S]_t = \frac{I}{v} \frac{Q_T - Q_t}{K_{exp}} \qquad (16)$$

where Q_T is the total heat quantity evolved by the culture and v is the volume of the culture. It can be seen (equations (15) and (16)) that it is possible to know A_c and $[S]_t$ at any time from the thermogram. Figure 6 shows A_c evolution versus the glucose concentration in the case of *Z. mobilis* growth and the double reciprocal plot obtained with both *Z. mobilis* and *S. cerevisiae* (Belaich et al., 1968).

Both the microcalorimetric technique and the method of calculation to determine K_m and V_{max} for entire cells and their energy substrate were employed for studies of bacteria and yeasts by Belaich et al. (1968, 1971), Murgier and Belaich (1971) and Belaich and Belaich (1976b). The kinetics obtained showed that in the case of anaerobic growth of *E. coli* at pH 7·8 (Belaich and Belaich, 1976b) with glucose, galactose and lactose as energy substrates the cellular rate of catabolic activity is a perfect hyperbolic function of the external sugar concentration at all concentrations used. This is in remarkable agreement with the earlier work of Monod (1942) who predicted in 1942, using primitive classical growth techniques, such types of kinetics. The Lineweaver–Burk plots of some other cases studied produced deviations in the straight line in the region of high substrate concentrations. It was previously observed that for *Z. mobilis* growing on glucose (Belaich et al., 1968) and *S. cerevisiae* growing on galactose (Murgier and Belaich, 1971) the cellular rate of catabolic activity becomes essentially

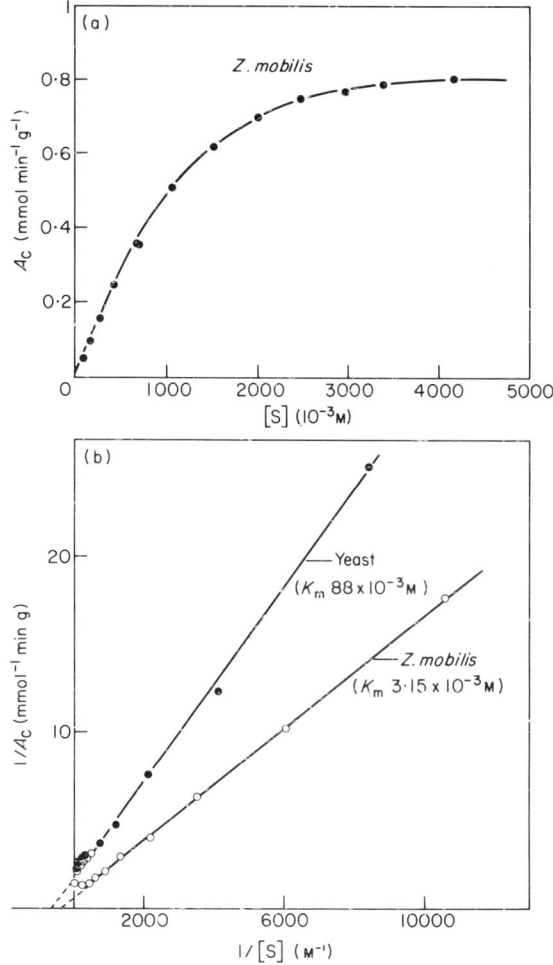

FIG. 6. (a) Cellular rates of catabolic activity (A_c) versus extracellular glucose concentration [S] in *Z. mobilis*. (b) Lineweaver–Burk plots of the curve represented in (a) and of a similar curve obtained with *S. cerevisiae*. (Belaich et al., 1968.)

constant and independent of energy substrate concentrations at high levels of concentration. In the case of *E. coli* fermenting fructose (Belaich and Belaich, 1976b) and *S. cerevisiae* growing on glucose (Belaich et al., 1968) the kinetics showed a marked inhibition at high substrate concentrations. Lastly, an allosteric kinetic relationship was reported for *S. cerevisiae* using maltose (Murgier and Belaich, 1971). Table 7 presents the K_m values and the V_{max} values obtained for all the cases studied. The measured K_m values vary by two orders of magnitude.

TABLE 7
Comparison of the kinetics determined by microcalorimetry for some microbial transport systems

Organism	Substrate	Shape[a] of the curve giving A_c	K_m (M)	V_{max} (extrapolated, mmol min^{-1} g^{-1})[b]	V_{max} (measured, mmol min^{-1} g^{-1})[c]
Z. mobilis[d]	Glucose	No	3.1×10^{-3}	1.818	0.816
S. cerevisiae[e]	Glucose	No	1.9×10^{-3}	0.597	0.597
	Fructose	Yes	4×10^{-3}	0.550	0.550
	Galactose	No	3.2×10^{-3}	0.550	0.180
	Maltose	No	—	—	0.180
E. coli[f]	Glucose	Yes	5.4×10^{-5}	0.27	0.27
	Galactose	Yes	1.1×10^{-3}	0.28	0.20
	Lactose	Yes	2.2×10^{-4}	0.12	0.09
	Fructose	No	7.5×10^{-5}	0.14	0.19

[a] Yes or no: A_c is (or is not) a hyperbolic function of the energy substrate concentration during all the growth.
[b] These values are extrapolated from the Lineweaver–Burk plots.
[c] These values are the maximum rates experimentally obtained.
[d] Belaich *et al.* (1968).
[e] Murgier and Belaich (1971).
[f] Belaich and Belaich (1976b).

The reciprocals of the K_m values reflect the affinities of whole cells for their energy substrates. It was assumed that, when the organism scrutinized has only one transport system for the energy substrate studied, the K_m and V_{max} obtained by microcalorimetry are the kinetic parameters for permease.

Another observation that it is possible to make from the results of Table 7 is that the V_{max} values are inversely proportional to the affinities. Thus, the lower and the higher affinities for glucose are presented by *Z. mobilis* and *E. coli* respectively. It is quite the contrary for the V_{max} values. The same observation can be made for the fermentation of fructose and galactose by *E. coli* and *S. cerevisiae*. It seems that a compensation phenomenon exists so that a low affinity of the transport system is compensated for by a high activity in the permease system.

As described above, the typical thermogram of an energy limited growth is presented by Fig. 4. Generally the thermogram returns to absolute zero when all the energy substrate has disappeared. However, in some cases a second thermal event has been seen during the stationary phase in anaerobiosis. Such an observation was reported practically simultaneously by Senez and Belaich (1963), who worked on the anaerobic growth of *E. coli*, and by Boivinet (1964) on *Aereobacter aerogenes*. The first interpretation given by Belaich and Senez (1963) and later adopted by Forrest (1969) was that this second heat evolution, which was not accompanied by any increase in biomass, was due to the degradation of some reserve material accumulated during the growth. No interpretation was given at this time by Boivinet. More recently A. and J. P. Belaich (1976a), studying the anaerobic glucose-limited growth of *E. coli*, showed that the existence of the second heat peak was dependent on the pH of the culture. It can be seen in Fig. 7 that the important second heat peak which occurs at pH 6·5 is not present at pH 7·8. The analysis of the fermentation products in both cases and the utilization of a hydrogen lyase-deficient strain of *E. coli* (Fig. 7(c)) showed that the second heat event occurring without growth, i.e. during the stationary phase, was related to the hydrogen–lyase activity of *E. coli*. Unfortunately, the cleavage of formic acid, which was shown to have accumulated during the growth phase, does not explain the importance of the secondary heat evolution. So it was concluded that a still unknown metabolic reaction, related to the hydrogen–lyase system, occurred during the second thermogenic peak. This type of metabolic study shows clearly how the microcalorimetric technique can lead to interesting physiological studies.

Aerobic growth was studied by Grangetto (1963) using a Calvet microcalorimeter. Aerobiosis was obtained either by the utilization of

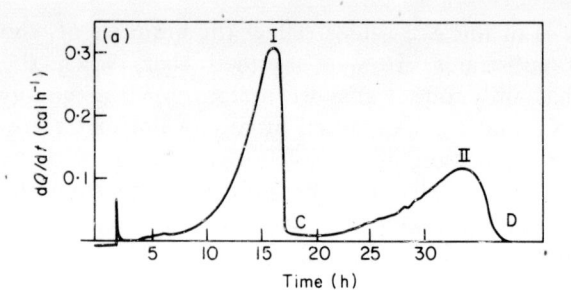

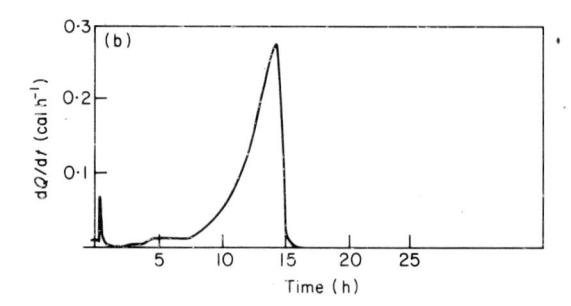

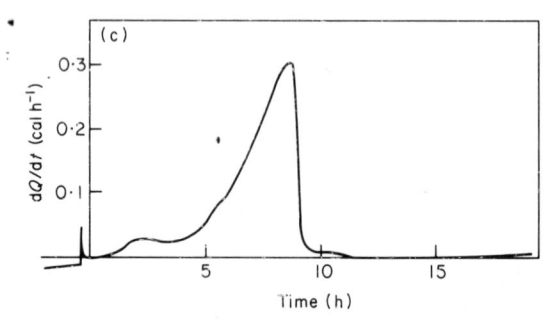

FIG. 7. (a) Thermogram of *E. coli* K–12 322 growing anaerobically in glucose-limited media at pH 6·5. Point C, glucose exhaustion. (b) Thermogram of *E. coli* K–12 322 growing anaerobically in glucose-limited media at pH 7·8. (c) Thermogram of *E. coli* 376 p4x growing anaerobically in glucose-limited media at pH 6·5. (A. Belaich and J. P. Belaich, 1976a.)

dissolved oxygen and a very low concentration of an energy and carbon source (the order of magnitude was 80 mg l^{-1}), or by bubbling wetted oxygen into the calorimetric cell. This last technique permitted the use of a glucose concentration of 500 mg l^{-1}. Thermograms of *A. aerogenes*, the growth of which was limited by succinate as the energy source, present the same characteristic shape as shown in Fig. 4. The heat quantity evolved during succinate oxidation was found to be very close to the theoretical enthalpy of succinate combustion, in good agreement with the discussion of section B. When *A. aerogenes* was grown on glucose as the energy and carbon source the thermogram obtained showed two well-separated heat peaks (Fig. 8). Only the first heat event was found associated with growth. The second heat evolution was due to degradation of acetic acid which was accumulated during the growth. The integration of the total thermogram obtained under these conditions

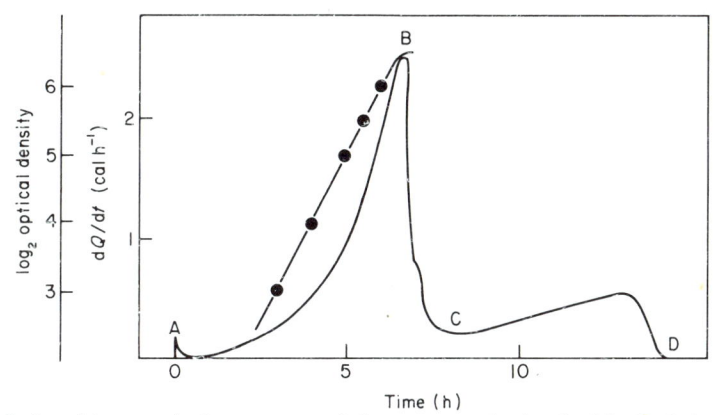

FIG. 8. Aerobic growth thermogram of *A. aerogenes* obtained with the Calvet batch calorimeter (Grangetto, 1963).

showed a large deficit in the heat quantity associated with the metabolism of glucose. It was concluded that some intermediary products of metabolism must be accumulated by the bacteria. Ericksson and Wadsö (1971), using a flow LKB apparatus, observed exactly the same phenomenon (Fig. 9) which occurs when *E. coli* is growing aerobically on glucose. Here, as in the case of anaerobic growth studies, the microcalorimetric technique permits direct observation of an interesting example of metabolic regulation.

3. GROWTH LIMITED BY A FACTOR OTHER THAN THE ENERGY SOURCE

When the energy substrate is the limiting factor, the growth is said to be well balanced which means that the energy which is delivered by catabolism is immediately used to perform biosynthesis. The rate of synthesis of macromolecules, of degradation of the energy substrate and of heat production are all generally described by the same function. However, as shown above, examples of some peculiar cases do exist

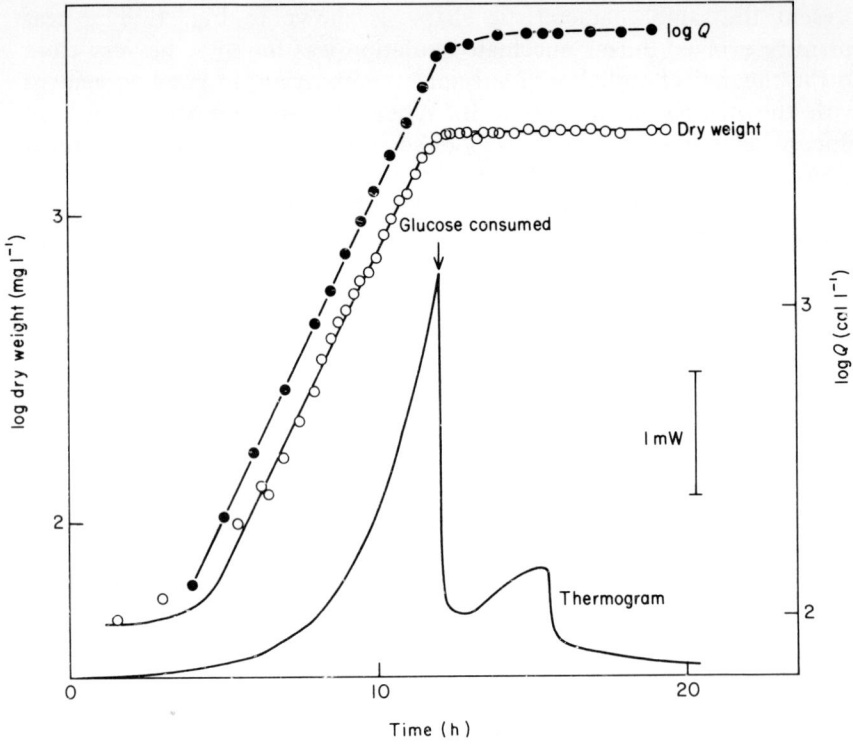

FIG. 9. *E. coli* growth curves for *E. coli* under aerobic conditions with glucose as substrate. ●, Integrated heat effect; ○, dry weight. Thermogram obtained with a flow calorimeter. (Eriksson and Wadsö, 1971.)

where thermogenic events occur during the stationary growth phase. On the contrary, when a nutritional factor other than the energy source is limiting, catabolism and anabolism are not well connected; the growth is said to be unbalanced. It has been shown that in such cases the rate of biosynthesis is much more repressed than is the rate of degradation of the energy source. The best example of this physiological condition is afforded by resting cells which are able to catabolize their energy substrate without growth at all. Microcalorimetric studies of such physiological situations were reported by Boivinet (1964), Belaich and Senez (1965a) and Belaich (1967). Belaich and Senez (1965a) reported the effect of phosphate and nitrogen limitation on catabolism in growing cells of *Z. mobilis* and *E. coli*. Figure 10 shows the results obtained in the case of *Z. mobilis* growing on minimal medium containing glucose as the sole energy source, ammonium chloride as the nitrogen source and pantothenic acid which is the only growth factor required for this

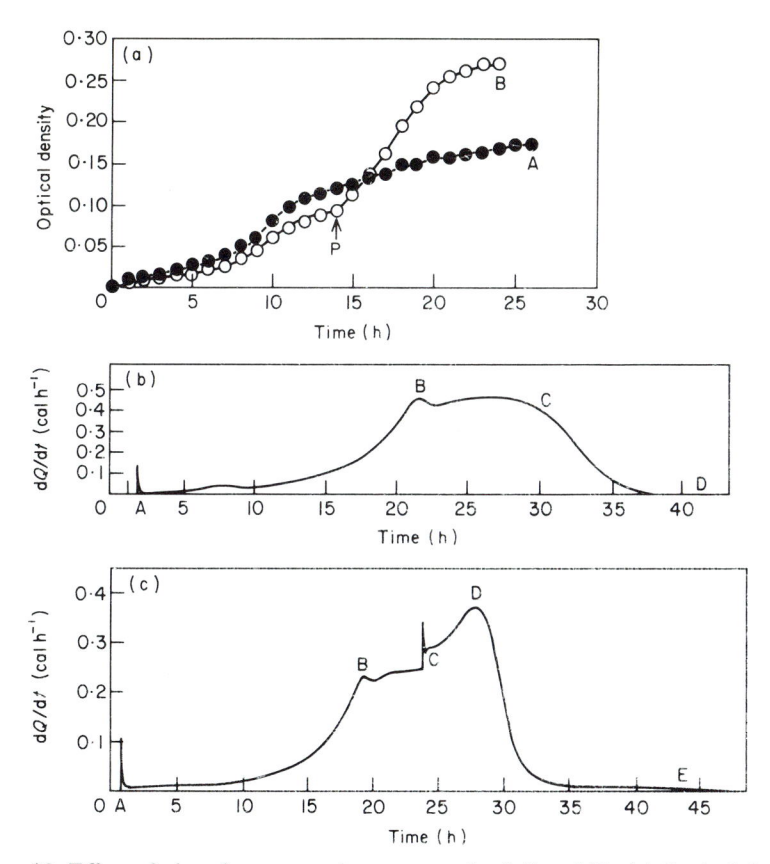

FIG. 10. Effect of phosphate starvation on growth of *Z. mobilis*. (a) Optical density versus time: ●, without HPO_4^{2-} addition; ○, with HPO_4^{2-} addition at point P (Belaich, 1967). (b) Thermogram of growth limited by phosphate without HPO_4^{2-} addition. Phosphorus becomes limiting at point B; BC corresponds to linear growth; at D all the glucose is exhausted. (Belaich and Senez, 1965a.) (c) Thermogram of growth limited by phosphate. At point C HPO_4^{2-} was added and the log phase restored (CD); at E all the glucose is exhausted. (Belaich and Senez, 1965a.)

organism (Belaich and Senez, 1965b). In this experiment phosphate was the limiting factor. As can be seen from Fig. 10, the growth is biphasic, first logarithmic, then linear when the phosphate concentration becomes limiting. The addition of phosphate at the beginning of the linear phase immediately restores the log phase. The thermogram obtained reflects this situation. Moreover, it can be seen that just after the addition of phosphate the thermal flux is markedly increased. The glucose growth yields, Y_g, were found to be equal to 4·8 during the log phase and 1·3 during the linear phase. The thermogram of Fig. 10 clearly shows that the diminution in the growth yield does not produce a corresponding decrease in the catabolic activity of *Z. mobilis*. The same results were reported for *E. coli* (Belaich and Senez, 1965a; Belaich, 1967).

Figure 11 shows the effect of nitrogen starvation on the metabolism of *Z. mobilis*. In this case, the exhaustion of nitrogen stopped the growth

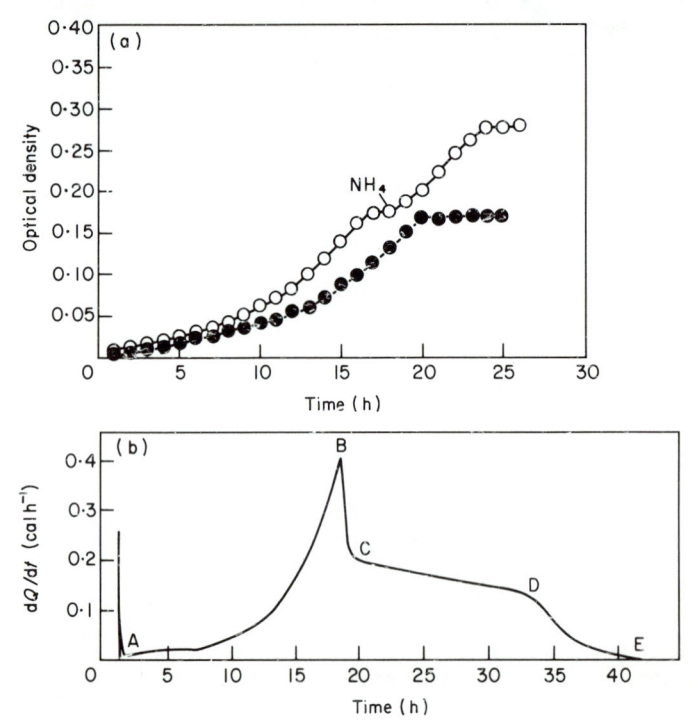

FIG. 11. Effect of nitrogen starvation on growth of *Z. mobilis*. (a) Optical density versus time: ●, without NH_4^+ addition; ○, with NH_4^+ addition (Belaich, 1967). (b) Thermogram of growth limited by nitrogen. At point B NH_4^+ is exhausted. No growth, nor accumulation of reserve product occurs during the period B to E. At point E all the glucose is exhausted. (Belaich and Senez, 1965a.)

which can be immediately restored by addition of NH_4^+ (Fig. 11(a)). The corresponding thermogram indicates an immediate decrease in the heat flux following the exhaustion of nitrogen. However 50 per cent of the catabolic activity remains and this fact demonstrates that catabolism is only partially controlled by anabolism. Boivinet (1964) reported that the same phenomenon occurs when the growth of *A. aerogenes* is starved of the nitrogen source. The main interest of the microcalorimetric technique, in this case, is to show in detail the transition between the catabolic activity of the growth phase and the catabolic activity remaining after the exhaustion of the limiting factor.

The catabolic activity of resting cells was studied by Forrest et al. (1961), Forrest and Walker (1963) and by Boivinet and Grangetto (1963). Figure 12 shows the course of heat production by a resting cell suspension with glucose as the energy source. The rate of heat production follows zero-order kinetics due to the low K_m of this organism for glucose (Forrest et al., 1961). The enthalpy corresponding to the degradation of 1 mol glucose with such resting cells was found to be very close to the one obtained with growth cultures, showing experimentally that the enthalpy of anabolism (ΔH_{an}) cannot be determined by differential heat measurement between growing and resting cells. A very interesting observation was made by Forrest and Walker (1963) on the endogenous metabolism of cells. These authors reported that *S. faecalis* in starved suspensions is able to evolve some heat without the addition of an energy substrate. This heat evolution, which is not accompanied by any detectable respiration nor pH change, is markedly affected by environmental factors and occurs only with cells that have been grown with an excess of energy source. The most interesting fact seems to be that resting cells are able to catabolize glucose only if they have detectable endogenous metabolism. Resting cells coming from growth limited by the energy source have no detectable endogenous heat production and rapidly lose their ability to ferment glucose (Forrest, 1969). Thus Forrest (1972) pointed out that the energy of maintenance appears to be supplied by the endogenous metabolism. From Fig. 12 it can be calculated that the heat production of a washed suspension of *S. faecalis* in the presence of glucose is 2·8 cal min^{-1} g^{-1} dry weight and in its absence 0·14 cal min^{-1} g^{-1}. Thus, the magnitude of the endogenous metabolism can be estimated to be 5 per cent of that of exogenous metabolism. As Forrest pointed out recently (Forrest, 1969), microcalorimetry can be particularly useful for the estimation of small metabolic events such as endogenous metabolism. It is interesting to compare the cellular rate of catabolic activity of *S. faecalis* in growing and in resting cells. In resting cells the A_c is 2·8/29 000 $\simeq$ 0·1 mmol min^{-1} g^{-1} cell dry weight ($\Delta H_{met} =$ 29 kcal mol^{-1} glucose). In growing cells the A_c can be estimated from Fig. 3 which gives $Y_g = 15$ (this value is low for this bacterium) and a Naperian growth rate of 0·0177 min^{-1}, giving $A_c = \mu/Y_g = 1·18$ mmol min^{-1} g^{-1}. These calculations show that, for *S. faecalis*, the cellular rate of catabolic activity is ten times higher in growing than in resting cells. Belaich and Senez (1965a) found, using *Z. mobilis*, that the A_c in growing cells was only twice as high as in resting cells and this

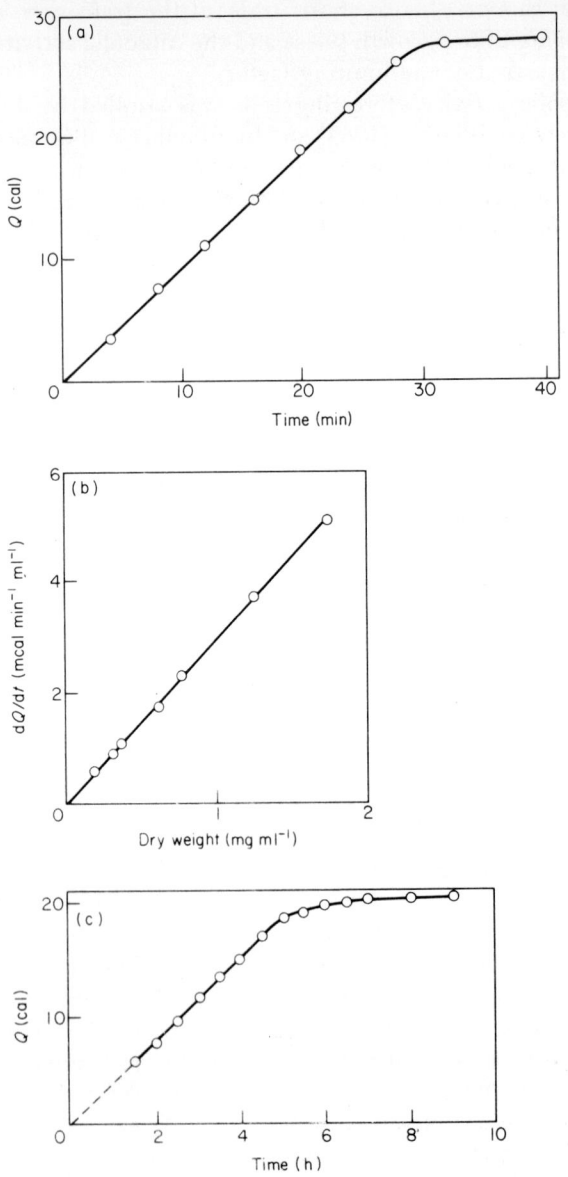

Fig. 12. (a) Heat production of a washed cell suspension of *S. faecalis* with glucose as the substrate (Forrest *et al.*, 1961). (b) Rate of fermentation of glucose by washed cell suspension of *S. faecalis* as a function of the concentration of bacteria (Forrest *et al.*, 1961). (c) Endogenous heat production by washed suspension of *S. faecalis* (450 mg of cells): gas phase, nitrogen. (Forrest and Walker, 1963.)

difference was confirmed by the study of nitrogen starvation in *Z. mobilis* (see Fig. 11(b)). It seems that *S. faecalis* is able to control its catabolic activity better than *Z. mobilis* does. Unfortunately, the starvation of growth of *S. faecalis* by a nitrogen source has not yet been studied; only such studies could prove or disprove the existence of the control of catabolic activity in *S. faecalis*.

Boivinet and Grangetto (1963) reported a study using resting cells of *A. aerogenes*. Figure 13 shows that in anaerobiosis, as in aerobiosis, the

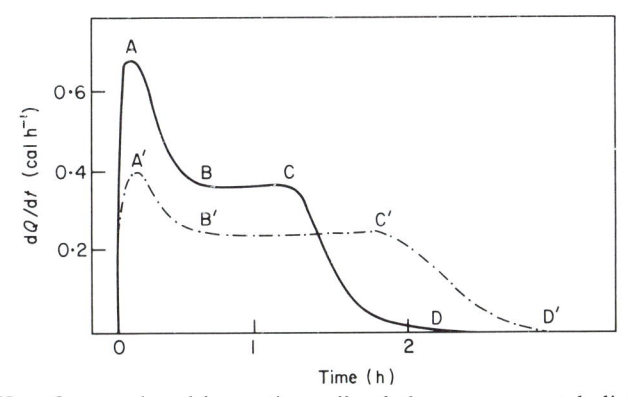

FIG. 13. Heat flux produced by resting cells of *A. aerogenes* metabolizing glucose anaerobically (———) or aerobically (- - - -) (Boivinet and Grangetto, 1963).

thermogram begins with a peak of about half an hour's duration; after this maximum the heat flux becomes stable and glucose is catabolized at a constant rate. From the growth parameters of *A. aerogenes* ($Y_g = 30$, $\Delta H_{met} = -29$ kcal) (Boivinet, 1964), the cellular rate of catabolic activities calculated in different physiological conditions is 0·247 mmol glucose min^{-1} g^{-1} dry weight when the bacteria are growing, 0·04 mmol min^{-1} g^{-1} when resting and about 0·125 mmol min^{-1} g^{-1} after nitrogen starvation. For *A. aerogenes*, as for *S. faecalis*, the A_c of resting cells is considerably lower than that of growing cells.

Another means for the study of uncoupling is to use antibiotics which specifically block biosynthesis. Figure 14 shows the effect of chloramphenicol on the growth of *Z. mobilis*. It has been observed that chloramphenicol at a final concentration of 50 μg ml^{-1} stops the growth immediately but it does not affect the cellular rate of catabolic activity, as shown by the constancy of the heat flux after the addition of the antibiotic (Lazdunski and Belaich, 1972).

All the experiments that have been described above are related through the uncoupling that occurs either when growth is limited and blocked by a deficiency in a structural nutrient (i.e. a nutrient such as phosphorus

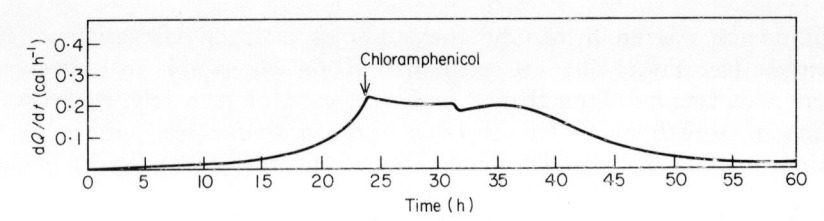

FIG. 14. Thermogram of growth blocked by chloramphenicol (chloramphenicol was added during exponential growth ↓) (Lazdunski and Belaich, 1972).

or nitrogen incorporated in relatively high percentage into the cell material), or when the growth is inhibited by chloramphenicol. The main feature of this type of study is that the cells pass, over varying times, from growth state conditions to a resting state. These experiments permitted the study of the transition from one state to another and showed clearly that catabolism is not under the strict control of anabolism. From a microcalorimetric point of view this is revealed by growth thermograms whose shapes are very different from those obtained when the growth is limited by the energy source and are characterized by the appearance of plateaux and by very slow returns to the base line.

4. UNCOUPLING IN GROWTH

Twenty years ago Le Gall and Senez (1960), Pichinoty (1960) and Rosenberg and Elsden (1960) simultaneously reported experiments showing that the cellular rate of catabolic activity is not dependent on the growth rate. This phenomenon was called uncoupling in growth. Le Gall and Senez (1960) performed their experiments using batch cultures of *Desulfo vibrio desulfuricans* and Pichinoty (1960) used *A. aerogenes*. The growth rate was found to be dependent on the nature of the nitrogen source. Rosenberg and Elsden (1960) used continuous cultures of *S. faecalis*, whose growth rate was dependent on the dilution of the culture medium. Belaich (1967) and Belaich *et al.* (1972) reported experiments with *Z. mobilis* which strengthened the concept of uncoupling in growth. As pointed out above, pantothenic acid is the only growth factor of *Z. mobilis*. Pantothenate starvation realized in batch culture resulted in a simultaneous decrease in both the glucose molecular growth yield and in the growth rate. In all these cases the cellular rate of catabolic activity calculated during the log phase by the relationship $A_c = \mu / Y_g$ remained constant and independent of the growth rate. In all cases of uncoupling the thermograms presented the characteristic shape of growth thermograms limited by energy source as shown in

Fig. 15. Starvation by pantothenate only diminishes the growth rate without change in the aspect of the growth kinetics. Thermograms were not observed, for example, which looked like those reported with nitrogen or phosphorus starvation.

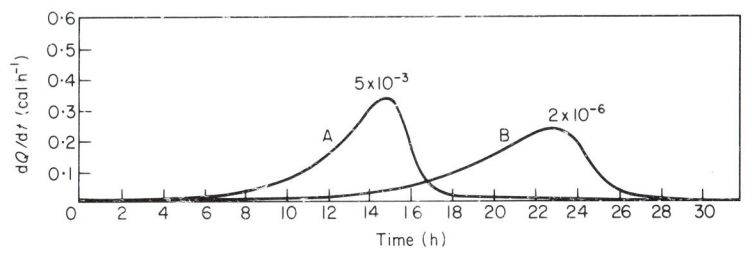

FIG. 15. Thermograms of *Z. mobilis* growing in defined medium (initial glucose concentration: 1 mg ml^{-1} in both cases). A, pantothenate concentration 5×10^{-3} mg ml^{-1}; B, pantothenate concentration 2×10^{-6} mg ml^{-1}. (Belaich et al., 1972.)

5. MIXED CULTURES

Calorimetry was applied to the study of natural biotopes which are able to metabolize mixtures of natural compounds with poorly defined compositions. A good example of such a study is given by Walker and Forrest (1964a, b) on the ruminal fermentation of sheep. The heat quantity evolved by a sample of 250 ml of rumen contents collected 16 h after feeding was as high as 685 cal and did not need a very sensitive calorimeter for its measurement. It was shown that heat production by rumen contents is a function of the percentage of the solid material contained in the sample and that the kinetics of heat production observed are of first order for long periods, as shown in Fig. 16. Walker (1965), using the same calculation techniques, showed

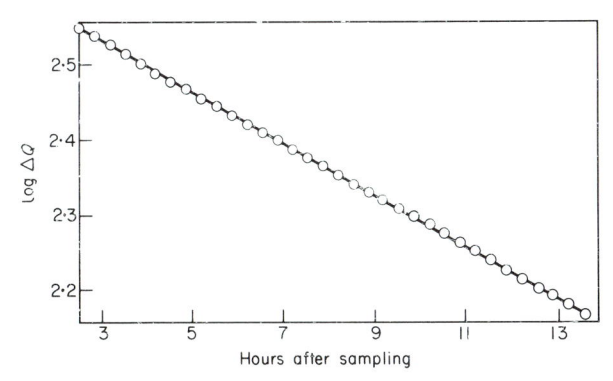

FIG. 16. Kinetic plot of heat production by rumen content (Forrest, 1969).

that when the sampling immediately follows the feeding, the kinetics are diphasic with a rapid primary degradation of soluble carbohydrates followed by a slower phase corresponding to that shown in Fig. 16. Figure 17 shows the effect of cellobiose added to a sample of rumen contents in active fermentation. From the experimental heat effect recorded, −56 cal per 2 mmol cellobiose, Forrest (1969) concluded that only 2 mmol volatile fatty acids was produced, corresponding to the

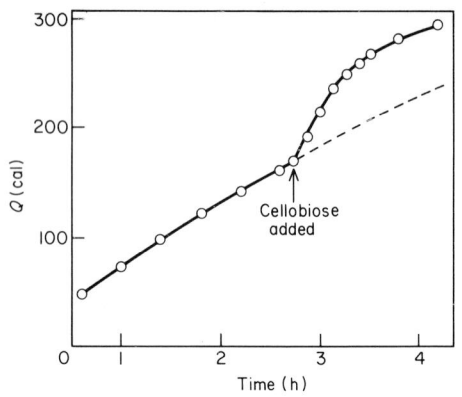

FIG. 17. Heat production by rumen content with added cellobiose (2 mmol) (Forrest, 1969).

fermentation of only 1 mmol glucose, and that 1·5 mmol of the glucose remaining was polymerized into a reserve polysaccharide. As Forrest (1969) pointed out, it seems that even in such complex systems, microcalorimetry can give valuable information.

D. Conclusions and prospects

Very often in scientific meetings I hear: "What is calorimetry good for?" I shall try to answer this formidable question, at least, in the field of microbiology. In my opinion, the most important thing for anyone who wishes to use this technique is to know what he is to measure by calorimetry or, more precisely, to know the physiological phenomena which are connected with the heat evolved by a growing or a resting microbial population. Such heat quantities are dependent on the nature of the culture medium and, more specifically, on the nature of the energy source and of the catabolic pathways used by the organisms studied. Section A shows how it is possible, by separating catabolism and anabolism, to estimate the heat quantities evolved during growth. This type of calculation, already performed by Battley (1960c), Forrest *et al.*

(1961), Belaich (1962, 1967), Boivinet (1964), and all the experimental results obtained (see Table 6), show that, when a sugar is the energy and carbon source, such heat quantities can, to a first approximation, be considered as the heat corresponding to the catabolic reactions. So, as Forrest (1972) wrote recently, there seems little possibility of using heat calorimetric measurements for the experimental determination of the enthalpy of growth. It seems also (see Table 2) that heats of combustion of bacterial dry weight presently available do not permit the calculation of ΔH_{an} with sufficient precision. For this purpose more elementary analyses and more precise measurements of heats of combustion, carried out in conjunction with direct calorimetric measurements, are needed for the determination of Δh_{an} in the few cases of organisms whose metabolism is well known (for example, in *S. faecalis*, *Z. mobilis* and *E. coli* in aerobics). When the energy source is oxidized to different degrees from glucose the calculations of the ΔH_{met} can be made, assuming that the cellular material is equivalent to glucose.

Bayne-Jones (1929a) and Bayne-Jones and Rhees (1929) reported that *A. aerogenes* and *S. aureus* cultures evolved more heat at the beginning than at the end of their growth, so that, for these authors, the young cells were thermally more active than the older ones. Later Stoward (1962), in a brief communication, pointed out that the heat quantity evolved per hour and per cell is lower at the end of the growth than at the beginning. From these observations Stoward (1962) concluded that bacterial cells obey Prigogine's hypothesis (1951) that living organisms tend towards a steady state of minimum entropy production. More recently McIlwaine and Langerman (1972), studying the growth of two luminescent bacteria on an ill-defined medium, published some observations which agree with the earlier work of Bayne-Jones and Rhees (1929). Stoward's results (1962) were disproved by Forrest and Walker (1962, 1965) who showed undeniably that the heat evolved per mg of cell by cultures of *S. faecalis* was constant during the whole of the logarithmic growth phase. Numerous reports by authors working with different organisms (Grangetto, 1963; Boivinet, 1964; Belaich, 1967; Ericksson and Wadsö, 1971) have confirmed Forrest's results. Moreover, some authors who have tried to demonstrate the applicability of Prigogine's theorem to growing bacteria seem to confuse the age of one cell with the age of the culture. It does not appear to be necessary to re-open discussion about the suggestion that living bacteria tend toward a steady growing state of minimum entropy per unit of mass unless there are further observations with, for example, continuous or synchronous cultures.

Section C shows that the shape of thermograms depends on the nature of the limiting growth factor and of the metabolic properties of the organism studied. In a complex medium containing more than one energy source, the growth curve can be diauxic and, as shown in Fig. 18,

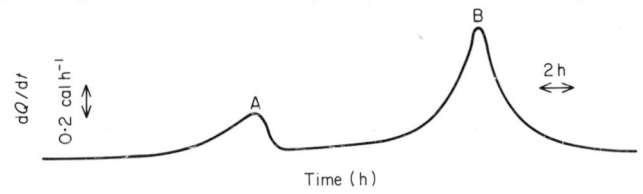

FIG. 18. Thermogram showing diauxic growth of Z. mobilis in the presence of glucose and saccharose (Belaich, 1967).

the thermograms present two well-separated thermal events. It is easy to imagine that, in an ill-defined medium, a pure culture may give thermograms with two or more heat peaks corresponding to different growth phases or to the effects of more than one limiting growth factor, as a nutritional factor or inhibition by metabolic products. Such complex curves (see Fig. 19) were obtained by earlier authors who immediately

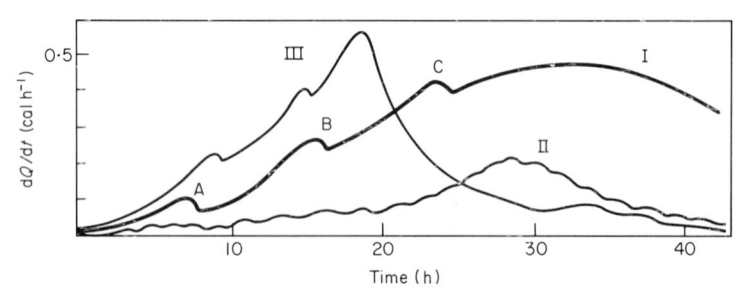

FIG. 19. Thermograms of different bacteria at 37 °C. I, *Salmonella Eberthella typhi*; II, *Salmonella paratyphi*; III, *Salmonella Schottmulleri*. (Prat et al., 1946.)

proposed the use of the particularities of thermograms for the identification of bacteria (Calvet *et al.*, 1945; Calvet and Fricker, 1946; Prat *et al.*, 1946; Prat, 1953). Recently a great deal of work has been done in this area (Boling *et al.*, 1973; Russell *et al.*, 1975a, b). The effects of antibiotics on the shape of thermograms can lead to interesting applications in the medical field. This has been shown by Mardh *et al.* (1976a, b) and Semenitz and Tiefenbrunner (1976). We can thus assume that thermograms can give interesting and reproducible information whose utilization seems promising in different medical areas. However, as pointed out recently by Beezer (1977), it is necessary to produce much work on the quantitative analysis of thermograms realized on complex media. The principal shortcoming at present is the lack of precise data which would permit the establishment of the formal relationships between the organism and the medium. More studies of the energetics of bacterial growth such as those described in section C (2, 3, 4) are needed.

Last but not least, microcalorimetry has promise in the field of natural mixed cultures. In section C a very demonstrative example is given by the studies on ruman fermentation. Another good example is offered by Mortensen et al. (1973) on soil microbiology. These studies demonstrate what it is possible to do with natural biotopes. Calorimetry allows work on insoluble and heterogeneous material and it is certainly possible to test rapidly the degradability of very complex substrates and the influence of chemical perturbations on natural biotopes.

Index of values

Values used for all the thermodynamic calculations in this paper. These values were taken from Forrest (1969) and from classical handbooks of thermochemistry.

	ΔH_f^0 (kcal mol^{-1})		ΔH_f^0 (kcal mol^{-1})
H_2O	− 68·31	Leucine	−152·5
NH_4^+	− 54·95	Threonine	−192·9
OH^-	− 31·75	Serine	−173·6
H_2SO_4	−207·5	Proline	−125·28
CO_2 aq.	− 98·69	Tyrosine	−160·5
Ethanol	− 68·85	Phenylalanine	−111·6
Lactic acid aq.	−164·02	Tryptophane	− 99·2
Glucose	−302·03	Methionine	−181·2
Lysine	−162·21	Cysteine	−126·7
Arginine	−149·0	Valine	−147·7
Aspartate	−232·6	Adenine	+ 23·2
Glutamate	−241·3	Guanine	− 45·09
Glycine	−128·4	Palmitic acid	−212·0
Alanine	−134·5	Ethanolamine	− 65·6

References

BATTLEY, E. H. (1960a). *Physiologia Pl.* **13**, 192.
BATTLEY, E. H. (1960b). *Physiologia Pl.* **13**, 628.
BATTLEY, E. H. (1960c). *Physiologia Pl.* **13**, 674.
BAYE-JONES, S. (1929). *J. Bact.* **17**, 105.
BAYE-JONES, S. and RHEES, H. S. (1929). *J. Bact.* **17**, 123.
BECK, K., GILL, S. J. and DOWNING, M. (1965). *J. Am. chem. Soc.* **87**, 901.

BEEZER, A. E. (1977). *In* "Application of Microcalorimetry in Life Sciences" (I. Lamprecht and B. Schaarschmidt, eds). Walter de Gruyter, Berlin and New York.
BELAICH, A. and BELAICH, J. P. (1967a). *J. Bact.* **125**, 14.
BELAICH, A. and BELAICH, J. P. (1976b). *J. Bact.* **123**, 19.
BELAICH, J. P. (1962). Thèse de spécialité, University of Marseilles.
BELAICH, J. P. (1963). *C. r. Séanc. Soc. Biol.* **157**, 16.
BELAICH, J. P. (1967). Thesis, University Aix-Marseille.
BELAICH, J. P. and SENEZ, J. C. (1965a). *Colln int. CNRS*, **156**, 381.
BELAICH, J. P. and SENEZ, J. C. (1965b). *J. Bact.* **89**, 1195.
BELAICH, J. P., SENEZ, J. C. and MURGIER, M. (1968). *J. Bact.* **95**, 1750.
BELAICH, J. P., MURGIER, M., BELAICH, A. and SIMONPIETRI, P. (1971). First European Biophysic Congress, Baden (Austria).
BELAICH, J. P., BELAICH, A. and SIMONPIETRI, P. (1972). *J. gen. Micobiol.* **70**, 179.
BENZINGER, T. H. and KITZINGER, C. (1963). *In* "Temperature—Its Measurement and Control in Science Industry", vol. 3, pp. 43–60. Rembold, New York.
BOIVINET, P. (1964). Thesis, University Aix-Marseille.
BOIVINET, P. and GRANGETTO, A. (1963). *C. r. hebd. Séanc. Acad. Sci.*, *Paris*, **256**, 2052–2054.
BOLING, E. A., BLANCHARD, G. C. and RUSSEL, W. J. (1973). *Nature, Lond.* **241**, 472.
BOUFFARD, A. (1895). *C. r. hebd. Séanc. Acad. Sci.*, *Paris*, **121**, 357.
BUNVILLE, L. G., GEIDUSCHEK, E. P., RAWITSCHER, M. A., and STURTEVANT, J. M. (1965). *Biopolymer*, **3**, 213.
BURKHEAD, C. E. and MCKINNEY, R. E. (1969). *Proc. Am. Soc. civ. Engrs*, *J. sanit. Engng Div.* **95**, 253.
BUZZELL, D. G. and STURTEVANT, J. M. (1952). *J. Am. Chem. Soc.* **74**, 1983.
CALVET, E. (1948). *C. r. hebd. Séanc. Acad. Sci.*, *Paris*, **226**, 1702.
CALVET, E. (1962). *In* "Experimental Thermochemistry", vol. 2, p. 385. Wiley, New York.
CALVET, E. and FRICKER, J. (1946). *C.r.hebd. Séanc. Acad. Sci.*, *Paris*, **226**, 1846.
CALVET, E. and PRAT, H. (1956). "Microcalorimetry". Masson, Paris.
CALVET, E., FRICKER, J. and PRAT, H. (1945). *C. r. hebd. Séanc. Acad. Sci.*, *Paris*, **226**, 1702.
CHAPMAN, D. (1973). *In* "Form and Function of Phospholipids", p. 127. Elsevier Scientific Publishing Company, Amsterdam.
DEBRUNFAUT, M. (1856). *C. r. hebd. Séanc. Acad. Sci.*, *Paris*, **42**, 945.
ERICKSSON, R. and WADSÖ, I. (1971). First European Biophysics Congress.
FORREST, W. W. (1961). *J. scient. Instrum.* **38**, 143.
FORREST, W. W. (1969). *In* "Biochemical Microcalorimetry" (H. D. Brown, ed.). Academic Press, New York and London.
FORREST, W. W. (1972). *In* "Methods in Microbiology", vol. 613, p. 285. Academic Press, London and New York.
FORREST, W. W. and WALKER, D. J. (1962). *Nature, Lond.* **196**, 190.

FORREST, W. W. and WALKER, D. J. (1963). *Biochem. biophys. Res. Comun.* **13**, 217.
FORREST, W. W. and WALKER, D. J. (1965). *Nature, Lond.* **205**, 46.
FORREST, W. W., WALKER, D. J. and HOPGOOD, M. F. (1961). *J. Bact.* **82**, 648.
GRANGETTO, A. (1963). Thèse de spécialité, Marseille.
GUNSALUS, I. C. and SCHUSTER, C. W. (1961). *In* "The Bacteria", vol. 2. Academic Press, New York and London.
GUNTFREUND, H. and STURTEVANT, J. M. (1953). *J. Am. Chem. Soc.* **75**, 5477.
HILL, A. V. (1912). *J. Physiol.* **43**, 261.
LAZDUNSKI, A. and BELAICH, J. P. (1972). *J. gen. Microbiol.* **70**, 187.
LAZNIEWSKI, M. (1959). *Bull. Acad. pol. Sci. Sér. Sci. Chim. Géol. Géographie*, **7**, 163.
LE GALL, J. and SENEZ, J. C. (1960). *C. r. hebd. Séanc. Acad. Sci., Paris*, **250**, 404.
LEHNINGER, A. (1965). *In* "Bioenergetics". Benjamin, New York.
LURIA, S. E. (1960). *In* "The Bacteria", vol. 1, pp. 1–34. Academic Press, New York and London.
MCCARTY, P. L. (1965). *Int. J. Ain. Wat. Poll.* **9**, 621.
MCILVAINE, P. and LANGERMAN, N. (1977). *Biophys. J.* **17**, 17.
MARDH, P. A., RIPA, T., ANDERSSON, K. E. and WADSÖ, I. (1976). *Antimicrob. Ag. Chemother.* **10**, 604.
MARDH, P. A., RIPA, T., ANDERSSON, K. E. and WADSÖ, I. (1976). *Scand. J. infect. Dis.* **9**, 12–16.
MEYERHOF, O. (1924). "Chemical Dynamics in Life Phenomena." Monograph, 110 pp.
MONK, P. and WADSÖ, I. (1968). *Acta chem. scand.* **22**, 1842.
MONOD, J. (1942). Thesis, University of Paris.
MORTENSEN, U., NOREN, B. and WADSÖ, I. (1973). *Bull. Ecol. Res. Commun.* **17**, 189.
MURGIER, M. and BELAICH, J. P. (1971). *J. Bact.* **105**, 573.
OESPER, P. (1951). *In* "Phosphorus Metabolism" (W. D. McElroy and B. Glass, eds).
O'REILLY, J. A., BAIR, H. E. and KARAZ, F. E. (1968). Unpublished data.
PICHINOTY, F. (1960). *Folio microbiologica*, **5**, 165.
PRAT, H. (1953). *Rev. Can. Biol.* **12**, 19.
PRAT, H., CALVET, E. and FRICKER, J. (1946). *Rev. Can. Biol.* **5**, 247.
PRIGOGINE, I. (1951). "Introduction to the Thermodynamics of Irreversible Processes". Thounds, Illinois.
PRIVALOV, P. R. and MONASELIDZE, D. R. (1963). *Biofizika*, **8**, 420.
PROCHAZKA, G. J., PAYNE, W. J. and MAYBERRY, W. R. (1970). *J. Bact.* **104**, 646.
RAVITSCHER, M., WADSÖ, I. and STURTEVANT, J. M. (1961). *J. Am. chem. Soc.* **83**, 3180.
ROBERTS, R. B., COWIE, D. B., ABELSOW, P. H., BOLTON, E. T. and BRITTEN, R. J. (1955). "Studies of Biosynthesis in *Escherichia coli*", pp. 607–648. Carnegie Institution, Washington, D.C.

Rosenberg, R. F. and Elsden, S. R. (1960). *J. gen. Microbiol.* **22**, 727.
Rubner, M. (1903). *Hyg. Rundsch.* **13**, 857.
Rubner, M. (1904). *Arch. Hyg.* **48**, 260.
Rubner, M. (1908). *Arch. Hyg.* **66**, 81.
Russel, W. J., Farling, S. R., Blanchard, G. C. and Boling, E. A. (1975a). "Interim Review of Microbial Identification by Microcalorimetry in Microbiology—1975" (D. Schlessinger, ed.). Am. Microbiol. Soc., Washington, D.C.
Russel, W. J., Zettler, J. F., Blanchard, G. G. and Boling, E. A. (1975b). "New Approaches to the Identification of Microorganisms" (G. G. Heden and I. Illeni, eds). John Wiley, New York.
Sedlaczek, L. (1964). *Acta microbiol. Polonica*, **13**, 101.
Semenitz, E. and Tiefenbrunner, T. (1977). *Chem. Arz.* **27**, 2247.
Senez, J. C. and Belaich, J. P. (1963). *Coll. int. CNRS*, no. **124**, 357.
Shearer, C. (1921). *J. Physiol.* **55**, 50.
Smullach, C. D., Van Wazer, J. R. and Irani, R. R. (1959). *J. Am. Chem. Soc.* **81**, 6347.
Sozaburo, O., Amanzuz, M., Takaschachi, K. and Hirani, K. (1963). *Kozokagabi Symposium*, **15**, 111.
Spink, C. and Wadsö, I. (1976). *In* "Methods in Biochemical Analysis", vol. 24, p. 1. John Wiley, New York.
Steiner, E. and Kitzinger, C. (1962). *Nature, Lond.* **194**, 1172.
Stoward, P. J. (1962). *Nature, Lond.* **194**, 977.
Sturtevant, J. M. (1953). *J. Am. chem. Soc.* **75**, 2016.
Sturtevant, J. M. (1954). *J. Phys. Chem.* **58**, 97.
Tangl, F. (1903). *Pflügers Arch. ges. Physiol.* **93**, 327.
Terroine, E. F. and Wurmser, R. (1922). *Bull. Soc. Chim. biol.* **4**, 519.
Tian, A. (1923). *Bull. Soc. Chim. biol.* **33**, 427.
Wadsö, I. (1970). *Q. Rev. Biophys.* **3**, 383.
Walker, D. J. (1965). "Physiology of Digestion of Ruminants" (R. W. Dougherty, ed.), pp. 296. Butterworth, London and Washington.
Walker, D. J. and Forrest, W. W. (1964a). *Austr. J. Agric. Res.* **15**, 199.
Walker, D. J. and Forrest, W. W. (1964b). *Austr. J. Agric. Res.* **15**, 299.

Growth and Metabolism in Yeasts

INGOLF LAMPRECHT

A. Introduction

1. THE YEAST

Yeasts belong to the simplest eucaryotic organisms and have always attracted the interest of scientists as model systems for higher degrees of evolution and for the study of general properties of metabolism and growth in animals and plants. But at the same time they are *per se* of great importance as essential factors in the production of bread, beer and wine, food and fodder, and as an excellent source of proteins and fats with a high concentration of vitamins and minerals. The ability to grow on inexpensive and otherwise waste carbon sources like spurt liquor and mash opens new dimensions for the fight of mankind against malnutrition and hunger.

From the pure scientific point of view yeasts are a group of fungi with special common characteristics, but they do not all belong to the same classes. Often they are botanically ill defined and only named by the purpose they are produced for: baker's yeast, brewer's yeast, growth yeast, or by their commercial form, e.g. pressed yeast. But the taxonomically well-defined yeasts, too, like *Saccharomyces cerevisiae* or *Candida utilis*, exist in a great variety of strains with a multiplicity of genetic properties making them an interesting object of basic research.

Many results concerning the metabolism and growth of yeasts under widely different conditions have been collected in the past, and it is impossible to deal even with a small part of them in this survey. The reader is referred to the excellent monographs that exist in the literature (see, for example, Cook, 1958; Reiff *et al.*, 1960; Mills, 1967; Rose and Harrison, 1969). Non-calorimetric experiments on metabolism and growth of yeasts will be cited only if they are necessary for the understanding of the topic, or if they have direct relevance (mainly energetic) to it.

Besides brewer's and baker's yeast, *Saccharomyces cerevisiae* is mainly used in calorimetric experiments; but sometimes *Candida utilis, Debaryomyces hansenii, Saccharomyces fragilis, Schizosaccharomyces pombe* and species of *Kluyveromyces, Rhodotorula* and *Hansenula* are used.

2. MICROCALORIMETRY

Because of the increasing application of calorimetry in the biological sciences, there exist some technically or instrumentally orientated reviews about calorimetry (see, for example, Skinner, 1962; McCullough and Scott, 1968; Sturtevant, 1972); and some papers on biological calorimetry have appeared (Wadsö, 1970; Boivinet, 1971; Forrest, 1972; Rouquerol and Boivinet, 1972; Sturtevant, 1974; Wadsö, 1975, 1976; Spink and Wadsö, 1976), together with proceedings of conferences concerned with the application of microcalorimetry in biology and medicine (Broda *et al.*, 1971; Peeters, 1973; Lamprecht and Schaarschmidt, 1977). They give surveys over the whole field of calorimetry, sometimes divided according to a biological scheme of increasing complexity of the systems under investigation (Boivinet, 1971). In reviews on bacterial calorimetry (Forrest, 1969), yeasts were subsumed under microbial or even bacterial calorimetry or indeed omitted completely. To our knowledge this paper is the first attempt to review the calorimetric literature on yeast.

This is quite astonishing as the first calorimetric experiments on microorganisms concerned the heat of fermentation developed in a huge brewer's vat (Dubrunfaut, 1856), and perhaps Lavoiser only failed to measure the heat of yeast growth during his intensive studies on heat and on fermentation, as he was bound to 0 °C by his ice calorimeter. Dubrunfaut's experiments would be termed "dynamical calorimetry" nowadays, and it was left to Rubner to start the first true calorimetric measurements on the development of heat during fermentation and growth of yeast (Rubner, 1903, 1904, 1906). Many subsequent "calorimetric" evaluations of energy metabolism were not calorimetric *sensu strictu*, as they did not measure directly the heat evolved during growth. With bombs of the Berthelot-type the heat content of the nutrient broth was determined at the beginning of an experiment and at the end, sometimes including the crop, sometimes after separation (Bouffard, 1895; Rubner, 1903; see also Tangl, 1908). The difference between the two derived figures equals the heat evolved during the growth period. The system thus obeys Hess' law of constant heat summation.

Even nowadays many data for the energy changes during an experiment are evaluated from a combination of bomb calorimetry and tabulated values of standard enthalpy and energy. But at the same time the progress in instrumentation and the variety of microcalorimeters commercially available enables one to measure some of these data directly under varying conditions. Although heat flow calorimeters operate with a sensitivity of 100 μV mW^{-1} and a resolution better than 1 μV, i.e. heat flows between 1 and 10 μW, they are far too insensitive to register the energy changes associated with *one* cell. At a maximum rate of heat production of 10^{-11} W per cell at least 10^5 cells are necessary to get a signal significantly greater than the noise level. Thus, cultures of $\sim 10^6$ to $\sim 10^8$ yeast cells are investigated. That implies a double integration in the method: all simultaneous heat productions in the cells are summed and then averaged over all cells regardless of which physiological state they are in. Thus knowledge of the synchrony (or lack of it) is sometimes of great importance when heat production in the different phases of a cell cycle is of interest.

Because of the excellent reviews about calorimetry that exist only a few remarks on instrumentation are given in section D. As necessary during the description of special experiments, more details will be cited in the text. For more information the reader is referred to the original literature.

B. Metabolism of yeast

The metabolism of the yeasts is very complex, as these organisms are able to maintain or grow on various substrates under widely differing conditions. Only a few introductory remarks will be given as far as they are necessary for the understanding of the calorimetric experiments and the interpretation of their results. Full accounts of yeast growth and metabolism are to be found in several monographs (Cook, 1958; Reiff *et al.*, 1960; Mills, 1967; Harrison, 1971). The chosen division into energy, anaerobic and aerobic metabolism, is somewhat arbitrary but offers advantages for the later discussions.

1. LEVELS OF METABOLISM

As Forrest and Walker (1964) pointed out, a microorganism may attain three different levels of metabolic activity, each corresponding to a steady state in a thermodynamic sense. These levels can be described by parameters which are independent of time or change very slowly compared with other time constants of biochemical reactions. During the growth in a batch culture the organism passes all three levels.

The lowest metabolic activity is due to the endogenous metabolism of the cell. It is defined as all the metabolic reactions that occur within the living cell when it is held in the absence of compounds or elements which may serve as specific exogenous substrates (Dawes and Ribbons, 1962, 1964). In this situation the organism is exclusively bound to its own reserve substances, mainly glycogen in the case of yeast (Goodwin *et al.*, 1974). This endogenous metabolism is chiefly aerobic, but some authors have reported anaerobic activities (Cook, 1958). Winzler and Baumberger (1938) observed, in the presence of oxygen, an endogenous heat production which decreased according to a first-order process. No heat flow was detected under anaerobic conditions, but it started with a heat burst like the "oxygen debt" in muscle tissue when oxygen was supplied. Therefore they supposed that an intermediate accumulated in the anaerobic phase which was readily oxidized after the addition of oxygen. Endogenous metabolism is observed as a tiny base line shift in thermograms derived from energy- and carbon-source limited batch cultures when growth has ceased because of depletion of exogenous substances.

The second level of metabolism is that of maintenance. Whereas endogenous metabolism is time limited because of depletion of reserve substances and is followed by the death of the cell, metabolism of maintenance keeps the cell alive for infinite periods as long as an exogenous substrate is available. Maintenance in yeast may be both anaerobic or aerobic following the external conditions. As no growth takes place in this situation, the external energy and carbon source—e.g. glucose—is only used for the energy supply of the cell. This energy is consumed in the different processes which keep the cell alive; steady resynthesis of macromolecules (RNA and protein, perhaps polysaccharides and cell wall materials); regulation of the pH value and the osmotic pressure in the cytoplasm; active transport into the cell and others.

It was demonstrated earlier that during the growth of the microorganism—in batch cultures—no energy of maintenance is necessary (Monod, 1942). However this energy of maintenance had to be invoked to explain deviations of the growth yield at low dilution rates in continuous flow cultures (Dawes and Ribbons, 1964; Pirt, 1966; see sections E.2 and E.3.b). The idea was that part of the energy source was consumed to maintain the cell in a proper state in the long periods between division and that this part of the energy source was lost for yield demands.

It will be shown below (section E.5) that the metabolism of maintenance is closely related to the capacity of yeast cells to repair the damage resulting from irradiation with u.v. light or X-rays.

The third level of metabolism corresponds to growth. In this state external energy, carbon and nitrogen sources are necessary as are trace elements and vitamins. Very often, energy and carbon source are the same, for example, a hexose for *S. cerevisiae*. But a great variety of other carbohydrates, hydrocarbons, amino acids or ill-defined mixtures like molasses are possible substrates. Intensity of growth is determined by external parameters, too, like kind and concentration of substrates and metabolites, pH value, aeration, temperature and concentration of cells.

The typical growth of a microbial batch culture, indicated by a sigmoidal relationship between a growth factor (number of cells, culture volume, dry weight, turbidity, heat or metabolite production) and time, may be divided into three to five distinct periods:

1. The lag or adaptation phase, when cells are transferred to a fresh medium. During this time no growth occurs as cells are occupied in establishing the necessary state of enzymes for catabolism.
2. The acceleration phase, when growth begins without obeying an exponential law.
3. The log or exponential phase. Here increase in cell number is exponentially dependent on time, or, after a mathematical transformation, the logarithm of cell number is a linear function of time.
4. The retardation phase when a low substrate or a high metabolite concentration limits the growth rate.
5. The stationary phase, when the exogenous parameters forbid further growth. This may be due to a depletion of substrate, to a shift in the pH value or to an unfavourable concentration of metabolites.

In the last phase the cells are fully alive. But, as an exogenous energy source is absent, they have to switch over to the endogenous metabolism which is time limiting. Therefore, often a sixth phase, the phase of slowly dying out, is included. All these phases are readily recognized in the growth thermograms of yeast cultures (see, for example, Figs 12 and 13). Modern calorimeters mainly determine the *rate* of heat production which is not proportional to the growth itself, but to the increment of growth. Therefore, most thermograms, as rate of heat production versus time, represent the derivative of the sigmoidal growth curve with respect to time and show a more or less bell-shaped form.

Only the exponential phase of growth corresponds to a thermodynamic steady state with constant specific values like increase of cell number or weight, rate of heat production or rate of catabolism (Forrest and Walker, 1964). The other parts in a growth curve are transition periods from one steady state to the next one with quickly changing parameters. When a microbial culture is growth limited, not by the energy but by the nitrogen or phosphate source, the steady state of growth is changed to a steady state of maintenance, when the limiting source is exhausted, and changes to the steady state of endogenous metabolism after depletion of the energy source. In this way the interesting questions of transition phases in the thermodynamics of irreversible processes can be readily studied in microbial cultures.

2. ENERGY METABOLISM

Metabolism of carbon and of energy are tightly coupled as both quantities are often derived from the same substrate. For instance, in a minimal medium for growth of yeasts, glucose can be the only source of carbon and of energy and one has to distinguish clearly between energy-yielding catabolism and carbon and energy consumption in cell biosynthesis or anabolism (Senez, 1962; Sols et al., 1971; Lagunas, 1973). Giaja (1920) has compared the heat production of yeast with that of horses and men and found a relation of 157 : 3 : 1 when based on the same amount of nitrogen. This enormous heat turnover does correspond to the energy actually needed by the organism for life, but is also a sign of its high catalytic activity. Only endogenous metabolism represents what was called "fundamental biologic energy" by Giaja.

Following Oura (1972), the energy demand of a yeast cell can be divided into various classes of differing importance:

   a. synthesis of monomers for cell material;
   b. polymerization of these monomers;
   c. formation of secondary and tertiary structures in proteins and nucleic acids;
   d. formation of supramolecular assemblies (membranes, ribosomes etc.);
   e. formation of cell organelles (nuclei, mitochondria, endoplasmic reticulum etc.) and the whole cell structure.

Besides these energy-consuming factors connected with growth there are two classes of reactions associated with the maintenance of viability of the cell:

   1. maintenance of the cell status (osmotic pressure, pH value, essential concentrations etc.);
   2. transport of substances to and from the cell and within the cell.

The amounts of energy used in the various processes are quite different, some even hard to calculate. It can be shown, for instance, that 75–85 per cent of the total energy requirement of the cell is due to cellular organization rather than for biosynthetic work. Under aerobic conditions the cell's structure is more complicated than that of the simpler anaerobic cell and the energy consumption increases by 40 per cent (Oura, 1973).

The distribution of energy for the different processes is shown in Table 1, presented as moles ATP per 100 g yeast dry matter.

It is well known nowadays that the energy requirements of cells can be expressed in terms of ATP which is directly, or by intermediates, involved in all energy-consuming reactions and couples catabolism with anabolism (Forrest and Walker, 1971). ATP plays the role of an energy

TABLE 1

Energy consumed for different cell functions as moles of ATP per 100 g yeast dry weight (Oura, 1973)

	Anaerobic	Aerobic with 10% oxygen
Cell material synthesis	1·12	1·12
Maintenance	0·13	0·04
Glucose uptake	0·75	0·23
Establishing of cell structure	6·53	8·58
TOTAL	8·53	9·97

carrier between energy-yielding (exergonic) and energy-requiring (endergonic) reactions. It is formed directly on the substrate level via the Embden–Meyerhof–Parnass pathway or by intermediates via the energy transfer chain. The low concentration of ATP in the yeast cell (~ 4 μmol g^{-1} dry weight) demonstrates that this compound is involved as an intermediate in coupled reactions for energy transfer and not for energy storage. Only 10^{-6} of the energy content of a cell stems from ATP or equivalent nucleotide triphosphates.

Many calorimetric and non-calorimetric experiments have been performed on the energy metabolism of microorganisms and especially of yeasts. More details can be found in the following papers: Gunsalus and Shuster (1961), Ingraham and Pardee (1967), Payne (1970), Stouthammer and Bettenhausen (1973), Goodwin *et al.* (1974), Verhoff and Spradlin (1976) and in several papers cited in section E.

3. ANAEROBIC METABOLISM

Yeasts belong to the organisms with the facility to perform anaerobic metabolism, i.e. to gain energy in the absence of oxygen. But this property varies significantly in the different types of yeasts. The Mycoderma yeasts are so strongly orientated towards aerobic catabolism that no anaerobic strain exists. In Torula yeasts fermentation is possible under anaerobic conditions but plays no role in the aerobiosis. The same holds true for baker's yeast where fermentation drops to one quarter if air is present in the culture. In contrast, fermentation in a brewer's yeast is not so influenced by the presence of air, so that in such cells 20–80 molecules of sugar are fermented for one molecule respired. These differences in metabolic behaviour are mirrored in the differences of the cytochrome spectra of these yeasts (Reiff *et al.*, 1960).

There have been intensive discussions on the existence of a true anaerobic metabolism. Some authors claimed that anaerobic growth is only possible for a few generations, perhaps at the expense of aerobically formed pools in the cell, and that traces of oxygen are necessary for the multiplication of yeasts (Battley, 1960a). But it has been shown that *S. cerevisiae* grows anaerobically for indefinite periods in the presence of ergosterol in the medium (Reiff *et al.*, 1960).

On the other hand, it is often very difficult to obtain totally anaerobic conditions in a culture, as traces of oxygen may be used by cells to shift the metabolism to aerobiosis. When calculating energy outputs under such conditions unexpectedly high enthalpy values are found which would be impossible for an exclusively anaerobic catabolism of the substrate (Hoogerheide, 1975). But there exist several—in the cell's view—external and internal tricks to avoid respiration. Many yeasts possess so-called petite- or respiration-deficient strains, mutants which lack certain respiratory enzymes belonging to the cytochrome system and which exhibit, even in the presence of oxygen, a true anaerobic metabolism (Cook, 1958). Several chemicals like KCN or azide may be counted as internal factors, as they have little or no influence on anaerobic fermentation, but block respiration completely (see, for example, Hoogerheide, 1975). More appropriate are factors which do not interfere with the enzyme system of the cell but keep the medium free of oxygen. Besides pure mechanical methods like flushing and deaerating the liquid with nitrogen or hydrogen of the highest degree of purity, substances which react with oxygen such as glucose-oxidase (Hoogerheide, 1975), amorphous ferrous sulphide (Brock and O'Dea, 1977) or pyrogallol may be placed in, or above, the medium to absorb residual traces of oxygen.

Pasteur suggested that fermentation would be life without air, i.e. the supply of free energy without the involvement of oxygen. The free energy change accompanying the metabolism of 1 mol glucose is only -253 kJ for anaerobic conditions compared with -2876 kJ under aerobic conditions (1 : 11, Wilhoit, 1969). The difference is not as striking, if one considers the percentage of the available free energy which is stored via the intermediate ATP: 40 per cent by fermentation and 69 per cent by oxidation (1 : 1·7) (Senez, 1962). The rest of the free energy, that is 60 per cent during fermentation and 31 per cent during oxidation, is dissipated as heat. But these are idealistic figures, as Hoogerheide (1975) pointed out; dependent upon the circumstances up to 100 per cent as heat can be distributed to the surroundings when the cell has no opportunity to use or store free energy (complete elimination of assimilatory processes).

The observed patterns of anaerobic metabolism on various substrates are indeed very complicated. Details can be found in the literature (Reiff *et al.*, 1960; Cook, 1968; Mills, 1968; Decker *et al.*, 1970).

4. AEROBIC METABOLISM

Aerobic metabolism is bound to the existence of free oxygen in the growth medium. This oxygen can act in three different ways in yeast cells: (a) it induces in a direct or indirect way the synthesis of mitochondrial respiratory complexes; (b) it participates as a reactant in several biosynthetic pathways, e.g. sterol or fatty acid synthesis; (c) and it functions as an electron acceptor in respiration (Rogers and Stewart, 1974). The concentration of oxygen in the medium and therefore the absorption into the cell determine the ratio of aerobic to anaerobic metabolism in the facultative anaerobic yeasts (Reiff *et al.*, 1960). Only at very high concentrations does it exhibit a poisoning effect on the cells (Oura, 1972, 1973, 1975). Besides oxygen the substrate and its concentration can also determine the type of metabolism in yeasts. If abundant glucose is available in the presence of oxygen, aerobic respiration, and a considerable aerobic fermentation, can exist at the same time in the cell although only respiration should be expected. This effect is well known as "catabolite repression", "negative Pasteur effect" or "Crabtree effect". It is understood as an excess of the breakdown products (ethanol and carbon dioxide) derived from glucose via the Embden–Meyerhof–Parnass pathway. They are unable to enter the tricarboxylic acid cycle and the energy transfer chain as these are saturated because of the limited capacities of the oxygen transfer system. When the glucose concentration drops below a value of ~ 3 mmol l^{-1} aerobic fermentation can be neglected while respiration stays at a maximum turnover level (Hoogerheide, 1975b).

The repressive effect of glucose influences the structure of the yeast cells. It especially represses the formation of mitochondria, the organelles in which respiration takes place (Oura, 1973c). Battley (1960a) gives two different compositions for cells grown anaerobically and aerobically which lead to slightly different heats of combustions (see section E.4).

It will be shown later on (see section E.3.a) that in calorimetric experiments true aerobic conditions are hard to achieve. By stirring the contents of a batch vessel, a homogeneous distribution of cells, substrates and metabolites is obtained but not a sufficient supply of oxygen because of the slow diffusion through the interface between air and liquid. Bubbling air through the medium is more effective but troublesome because of the very high heat of evaporation if the gas is not completely saturated with water. Flow calorimeters have the great advantage that the culture medium can be aerated and stirred (thus reducing the size of the air bubbles) at very high speeds without thermal interference within the calorimeter. However, the flow time of several minutes between the fermentor and the calorimeter can cause a considerable decrease in oxygen concentration in the culture, thus leading to a shift towards more anaerobic metabolism. Batch cultures on solid medium offer slight improvements over liquid cultures as oxygen is directly accessible to the cells growing on the surface. But only a few experiments have been performed in this manner (Lamprecht and Meggers, 1972; Ross, 1973).

In Fig. 1 the main steps of anaerobic and aerobic catabolism of glucose are summarized. The hexose enters the pathway and is phosphorylated in two steps at the expense of 2 mol ATP to a diphosphate which is further catabolized to 2 mol ethanol along with the formation of 4 mol ATP. This anaerobic part of hexose metabolism which renders a net gain of 2 mol ATP per mole of hexose is called the glycolytic or the Embden–Meyerhof–Parnass pathway. Via pyruvate and acetyl-coenzyme A, the two-carbon compound is fed into a cycle (the tricarboxylic acid cycle, citric acid cycle or Krebs cycle) in which most respiratory nucleotides are reduced. These nucleotides are then oxidized in the energy transfer chain if oxygen is available leading to 36 mole more of ATP per mole glucose. Therefore, by the complete oxidation of 1 mole glucose 38 mole ATP are formed. Giving simplified energetic balance equations for anaerobic and aerobic catabolism of glucose, one can write for fermentation:

$$C_6H_{12}O_6 + 2\,ADP + 2\,P_i \longrightarrow 2\,C_2H_5OH + 2\,CO_2 + 2\,ATP$$

and for respiration:

$$C_6H_{12}O_6 + 6\,O_2 + 38\,ADP + 38\,P_i \longrightarrow 6\,CO_2 + 6\,H_2O + 38\,ATP$$

These two equations demonstrate the great energetic advantage of respiration over fermentation and indicate that endogenous metabolism should be aerobic from an energy economy point of view.

5. SOURCES OF ENERGY AND CARBON

Carbohydrates are most frequently used in the production of yeasts. They simultaneously serve as sources of carbon, hydrogen and energy for the

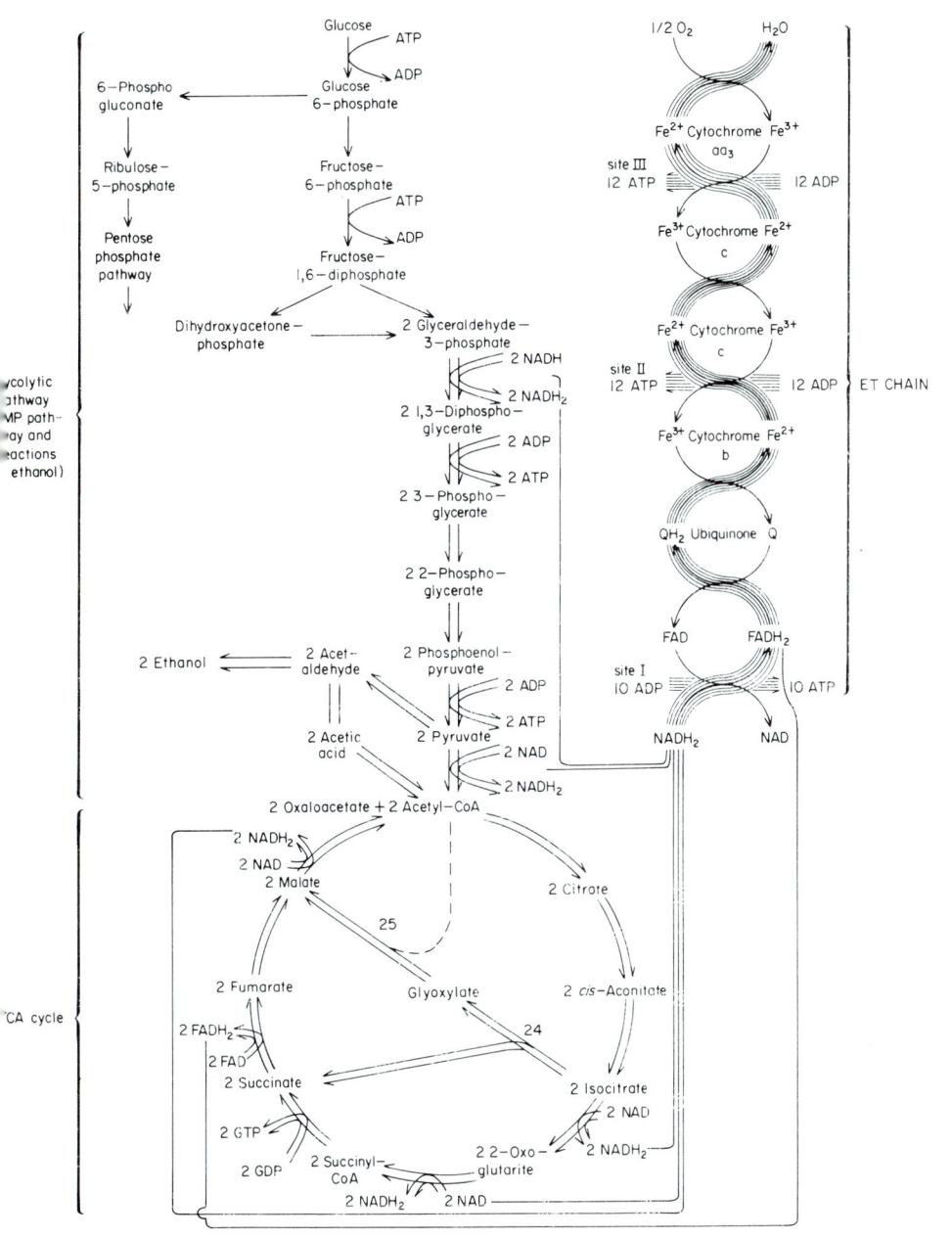

FIG. 1. Catabolism of glucose in yeast cells via the Embden–Meyerhof–Parnass pathway (EMP), tricarboxylic acid cycle (TCA) and energy transfer chain (ET chain) (Oura, 1973).

cell. These compounds, mainly of botanical origin, may contain sugars in an immediately available form such as glucose, fructose or other mono- and disaccharides, they may be starch-containing cereals, roots and tubers and all sorts of molasses. In addition, sulphite processes for wood pulp manufacture make available large amounts of carbohydrates for commercial yeast production (Harrison, 1971). These spent liquors are rich in pentoses, which are normally more difficult for the yeast to metabolize than hexoses. The degree of transformation from substrate to cell dry mass decreases and the dissipation of heat increases. Thus in spent liquor derived from beech trees with a pentose content of ~ 90 per cent the heat output per gram yeast is 30 per cent greater than that in pine-tree spent liquor which contains only 15 per cent pentose (Schmidt, 1947).

But for industrial purposes many other compounds besides the carbohydrates are of interest as energy and carbon sources for the production of yeasts. Hydrocarbons are among the most promising substrates for industrial production of proteins by yeasts. Yeast cells can even attack liquid hydrocarbons and they show growth yields of 0·8–1·1 g dry weight per gram substrate (Wang, 1968). The disadvantage of such substrates is the high dissipation of heat which is double that of carbohydrates for n-paraffins and increases more than five-fold for methane; 200 g of carbohydrates produce ~ 100 g yeast and dissipate 1600 kJ of heat, whilst 100 g of hydrocarbons render 100 g yeast and 3270 kJ heat (Guenther, 1965). But it seems possible that the growth yield can be improved, in the latter case, to 150 per cent, thus reducing the cost three-fold: only one-third of the raw material is necessary, the oxygen consumption is only slightly increased, the heat dissipation (and thus the cooling requirement) is equivalent (always compared with the same amount of yeasts grown on carbohydrates (Guenther, 1965)). Cooling is an essential factor in the industrial production of yeasts, as these organisms are limited to a small temperature region for growth, the optimum temperature being 30 °C. In large fermentors working aerobically on concentrated sugar solutions as much as 2·3 MW must be removed from the solution to keep the proper temperature. This figure increases with growth on hydrocarbons, so that cooling by sea water has been proposed (Harrison, 1971).

Considerable amounts of heat are also dissipated during aerobic ensilage, (McDonald et al., 1973). Under anaerobiosis less opportunity exists for heat to escape so that more energy is stored in the products, up to 99 per cent. In these reactions yeasts are involved together with bacteria. Ensilage is mentioned in this context as an example of an ill-defined substrate for energy metabolism of yeasts.

C. Growth of yeast

The growth of yeast is related to the presence of sources of carbon, energy and of nitrogen. All the other necessary components are required in such small concentrations that they can be neglected in these considerations.

When there is more than one source of energy in the medium either simultaneously or when metabolites of the first source can act as a new energy compound after the depletion of the first one, the phenomenon of diauxic metabolism is observed (Dedem and Moo-Young, 1975; Schaarschmidt and Lamprecht, 1977a, b). A typical example is the consecutive pattern of heat production in the aerobic fermentation of glucose at high concentrations: despite the presence of oxygen the glucose is primarily fermented via the Embden–Meyerhof–Parnass pathway to ethanol which is fed to the Krebs cycle and the energy transfer chain after the depletion of glucose. Between the two different metabolic processes there is a distinct lag phase and a period of transition which sometimes shows clear oscillations in the heat output (Brettel, 1977a). Such behaviour is shown in Fig. 15. Diauxic properties in a culture lead to structure in the growth curve and more directly in the thermogram and allows one to observe characteristic "finger-print" thermograms for growing yeasts (see section E.1 and the contribution by Newell, p. 163). But the simultaneous availability of energy from different energy sources does not always lead to diauxic metabolism; if, for instance, glucose and fructose are present in an anaerobic growth medium, only one phase is observed.

In section B.1, the different phases of a growing culture were mentioned. However, there exists another integral division of time for a culture of cells, the time between two cell divisions—the interphase—which itself can be divided into three distinct periods:

1. G1: the first gap phase, when only RNA and proteins are synthesized;
2. S: the synthetic phase, when DNA and histones are constructed;
3. G2: the second gap phase, when again mainly RNA and proteins are synthesized.

After completion of G2 the cell is ready for mitotic division. Normally one cannot observe these phases in a growing culture of yeasts as the cells are in different parts of the mitotic cell cycle. Only by synchronizing the culture, or, after the appearance of self-synchronized growth, can the phases be separated and the specific synthetic properties and the energetic turnover be distinguished. Few calorimetric experiments on synchronous cultures of yeasts have been performed (Poole *et al.*, 1973; Brettel, 1977a, b; Poole, 1977), as long-lasting synchrony in yeasts is hard to obtain. But they offer the possibility of observing the energy requirement of a cell for special synthetic processes and for the mechanical division of the two newly formed cells.

One has to distinguish between two different methods for cultivating yeast cells:

1. batch culture;
2. continuous culture.

In batch culture growth is possible only for a finite period or for a finite number of generations, as the energy source is depleted over time. In such a set-up one observes the various growth phases like lag, log and stationary phase and it is only in these cultures that diauxic growth takes place. As Monod (1942) has demonstrated the growth rate of a microorganism is determined either by internal or by external growth factors. At high substrate concentrations cells are under internal control, while they are externally controlled when the growth rate depends on a growth factor externally supplied (see, for example, Fig. 9). In contrast to a "closed" batch culture a continuous culture is an "open" system, with a steady supply of fresh medium. Following Kubitschek (1970) one may define a continuous culture as a system in which individual cells are suspended in a (nearly) constant volume, at or near a steady state of growth established by the continual addition of fresh growth medium, and the continual removal of part of the culture. They can be run as a turbidostat with an abundant substrate supply (internal control) or as a chemostat at lower substrate concentrations (external control). In these cultures growth rates vary widely with change in the substrate concentration and the dilution rate (the fractional replacement rate of nutrient medium). It is readily shown that under these chemostat conditions the growth rate equals the dilution rate (Kubitschek, 1970; Herbert, 1959). A continuous culture can be kept for infinite time in a quasi-steady state as long as the externally controlling growth factor is not changed and no mutation of the yeast or infection by bacteria (which grow more quickly) appears. Calorimetry of continuous cultures is best performed in a combination of separate fermentor and flow microcalorimeter (Brettel et al., 1971; Brettel, 1974, 1977a, b). For more details about continuous cultures the reader is referred to the literature (see, for example, Malek et al., 1964; Malek and Fencl, 1966; van Uden, 1967, 1971; Mor and Fiechter, 1968; Kubitschek, 1970; Leuenberger, 1971, 1972).

As pointed out above, growth can be followed continuously by the increase in dry weight or the number of the cells, the turbidity of the culture or the rate of heat production. Frequently it is of interest to know the final yield of the culture. Bauchop and Elsden (1960) claimed that the ratio of the cell yield (dry weight) to the moles ATP formed in the specific pathway, the Y_{ATP} value, should be equal to approximately 10·5 g dry weight mol^{-1} ATP for all microorganisms. It could be shown later that this value is a good approximation for many organisms, but it can vary considerably with the conditions of the culture (Stouthamer and Bettenhausen, 1973). Higher Y_{ATP} values are found in richer growth media. Another way to present growth yields is the Y-value, the amount of dry cell matter formed per mole of substrate (see, for example, Forrest and Walker, 1971). For fermentating yeasts Y values around 20 are cited (Bauchop and Elsden, 1960). Different values for Y are presented in Table 8.

The mathematical description of growth phenomena has attracted much interest, in order to get an exact evaluation of specific metabolic parameters and thus to determine the effects of small changes in the external conditions. Three classes of mathematical models for biological growth can be distinguished: the deterministic, the stochastic and the parametric (Walter and Lamprecht, 1976). Nearly all growth relationships incorporate the Malthus equation which properly describes the exponential phase of microbial growth. Many of these models are adapted to special problems. Dedem and Moo-Young (1975) deduced parameters for continuous and batch cultures fermenting on two different carbon and energy sources, van Uden (1967) described the transport-limited growth in a chemostat, Peringer *et al.* (1974) presented a generalized model for growth kinetics of *S. cerevisiae* with regard to incorporation of substrate into biomass and to maintenance energy, and Lamprecht and Meggers (1971) adapted an expanded logistic growth function to thermograms of stirred and unstirred batch cultures. To describe and explain growth often only a polynomial of the *n*th order fitted to the experimental results is wanted—and obtained by computers (Heinmets, 1966; Edwards and Wilke, 1968). Many attempts in this direction have been made and the results may be found in the literature.

D. Calorimetry of metabolism

The term "calorimetric" determination is not clearly defined in the literature and refers to true calorimetric experiments by calorimeters as well as to estimations of heat production from temperature rises and heat flows in fermentors, and also to mere calculations of the rate of heat production from thermodynamic standard values. Moreover, the term "indirect calorimetry" is often used in discussions of cultures of microorganisms though more usually referred to in research studies on higher animals. To avoid confusion a brief introduction to the various techniques for the determination of heat production, rate of heat production and heat content is appropriate.

1. DIRECT CALORIMETRY

This term will be reserved for all those experiments which are performed by means of calorimeters of the adiabatic or the isothermal type. Although yeasts are able to grow over a broad range of temperature, they prefer an optimal temperature of around 30 °C. And as kinetic data are more easily obtained at constant temperature, isothermal instruments are the calorimeters of choice. Only a few experiments have been performed recently with adiabatic or quasi-adiabatic set-ups (Terveen and Reis, 1977). But besides adiabatic and isothermal instruments one has to distinguish between batch and flow calorimeters.

a. *Batch calorimeters*

The most popular batch calorimeters are those of the Tian–Calvet type (Calvet and Prat, 1956, 1963), which are commercially available or "own-built" instruments with adaptations to suit special problems. The more modern constructions use Peltier elements instead of thermopiles. The volume of the calorimetric vessels[1] varies between a few millilitres (Fujita *et al.*, 1976) to more than 100 millilitres (Takahashi, 1973). Even larger amounts of culture medium can be used in a gradient-layer calorimeter resembling that of Benzinger and Kitzinger constructed for metabolic experiments on babies and later applied to microbial growth (Reis and Eiber, 1967). In this instrument the simultaneous measurement of the rates of heat production and of oxygen consumption is possible.

The great disadvantage of batch calorimeters lies in the fact that the contents of the calorimetric vessel are not easily accessible, that manipulations are difficult to perform, and that either of these is always connected with a thermal disturbance of the system. It was mentioned above (see section B.3) that true aerobic conditions are nearly impossible to achieve in batch calorimeters.

One barrier for the penetration of oxygen into the medium is presented by the interface between the liquid and the gaseous phases, the other the long diffusion path from the outside to the vessel. The mode of increasing oxygen concentration by bubbling air through the medium introduces other problems since air bubbles promote the evaporation of the medium resulting in large endothermic effects. In spite of these problems much

---

[1] Dealing with biological calorimetry the expression "cell" should be avoided and replaced by "vessel" to avoid confusion.

equipment has been constructed to gain more information about batch cultures. A recent detailed description can be found in the article by Spink and Wadsö (1976).

b. *Flow calorimeters*
Many of the problems with batch instruments are avoided with flow calorimeters (Monk and Wadsö, 1968; Picker *et al.*, 1969). When applied in microbiology they are connected to external fermentors in which the culture grows (Eriksson and Wadsö, 1971; Brettel *et al.*, 1972). All the necessary manipulations are performed in the fermentor so that the thermal equilibrium of the calorimeter is not disturbed. Simultaneous measurements on the culture are readily carried out, for instance continuous determination of turbidity (Brettel, 1977a; compare with Schaarschmidt and Lamprecht, 1973), of pH value, oxygen concentration, and of dry matter content (Eriksson and Wadsö, 1971), rate of oxygen uptake and of carbon dioxide production (Brettel, 1974, 1977a). Moreover, addition of fresh medium or of special substances like drugs or inhibitors is facilitated and continuous culture made possible.

Some difficulties, especially with respiring cultures of high cell density, arise because of the time involved in pumping from fermentor to calorimeter. In this time a substrate of the medium, often the oxygen, can become limiting so that the metabolic situation in the calorimeter does not correspond to that in the fermentor. Ackland *et al.* (1976) have tried to overcome the problem by placing a small fermentor vessel in the air bath of an LKB flow calorimeter and omitting the heat exchangers. In this set-up they reduced the flow time of normally 4–5 min to about 1 min. But at higher cell concentrations even this time can be too long.

Brettel (1977b) and Cardaso-Duarte *et al.* (1977) have given a mathematical approach to calculate the real rate of heat production in the fermentor from the figures measured in the calorimetric vessel. From one point of view it is an elegant attempt to solve the problem of the time of culture flow, but unsatisfactory from another point of view as assumptions about the metabolism during the flow have to be made. The best approach lies in a submersible calorimeter to be placed in the medium in the fermentor. Allowing for complete homogeneity in the fermentor such a calorimeter would render a true picture of the heat production in the culture medium. Although efforts in this direction are made in many laboratories, only one citation without further details could be found in the literature (Chirkov, 1976).

As the density and the volume of yeasts are higher than those of bacteria and, in addition, many yeasts aggregate in the later phases of growth, the cells may tend to settle in the line and in the calorimetric vessel. A sedimentation in the flow line would only increase the flow resistance and thus diminish the flow rate, while a sedimentation in the calorimetric vessel leads to an otherwise undetected higher cell concentration and to an increased heat output. To overcome this problem the medium flow to the calorimeter is often mixed with a flow of air bubbles. The interfaces between bubbles and medium have a scrubbing effect on the walls of the flow line and the vessel and prevent sedimentation (Eriksson and Wadsö, 1971; Brettel et al., 1972). At the same time the bubbles increase the oxygen concentration in the medium. However, the flow time is too short and the oxygen diffusion rate into the liquid phase too small to provoke a significant effect (James, personal communication). If the calorimetric "vessel" is formed like a spiral as part of the flow line the intermittent flow of air and medium can overcome the sedimentation in the vessel. Attempts have been made to improve flow conditions in the normal vessel (Eriksson and Wadsö, 1971) by altering its shape and the length and position of the inlet and outlet tubes. Gustafsson and Lindman (1977) describe a vessel with a rounded bottom, a shaped inlet tube, and a funnel-shaped top for the outlet which facilitated the removal of air bubbles. Besides counting the cell numbers the occurrence or non-occurrence of sedimentation was checked in a "metabolic" way: in the middle of the exponential phase the vessel was emptied, washed and connected again to the fermentor. If sedimentation had taken place, a shift in the thermogram would appear, with no sedimentation the thermogram would continue undisturbed.

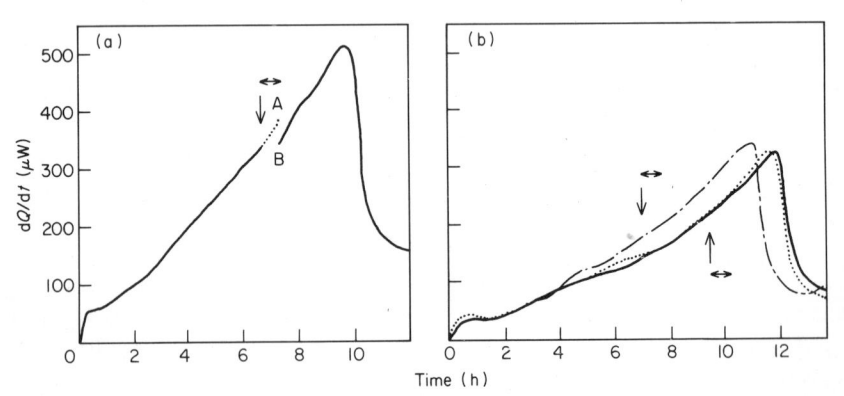

FIG. 2. Thermograms of growth of *D. hansenii* obtained with two different flow vessels: (a) commercial flow vessel, and (b) model LB. At the vertical arrow the vessel was emptied and again connected to the fermentor. After a period of re-equilibration of the calorimeter (double arrow) the thermogram was recorded again. In (b) three independent experiments are shown. (Gustafsson and Lindman, 1977.)

Figure 2 shows both cases for a commercial flow vessel and the model LB of Gustafsson and Lindman (1977).

2. BOMB CALORIMETRY

Some of the first experiments on the metabolism of microorganisms were carried out with bomb calorimeters of the Berthelot type (Rubner, 1903, 1906; Tangl, 1903). Nowadays more sensitive batch and flow calorimeters with good long time base line stability are commercially available and bomb calorimeters are less often used. But they still have their importance in determining the heat content of the crop obtained when yeast is grown under different external conditions on varying substrates (Reiff *et al.*, 1960; Guenther, 1965; Lamprecht, 1969; Prochazka, 1973) and for the storage of energy in cells (Wang, 1968; McDonald *et al.*, 1973).

3. DYNAMICAL CALORIMETRY

Since no "inverse" calorimeter, which is submerged in the growth medium instead of placing the medium in the calorimeter, is available at the moment, other techniques are of interest in large commercial fermentor units; Cooney *et al.* (1968) and Mou and Cooney (1976) adapted a "dynamical calorimetry" for monitoring fermentation processes in fermentors of ~ 1 litre volume. This method consisted of measuring the temperature increase after switching off the temperature control system and taking into account several heat losses and gains. In total one can construct an energy balance equation:

$$Q_H = Q_{acc} - Q_{ag} + Q_{surr} + Q_{evp} + Q_{sen}$$

with Q_H being the heat of fermentation which is to be determined, Q_{acc} the heat accumulated in the system by the rise of temperature, Q_{ag} the heat of agitating the medium, Q_{surr} the heat loss to the surroundings, Q_{evp} the heat loss due to evaporation of water in the air stream and Q_{sen} the heat loss to the air stream. By specific experimental set-ups Q_{evp} and Q_{sen} can be made negligibly small and $Q_{ag} - Q_{surr}$ determined by calibration before the experiment. The remaining Q_{acc} is evaluated by a sensitive thermistor circuit. The thermograms obtained are more complex and sometimes hard to compare with those from more frequently used microcalorimeters. The procedure was tested for several microorganisms, among them the yeast *Candida intermedia*. Like flow calorimetry with a separate fermentor, manipulations on the growth medium and simultaneous determinations of other parameters like pH value, oxygen uptake, carbon dioxide, metabolite and cell mass production and consumption of substrate are simpler than in a batch calorimeter. The drawbacks of the method lie in the possible changes of fluid properties which can alter the power drawn into the medium (Mou and Cooney, 1976). However, differences of only a few per cent between the calculated yields and actual yields demonstrate the usefulness of this method for larger fermentors. It was used by other workers and adapted for gaseous hydrocarbon fermentation by the recently isolated fungus *Graphium* sp. (Volesky and Thambimuthu, 1976). In Fig. 3 is shown a thermogram, together with other data, for the fungus *Trichoderma viride* growing in a medium containing insoluble cellulose; the fluctuations in the thermogram only partly correspond to changes in external parameters.

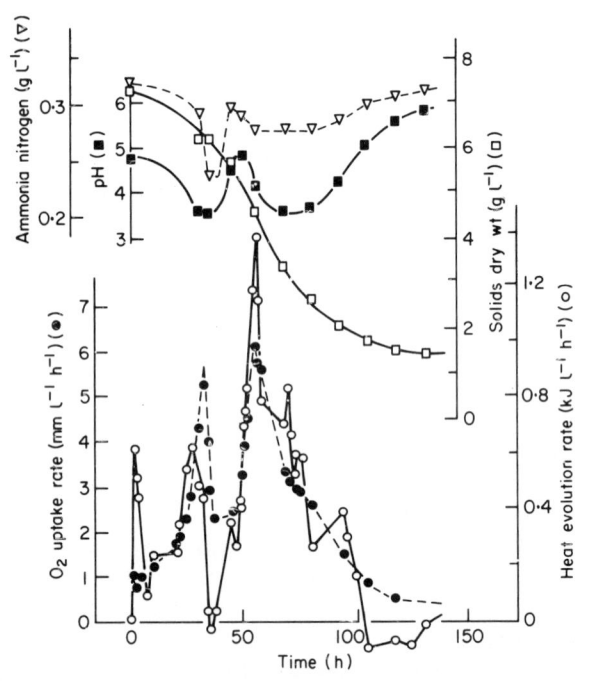

FIG. 3. Growth of the hyphomycete *Trichoderma viride* in a medium containing insoluble cellulose. Rate of heat production (○) and of oxygen uptake (●) are compared with the nitrogen concentration (▽), pH value (■) and concentration of solids (□) in the medium. (Mou and Cooney, 1976.)

Another type of dynamical calorimeter which may be traced back to the labyrinth flow calorimeter of Swietoslawski (1946) measures the temperature difference between inlet and outlet of the cooling water under constant temperature conditions (Goma *et al.*, 1976). After appropriate calibration, the flow rate of water and temperature increase determine the rate of heat production in the system. Especially in large fermentors with a favourable surface-to-volume ratio this type of calorimetry can be successfully applied (Mou and Cooney, 1976).

In a broad sense the first calorimetric experiments of Dubrunfaut (1856), Bouffard (1895), A. Brown (1901) and H. T. Brown (1914) and later quasi-adiabatic calorimetric measurements are of this type.

4. INDIRECT CALORIMETRY

Indirect calorimetry—a term frequently used in metabolic experiments on animals and men—derives its heat quantities by calculation from the gas exchange of an organism. By determination of the rate of oxygen consumption and carbon dioxide production, the RQ value (the ratio of these two figures) is evaluated. This is an indication of the nature of the energy source that is metabolized. One obtains the following RQs: carbohydrates (1·000), fats (0·707), proteins (0·801). As the heat of combustion per mole of these compounds is known, carbohydrates (17·2 kJ g^{-1}), fats (38·9 kJ g^{-1}), proteins (22·6 kJ g^{-1}), the RQs are readily transformed into energetic data.

Warburg or manometric techniques have been most frequently applied in this field of calorimetry, and more recently polarographic methods. It has been shown by several authors (see, for example, Schaarschmidt et al., 1975, 1977; Zotin and Lamprecht, 1976) that simultaneous measurements of heat production by direct and indirect calorimetry yields results with considerable differences. Attempts were made to explain these differences by the theory of bound dissipation functions (Zotin, 1972). Therefore, it is not recommended to draw conclusions about heat production from measurements of gas metabolism only. Simultaneous measurements yield interesting results for the thermodynamics of irreversible processes. Such experiments are readily carried out with flow calorimeters or by dynamic calorimetry. Lamprecht (1976) adapted a microchip with a pressure transducer to a 100 ml vessel in a Calvet calorimeter, thus enabling both heat and pressure measurements.

Another form of indirect calorimetry frequently used consists of the application of standard thermodynamics data for the substances involved in metabolism as substrates or metabolites. Using Hess' law of constant heat summation (and believing in the first law of thermodynamics) the heat changes accompanying the metabolic process may be calculated. A prerequisite for such determinations is an exact biochemical analysis of growth and the establishment of growth or heat balance equations which include a chemical formulation of the grown cell (e.g. $C_{4·20}H_{7·36}N_{0·84}O_{1·90}$; Battley, 1960c). In this way much metabolic data could be obtained (see, for example, Battley, 1960a, b, c, 1971; Mayberry et al.,

TABLE 2
Aerobic growth of yeast in a glucose medium (Oura, 1973)

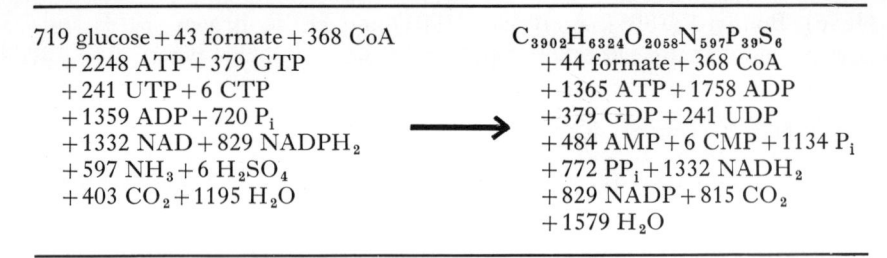

719 glucose + 43 formate + 368 CoA + 2248 ATP + 379 GTP + 241 UTP + 6 CTP + 1359 ADP + 720 P_i + 1332 NAD + 829 $NADPH_2$ + 597 NH_3 + 6 H_2SO_4 + 403 CO_2 + 1195 H_2O ⟶	$C_{3902}H_{6324}O_{2058}N_{597}P_{39}S_6$ + 44 formate + 368 CoA + 1365 ATP + 1758 ADP + 379 GDP + 241 UDP + 484 AMP + 6 CMP + 1134 P_i + 772 PP_i + 1332 $NADH_2$ + 829 NADP + 815 CO_2 + 1579 H_2O

1968; Minkevich and Eroshin, 1973; Oura, 1973a, b, 1975). Table 2 shows an example of such a balanced equation for aerobic growth of yeast.

E. Microcalorimetric investigations

Microcalorimetry is not exclusively used for quantitative measurements of heat production during the metabolism of yeasts; it is also used as an analytical tool. Newell (p. 163) shows the applicability of microcalorimetric finger-printing for the speciation of different organisms. Some examples of diauxic growth will be given later (see Figs 15 and 17), where the calorimeter serves to detect the temporal order of succession of alternating metabolic pathways. In a semi-quantitative way the effect of antibiotics, drugs or uncouplers on the cell metabolism may be tested calorimetrically. These experiments can be carried out from two points of view: (a) Does a substance interfere with the metabolism? (b) To what extent can the metabolism be disturbed by a special substance? The more analytical approach is dealt with in the next section.

1. QUALITATIVE CALORIMETRIC MEASUREMENTS

Yeasts are normally stored in stock cultures on agar slants at temperatures between 4 and 6 °C. From time to time they are genetically controlled and transferred to fresh slants although the metabolism is very reduced at these temperatures. Precultures are taken from the agar and grown in the right medium and at the correct temperature to the required phase. Very often, cells are harvested in the late logarithmic or the early stationary phase and then prepared for the main experiment following a detailed schedule. But in spite of extreme care in the preculture the inocula thus obtained vary slightly from day to day. As far as growth experiments are concerned these differences are of no significance. Under batch conditions of medium glucose concentrations the cell numbers increase by a factor of 100 or more so that the 1 per cent inoculum can be neglected. The situation changes when high initial cell concentrations are applied in research on maintenance or for analytical problems. Under these conditions the inoculum has to be as reproducible as possible. Beezer *et al.* (1976) have proposed a simple method for freezing up to 150 samples of yeast inocula at liquid nitrogen temperature. They tested the viability of cells of *S. cerevisiae* after a storage time of 1–500 days by means of a flow microcalorimeter and found a viable recovery factor of 93·7 per cent as compared with the unfrozen sample. Such a procedure is always recommended when the inoculum enters into the experiment as a "substrate", the concentration of which is an essential part of the analysis.

This method of preparing defined inocula was applied to the microcalorimetric bioassay of antifungal drugs (Beezer, 1977; Beezer *et al.*, 1977; Chowdry, 1977; Newell, 1977). *Saccharomyces cerevisiae* cells (*Candida albicans* respectively, Newell, 1977) were incubated (after thawing) in glucose buffer under anaerobic conditions. They produced a metabolism of maintenance showing zero-order kinetics as demonstrated by a thermogram parallel to the base line. Following an exact time schedule varying amounts of Nystatin bulk material or other drugs were added, changing the thermogram significantly (Fig. 4). The decrease in heat output was checked in different ways and the most appropriate then used for the test.

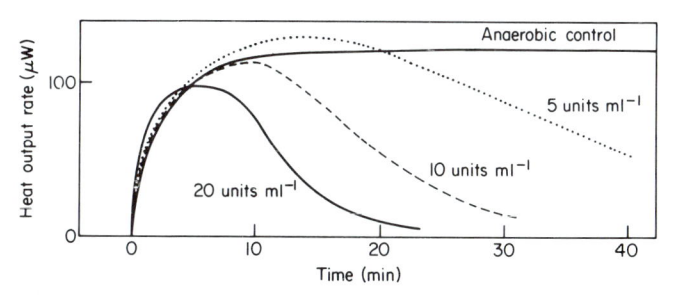

FIG. 4. Thermogram of anaerobic metabolism of maintenance of *S. cerevisiae* and the influence of varying amounts of Nystatin bulk material (Beezer *et al.*, 1977).

Compared with the normally used plate assay for Nystatin, this procedure is more sensitive, so that lower concentrations can be employed, and the reproducibility of ±5 to ±10 per cent is improved to ±3·5 per cent. The shortening of assay time from 16 to 1 h would be only advantageous when using multi-channel calorimeters to compete with the plate technique.

Reis and Büger (1967) and Reis (1975) have developed a calorimetric test for pharmacological and toxic substances. When poisons with a weak influence on respiration are applied, the yeast cell may regulate its metabolism to a new steady state with an unchanged rate of heat production so that the action of the drugs is veiled. The authors, therefore, take the oxygen calorie equivalent (oce) as the criterion of the toxicity. The oce-value is obtained by simultaneous measurement of the rate of heat production and oxygen consumption. Figure 5 demonstrates the influence of a strong poison (KCN) on the rate of respiration and heat production in non-growing yeasts.

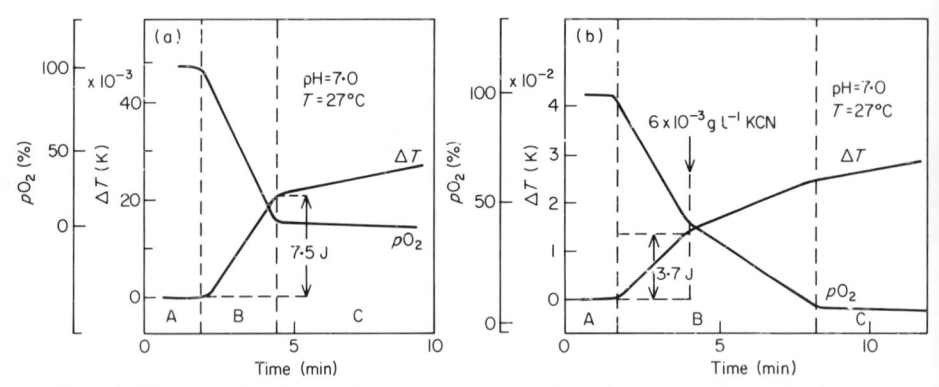

FIG. 5. Heat production and oxygen consumption of non-growing baker's yeast in a quasi-adiabatic calorimeter. After phase A, cells are inoculated. At the end of phase B the physically dissolved oxygen is consumed so that only anaerobic fermentation remains in phase C. (a) Undisturbed culture; (b) change of metabolism after addition of KCN. (Reis, 1975.)

From both rates the data of Fig. 6 are obtained, the change of the oce-value after addition of various modifiers. Line 5 represents the normal linear relationship between oxygen consumption and heat production (2930 J l^{-1} O$_2$ after correction). Histotropic drugs like insulin or cortineurin increase the oce-value, ergotropic drugs like Novadral or Endoxan decrease it.

The experiments were carried out first with a gradient layer calorimeter of the Benzinger–Kitzinger type and later on with a quasi-adiabatic instrument which incorporated oxygen electrodes. These results were then related to human metabolism. Thus, this toxicity test offers a very quick method of surveying various substances. Perhaps the most advantageous application lies in the field of water purity control.

Belaich *et al.* (1968, 1971) and Murgier and Belaich (1971) applied microcalorimetric methods to determine the influence of extracellular glucose concentration on the rate of catabolism in a respiratory-deficient mutant of *S. cerevisiae*, and got Michaelian kinetics which were apparently related to glucose transport. They discovered two-phase kinetics with constant V_{max} values. The two K_m values could be due to two different processes governing sugar transport and utilization at low and high external glucose concentration. As these experiments are extensively described in the contribution by Belaich on bacterial metabolism (p. 1) no further details will be given here.

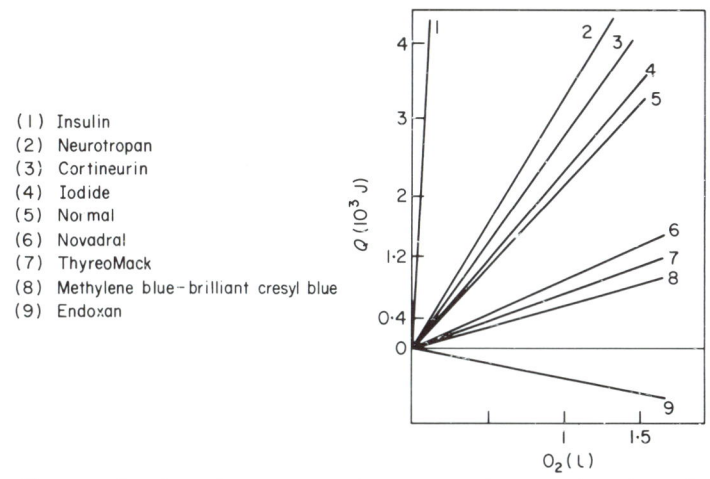

(1) Insulin
(2) Neurotropan
(3) Cortineurin
(4) Iodide
(5) Normal
(6) Novadral
(7) ThyreoMack
(8) Methylene blue-brilliant cresyl blue
(9) Endoxan

FIG. 6. Heat production of a non-growing yeast culture as function of oxygen consumption under the influence of histotropic and ergotropic drugs (Reis and Büger, 1967).

In a similar way the growth of yeast on different saccharides was investigated and the appearance of diauxy analysed calorimetrically (Schaarschmidt and Lamprecht, 1977a, b). In mixtures of monosaccharides with disaccharides or of disaccharides with each other biphasic thermograms were always observed which followed, after transformation, complex Michaelis kinetics. The K_m and V_{max} constants confirm results obtained by biochemical and genetic methods.

2. METABOLISM OF MAINTENANCE

It was shown in Table 1 that only 0·4 per cent of the energy consumption in a growing yeast culture is used for maintenance, and it may be supposed that this value of 0·13 mol ATP per 100 g dry weight is the true energy demand required to keep a cell alive. Thus Giaja's (1920) statement that the large energy turnover during maintenance is a sign of the high catalytic activity of the cell and not the "fundamental biological energy" as in higher organisms seems reasonable.

Some of the first calorimetric experiments on yeast concerned the metabolism of maintenance since, although not mentioned explicitly, the fermentative conditions make growth negligible. For historical reasons and to show the development of "micro"-calorimetry the data of Dubrunfaut (1856) are compiled in Table 3 from his original paper.

TABLE 3

Heat of fermentation observed in a fermentor of 21 400 litres (Dubrunfaut, 1856)

Increase in temperature: 14·05 °C	
$21·4 \times 10^6$ g wine =	1258×10^3 kJ
Heat stored in the tank =	$30·5 \times 10^3$ kJ
$1·156 \times 10^6$ g CO_2 liberated =	$25·5 \times 10^3$ kJ
$19·236 \times 10^3$ g H_2O evaporated =	$45·5 \times 10^3$ kJ
Mechanical work of CO_2 =	$60·8 \times 10^3$ kJ
TOTAL	$= 1420 \times 10^3$ kJ
$2·559 \times 10^6$ g sugar fermented =	552 Jg^{-1}

Half a century later A. Brown (1901) and H. T. Brown (1914) used even larger industrial fermentors (up to 53 000 litres) to determine the heat of fermentation of maltose by brewer's yeast. H. T. Brown illustrated the high fermentative activity of yeasts by the calculation that the temperature of a cell at 30 °C would increase by 75·5 °C if all the heat evolved could be stored within it.

The metabolism of maintenance may be determined calorimetrically under two experimental conditions. If the substrate concentration is high and growth is excluded by lack of an essential component of the medium—e.g. the nitrogen source—one observes a thermogram with zero-order kinetics (see Fig. 4, the thermogram without Nystatin). These forms of thermogram are only obtained with flow calorimeters. More distinct curves are found in batch cultures when at time zero glucose is added to a yeast/buffer suspension (see Fig. 10). At high substrate concentrations the thermogram reveals distinct phases: a mere physico-chemical heat of dilution after addition of glucose, a peak due to a very rapid uptake of glucose by the cell, a broad maximum due to the construction of reserve pools, a steady state of maintenance metabolism following zero-order kinetics, and a return to the steady state of endogenous metabolism. If the initial substrate concentration is low the thermogram is distinctly less structured. These two forms will be discussed separately.

In a series of flow-calorimetric experiments Hoogerheide (1974, 1975a, b) analysed the anaerobic and aerobic metabolism of "resting" cells of baker's yeast. He calculated the minimum and the maximum heat that can be dissipated in a fermentation following the classical Gay–Lussac formula:

$$C_6H_{12}O_6 + 2\,ADP + 2\,P_i \longrightarrow 2\,C_2H_5OH + 2\,CO_2 + 2\,ATP$$

If the energy content of 2 mol ATP (2×50 kJ) is conserved in the cell and if an entropy decrease of the Gibbs energy of 10 per cent occurs (Wilhoit, 1969), the original Gibbs energy decreases to $253 - 25 - 100$ kJ $= 128$ kJ. This is the minimum heat produced in the reaction. To avoid disturbances of the anaerobic conditions by traces of oxygen which diffused into the medium via the connecting teflon tubes between the fermentor and the calorimeter, Hoogerheide used the enzyme glucose oxidase as oxygen sink, and KCN or azide as poisons of the respiratory enzyme system, or respiratory-deficient mutants. Figure 7 shows the effect of azide on the metabolism of maintenance ("+ glucose") and the endogenous metabolism ("− glucose") of baker's yeast.

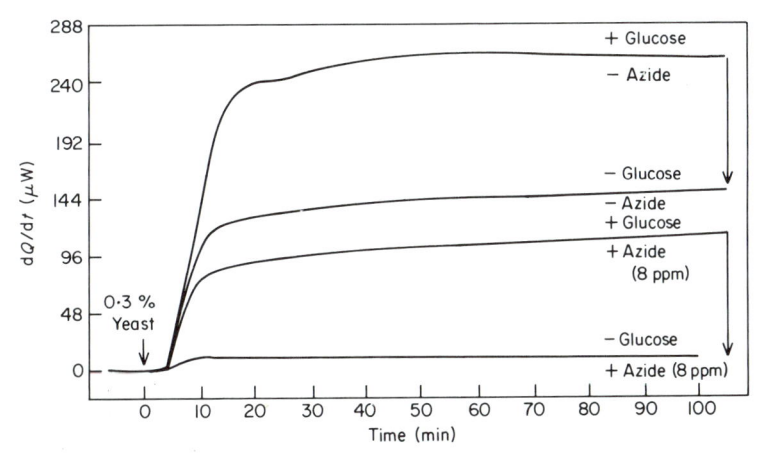

FIG. 7. Heat development of a baker's yeast suspension (0·42 mg dry weight) in buffer in the presence and absence of glucose. An amount 8 ppm azide was added to obtain "anaerobic" fermentation by inactivating the respiratory enzyme system. (Hoogerheide, 1975a.)

The results were not calculated in energy per mole of glucose but per mm³ CO_2 produced during fermentation, since it is in this way that the fraction of the glucose which does not participate in the Gay–Lussac reaction may be ignored. The author recorded a difference of 20 per cent in the glucose that is utilized for synthetic processes between the two

TABLE 4

Heat production during anaerobic and aerobic metabolism of maintenance and growth of baker's yeast, obtained by a flow microcalorimeter (Hoogerheide, 1975a, b)

Substrate	ΔH_{min} kJ mol^{-1}	ΔH_{max} kJ mol^{-1}	Δh_{min} µJ mm^{-3} gas	Δh_{max} µJ mm^{-3} gas	Δh_{obs} µJ mm^{-3} gas	Additions
a. Resting cells, anaerobic						
glucose	126	253	2813	5651	2913	60 ppm KCN
					2900	8 ppm azide
					2637	Glucose-oxidase
					3491	Petite strain
b. Resting cells, aerobic						
glucose	963	2872	7175	21349	18418	—
lactic acid	460	1365	6865	20260	19297	—
acetic acid	322	875	7200	19549	19172	—
ethanol[a]	456	1360	6823	20260	23735	—
ethanol[b]	138	490	6112	21809	23735	—
glucose	963	2872	7175	21349	18879	8 ppm DNP
c. Growing cells, anaerobic						
glucose	126	253	2813	5651	2884	60 ppm KCN
d. Growing cells, aerobic						
ethanol[c]	(1 h) 252	803	6368	21251	18084	—

ΔH_{min}, enthalpy change with complete storage.
ΔH_{max}, enthalpy change without storage.
Δh_{min}, specific enthalpy change with complete storage.
Δh_{max}, specific enthalpy change without storage.
[a] For complete oxidation as $C_2H_5OH + 3\,O_2 = 2\,CO_2 + 3\,H_2O$.
[b] For oxidation as $C_2H_5OH + O_2 = CH_3COOH + H_2O$.
[c] For RQ = 0.24.

conditions, notwithstanding the absence of an assimilable nitrogen source in the medium. This result is supported by the fact that under all conditions of anaerobiosis mentioned above the heat production per mm³ CO_2 is very close to the minimum value. The high rate of fermentation is thus not due to the ATP-ase activity, but to a consumption of ATP in assimilatory processes.

The inverse holds good for aerobic conditions (Hoogerheide, 1975b). The observed heat production related to the consumption of oxygen is very close to the maximum value for all substrates. This indicates that practically all the ATP produced in respiratory processes is dephosphorylated with heat dissipation and is not used for synthetic reactions. Thus, 2-4-DNP as an uncoupler of phosphorylation from respiratory processes could not increase the heat development. The results of Hoogerheide are compiled in Table 4.

Batch experiments on the metabolism of maintenance by Fujita and Nunomura (1977) exhibited quite a different behaviour (Fig. 8). Immediately after addition of glucose a high heat output is registered which

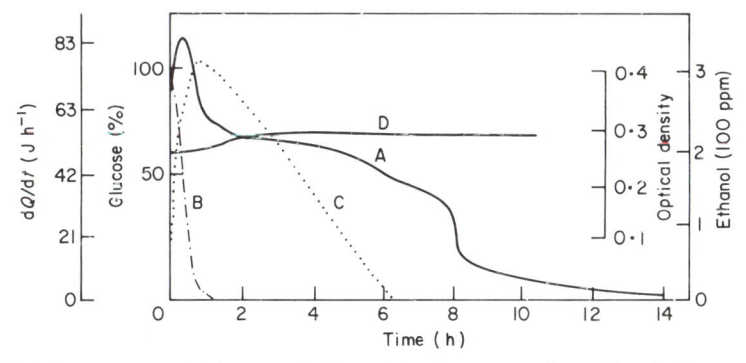

FIG. 8. Time course of the metabolism of maintenance by cells of *S. cerevisiae* under aerobic conditions. a, thermogram; b, glucose concentration; c, ethanol concentration; d, turbidity. (Fujita and Nunomura, 1977.)

decreases to a steady state for a short period and then falls to the base line. By chemical tests the authors demonstrated that the first phase of the thermogram was mainly associated with the uptake of glucose via a mechanism coupled to endogenous metabolism. The second phase, of ethanol consumption, may begin at nearly the same time as the glucose uptake. It can be depressed by completely anaerobic conditions or by addition of KCN. Equivalent thermograms for *Aerobacter aerogenes* were found by Boivinet and Grangetto (1963).

The thermograms observed by Schaarschmidt (1972), Lamprecht *et al.* (1973) and Schaarschmidt *et al.* (1973) show similarities with those described above at low substrate concentration (Fig. 9), but were explained in quite a different way. At higher glucose concentrations one gets a multiphase thermogram as described at the beginning of this contribution.

The steady state of horizontal deflection, $\Delta^{\bullet}H_{stat}$, corresponds to the real metabolism of maintenance, which remains at the same level as long as the glucose concentration is higher than 50×10^{-12} g per cell.

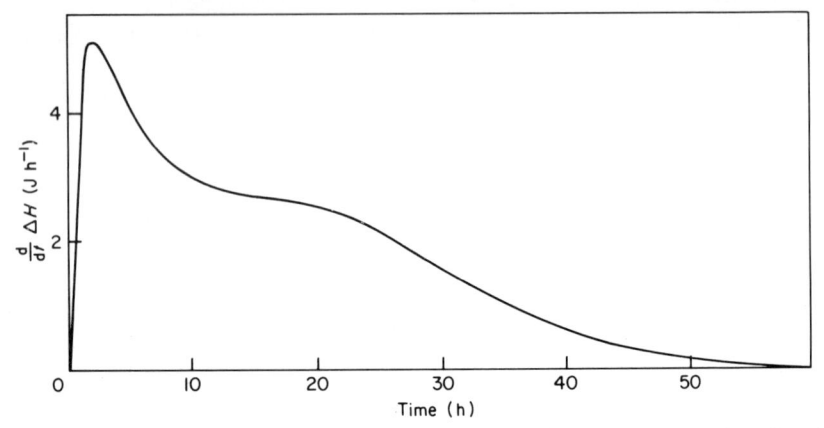

FIG. 9. Thermogram of the maintenance at low glucose concentration for *S. cerevisiae* (Lamprecht et al., 1973).

$\Delta^{\bullet}H_{stat}$ is proportional to the number of cells in the culture, but varies with the yeast strain under investigation. In an analysis of a series of polyploid and isogenic strains of *S. cerevisiae* $\Delta^{\bullet}H_{stat}$, given in watts per cell, increased in a sigmoidal form with increasing ploidy, i.e. cell weight or volume (Lamprecht et al., 1976). But related to the cell mass there are two distinct levels of maintenance, the lower one for haploid to triploid strains, the higher one, increased over the lower value by a factor two, for tetraploid to hexaploid strains. This result may be interpreted by suggesting that more than a triple set of chromosomes represents an unfavourable energetic situation, so that per gram dry matter a more intensive metabolism is required to maintain vital functions.

The hatched area (ΔH_{depot}) in Fig. 10 corresponds to the formation of depots in the cell since:

a. if the cells incubated in buffer were taken from different growth phases ΔH_{depot} was a minimum for the logarithmic phase when the depots should be filled and a maximum for the lag and stationary phases;
b. if no glucose is added, no maximum appears;
c. by radioactive labelling an increased incorporation is found in the first few hours;

d. the energy values obtained correspond well with data on glucose incorporation (van Niel and Anderson, 1941; Pohlit et al., 1966).

The heat of depot formation amounts to 3.9×10^{-8} J per cell or 7.3×10^{-12} g glucose per cell of a diploid strain and exhibits the same dependence on ploidy as the stationary heat production.

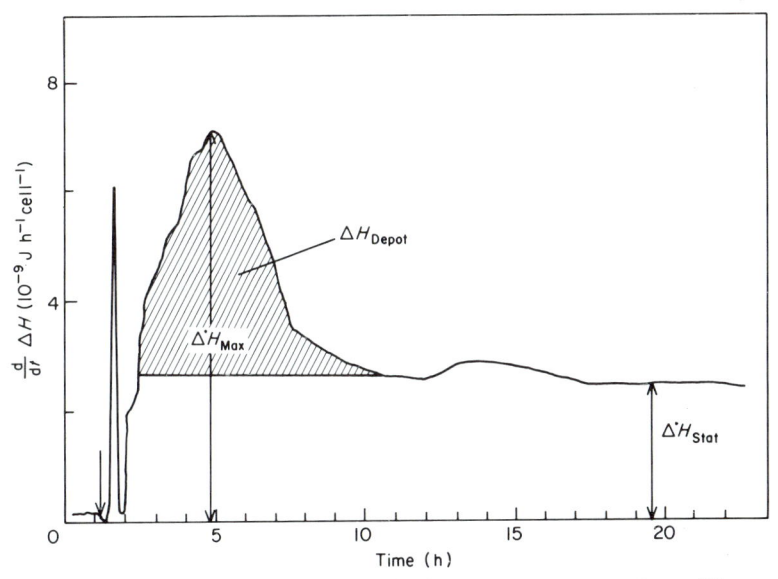

FIG. 10. Thermogram of maintenance at high glucose concentrations. Glucose is added after a period of endogenous metabolism (arrow). (Lamprecht et al., 1973.)

The short exothermic peak a few minutes after addition of glucose corresponds to heat production of $\sim 2.5 \times 10^{-12}$ W per cell and can be explained by a more phasic catabolism of glucose, due to a regulation by phosphofructokinase (Kopperschläger, 1968).

The metabolism of maintenance in batch cultures is strongly influenced by small amounts of drugs which block respiratory enzymes (sodium azide) or interfere with the DNA (dyes) (see p. 66). The resulting thermograms are even less distinct than those of respiratory-deficient strains. Addition of acriflavine to a non-growing culture in a steady state leads to a small exothermic reaction immediately after injection which can be ascribed to the binding of the dye to cell components. After 1 h a steady state is achieved (Schaarschmidt, 1972). Results on some yeasts and different dyes are shown in Table 5. If the cells are incubated in the presence of the drug, the initial maxima (ΔH_{depot}) almost disappear and only a rather low, constant, heat flow results.

TABLE 5

The influence of respiration repressors on the metabolism of maintenance of different strains of *S. cerevisiae* (Schaarschmidt, 1972)

		\multicolumn{4}{c}{Strain}
		211	2012	S 2095	298
Repressor		Thio-pyronine $5\ \mu g\ ml^{-1}$	Acridine yellow $7.1\ \mu g\ ml^{-1}$	Acriflavine $5\ \mu g\ ml^{-1}$	Sodium azide $7.8\ \mu g\ ml^{-1}$
$\Delta H_{depot} \times 10^{-8}$ J per cell	with without	0·42 4·04	2·05 4·73	0·42 4·31	0·63 3·98
$\frac{d}{dt}\Delta H_{max} \times 10^{-12}$ W per cell	with without	0·55 2·05	0·20 2·34	0·14 1·86	0·23 1·81
$\frac{d}{dt}\Delta H_{stat} \times 10^{-12}$ W per cell	with without	0·55 1·08	0·13 1·29	0·14 1·36	0·21 1·56

The ability of yeast cells to store glucose was compared with their recovery rates after u.v. irradiation and "liquid-holding" in buffer (Schaarschmidt and Lamprecht, 1974). The survival rates were found to be proportional to the intracellular glycogen reservoirs as observed by microcalorimetry in accordance with the suggestion that the capacity of repair is primarily determined by the size of intracellular reserves and the turnover rates of energy sources (Kiefer, 1971; Jain and Pohlit, 1972; Schaarschmidt *et al.*, 1973). However, there was no difference in the steady-state metabolism of cells which were unirradiated or irradiated up to doses resulting in 1 per cent survival. Growth is a more sensible index to radiation injuries than the metabolism of maintenance. The same held good when X-rays were used instead of u.v. irradiation.

During experiments to increase the oxygen concentration in batch cultures by higher pressures of oxygen in the upper part of the calorimetric vessel and thus shifting the metabolism of maintenance to aerobiosis, strong oscillations in the rate of heat production were observed (Schaarschmidt *et al.*, 1973). Simultaneous measurements of the optical density in the upper layer of the medium exhibited a circulation of the cells through the vessel which could be brought to zero by increasing the specific density of the liquid to that of the cells. The explanation of the phenomenon lies in the fact that when cells leave a layer of adequate oxygen concentration and switch their metabolism to anaerobiosis. Production of carbon dioxide within the cells causes an influx of water into the cell and a decrease in specific density. The cells then return to the upper layers and to aerobic catabolism, release carbon dioxide and water, and are then ready to repeat the cycle. This example of external control over metabolic pathways demonstrates how difficult it can be to adjust external parameters properly in batch cultures.

Table 6 compiles the data found in the literature relating to metabolism of maintenance (some could only be calculated approximately from the data given). All the figures show great variation which demonstrates that the conditions were ill defined. Sometimes true anaerobic or aerobic conditions were not achieved, sometimes the glucose incorporated into the depots of starved cells may have played an essential role. The $-d\Delta H/dt$ for *S. cerevisiae* (Schaarschmidt, 1972) is that for the widely used wild strain 211. The corresponding values for *S. cerevisiae* strains with special genetic properties like sensitivity or resistance against u.v. irradiation or for respiratory-deficient strains varied considerably (see Table 5).

Only a few data are available on endogenous metabolism, i.e. catabolism without an exogenous energy source. Rubner (1904) was the first to calculate this energy state of the cell from the heat of combustion. When yeast was stored for 2 days in a buffer without substrate the heat of combustion decreased by a value that corresponded to a rate of heat production of 4·0 mW g^{-1}. By direct calorimetry he obtained a figure of 0·40 mW g^{-1} dry weight. These rates are smaller by a factor of 10 to 100 than the usual rates described for the metabolism of maintenance (see Table 6). Fujita and Nunomura (1977) observed an endogenous metabolism lasting only a few hours under anaerobic or aerobic conditions without adjustment to a steady state with a total heat production of 780 J g^{-1} dry weight (aerobically) and 336 J g^{-1} (anaerobically). The thermograms exhibited a "heat burst" in the first hour of aerobic metabolism and are less featured under anaerobic conditions. The maximum rate of heat production corresponded to $\sim$130 mW g^{-1} dry weight (aerobically) and 27 mW g^{-1} dry weight (anaerobically). These rates of heat production seem extremely high when compared with the metabolism of maintenance (Table 6).

If one assumes that endogenous sugar in the form of glycogen was catabolized in this reaction only 0·3 per cent of the cell dry mass was used up, whereas the depots amount to 10–30 per cent of the mass. Thus, these thermograms can be taken as the first part of a transition phase to a new

TABLE 6

Metabolism of maintenance in yeasts obtained by calorimetric experiments in anaerobic batch cultures, aerobic batch cultures and continuous cultures

Yeast	Substrate	$-\Delta H$ kJ mol^{-1}	$\frac{d \Delta H^a}{dt}$ mW g^{-1}	$\frac{d S^a}{dt}$ g h^{-1} g^{-1}	Reference
Anaerobic batch cultures					
Brewer's yeast (?)	Dextrose	98·4	—	—	Bouffard (1895)
Brewer's yeast	Maltose	2 × 90·0	—	—	Brown (1901)
Brewer's yeast	Cane sugar	2 × 112	116^a	—	Rubner (1904)
Baker's yeast	Cane sugar	2 × 110	87^a	0·45^a	Hill (1911)
Brewer's yeast	Maltose	2 × 94·2	—	0·045	Brown, H. T. (1914)
Yeast	Glucose	98·0	—	—	Coon and Daniels (1933)
Baker's yeast	Glucose	71·2	74	0·68	Ohlmeyer and Fritz (1965)
S. cerevisiae (211)	Glucose	111	48	0·25	Lamprecht (1969)
S. cerevisiae (211)	Glucose	112	48 (157)^c	—	Schaarschmidt (1972)
S. cerevisiae	Glucose	101	40	—	Lamprecht *et al.* (1973)
Baker's yeast	Glucose	153	310	—	Hoogerheide (1975a)
Aerobic batch cultures					
Baker's yeast	Glucose	365^b	130^a	—	Grainger (1968)
Baker's yeast	Glucose	—	454	0·009	Ohlmeyer and Fritz (1965)
Baker's yeast	Glucose	—	594	0·149	Hoogerheide (1975b)
Continuous cultures					
S. cerevisiae	Glucose	531^b	55	—	Brettel (1974)
S. cerevisiae	Glucose	470^b	100	—	Brettel (1977b)

^{a} Calculated.
^{b} Per mole oxygen.
^{c} Stirred culture.
^{d} The differentials on the enthalpy and entropy functions are quoted per gram dry weight.

low steady state of endogenous metabolism, which is too small to be detected in these experiments. If the 336 J g^{-1} were dissipated during 2 days (as in Rubner's experiments), a rate of heat production of 1·9 mW g^{-1} can be calculated (comparable to the data of Rubner and to a value of 13·2 mW g^{-1} for *S. cerevisiae* (Schaarschmidt, personal communication) and 9·67 mW g^{-1} for *Streptococcus facealis* (Forrest and Walker, 1963)). From Fig. 4 one can deduce the extremely high value of endogenous metabolism in Hoogerheide's experiments (Fig. 4: "-glucose, -azide"). The rate of heat production under "anaerobic" conditions is calculated to be ∼360 mW g^{-1}. Under aerobic conditions a short heat burst appeared as in the experiments of Fujita and Nunomura with a maximum rate of 340 mW g^{-1} and a steep decrease to a lower level of ∼60 mW g^{-1} after 1 h. The total heat developed in this period amounted to 470 J g^{-1}. Figure 11 shows the thermograms for "anaerobic" and aerobic endogenous metabolism. Possible explanations for the high "anaerobic" metabolism are (a) traces of oxygen diffusing in through the

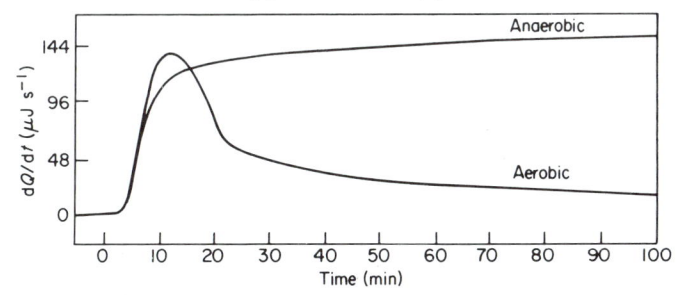

FIG. 11. Endogenous metabolism of 0·42 mg dry weight baker's yeast "anaerobic" and aerobic conditions in a flow microcalorimeter (Hoogerheide, 1975a).

Teflon tubes and (b) a continuous presentation of fresh anaerobically maintained cells to aerobic conditions establishing a permanent heat burst situation. When in these experiments the respiratory enzyme system was poisoned by azide or KCN, the endogenous metabolism dropped to approximately zero (Hoogerheide, 1975a).

3. METABOLISM OF GROWTH

In experiments on the growth of microorganisms a calorimeter can be used as an analytical or a quantitative tool. Without quantifying the value of heat production the signal may be taken as an indication of growth as such, on different metabolic pathways, on different metabolic situations in the culture and on the ease of using a chemical compound as substrate. The combination of these answers is used in "finger-printing" to identify microorganisms from their thermograms (see contribution by Newell, p. 163) or to evaluate the applicability of special microbial strains to fermentation problems. Section (a) is more dedicated to the calorimetric determination of the rate of heat production and the total dissipated heat during growth in batch cultures. Continuous cultures are examined in section (b).

Growth means formation of new cellular material from inanimate substances at the expense of free energy. This is mainly needed for the synthesis of proteins, nucleic acids, lipids and cell wall material. In autotrophic organisms synthesis from the low level of carbon dioxide at a high cost of energy and time is possible, whereas heterotrophs use building blocks already present in the medium. Between these two extremes normal metabolism takes place. Yeasts, as do many other organisms, grow on carbohydrates as the sole source of energy and carbon. Forrest and Walker (1971) demonstrated that the energy required for the synthesis of cell monomers from hexose does not exceed 0.76×10^{-3} mol ATP g^{-1} of cells. This corresponds to an energy maximum of 40 J g^{-1} as compared with the combustion heat of 20 930 J g^{-1} or a heat released during growth of more than 5000 J g^{-1} under anaerobic conditions. Thus it seems plausible that in some growth experiments the measured heat equals the calculated figure for the fermentation, as the real enthalpy of growth is too small to be detected and the carbon source is assimilated at the same level of oxidation as the cellular material (Murgier and Belaich, 1971). But it will be calculated in section E.4 that the mean composition of a yeast cell may be symbolized by $CH_{1.68}O_{0.49}$ instead of CH_2O.

Calorimeters determine primarily the rate of heat production or the heat produced. Operating under constant pressure and at constant volume the heat produced equals the enthalpy change. This is the case if no gases are involved in the metabolism, as in the lactic acid fermentation, if the volume of all gases stays constant, as in the aerobic fermentation of glucose (RQ = 1) or if the term $p\Delta V$ is negligible. But it is not correct to omit the entropy term and to present the heat values as changes in free energy.

a. *Growth in batch cultures*
In batch cultures the form of the thermogram is very dependent on the conditions in the calorimetric vessel although the type of metabolism remains unchanged. Because of the higher specific density of yeast cells they tend to sediment in the culture and to grow at the bottom of the vessel. The rate of heat production is then determined by the diffusion rates of substrate and metabolites and not by the substrate concentration.

Stirring the medium avoids sedimentation and changes the slope of the thermogram (Fig. 12) (Lamprecht and Meggers, 1969). The steep decay following the maximum corresponds initially to the Michaelis–Menten kinetics (Lamprecht and Meggers, 1971) used by Belaich et al. (1968) for the microcalorimetric determination of glucose permeation in bacterial and yeast cells. This type of thermogram is often observed in bacterial cultures where no sedimentation occurs.

Figure 13 shows a thermogram typical for growth under nearly anaerobic conditions in a batch calorimeter. After a short lag phase with no detectable heat production (because of the small number of cells present initially when compared with the final number), glucose fermentation and

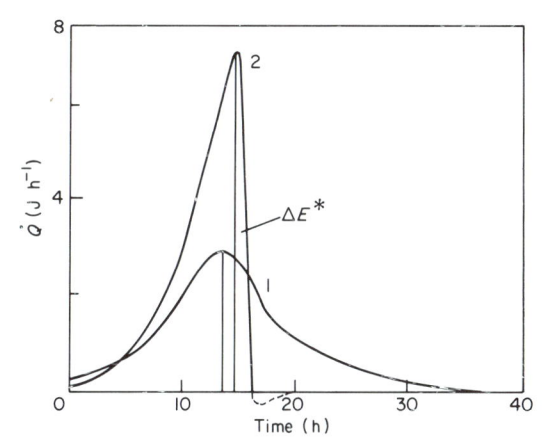

FIG. 12. Thermograms of two identical anaerobic cultures of *S. cerevisiae* in a simple medium. 1, Unstirred; 2, stirred. (Lamprecht and Meggers, 1969.)

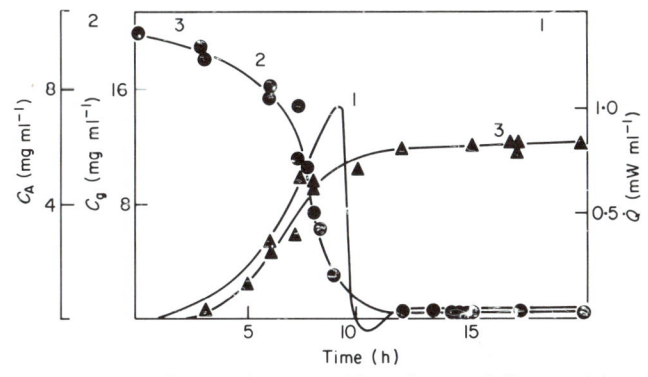

FIG. 13. Thermogram of a nearly anaerobic culture of *S. cerevisiae*. 1, Rate of heat production $\dot{Q}$; 2, glucose concentration C_g; 3, alcohol concentration C_A as functions of time. (Schaarschmidt et al., 1975.)

heat production increase to a maximum when the glucose concentration becomes growth limiting. The rate of heat production drops towards a short endothermic phase which is caused by evaporation of carbon dioxide from the liquid. This effect is more pronounced at high glucose concentrations and vanishes for small ones when all the carbon dioxide remains dissolved in the medium. The third phase is that of a slow aerobic respiration of the formed ethanol. The turnover is limited by the diffusion rate of oxygen to the calorimetric vessel. Thermograms of stirred cultures with varying glucose concentrations are similar in a mathematical sense, so that the maximum heat flow is proportional to the square root of the glucose concentration (Lamprecht et al., 1971).

Experiments are often performed in which the energy source is growth limiting. By changing the substrate concentration and keeping the other components in the medium constant one obtains the enthalpy change per mole of substrate. But at high concentrations another compound in the medium can become limiting so that the total heat output is determined by the concentration of the energy source, while growth is limited by another factor and a period of maintenance metabolism appears. Such behaviour is demonstrated in Fig. 14 where after 10 h the nitrogen

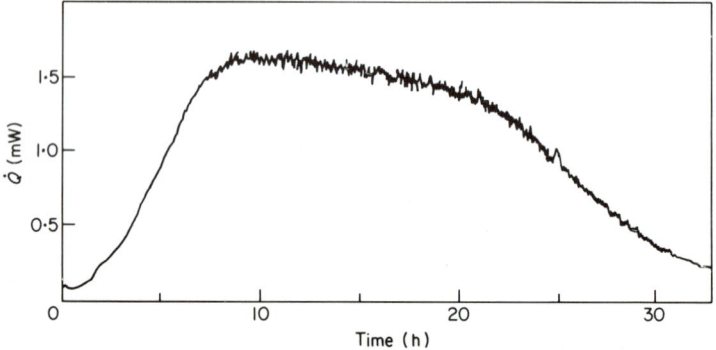

FIG. 14. Unstirred culture of *S. cerevisiae* in minimal medium at high glucose concentration (Lamprecht and Meggers, 1969).

source is depleted and after 20 h the glucose concentration becomes limiting for the metabolism of maintenance. The situation becomes more pronounced when a stirred culture without diffusion problems is used.

It was pointed out above that true aerobic conditions are hard to obtain in batch calorimeters. The small deviation from the base line in Fig. 13 after the depletion of glucose exhibits the respiration of the previously formed ethanol. The situation changes drastically when a flow calorimeter is used in combination with a fermentor with intensive aeration and a low cell concentration so that the oxygen is not rate limiting during the flow to the calorimeter. Figure 15 demonstrates the biphasic thermogram of such a culture (Schaarschmidt and Lamprecht, 1977). An initial, exponential, growth phase takes place at the expense of aerobic fermentation of glucose as shown by the decrease in glucose and the increase in ethanol concentration. After depletion of glucose a lag phase of ~1 h appears and a second exponential growth phase resulting from respiration of ethanol. This energetic behaviour is reflected in the growth of biomass (X) too, but its slope is less pronounced and informative.

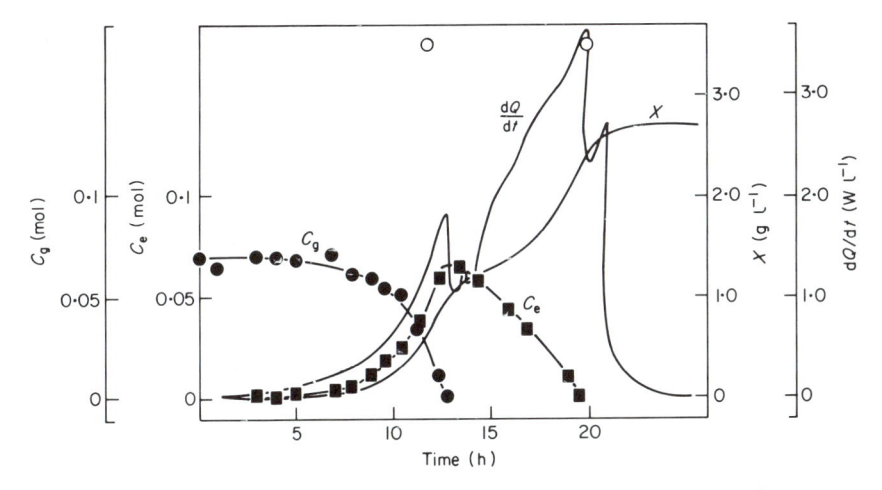

FIG. 15. Aerobic culture of *S. cerevisiae*. Rate of heat production (dQ/dt), dry weight (X), glucose concentration (C_g) and ethanol concentration C_e as functions of time. (Schaarschmidt and Lamprecht, 1977.)

The distinct oscillations in the second lag phase appear when the cells have to adapt from aerobic fermentation of glucose to respiration of ethanol. They have been observed also by Mochan and Pye (1973) during respiration experiments on the yeast *Saccharomyces carlsbergensis* and should not be confused with higher frequency oscillations in glycolysis when changing from aerobic to anaerobic conditions (Winfree, 1972). These slow periodic changes in the respiration rate and heat production are expressions of a control mechanism at the level of enzyme activation and inhibition within the oxidative phosphorylation or the tricarboxylic acid cycle system. Thermodynamically the transition from the steady state of aerobic fermentation to the new one of respiration seems to follow a limit-cycle behaviour (Mochan and Pye, 1973).

In Fig. 16 two identical aerobic cultures of *S. cerevisiae* are compared, one grown in a stirred batch calorimeter, the other in an intensively agitated and aerated fermentor and monitored by a flow calorimeter.

The enormous differences even during the aerobic fermentation of glucose results from the effect that additional respiration at a low rate takes place in this part of growth (Pohlit et al., 1966) and increases the heat output strongly (Brettel, 1977b).

A thermogram corresponding to those in Figs 15 and 16 was obtained for *S. cerevisiae* by Fujita et al. (1976) in a batch calorimeter with a favourable ratio between liquid and gas phases. By rotating the instrument a thorough agitation of the culture is achieved and as a consequence

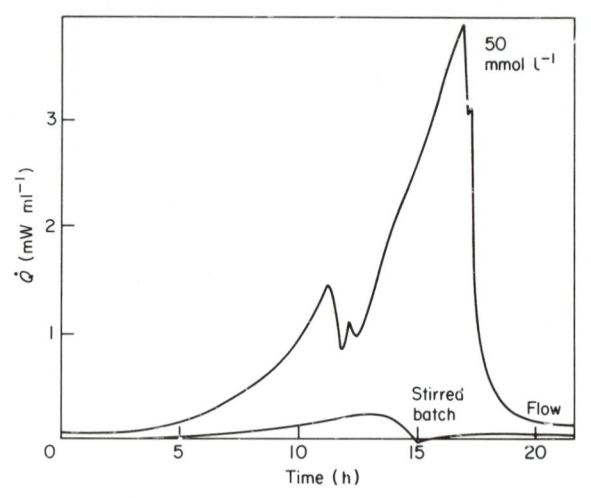

FIG. 16. Comparison of two aerobic cultures of *S. cerevisiae*, one grown in a batch calorimeter, the other in a separate fermentor and monitored by a flow calorimeter (Brettel, 1977b).

aerobic conditions are obtained. Intensively structured thermograms in a flow calorimeter occur during the growth of a salt-tolerant yeast *Debaryomyces hansenii* in a synthetic medium (Gustafsson and Norkrans, 1976).

The biphasic growth shown in Figs 15 and 16 is a special case of diauxism, the growth on two or more energy and carbon sources. In the first calorimetric experiments on yeast such a behaviour was observed (Rubner, 1904; Hill, 1911). It is calorimetrically used for "finger-printing" the different strains. Examples are given by Beezer (1977) with up to five different sugars in a growth medium for *Klyveromyces fragilis*. In the brewing industry even more sugars are used to define the properties of strains for fermentation. Van Dedem and Moo-Young (1975) have developed a model to predict the behaviour of continuous and batch cultures when growing on two different energy and carbon sources. The basic assumption made was that the permease of the favoured substrate is

GROWTH AND METABOLISM IN YEASTS 83

constitutive in the cells, whereas the permease of the second substrate is subject to induction and to catabolite repression (see section B).

Schaarschmidt and Lamprecht (1977, 1978) investigated, calorimetrically, the diauxic performance of *S. cerevisiae* when growing on combinations of the monosaccharides fructose, galactose, glucose and mannose and the disaccharides maltose and saccharose. The biphasic behaviour could be observed in two distinct peaks in the thermogram (Fig. 17), a stepwise increase of the biomass or biphasic kinetics in the sugar uptake (Fig. 18). Diauxy always occurs in mixtures of mono- with disaccharides or of two disaccharides, but in mixtures of two monosaccharides only when galactose is concerned. In all sugar combinations a heat production of 446 ± 17 J g^{-1} under nearly anaerobic conditions is observed. Taking the disaccharides

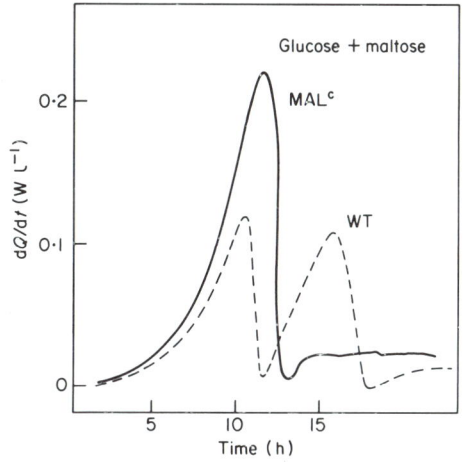

FIG. 17. Thermogram of a maltose constitutive mutant MAL^c of *S. cerevisiae* and of the corresponding wild type (WT) in a medium with glucose and maltose as energy and carbon source (Schaarschmidt and Lamprecht, 1978).

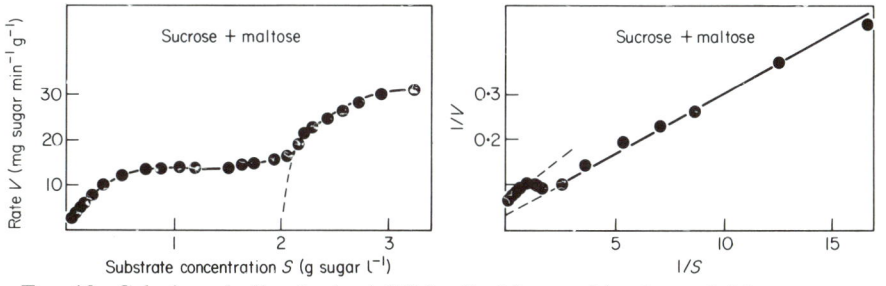

FIG. 18. Calorimetrically obtained Michaelis–Menten kinetics and Lineweaver–Burk plot for growth in a sucrose–maltose medium (Schaarschmidt and Lamprecht, 1978).

as equivalent to 2 mol monosaccharides this transforms to 80·3 kJ mol^{-1} a rather low value when compared with the theoretical value of 135 kJ mol^{-1} (Belaich et al., 1968).

Whereas most authors studied the whole growth period up to the exhaustion of the substrate, Hoogerheide (1975a, b) concentrated on the lag phase of anaerobic and aerobic yeast cultures. During the first hour of the experiment no growth took place, but a considerable heat development. After the end of this period heat production and growth increased at the same rate, i.e. the rate of heat dissipation per cell stayed constant (Fig. 19). After 5 h the cell density had increased so much that the oxygen concentration in the flow line became limiting and the heat production dropped dramatically as a result of the change from aerobic to anaerobic conditions where ethanol cannot be degraded further. Growth, as measured by turbidity in the fermentor, increased steadily demonstrating that "parallel" experiments in a calorimeter and in a corresponding vessel may not be parallel at all.

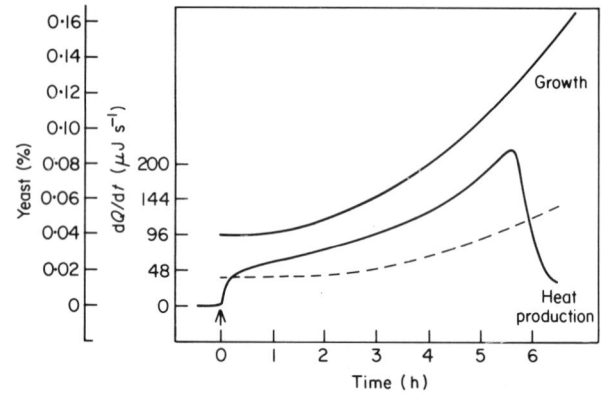

FIG. 19. Aerobic growth and heat development of baker's yeast on ethanol and corresponding heat production of the same concentration of non-growing cells (dashed line) (Hoogerheide, 1975b).

Under anaerobic growth conditions Hoogerheide (1975a) observed an increase in the rate of heat and carbon dioxide production in the first 2 h after inoculation which was not due to growth but to an increased fermentative activity of the single yeast cell. The values of heat production per volume carbon dioxide are very close to those found in the metabolism of maintenance and demonstrate that under anaerobic conditions the yeast cells utilize the ATP produced during fermentation quite efficiently for assimilation, whereas under aerobic conditions the cells are extremely wasteful with the ATP energy.

During growth part of the substrate energy is stored via assimilation in the cells and not dissipated as heat. Some of the differences between the enthalpy changes in Table 8 may be due to the fact that some enthalpies are calculated per mole substrate globally consumed and some per mole substrate *not* assimilated to cell material. Battley (1960a, b) gives a thorough account of the conservative and non-conservative reactions in the growth of *S. cerevisiae* (Hansen). In a "conservative" reaction energy is stored in products which are different from those of the relative "non-conservative" reaction, i.e. in the substance of the cells. To obtain a completely non-conservative reaction the author used DNP as uncoupler of assimilation. During metabolism of maintenance, although by definition without growth, one observes some assimilation (see Table 4). During anaerobic growth on glucose and aerobic growth on glucose, ethanol and acetic acid the enthalpy change in a conservative reaction averages 71·2 per cent of that in the relative non-conservative reaction. Thus 28·8 per cent of energy is stored in the cellular products under Battley's experimental conditions. Calculation of the possible free energy changes during aerobic and anaerobic catabolism of glucose results in 2922 kJ mol^{-1} and 218 kJ mol^{-1}, respectively, i.e. 13·4 times as much energy is available for aerobic growth than is available for anaerobic growth. Nevertheless, the observed increase by only 3·4 demonstrates that the energy recovery is not complete. The formation of glycerol in the first phase of aerobic fermentation of glucose provides an explanation since glycerol is not metabolized by this yeast and its chemical potential is wasted as heat and entropy (Battley, 1960a).

Multi-channel calorimeters are constructed for parallel measurements on many cultures under varying conditions, especially for the identification of microorganisms (Russell *et al.*, 1975). Takahashi (personal communication) changed his twin instrument of 200 ml working volume which was successfully tested with growth experiments of *S. cerevisiae* (Takahashi, 1973) to a calorimeter with six reaction vessels centred around one reference. The influence of various sugars and of the size of the inoculum on the form of the thermograms were determined. In the latter case only the lag phase was prolonged if the amount of inoculum decreased while the form of the thermogram, the maximum heat flow and the total heat output stayed constant. These results are in accordance with findings of Lamprecht (1969) who got unchanged thermograms as long as a lag phase appeared. With an inoculum of $>1.5 \times 10^8$ cells the thermogram lost its typical shape as no real logarithmic phase could be established in the culture.

In industrial fermentations liquid media are always used. Solid media— mainly provided in Petri dishes or as agar slants—are, however, frequently used in microbiology and microbial and radiation genetics. Astonishingly only a few calorimetric experiments with growth on solid media are known. Ross (1973) reported results using a modified NIH Calorimeter which accommodated normal Petri dishes as the calorimetric vessels. Lamprecht and Meggers (1972) formed a solid layer of medium along the walls and on the bottom of normal calorimetric vessels, so that a large surface of between 27 and 19 cm² (3·7–10·3 ml medium) was obtained. Thus good aeration of the cells and release of carbon dioxide to the gas phase was possible. Under these conditions aerobic metabolism is only limited by the supply of oxygen to the calorimeter vessel and not by diffusion through a small layer between gas and liquid phase or through the whole

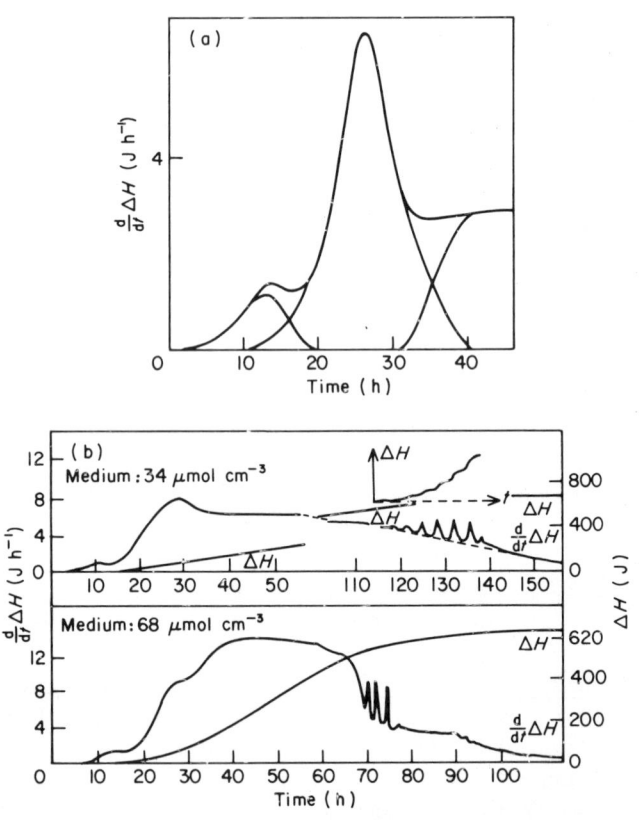

FIG. 20. First part of aerobic fermentation of a typical thermogram from cultures growing on a solid medium (a). The graphs show two complete thermograms of growth on solid medium under aerobic conditions. The glucose concentration and the amount of medium were varied in the experiments (b). (Lamprecht and Meggers, 1972.)

liquid medium. Diffusion only becomes limiting for low substrate concentrations at the end of growth and leads to the oscillations shown in Fig. 20. Diffusion of glucose is high enough so that the whole energy source in the medium can be used, not only that of an upper layer. Anaerobic glucose fermentation results in a quite low enthalpy of 67·8 kJ mol^{-1}, and aerobic respiration of 1674 kJ mol^{-1} agreeing with values obtained in liquid media (Table 7). The polysaccharide forming the agar is not catabolized by the cells under these conditions.

To overcome the problem of insufficient oxygen supply in batch calorimeters higher gas pressures were applied to liquid media (Lamprecht and Schaarschmidt; 1973; Schaarschmidt et al., 1974).

To evaluate the effect of pressure *per se* metabolically inert gases like nitrogen or argon were compared with air or oxygen. Figure 21 shows thermograms obtained under different conditions in liquid cultures of *S. cerevisiae*. The poisoning of metabolism at higher oxygen concentrations is quite obvious from curve 4. Under nitrogen, an anaerobic fermentation is established whose heat production is independent of agitation

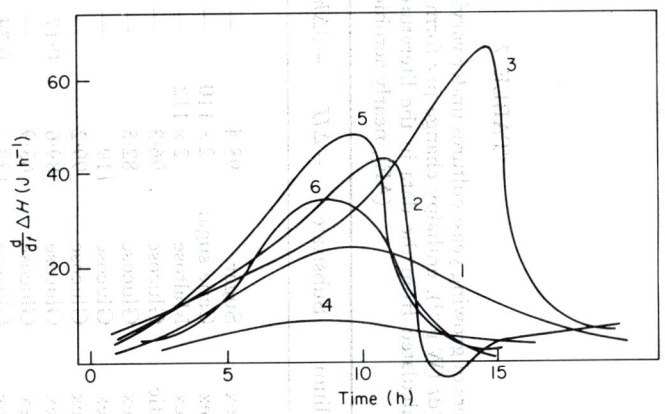

FIG. 21. Thermograms of growing cultures in a synthetic medium at various conditions of gas pressure. 1, Air atmosphere, unstirred; 2, air atmosphere, stirred; 3, oxygen ($p=0$), stirred; 4, oxygen ($p=3$), stirred; 5, nitrogen ($p=5$), stirred; 6, nitrogen ($p=5$), unstirred. The pressure p is given in kp cm^{-2} as pressure in excess of atmospheric pressure.

and of the pressure up to 10 kp cm^{-2} (98 kJ mol^{-1} glucose). The heat production under oxygen in stirred cultures is maximal just above 1 kp cm^{-2} absolute pressure and exhibits an essentially higher enthalpy of 490 kJ mol^{-1} glucose instead of the 80 kJ mol^{-1} achieved with air. However, it drops to very low values at higher pressures because of the well-known toxic effect of oxygen. If not stirred, higher oxygen pressures are tolerable because of the low absorption and diffusion coefficients of oxygen. A

TABLE 7

Calorimetrically determined parameters of growing yeast cultures under varying conditions. Enthalpy changes ΔH (kJ mol^{-1} substrate), specific rate of heat production $d\Delta h/dt$ (W g^{-1}), enthalpy change per formed biomass $\Delta H'$ (kJ g^{-1}), growth yield Y (g dry weight per g substrate). Some values are calculated from the data in the literature. AN, anaerobic; (AN), nearly anaerobic; AE, aerobic; (AE), nearly aerobic

Yeast	Condition	Medium	Substrate	$-\Delta H$	$-d\Delta h/dt$	$-\Delta H'$	Y	Reference
a. Anaerobic batch cultures								
Brewer's yeast	AN	Complex	Sugar	98·4	—	—	—	Bouffard (1895)
Yeast (undefined)	(AN)	Complex	Cane sugar	2 × 110	—	2 × 27·9	0·0218	Rubner (1904)
Yeast (undefined)	(AN)	Complex	Maltose	2 × 112	—	2 × 32·1	0·0193	Rubner (1904)
S. cerevisiae	AN	Synthetic	Glucose	96·3	—	6·28	0·085	Battley (1960b)
Baker's yeast	AN	Complex	Glucose	82·5	—	1·93	0·237	Ohlmeyer and Fritz (1966)
S. cerevisiae Y Fa	AN	Complex	Glucose	119	—	5·25	0·126	Belaich *et al.* (1968)
S. cerevisiae 211	(AN)	Complex	Glucose	99·2	—	4·23	0·131	Lamprecht (1969)
S. cerevisiae 211	(AN)	Complex	Glucose	89·6	0·17	4·48	0·111	Lamprecht *et al.* (1971)
S. cerevisiae	(AN)	Complex	Glucose	63·9	—	—	—	Takahashi (1973)
Baker's yeast	AN	Complex	Glucose	129	0·34	—	—	Hoogerheide (1975a)
S. cerevisiae	AN	Complex	Fructose	124	—	4·78	0·144	Murgier and Belaich (1971)
S. cerevisiae	AN	Complex	Galactose	125	—	4·82	0·144	Murgier and Belaich (1971)
S. cerevisiae	AN	Complex	Maltose	2 × 126	—	2 × 6·31	0·111	Murgier and Belaich (1971)
S. cerevisiae	(AN)	Complex	Sugars	80·4	—	1·72	0·26	Schaarschmidt and Lamprecht

S. cerevisiae	AE	Complex	Glucose	2005	—	40.0	0.278	Battley (1960b)
Rhodotorula sp.	AE	Complex	Glucose	536	—	1.05	7.61	Ohlmeyer and Fritz (1966)
C. intermedia	(AE)		Glucose	1695	—	11.4	0.826	Cooney et al. (1968)
S. cerevisiae 211	AE	Solid medium	Glucose	1674	—	—	—	Lamprecht and Meggers (1972)
Baker's yeast	AE		Glucose	—	1.97	—	—	Grainger (1973)
Kl. aerogenes	AE	Synthetic	Glucose	—	3.0–4.2	—	0.19	Few et al. (1976)
D. hansenii	AE	Synthetic	Glucose	1130–1690	0.4	12.6–18.8	0.5	Gustafsson and Norkrans (1976)
S. cerevisiae	AE	Complex	Ethanol	854	—	33.7	0.55	Battley (1960b)
Baker's yeast	AE	Complex	Ethanol	—	1.23	—	—	Hoogerheide (1975b)
Hansenula sp.	AE		Ethanol	544	2.09	16.7	0.61	Waki et al. (1976)
S. cerevisiae	AE	Complex	Acetic acid	678	—	45.2	0.25	Battley (1960b)

c. *Continuous cultures*

S. cerevisiae	AN	Complex	Glucose	139	0.20	10.4	0.074	Cardoso-Duarte et al. (1977)
S. cerevisiae	AE	Complex	Glucose	2663	2.15	14.5	1.017	Brettel (1974)
Kl. aerogenes	AE	Synthetic	Glucose	—	0.13	—	—	Ackland et al. (1976)
S. cerevisiae	AE	Synthetic	Glucose	1070	1.54	10.4	0.572	Brettel (1977b)
S. cerevisiae	AE	Synthetic	Ethanol	616	—	28.3	0.474	Brettel (1974)
Hansemula sp.	AE	Synthetic	Ethanol	607	—	18.0	0.73	Waki et al. (1976)

maximum of 188 kJ mol^{-1} glucose is observed at 5 kp cm^{-2}. These experiments show that aerobic catabolism is most strongly increased by replacing air by oxygen in stirred batch cultures. But of course one is still far from the maximum possible heat production in non-assimilating cultures (2872 kJ mol^{-1} glucose).

The question has been raised: In which phase of growth does the cell produce most heat? It has been shown from continuous culture studies (see following section) that parallel to growth a metabolism of maintenance takes place. The results shown in Figs 9 and 10 imply that during the lag phase of a culture an increased rate of heat production should occur. Gustafsson and Norkrans (1976) observed a drastic decrease of this rate in growing aerobic cultures of *Debaryomyces hansenii* at high salt concentrations. Just before the acceleration phase they obtained 2000 μW mg^{-1} and 200 μW mg^{-1} in the late exponential phase. The differences were less pronounced at lower salt concentrations: 500 μW mg^{-1} to 400 μW mg^{-1}. Hoogerheide (1975b) and other authors came to the same conclusion (Pohlit *et al.*, 1966; Jain, 1972; Kemp and Ross, 1975). Lagunas (1973) found a maximum rate of fermentation at the early and middle part of the logarithmic phase. Schaarschmidt *et al.* (1975, 1977) reported a pronounced maximum acceleration of cell doubling and specific heat output at the beginning of the exponential phase and a semilogarithmic decrease of specific rate of heat production as a function of the culture biomass. Thus the results of Parnas and Cohen (1976) could not be proved, i.e. that maximum synthesis of storage materials and therefore maximum weight increase occur at the end of the exponential phase when the decrease in substrate concentration is a signal to the cell of impending starvation.

Poole *et al.* (1973) and Poole (1977a, b) analysed the energetic situation in a growing batch culture of the fission yeast *Schizosaccharomyces pombe* after induction of synchrony by velocity sedimentation and by deoxyadenosine, respectively. While strong oscillations appeared in the rate of oxygen uptake and the cell number increased stepwise, the heat dissipation grew at a constant rate without any sign of oscillations. But when the cells were incubated in the presence of 16 μM CCCP as uncoupler of the oxidative phosphorylation an overall exponentially increasing but discontinuous calorimetric response was obtained. The maximum of heat dissipation occurred slightly before the next doubling of the cell. This is in good agreement with the results of Brettel (1977a) presented in Figs 26 and 27 for a synchronous continuous culture. Poole *et al.* (1973) interpreted their observations by assuming that the oscillatory component of respiration is involved in energy conversion and that two components of respiration exist. The exponentially increasing one is difficult to uncouple and mainly blocked by inhibitors of the electron transport. In contrast,

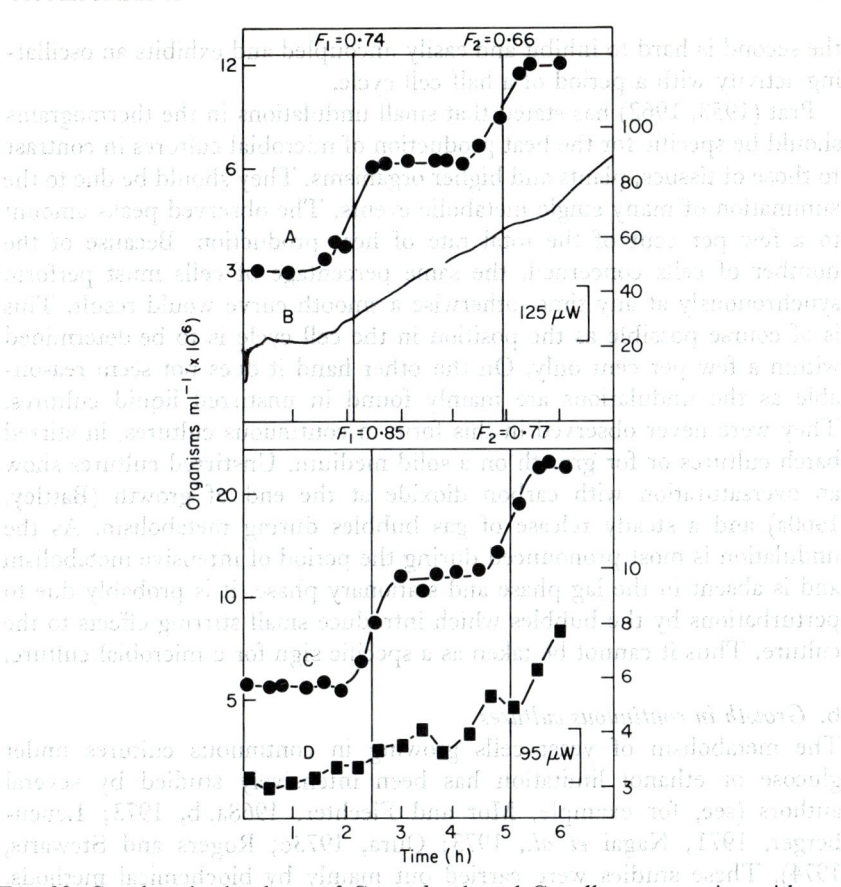

FIG. 22. Synchronized culture of *S. pombe*. A and C, cell concentration without and with 16 μM CCCP; B and D, corresponding rate of heat production. (Poole et al., 1973.)

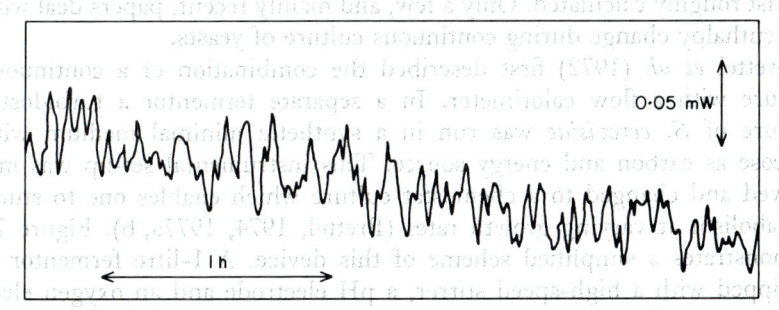

FIG. 23. Undulations in an enlarged part of a thermogram from a batch culture of *S. cerevisiae* at high glucose concentration. The total rate of heat dissipation amounts to 2·9 mW from approximately 5×10^8 cells.

the second is hard to inhibit and easily uncoupled and exhibits an oscillating activity with a period of a half cell cycle.

Prat (1953, 1962) has stated that small undulations in the thermograms should be specific for the heat production of microbial cultures in contrast to those of tissues, plants and higher organisms. They should be due to the summation of many single metabolic events. The observed peaks amount to a few per cent of the total rate of heat production. Because of the number of cells concerned, the same percentage of cells must perform synchronously at any time, otherwise a smooth curve would result. This is of course possible as the position in the cell cycle is to be determined within a few per cent only. On the other hand it does not seem reasonable as the undulations are mainly found in unstirred liquid cultures. They were never observed in this form in continuous cultures, in stirred batch cultures or for growth on a solid medium. Unstirred cultures show an oversaturation with carbon dioxide at the end of growth (Battley, 1960a) and a steady release of gas bubbles during metabolism. As the undulation is most pronounced during the period of intensive metabolism and is absent in the lag phase and stationary phase, it is probably due to perturbations by the bubbles which introduce small stirring effects to the culture. Thus it cannot be taken as a specific sign for a microbial culture.

b. *Growth in continuous cultures*

The metabolism of yeast cells growing in continuous cultures under glucose or ethanol limitation has been intensively studied by several authors (see, for example, Mor and Fiechter, 1968a, b, 1973; Leuenberger, 1971; Nagai *et al.*, 1973; Oura, 1973c; Rogers and Stewarts, 1974). These studies were carried out mainly by biochemical methods. In spite of the importance of the heat of fermentation and the demand for cooling in large-scale fermentors, the evolution of heat was often neglected or just roughly calculated. Only a few, and mainly recent, papers deal with the enthalpy change during continuous culture of yeasts.

Brettel *et al.* (1972) first described the combination of a continuous culture with a flow calorimeter. In a separate fermentor a turbidostat culture of *S. cerevisiae* was run in a synthetic minimal medium with glucose as carbon and energy source. This instrumental set-up was improved and changed to a chemostat culture which enables one to study metabolism at varying growth rates (Brettel, 1974, 1977a, b). Figure 24 demonstrates a simplified scheme of this device. A 1-litre fermentor is equipped with a high-speed stirrer, a pH electrode and an oxygen electrode, a sampling and an outflow tube and an inlet for fresh medium. Moistened and exactly thermostatted air is supplied from below and the bubbles broken by the stirrer to obtain sufficient aeration. The cell suspension from the fermentor is continuously pumped through a light beam to monitor turbidity and then through the calorimeter. As high flow rates of 200 ml h^{-1} are preferred the medium is recycled to the fermentor, so that absolutely sterile handling is essential but sometimes difficult to achieve in cultures of some weeks' duration. To avoid sedimentation in the flow lines and in the tubular flow calorimetric vessel the

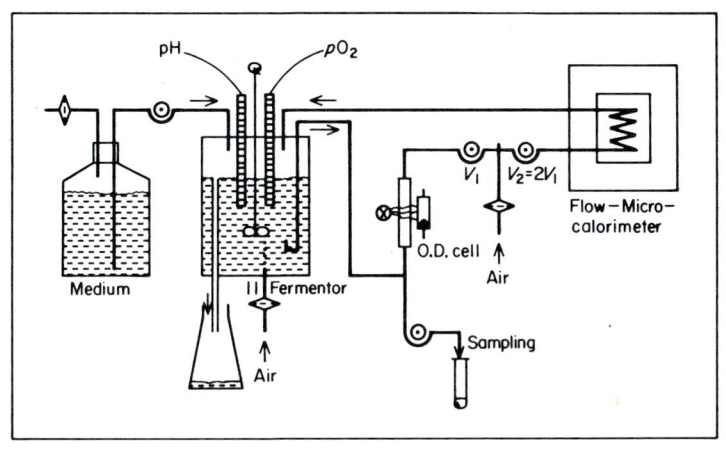

FIG. 24. Simplified scheme of the combination of a continuous chemostat culture with a flow calorimeter (Brettel, 1977a).

flow stream is mixed half and half with air. This arrangement was sufficient for oxygen consumptions smaller than 40 mmol O_2 l^{-1} h^{-1}. The noise introduced to the calorimeter signal by the bubbles was damped by a capacitor.

The results of the experiments of Brettel (1974, 1977a, b) are partly given in Fig. 25. The graph of the dry mass concentration X as function of the dilution rate and hence the specific growth rate (see section C) exhibits the typical biphasic behaviour of a yeast chemostat culture which is caused by the repression of the respiratory enzyme system at high glucose concentrations, that is above approximately $D = 0.3$ h^{-1}. Below this dilution rate glucose is mainly catabolized aerobically. For very small growth rates the dry mass concentration decreases, too, as essential parts of the glucose are consumed in the metabolism of maintenance (see section E.2).

In the dilution range from 0·05 h^{-1} to 0·4 h^{-1} the specific rate of heat production was a linear function of the growth rate. In a complex medium it followed that

$$\dot{Q}/X = 0{\cdot}056 \text{ W g}^{-1} + \mu \times 9{\cdot}41 \text{ kJ g}^{-1} \quad \text{(Brettel, 1974)}$$

and in a synthetic medium with glucose as sole carbon and energy source and growth-limiting substance:

$$\dot{Q}/X = 0.10 \text{ W g}^{-1} + \mu \times 9.07 \text{ kJ g}^{-1} \quad \text{(Brettel, 1977b)}$$

where $\mu = D$ is the specific growth rate. The division into a rate-dependent (second) term and a constant heat production term demonstrates that parallel to growth a metabolism of maintenance takes place (see Table 6).

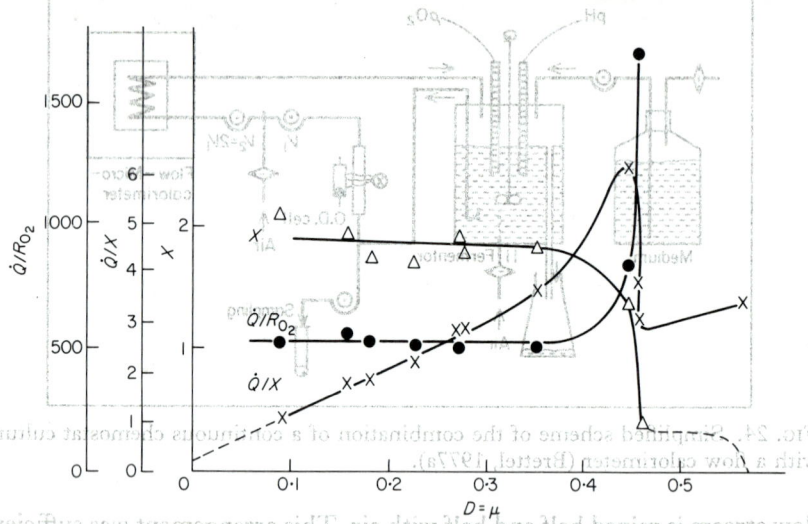

FIG. 25. Performance of a chemostat culture of *S. cerevisiae* as function of the dilution rate $D(\text{h}^{-1})$. X, cell density (g dry weight l^{-1}); $\dot{Q}/R_{O_2}$, heat production per mole of oxygen consumed (kJ $\text{mol}^{-1} O_2$); $\dot{Q}/X$, specific rate of heat production (kJ $\text{h}^{-1} \text{g}^{-1}$). (Brettel, 1974.)

Ackland *et al.* (1976, 1977) performed similar experiments on *Klebsiella aerogenes* in an inverse conical fermentor which was partly placed inside the air bath of a flow calorimeter to shorten the flow time from the culture to the calorimetric flow vessel. But they only describe the specific rate of heat production between $\mu = 0.3 \text{ h}^{-1}$ and the washout at $\mu = 0.76 \text{ h}^{-1}$. In this range they observe a linear decrease with μ:

$$\dot{Q}/X = 0.22 \text{ W g}^{-1} - \mu \times 104 \text{ kJ g}^{-1} \quad \text{(Ackland et al., 1976)}$$

which cannot be compared with the results of Brettel. At a dilution rate of $D = 0.3 \text{ h}^{-1}$ which is covered by all three experiments one gets specific rates of heat production of 0.840 W g^{-1} and 0.856 W g^{-1} for *S. cerevisiae* and 0.133 W g^{-1} for *Klebsiella*. Cardoso-Duarte *et al.* (1977) obtained 0.20 W g^{-1} for a respiration deficient mutant of *S. cerevisiae* at a dilution rate of 0.079 h^{-1}.

The complete oxidation of glucose liberates 470 kJ mol^{-1} oxygen. At low growth rates where the metabolism is almost completely aerobic, the heat production of 470 kJ mol^{-1} oxygen in a synthetic medium (Table 6; Brettel, 1977b) is identical with this value, whereas the 531 kJ mol^{-1} oxygen in a complex medium is close to the data found by Hoogerheide (1975b) for non-growing cultures of baker's yeast (Table 4). This indicates, in agreement with the results of Hoogerheide (see section E.2 and Table 4), that most of the free energy derived from the oxidative degradation of glucose is wasted as heat. At higher dilution rates fermentation contributes considerably to the energy turnover. The value of $\dot{Q}/R_{O_2}$ increases to 1700 kJ mol^{-1} oxygen as no oxygen is consumed in the glycolytic pathway. This means that about 70 per cent of heat production stems from the Embden–Meyerhof–Parnass pathway (Brettel, 1974).

Brettel (1977b) established the energy balance of the continuous culture between glucose supply and the formation of cells and metabolites and the liberation of heat. The energy recovery amounted to 99·2 ± 8·8 per cent. In the range of pure aerobic metabolism 60 per cent of the energy content of the glucose was converted into biomass and the remainder liberated as heat. At higher dilution rates with considerable fermentation energy storage in biomass it dropped to less than 30 per cent as the concentration of metabolites—mainly ethanol—increased.

Cardoso-Duarte et al. (1977) analysed the glucose decay in the culture flowing from a fermentor to a calorimeter and found an exponential slope. By varying the flow rate to the calorimeter and calculating the rate of glucose catabolism in the calorimetric vessel they could determine the enthalpy change when 1 mol glucose is consumed by the population. The authors obtained $\Delta H = -139$ kJ mol^{-1} glucose for a respiration-deficient mutant of S. cerevisiae under aerobic conditions in a chemostat.

Brettel (1977b) used a different way to overcome the problem of substrate limitation during the flow time. When in a continuous culture after the establishment of a steady state the substrate supply was cut off, the substrate concentration in the fermentor must decrease in the same manner as in the flow line. After a period corresponding to the flow time between fermentor and calorimetric vessel the previously constant signal should diminish. Brettel found the expected exponential decrease and from it the time constant of the process. Thus it becomes possible to calculate the true rate of heat production in the fermentor from the thermogram.

Figure 26 shows the heat production in a partly synchronous continuous culture of S. cerevisiae (Brettel, 1977a). This type of budding yeast was chosen as the growth of a bud is a convenient visual marker of the position in the cell cycle. Synchronously dividing cells can be prepared by selection and induction (e.g. by starvation, substrate shocks or

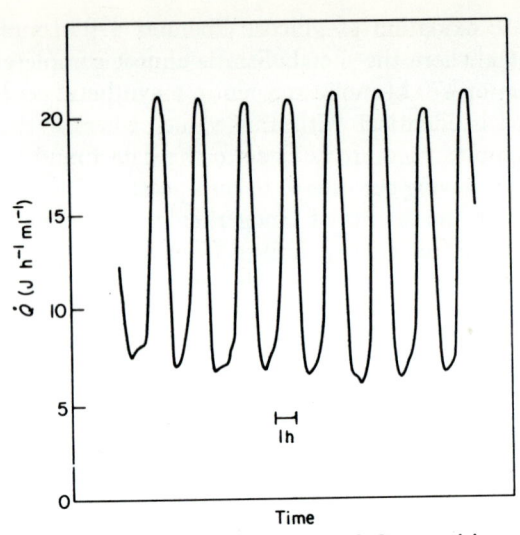

FIG. 26. Thermogram of a continuous culture of *S. cerevisiae* with partly synchronous growth (Brettel, 1977a).

temperature changes) or they may occur spontaneously. In Brettel's experiments 30–50 per cent synchrony appeared spontaneously in glucose or ethanol-limited chemostat cultures. Figure 27 shows the results for various parameters of the culture as a function of the cell cycle time. The ratio N_{IBC}/N gives the percentage of initially budding cells which shows a marked increase to about 30 per cent. A short time before the appearance of the bud the rate of heat production $\dot{Q}$ increases considerably—by a factor of 2·5. The rate of oxygen consumption grows, too, but is less pronounced and without the small maximum seen in $\dot{Q}$. This proves that in this phase of the cell cycle the cell has to achieve a high rate of biosynthesis and structural changes to prepare for the cell division. The required energy is taken from the degradation of reserve carbohydrates, as shown by Küenzi and Fiechter (1969). These reserve carbohydrates are catabolized aerobically as is demonstrated by the increase in the R_{O_2} value. But during the time of maximum turnover the respiratory enzyme system becomes saturated so that a fraction of the reserve substances is fermented to ethanol. This is demonstrated by the increased ratio $\dot{Q}/R_{O_2}$, which is 0·47 J μmol^{-1} O_2 for a pure oxidative degradation of carbohydrates.

It is clear from the above that the main part of the energy turnover takes place in one-third of the cell cycle. A rough calculation from the data in Fig. 27 reveals two metabolic levels in the cell cycle: the high one of 296 per cent around the budding time, and in the rest of the cycle a low level of 16 per cent of the mean rate of heat production during the whole cycle. 89 per cent of the heat is thus distributed throughout the active phase and only 11 per cent elsewhere. Brettel confirmed the observation of other authors (e.g. Meyenburg, 1968) that the period of increased metabolism has an almost constant length, which is independent of the specific growth rate and the duration of the complete cell cycle.

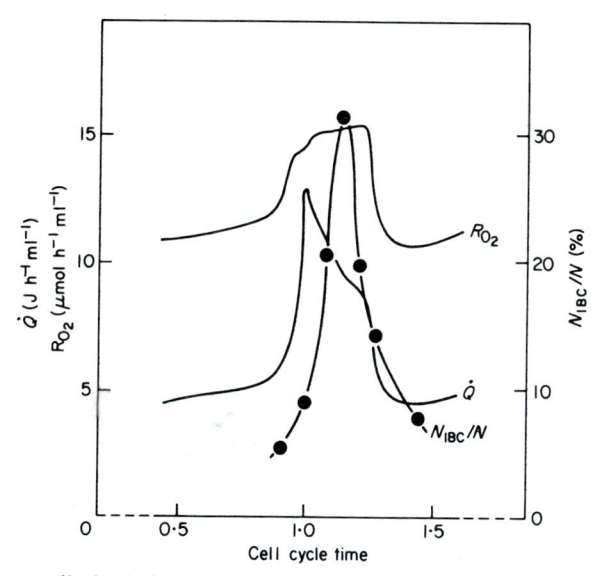

FIG. 27. Glucose-limited chemostat culture of *S. cerevisiae*. Rate of heat production $\dot{Q}$ and of oxygen consumption R_{O_2} and percentage of initially budding cells N_{IBC}/N as function of the position in the cell cycle. (Brettel, 1977a.)

The question of the changing energy content of yeast cells is raised in section E.4. Grainger (1973) demonstrated the influence of the growth temperature on the heat of combustion of baker's yeast. In his batch experiments he could not distinguish between the effect of temperature and that of a changed growth rate occurring simultaneously. In a chemostat culture one could fix the temperature at a desired value and vary the growth rate independently. Brown and Rose (1969) showed that some cell components of *Candida utilis* varied more when the cells were grown at a fixed temperature but at different rates. Experiments on yeast cells from chemostat cultures should be performed to test the results of Grainger (1973) and to compare them with those of Brown and Rose (1969). Moreover, under these conditions growth rates on different substrate can be fixed to the same value so that the energy content of the resulting cells is easier to compare than when harvested from batch cultures.

4. ENERGY CONTENT OF YEASTS

In large-scale industrial fermentation it is often easier, or indeed the only way, to determine the energy content of yeasts from the heat content of substrate, product and crop and calculation of the dissipated heat from the differences. Wang *et al.* (1976) have shown that the rate of heat production during fermentation is given by

$$\Delta^{\cdot}H_{\text{ferm}} = K_1 - (K_2/Y)\mu X$$

with μX the cell growth rate (g h^{-1}), Y the cellular yield coefficient (g cells g^{-1} substrate), K_1 and K_2 are constants proportional to the heat of combustion ΔH_c of cells and substrate, respectively. Other authors have presented similar equations. μX and Y are readily determined during a pilot experiment and K_2 can be calculated from thermogram data. Thus it remains to evaluate the heat of combustion of the crop which is grown during the process.

Direct and indirect methods for the determination of ΔH_c are given in the literature. The classical method of combustion under high oxygen pressure in a Berthelot bomb calorimeter transforms the biomass into H_2O, CO_2 and N_2 and renders direct values of ΔH_c. Table 8 compiles data from the literature on different, often undefined yeasts. One observes that even for the same organisms large deviations (of several per cent) are possible which can be important in industrial fermentation. Grainger (1973) has reported the variations in the heat content of baker's yeast resulting from different temperatures of growth. These values are more impressive when calculated as heat content per cell at the different temperatures: 0·536 J per 10^6 cells (20·0 °C), 0·294 J per 10^6 cells (30·0 °C) and 0·703 J per 10^6 cells (40·0 °C). They are mainly derived from changes in the cell mass under the varying growth conditions. Another example of the influence of varying growth conditions is found in Table 9 in the data of Battley (1960a). Although the substrate and the kind of metabolism changes considerably, the heat content varies by no more than 3 per cent. This underlines the observation of Mennett and Nakayama (1971) that cells operate within a caloric parameter. The observation of Oura (1973b) that aerobically grown cells contain substantially more ordered structure than those grown anaerobically and that correspondingly more heat is produced under aerobic conditions finds no equivalent in the heat of combustion. The mean of the heat contents of the three aerobically grown yeasts is exactly the same as the anaerobic figure.

The Dulong formula which is frequently employed to calculate the heat of combustion of coal samples was altered by Wang *et al.* (1976) to fit the heat of combustion of microorganisms. They suggest

$$\Delta H_c = 33\cdot78C + 144\cdot1(H - \tfrac{1}{8}O) \text{ kJ g}^{-1}$$

TABLE 8

Experimentally determined heats of combustion of yeasts grown under different conditions

Yeast	Condition	ΔH_c kJ g^{-1}	Reference
Brewer's yeast Bottom fermenting		18·73	Rubner (1904)
Torula Top fermenting		19·06	
Brewer's yeast		20·13–21·22	Fink and Just (1938)
Yeast	Pressed yeast	19·09–20·26	
Baker's yeast	20·0 °C aerobic;	15·20	Guenther (1965)
	30·0 °C complex	18·06	Grainger (1973)
	40·0 °C medium	17·70	
		19·14	
Yeast, different strains	Various media	21·56	Oura (1975)
Yeast[a]		21·22–24·12	Prochazka et al. (1973)
S. cerevisiae[a]		22·22	
Candida sp.	Cellulose hydrolysate	17·02–19·47	Minkevich and Eroshin (1973)
S. cerevisiae 211	Anaerobic, complex medium	18·42	Tachoire (personal communication)
S. cerevisiae 211	Various media	20·51–21·77	Lamprecht (1969)
S. cerevisiae	Various strains	17·1 –17·6	Lamprecht (1969)
S. cerevisiae	Various media	16·64–18·35	Brettel (1977b)
	Mean	19·33 ± 2·09	

[a] Per gram ash-free dry weight.

TABLE 9

Chemical composition of different yeasts and calculated heat of combustion per gram (ΔH_c) or per gram ash-free dry weight (ΔH_c^A) (if no ash content is given, ΔH_c^A is calculated with the mean)

Yeast	Given in the literature	Reduced to $C_6H_xO_yN_z$	CH_uO_v	ΔH_c kJ g^{-1}	ΔH_c^A kJ g^{-1}	Ash content %	Reference
Yeast	$C_{170}H_{258}O_{88}N_{26}$	$C_6H_{9.11}O_{3.11}N_{0.92}$	$CH_{1.52}O_{0.52}$	21.8	23.7	—	Reiff et al. (1960)
S. cerevisiae (Hansen)[a]	$C_{4.20}H_{7.36}O_{1.90}N_{0.84}$	$C_6H_{10.50}O_{2.71}N_{1.20}$	$CH_{1.75}O_{0.45}$	25.3	27.7	8.6	Battley (1960a)
S. cerevisiae (Hansen)[b]	$C_{4.26}H_{7.31}O_{1.96}N_{0.73}$	$C_6H_{10.30}O_{2.76}N_{1.45}$	$CH_{1.72}O_{0.46}$	24.6	27.0	8.8	Battley (1960a)
S. cerevisiae (Hansen)[c]	$C_{4.48}H_{7.69}O_{1.83}N_{0.67}$	$C_6H_{10.30}O_{2.45}N_{0.90}$	$CH_{1.72}O_{0.41}$	26.4	28.8	8.3	Battley (1960a)
S. cerevisiae (Hansen)[d]	$C_{4.38}H_{7.16}O_{1.94}N_{0.66}$	$C_6H_{9.81}O_{2.66}N_{0.90}$	$CH_{1.64}O_{0.44}$	24.9	26.6	6.4	Battley (1960a)
Torulopsis	$C_{4.16}H_{7.30}O_{2.18}N_{0.57}$	$C_6H_{10.53}O_{3.14}N_{0.82}$	$CH_{1.76}O_{0.52}$	23.1	25.0	7.75	Chen (1964)
Baker's yeast	$C_{3.92}H_{6.50}O_{1.94}$	$C_6H_{9.95}O_{2.97}$	$CH_{1.65}O_{0.49}$	23.0	25.0	—	Guenther (1965)
Yeast	$C_6H_9O_3N$	$C_6H_{9.00}O_{3.00}N_{1.00}$	$CH_{1.50}O_{0.50}$	22.2	24.1	8.0	Bronn (1970)
Yeast	$C_6H_{10}O_3N$	$C_6H_{10.00}O_{3.00}N_{1.00}$	$CH_{1.67}O_{0.50}$	23.2	25.2	—	Harrison (1971)
S. cerevisiae H 1022[c]	$C_{4.2}H_{7.6}O_{2.2}N_{0.6}$	$C_6H_{10.86}O_{3.14}N_{0.86}$	$CH_{1.81}O_{0.52}$	23.4	25.5	—	Nagai et al. (1973)
Yeast	$C_{3902}H_{6324}O_{2058}N_{597}$	$C_6H_{9.72}O_{3.16}N_{0.92}$	$CH_{1.62}O_{0.53}$	21.9	23.8	—	Oura (1973, 1975)
Baker's yeast	—	$C_6H_{10.90}O_{3.06}N_{1.03}$	$CH_{1.82}O_{0.51}$	23.7	25.9	8.6	Wang et al. (1976)
	MEAN	$C_6H_{10.08}O_{2.93}N_{1.00}$	$CH_{1.68}O_{0.49}$	23.6 ±1.4 (6.0%)	25.7 ±1.6 (6.2%)	8.1 ±0.8 (10.2%)	

Mean composition in weight per cent: carbon 50.4 ± 1.6; hydrogen 7.1 ± 0.4; oxygen 32.8 ± 2.6; nitrogen 9.8 ± 1.8.

[a] Anaerobic, glucose.
[b] Aerobic, glucose.
[c] Ethanol.

where C, H and O are the weight fractions of carbon, hydrogen and oxygen in the dry weight. In balanced equations of microbiological processes a symbolic cell formula is given. Using these formulae and the equation above, one can calculate the heat of combustion of microorganisms. Table 9 shows different compositions of yeasts taken from the literature and calculated ΔH_c values on a gram of dry matter, and on a gram of ash-free dry matter basis. Ash contents between 1·9 and 10 per cent are reported in the literature, with most frequent values between 5 and 9 per cent. Therefore, a figure of 7 per cent was used, when no ash content was given for the specific yeast. The mean for all data is $23·6 \pm 1·4$ kJ g^{-1} dry weight or $25·7 \pm 1·6$ kJ g^{-1} ash-free dry weight with the cited standard deviation. Molar combustion heats of organic compounds are nearly proportional to the amount of oxygen used during the combustion (Kharasch, 1929). Calculated on a gram-equivalent of oxygen consumed one obtains a mean of $112·7 \pm 4·5$ kJ. Practically the same value ($112·8 \pm 4·1$ kJ) is found for 12 different cultures of a *Candida* species, which transforms to $17·74 \pm 0·64$ kJ g^{-1} dry mass (Minkevich and Eroshin, 1973). Senez (1962) has stated that the heat of combustion calculated per gram-atom carbon is 469 kJ for microbial biomass and 473 kJ for glucose. Thus, the carbon in the cell is at the same level of reduction as in the glucose and cell material can be symbolized by (CH_2O). Table 9 presents the cell compositions given by the authors, transformed to a glucose-like formula and to a one-carbon compound omitting the nitrogen. This latter column reveals a mean of ($CH_{1·68}O_{0·49}$) with a standard deviation of 6·0 per cent for H and 12·2 per cent for O. Thus, the level of reduction of cell material for yeasts is not as close to that of glucose as stated by Senez (1962).

Another way of calculating the heat content of yeasts is to determine the amounts of proteins, carbohydrates, nucleic acids, fats and ash and multiply them by the heats of combustion of these compounds. Mean values are given in the literature (Mennett and Nakayama, 1971): proteins 22·6 kJ g^{-1}; carbohydrates 17·2 kJ g^{-1}; nucleic acids 14·7 kJ g^{-1}; fats 38·9 kJ g^{-1}. For yeasts, the following composition is found (Reiff et al., 1960): 45-60 per cent raw proteins; 25-35 per cent total carbohydrates; 4-7 per cent raw fats; 7 per cent nucleic acids; 6-9 per cent ash. These values are main values for different yeasts under various conditions. Of course, in special strains large deviations may be observed. Taking a mean composition of 50 per cent proteins, 30 per cent carbohydrates, 6 per cent fats, 7 per cent nucleic acids and 7 per cent ash, one obtains a heat of combustion of 19·9 kJ g^{-1} dry matter or 21·4 kJ g^{-1} ash-free dry matter. The same value of 19·9 kJ g^{-1} can be calculated from the yeast composition given by Oura (1973). From the data of Brown and Ross (1969) one can calculate a heat content of 21·1 kJ g^{-1} dry matter and from those of Paredes-Lopez et al. (1976) 19·6 kJ g^{-1} dry matter for *Candida utilis*.

Comparing the three methods of determining the heat content of the cells one gets Table 10. Of course it is questionable to calculate the means as very different organisms are concerned. However, they do give a rough impression of the heat of combustion to be expected and of the deviation found between calorimetric determinations and computation.

TABLE 10

Mean values of the heat content of yeast cells as determined by different methods

Determined by	ΔH_c kJ g^{-1}	± Standard deviation kJ g^{-1}	Range kJ g^{-1}
Combustion	19·3	2·1	15·2–24·1
Calculation from the elements	23·6	1·4	21·8–26·4
Calculation from the composition	20·1	0·6	19·6–21·1

F. Some thermodynamic considerations from calorimetric experiments

The first law of thermodynamics on the conservation of energy has never been really doubted in its application to living matter and many scientists even thought it a waste of time to prove this law. On the other hand, calorimetry has been used to check its universal applicability not only for large animals but for microorganisms (Rubner, 1902). By bomb and direct calorimetry Rubner (1903, 1904) found a complete recovery of energy from the growth medium. Cooney et al. (1968) observed an energy recovery with a deviation between +1·1 and −4·5 per cent with a mean of −0·75 per cent using dynamic calorimetry (confirmed by Mou and Cooney (1976) on other organisms). And it could be calculated from batch experiments that the heat evolved during microbial growth is practically equal to the enthalpy change of the catabolic reaction (Forrest, 1969, 1970; Murgier and Belaich, 1971). In a combination of a flow calorimeter with a continuous culture Brettel (1977b) got an energy recovery of 99 ± 8 per cent over a large range of dilution rates, for pure aerobic catabolism as well as for aerobic fermentation. On the basis of the first law the determination of heat generated during growth is possible by means of heat balances (Goma et al., 1976), dynamical calorimetry (Volesky and Thambimuthu, 1976) and combustion heats (Chirkov, 1976).

Although the validity of the second law of thermodynamics in living systems is less evident and was doubted for a long time—not only by the vitalistic school—it is generally accepted today in its more universal form for closed and for open systems. But the interesting question remained: Within what limits is the second law fulfilled? Battley (1960c) has shown for anaerobic growth that the entropy difference between a non-conservative reaction with growth is ~ 29 per cent of the entropy change in the non-conservative reaction and ~ 41 per cent in the conservative one. The counterbalance suffices in any case to satisfy the second law. This example demonstrates that cells act as a retarding agency in the decrease of free energy and increase of entropy. Gorski (1966) has presented a thorough investigation of the different entropy changes accompanying growth of the fungus *Aspergillus oryzae* and the growth of *S. cerevisiae* from the data of Battley. Taking into account the entropy change of the medium, of the newly formed biomass and its structural entropy, and comparing it to the entropy increase in the surroundings by the heat production during catabolism, the overall entropy change in the closed system: calorimetric vessel + thermostat is always positive, as expected from the second law. Under aerobic conditions the entropy change due to growth can be neglected in relation to the total entropy change of the system (mean 4·6 per cent). This corresponds well with the statement of Forrest (1970) that the entropy outflow from a microbial culture is almost entirely due to the degradation of the energy source. Because of the smaller amount of heat dissipation during anaerobic fermentation the contribution of the growth of cells amounts to 7 per cent, and the whole entropy change in the batch culture to 64 per cent. A figure of 59 per cent is given by Lamprecht (1969) for an anaerobic culture of *S. cerevisiae* under different conditions. However, it has to be pointed out that no entropy data on growth can be obtained from calorimetric measurements alone. Additional calculations on the composition and the biochemical structure of the cells have to be included.

As with yeasts (see section B.1), Forrest and Walker (1964) observed three distinct metabolic levels in cultures of *Streptococcus faecalis* each corresponding to a steady state of entropy production. The second law is obeyed as long as the change in internal entropy, calculated as the difference between the total entropy production and the entropy outflow from the system, is larger than zero. In a steady state the rate of total entropy production becomes zero so that the internal change of entropy is counterbalanced by the outflow. A steady state can only be achieved with constant external parameters. In batch cultures this condition is hard to achieve as the medium changes during catabolism. Therefore, no constant entropy production could be observed in yeast batch cultures (see Fig. 28). But continuous cultures are in a steady state at all dilution rates. A continuous culture of *S. cerevisiae* lasting several weeks showed no change in the rate of heat production, once the new steady state was established (Brettel, 1977b). It is then well adapted to its growth conditions and all processes are believed to be balanced, so that the biochemical reactions work close to thermodynamic equilibrium (Schaarschmidt *et al.*, 1975).

The product of temperature and rate of entropy development, specific to the volume or mass concerned, is called the specific dissipation function, ψ. Following a theory of Zotin (1972), this function can be divided into two parts when an open system is far from thermodynamic equilibrium. One part, the external dissipation function ψ_d, leaves the system, while the bound dissipation function ψ_u stays in the system and may be used for some irreversible processes. Thus $\psi = \psi_d + \psi_u$. Near equilibrium, ψ equals ψ_d and all the heat formed in the system is liberated to the environment. The amount of dissipation heat fixed in the system increases with the distance from equilibrium. The metabolism of an organism determines all dissipative processes, so that ψ is given by the rate of respiration and glycolytic catabolism. On the other hand, the external dissipation function

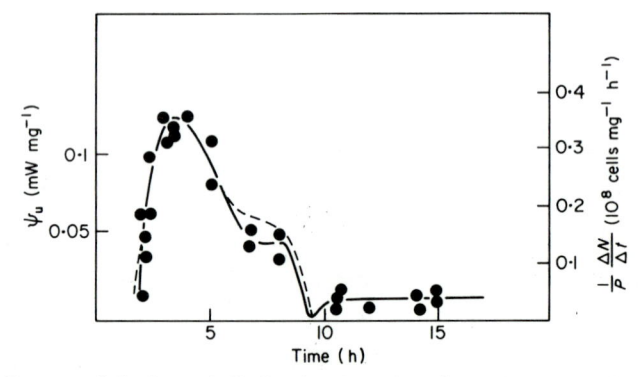

FIG. 28. Course of the bound dissipation function ψ_u and the specific rate of cell production (dotted line) of an anaerobic batch culture of *S. cerevisiae* as calculated from the metabolism during growth. (Schaarschmidt *et al.*, 1975).

ψ_d corresponds to the specific rate of heat production. Therefore, the bound dissipation function ($\psi_u = \psi - \psi_d$) equals the difference between the metabolism intensity and the heat production intensity, the first being measured manometrically, the second calorimetrically. Figure 28 demonstrates the slope of the ψ_u function in an anaerobic batch culture. Just at the beginning of the exponential phase a maximum appears which is due to the new thermodynamic state of high activity which the population achieves. In a continuous culture of steady-state growth the ψ_u function is constant and very close to zero at all dilution rates (Schaarschmidt et al., 1975; Schaarschmidt and Brettel, 1976).

In contrast to Forrest and Walker (1964) who could not observe any transition states when the culture changed from one steady state to the next, the oscillations found in the transition from aerobic to anaerobic conditions (Winfree, 1972) or in the adaptation from aerobic fermentation to respiration (Mochan and Pye, 1973; Schaarschmidt and Lamprecht, 1977b; see Figs 15 and 16) demonstrate that a monotonous progression to a new steady state does not always occur, but limit-cycle behaviour can be included.

References

ACKLAND, P. J., PRICHARD, F. E. and JAMES, A. M. (1976). *Microbios Lett.* **3**, 21–24.
ACKLAND, P. J., PRICHARD, F. E. and JAMES, A. M. (1977). Energy changes during the growth of *Klebsiella aerogenes* in continuous culture. Third Int. Symp. on Microcalorimetry in Microbiology, Chelsea College, London.
BATTLEY, E. H. (1956). A contribution to the study of the thermodynamics of growth of microorganisms. Dissertation, Stanford University, Stanford, California.
BATTLEY, E. H. (1960a). *Pl. Physiol.* **13**, 192–203.
BATTLEY, E. H. (1960b). *Pl. Physiol.* **13**, 628–640.
BATTLEY, E. H. (1960c). *Pl. Physiol.* **13**, 674–686.
BATTLEY, E. H. (1971). A calculation of the free-energy changes accompanying the growth of *Saccharomyces cerevisiae* on several substrates. In "Proc. First Eur. Biophys. Congr., Vienna" (E. Broda, A. Locker and H. Springer-Lederer, eds), pp. 299–305.
BAUCHOP, T. and ELSDEN, S. R. (1960). *J. gen. Microbiol.* **23**, 457–469.
BEEZER, A. E. (1977). Microcalorimetric studies of micro-organisms. In "Application of Calorimetry in Life Sciences" (I. Lamprecht and B. Schaarschmidt, eds), pp. 109–118. de Gruyter, Berlin.
BEEZER, A. E., NEWELL, R. D. and TYRRELL, H. J. V. (1976). *J. appl. Bact.* **41**, 197–207.
BEEZER, A. E., NEWELL, R. D. and TYRRELL, H. J. V. (1977). *Analyt. Chem.* **49**, 34–37.
BELAICH, J.-P., SENEZ, J. C. and MURGIER, M. (1968). *J. Bact.* **95**, 1750–1757.
BELAICH, J. P., MURGIER, M., BELAICH, A. and SIMONPIETRI, P. (1971). Application de la microcalorimétrie à la mesure de l'affinité de cellules microbiennes pour leur substrat énergétique. In "Proc. First Eur. Biophys. Congr., Vienna" (E. Broda, A. Locker and H. Springer-Lederer, eds), pp. 289–297.
BIJKERK, A. H. and HALL, R. J. (1977). *Biotechnol. Bioengin.* **19**, 267–296.
BOIVINET, P. (1971). Contributions de la calorimétrie aux recherches biologiques actuelles. In "Proc. First Eur. Biophys. Congr., Vienna" (E. Broda, A. Locker and H. Springer-Lederer, eds), pp. 277–288.

BOIVINET, P. and GRANGETTO, A. (1963). *C. r. Séanc. Soc. Biol.* **256**, 2052–2053.
BOUFFARD, A. (1895). *C. r. Séanc. Soc. Biol.* **121**, 357–360.
BRETTEL, R. (1974). Microcalorimetric measurements of the energy utilization in chemostat cultures of *Saccharomyces cerevisiae*. *In* "Proc. Fourth Int. Symp. Yeasts, Vienna" (H. Klaushofer and U. B. Sleyter, eds), Part I, **B6**, pp. 87–88. University of Agriculture, Wien.
BRETTEL, R. (1977a). Microcalorimetric measurements of the heat production in partially synchronous cultures of baker's yeast. *In* "Application of Calorimetry in Life Sciences" (I. Lamprecht and B. Schaarschmidt, eds), pp. 129–138. de Gruyter, Berlin.
BRETTEL, R. (1977b). Mikrokalorimetrische Untersuchungen an kontinuierlichen und statischen Kulturen der Hefe *Saccharomyces cerevisiae*. Thesis, Free University of Berlin.
BRETTEL, R., CORTI, L., LAMPRECHT, I. and SCHAARSCHMIDT, B. (1972). *Stud. Biophys.* **34**, 71–76.
BROCK, T. D. and O'DEA, K. (1977). *Appl. Environ. Microbiol.* **33**, 254–256.
BRODA, E., LOCKER, A. and SPRINGER-LEDERER, H. (eds) (1971). Biological calorimetry. *In* "Proc. First Eur. Biophys. Congr., Baden", vol. IV, part IX, pp. 277–395.
BRONN, W. K. (1970). Probleme der mikrobiellen Wärmebildung und adäquaten Fermenterkühlung. *In* "2. Symp. Techn. Mikrobiol., Berichte" (H. Dellweg, ed.). Inst. Gärungsgew., Berlin.
BROWN, A. (1901). *J. Fed. Inst. Brewing*, **7**, 93–103.
BROWN, C. M. and ROSE, A. H. (1969). *J. Bact.* **97**, 261–272.
BROWN, H. T. (1914). *Ann. Bot.* **28**, 197–226.
CALVET, E. and PRAT, H. (1956). "Microcalorimétrie—Applications physicochimiques et biologiques." Masson et Cie, Paris.
CALVET, E. and PRAT, H. (1963). "Recent Progress in Microcalorimetry". Pergamon Press, Oxford, London, New York, Paris.
CARDASO-DUARTE, J. M., MARINKO, M. J. and UDEN, N. van (1977). Flow microcalorimetry of the chemostat. *In* "Continuous Culture", vol. 6 "Applications and New Fields" (A. Dean, ed.), pp. 40–48. Plenum Press, New York.
CHEN, S. L. (1964). *Nature, Lond.* **202**, 1135–1136.
CHIRKOV, I. M. (1976). Thermophysical measurements in studies of microbiological synthesis processes. Fifth Int. Fermentation Symp., Berlin, pp. 21–22.
CHOWDRY, B. Z. (1977). Bioassay of antifungal drugs and use of microcalorimetry in interaction studies. Third Int. Symp. on Microcalorimetry in Microbiology, Chelsea College, London.
COOK, A. H. (1958). "The Chemistry and Biology of Yeasts." Academic Press, New York.
COON, E. D. and DANIELS, F. (1933). *J. Phys. Chem.* **37**, 1–12.
COONEY, C. L., WANG, D. I. C. and MATELES, R. I. (1968). *Biotechnol. Bioengin.* **11**, 269–281.
DAWES, E. A. and RIBBONS, D. W. (1962). *A. Rev. Microbiol.* **16**, 241–264.

DAWES, E. A. and RIBBONS, D. W. (1964). *Bact. Rev.* **28**, 126–149.
DECKER, K., JUNGERMANN, K. and THAUER, R. K. (1970). *Angew. Chem.* **82**, 153–173.
DEDEM, G. VAN and MOO-YOUNG, M. (1975). *Biotech. Bioeng.* **17**, 1301–1312.
DUBRUNFAUT, M. (1856). *C. r. Séanc. Soc. Biol.* **42**, no. 2, 945–948.
EDWARDS, V. H. and WILKIE, C. R. (1968). *Biotech. Bioeng.* **10**, 205–232.
ERIKSSON, R. and WADSÖ, I. (1972). Design and testing of a flow microcalorimeter for studies of aerobic bacterial growth. *In* "Proc. First Eur. Biophys. Congr., Vienna" (E. Broda, A. Locker and H. Springer-Lederer, eds), vol. 4, pp. 319–327.
FEW, G. A., YAU, A. O. P., PRICHARD, E. and JAMES, A. M. (1976). *Microbios.* **16**, 37–48.
FINK, H. and JUST, F. (1938). *Biochem. Z.* **300**, 84–88.
FORREST, W. W. (1969a). Bacterial calorimetry. *In* "Biochemical Microcalorimetry" (H. D. Brown, ed.), pp. 165–180. Academic Press, New York.
FORREST, W. W. (1969b). Energetic aspects of microbial growth. *In* "Microbial Growth. 19th Symp. Soc. Gen. Microbiol.", pp. 65–86. The University Press, Cambridge.
FORREST, W. W. (1970). *Nature, Lond.* **225**, 1165–1166.
FORREST, W. W. (1972). Microcalorimetry. *In* "Methods in Microbiology" (J. R. Norris and D. W. Ribbons, eds). Academic Press, London and New York.
FORREST, W. W. and WALKER, D. J. (1963). *Biochem. Biophys. Res. Commun.* **13**, 217–222.
FORREST, W. W. and WALKER, D. J. (1964). *Nature, Lond.* **201**, 49–52.
FORREST, W. W. and WALKER, D. J. (19' 1). *Adv. Microb. Physiol.* **5**, 213–274.
FUJITA, T. and NUNOMURA, K. (1977). Calorimetric studies of yeast metabolism under non-growing conditions. *In* "Application of Calorimetry in Life Sciences" (I. Lamprecht and B. Schaarschmidt, eds), pp. 119–127. de Gruyter, Berlin.
FUJITA, T., NUNOMURA, K., KAGAMI, I. and NISHIKAWA, Y. (1975). *J. gen. appl. Microbiol.* **22**, 43–50.
GIAJA, J. (1920). *C. r. Séanc. Soc. Biol.* **83**, 1479–1480.
GOMA, G., RIBET, D. and POURCIEL, J. B. (1976). Heat balance, analytical tool in fermentation. Fifth Int. Fermentation Symp., Berlin, p. 19.
GOODWIN, CH. D., BALCAVAGE, W. X. and MATTOON, J. R. (1974). *Arch. Biochem. Biophys.* **165**, 413–420.
GORSKI, F. (1966). "Plant Growth and Entropy Production". Zaklad Fizjologii Roslin Pan, Krakov.
GRAINGER, J. N. R. (1968). The relation between heat production, oxygen consumption and temperature in some poikilotherms. *In* "Quantitative Biology of Metabolism. Third Int. Symp." (A. Locker, ed.), pp. 86–90. Springer-Verlag, Berlin.
GRAINGER, J. N. R. (1973). The measurement of the efficiency of growth at different temperature and its significance. *In* "Effects of Temperature on Ectothermic Organisms" (W. Wieser, ed.), pp. 209–215. Springer-Verlag, Berlin.

GUENTHER, K. R. (1965). *Biotech. Bioeng.* **7**, 445–446.
GUNSALUS, I. C. and SHUSTER, C. W. (1961). Energy-yielding metabolism in bacteria. *In* "The Bacteria" (I. C. Gunsalus and R. Y. Stanier, eds), vol. II, pp. 1–58. Academic Press, New York.
GUSTAFSSON, L. and LINDMAN, B. (1977). *FEMS Microbiology Letters*, **1**, 227–230.
GUSTAFSSON, L. and NORKRANS, B. (1976). *Arch. Mikrobiol.* **110**, 177–183.
HARRISON, J. S. (1971). *Prog. ind. Microbiol.* **10**, 129–177.
HEINMETS, F. (1966). *Helgolaend. wiss. Meeresunters.* **14**, 168–194.
HERBERT, D. (1959). *In* "Recent Progress in Microbiology" (G. Tunevall, ed.), pp. 381–402. C. C. Thomas, Springfield, Illinois.
HILL, A. V. (1911). *J. Physiol.* **43**, 261–285.
HOOGERHEIDE, J. C. (1974). Studies on the energy metabolism during anaerobic fermentation of glucose by "resting" cells of baker's yeast. *In* "Proc. Fourth Int. Symp. Yeasts, Vienna" (H. Klaushofer and U. B. Sleyter, eds), Part I, A9, 17. University of Agriculture, Wien.
HOOGERHEIDE, J. C. (1975a). *Rad. Environ. Biophys.* **11**, 295–307.
HOOGERHEIDE, J. C. (1975b). *Rad. Environ. Biophys.* **12**, 281–290.
INGRAHAM, L. L. and PARDEE, A. B. (1967). Free energy and entropy in metabolism. *In* "Metabolic Pathways" (D. M. Greenberg, ed.), 3rd ed., vol. 1, pp. 2–46. Academic Press, New York.
JAIN, V. K. and POHLIT, W. (1972). *Biophysik*, **8**, 254–263.
KEMP, R. B. and ROSS, P. D. (1975). Microcalorimetry Research Report 5, Bedford College, London, 25–26 March. Report on "Second International Symposium on Microcalorimetry in Microbiology".
KHARASCH, M. S. (1929). *Bur. Stand. J. Res.* **2**, 359.
KIEFER, J. (1971). The importance of energy metabolism in post-irradiation cellular processes. *In* "Proc. First Eur. Biophys. Congr., Vienna" (E. Broda, A. Locker and H. Springer-Lederer, eds), vol. II, pp. 209–212.
KIEFER, J. (1974). *Int. J. Radiat. Biol.* **26**, 167–179.
KOPPERSCHLÄGER, G. (1968). *Wiss. Z. Karl-Marx-Univ., Leipzig*, **17**, 623–629.
KUBITSCHEK, H. E. (1970). "Introduction to Research with Continuous Culture". Prentice-Hall, Englewood Cliffs.
KÜENZI, M. T. and FIECHTER, A. (1969). *Arch. Mikrobiol.* **64**, 396–407.
LAGUNAS, R. (1973). Energy metabolism of *S. cerevisiae* in relation with the carbon energy source. *In* "Proc. 3rd Int. Spec. Symp. Yeasts, Helsinki", part I, p. 42.
LAMPRECHT, I. (1969). Mikrokalorimetrische Untersuchungen an Hefekulturen und Beiträge zur theoretischen Deutung. Thesis, Free University, Berlin.
LAMPRECHT, I. (1976). *Biochem. Soc. Trans.* **4**, 565–569.
LAMPRECHT, I. and MEGGERS, C. (1969). *Z. Naturf.* **24b**, 1205–1207.
LAMPRECHT, I. and MEGGERS, C. (1971). Mathematische Beschreibung Mikrokalorimetrisch Ermittelter Wachstumskurven von Saccharomyces. *In* "Proc. First Eur. Biophys. Congr., Vienna" (E. Broda, A. Locker and H. Springer-Lederer, eds), vol. 4, pp. 335–339.
LAMPRECHT, I. and MEGGERS, C. (1972). *Biophysik*, **8**, 316–325.

LAMPRECHT, I. and SCHAARSCHMIDT, B. (1973). *Bull. Soc. chim. Fr.* **4**, 1200–1201.
LAMPRECHT, I. and SCHAARSCHMIDT, B. (eds) (1977). "Application of Calorimetry in Life Sciences". de Gruyter, Berlin.
LAMPRECHT, I., MEGGERS, C. and STEIN, W. (1971). *Biophysik*, **8**, 42–52.
LAMPRECHT, I., SCHAARSCHMIDT, B. and STEIN, W. (1973). *Biophysik*, **10**, 177–186.
LAMPRECHT, I., SCHAARSCHMIDT, B. and WELGE, G. (1976). *Rad. Environ. Biophys.* **13**, 57–61.
LEUENBERGER, H. G. W. (1971). *Arch. Mikrobiol.* **79**, 176–186.
LEUENBERGER, H. G. W. (1972). *Arch. Mikrobiol.* **83**, 347–358.
MALEK, I. and FENCL, Z. (eds) (1966). "Theoretical and Methodological Basis of Continuous Culture of Microorganisms". Academic Press, New York.
MALEK, I., BERAN, K. and HOSPODKA, J. (eds) (1964). "Continuous Cultivation of Microorganisms". Academic Press, New York.
MAYBERRY, W. R., PROCHAZKA, G. J. and PAYNE, W. J. (1967). *J. appl. Microbiol.* **15**, 1332–1338.
MAYBERRY, W. R., PROCHAZKA, G. J. and PAYNE, W. J. (1968). *J. Bact.* **96**, 1424–1426.
MCCULLOUGH, J. P. and SCOTT, D. W. (eds) (1968). "Experimental Thermodynamics". Butterworth, London.
MCDONALD, P., HENDERSON, A. R. and RALTON, I. (1973). *J. Sci. Fd. Agric.* **24**, 827–834.
MENNETT, R. H. and NAKAYAMA, T. O. M. (1971). *Appl. Microbiol.* **22**, 772–776.
MEYENBURG, H. K. v. (1968). *Path. Microbiol.* **31**, 117–127.
MILLS, A. K. and KREBS, H. (1965). "Aspects of Yeast Metabolism". Blackwell Scientific Publications, Oxford.
MINKEVICH, I. G. and EROSHIN, V. K. (1973). *Folia microbiol.* **18**, 376–385.
MOCHAN, E. and PYE, E. K. (1973). *Nature, New Biology*, **242**, 177–179.
MONK, P. and WADSÖ, I. (1968). *Acta chem. scand.* **22**, 1842–1852.
MONOD, J. (1942). "Recherches sur la Croissance Bactérienne". Masson et Cie, Paris.
MOR, J. R. and FIECHTER, A. (1968a). *Biotech. Bioeng.* **10**, 159–176.
MOR, J. R. and FIECHTER, A. (1968b). *Biotech. Bioeng.* **10**, 787–803.
MOR, J. R., ZIMMERLI, A. and FIECHTER, A. (1973). *Analyt. Biochem.* **52**, 614–624.
MOU, D.-G. and COONEY, CH. L. (1976). *Biotech. Bioeng.* **18**, 1371–1392.
MURGIER, M. and BELAICH, J. P. (1971). *J. Bact.* **105**, 573–579.
NAGAI, SH., NISHIZAWA, Y. and AIBA, S. (1973). *J. gen. appl. Microbiol.* **19**, 221–232.
NEWELL, R. (1977). Growth of, and drug interactions with, *Candida albicans*. Third Int. Symp. on Microcalorimetry in Microbiology, Chelsea College, London.
NIEL, C. VAN and ANDERSON, E. (1941). *J. cell. comp. Physiol.* **17**, 49–56.
OHLMEYER, P. and FRITZ, U. (1966). *Z. Naturf.* **216**, 175–180.
OURA, E. (1972). The effect of aeration on the growth energetics and biochemical composition of baker's yeast. Thesis, Helsinki.

Oura, E. (1973a). Die Verwendung einer biochemischen Modellgleichung bei der Betrachtung des Hefewachstums. 3. Symp. Techn. Mikrobiol., Berlin, pp. 355–360.

Oura, E. (1973b). Heat formation during the anaerobic and aerobic growth of baker's yeast. Proc. First Nat. Meet. Biophys. Biotechn. Finland, Helsinki, pp. 142–144.

Oura, E. (1973c). Proc. Third Int. Spec. Symp. Yeasts, Helsinki, Part II, pp. 216–230.

Oura, E. (1973d). Energetics of yeast growth under different intensities of aeration. Biotechnol. Bioeng. Symp. No. 4, pp. 117–127.

Oura, E. (1975). Enthalpy changes of yeast growth under different intensities of aeration. In "Fifth Int. Biophys. Congr., Copenhagen" (Alkon Keskuslaboratorio Report 6296/75).

Paredes-López, O., Camargo-Rubio, E. and Ornelas-Vale, A. (1975). Appl. environ. Microbiol. 31, 487–491.

Parnas, H. and Cohen, D. (1976). J. theor. Biol. 56, 19–55.

Payne, W. J. (1970). Ann. Rev. Microbiol. 24, 17–52.

Peeters, H. (ed.) (1973). "Protides in the Biological Fluids". Section C: Microcalorimetry in biological systems, vol. 20. Pergamon Press, Oxford.

Peringer, P., Blachare, H., Corrieu, G. and Laue, A. G. (1974). Biotech. Bioeng. 16, 431–454.

Picker, P., Jolicœur, C. and Desnoyers, J. E. (1969). J. chem. Thermodyn. 1, 469–483.

Pirt, S. J. (1966). Proc. Roy. Soc. B163, 224–231.

Pohlit, W., Laskowski, W. and Lochmann, E.-R. (1966). Z. Naturf. 21b, 1089–1096.

Poole, R. K. (1977). Respiration and heat evolution in deoxyadenosine-synchronised cultures of the yeast Schizosaccharomyces pombe. Third Int. Symp. on Microcalorimetry in Microbiology, Chelsea College, London.

Poole, R. K. (1977). J. gen. Microbiol. 103, 19–27.

Poole, R. K., Lloyd, D. and Kemp, R. B. (1973). J. gen. Microbiol. 77, 209–220.

Prat, H. (1953). Rev. can. Biol. 12, 19–34.

Prat, H. (1962). Biochemical and zoological thermogenesis. In "Experimental Thermochemistry" (H. A. Skinner, ed.), vol. 2, pp. 411–425. Interscience Publishers, New York.

Prochazka, G. J., Payne, W. J. and Mayberry, W. R. (1973). Biotech. Bioeng. 15, 1007–1010.

Reiff, F., Kautzmann, R., Lüers, H. and Lindemann, M. (1960). "Die Hefen", vols I and II. Hans Carl Verlag, Nürnberg.

Reis, A. (1975). G-I-T Fachz. Labor, 758–765.

Reis, A. and Büger, P. (1967). Elektromedizin, 12, 183–187.

Reis, A. and Eiber, H. (1967). Elektromedizin, 12, 75–78.

Rogers, P. J. and Stewart, P. R. (1974). Arch. Microbiol. 99, 25–46.

Rose, A. H. and Harrison, J. S. (eds) (1969). "The Yeasts". Academic Press, London and New York.

Ross, P. D. (1973). Calorimeter for the study of cells grown in tissue cultures. Symp. "Techniques of Microcalorimetric Investigations", Lund, 1973.
Rouquerol, I. and Boivinet, P. (1972). Calorimetric measurements. In "Differential Thermal Analysis" (R. C. Mackenzie, ed.) vol. 2. Academic Press, London and New York.
Rubner, M. (1903). Arch. Hyg. 48, 260–311.
Rubner, M. (1904). Arch. Hyg. 49, 355–418.
Rubner, M. (1906). Arch. Hyg. 57, 193–243.
Rubner, M. (1906). Arch. Hyg. 57, 244–268.
Russell, W. J., Zettler, J. F., Blanchard, G. C. and Boling, E. A. (1975). Bacterial identification by microcalorimetry. In "New Approaches to the Determination of Micro-organisms" (C.-G. Hedén and T. Illéni, eds), pp. 103–121. John Wiley, New York.
Schaarschmidt, B. (1972). Mikrokalorimetrische Untersuchungen zum Betriebsstoffwechsel UV-bestrahlter Hefemutanten. Thesis, Free University, Berlin.
Schaarschmidt, B. and Brettel, R. (1976). The change of the ψ_u-function during the growth of microbial cultures. In "Thermodynamics of Biological Processes" (A. I. Zotin and I. Lamprecht, eds), pp. 124–131. Nauka Publisher, Moscow.
Schaarschmidt, B. and Lamprecht, I. (1971). Mikrokalorimetrische Bestimmung des Betriebsstoffwechsels Strahlensensibler Mutanten von Saccharomyces vor und nach uv-bestrahlung. In "Proc. First Eur. Biophys. Congr., Vienna" (E. Broda, A. Locker, H. Springer-Lederer, eds), vol. 4. pp. 329–333.
Schaarschmidt, B. and Lamprecht, I. (1973). Experientia, 29, 505–506.
Schaarschmidt, B. and Lamprecht, I. (1974). Int. J. Radiat. Biol. 25, 167–173.
Schaarschmidt, B. and Lamprecht, I. (1977). Rad. Environ. Biophys. 14, 153–160.
Schaarschmidt, B. and Lamprecht, I. (1978). Thermochim. Acta. 22, pp. 333–338.
Schaarschmidt, B., Lamprecht, I. and Stein, W. (1973). Biophysik 9, 349–355.
Schaarschmidt, B., Umlauf, C. and Lamprecht, I. (1973). Int. J. Radiat. Biol. 24, 433–441.
Schaarschmidt, B., Lamprecht, I. and Welge, G. (1974). Rad. Environ. Biophys. 11, 53–61.
Schaarschmidt, B., Zotin, A., Brettel, R. and Lamprecht, I. (1975). Arch. Mikrobiol. 105, 13–16.
Schaarschmidt, B., Zotin, A. I. and Lamprecht, I. (1977). Quantitative relation between heat production and weight during growth of microbial cultures. In "Application of Calorimetry in Life Sciences" (I. Lamprecht and B. Schaarschmidt, eds), pp. 139–148. de Gruyter, Berlin.
Schmidt, E. (1947). Angew. Chem. 59, 16–20.
Senez, J. C. (1962). Bact. Rev. 26, 95–107.
Skinner, H. A. (ed.) (1962). "Experimental Thermochemistry". Interscience Publishers, New York.

SOLS, A., GANCEDO, C. and DELAFUENTE, G. (1971). *In* "The Yeasts" (A. H. Rose and J. S. Harrison, eds), vol. 2. Academic Press, London and New York.
SPINK, C. and WADSÖ, I. (1975). Calorimetry as an analytical tool in biochemistry and biology. *In* "Methods in Biochemical Analysis" (D. Glick, ed.), vol. 23. Wiley–Interscience, New York.
STOUTHAMER, A. H. and BETTENHAUSSEN, C. (1973). *Biochim. biophys. Acta*, **301**, 53–70.
STURTEVANT, J. M. (1972). *Meth. Enzym.* **26**, 227–253.
STURTEVANT, J. M. (1974). *Ann. Rev. Biophys. Bioeng.* **3**, 35–51.
SWIETOSLAWSKI, W. (1946). "Microcalorimetry". Reinhold, New York.
TAKAHASHI, K. (1973). *Agric. Biol. Chem.* **37**, 2743–2747.
TAKAHASHI, K. and HASHIMOTO, Personal communication.
TANGL, F. (1903). *Arch. ges. Physiol.* **98**, 475–489.
TERVEEN, A. and REIS, A. (1977). *G-I-T Fachz. Labor*, |393–397.
UDEN, N. VAN (1967). *Arch. Mikrobiol.* **58**, 145–154.
UDEN, N. VAN (1971). *Z. allg. Mikrobiol.* **11**, 541–550.
VERHOFF, F. H. and SPRADLIN, J. E. (1976). *Biotech. Bioeng.* **18**, 425–432.
VOLESKY, B. and THAMBIMUTHU, K. V. (1976). Study of heat of selected aerobic fermentations. Fifth Int. Fermentation Symp., Berlin, p. 20.
WADSÖ, I. (1970). *Q. Rev. Biophys.* **3**, 383–427.
WADSÖ, I. (1975). Microcalorimetry and its application in biological sciences. *In* "New Techniques in Biophysics and Cell Biology" (R. H. Pain and B. J. Smith, eds), pp. 85–126. John Wiley, London.
WADSÖ, I. (1976). *Biochem. Soc. Trans.* **4**, 561–565.
WAKI, T., SUGA, K., HAMURO, K. and ICHIKAWA, K. (1976). Dynamic analysis and control of yeast production process. Fifth Int. Fermentation Symp., Berlin, p. 18.
WALTER, R. and LAMPRECHT, I. (1976). Modern theories concerning the growth equation. *In* "Thermodynamics of Biological Processes" (A. I. Zotin and I. Lamprecht, eds), pp. 98–112. Nauka Publisher, Moscow.
WANG, D. I. C. (1968). *Chem. Eng.* 99–108.
WANG, H. Y., MOU, D.-G. and SWARTZ, J. R. (1976). *Biotechn. Bioeng.* **18**, 1811–1814.
WILHOIT, R. C. (1969). Thermodynamic properties of biochemical substances. *In* "Biochemical Microcalorimetry" (H. D. Brown, ed.), pp. 33–81. Academic Press, New York.
WINFREE, A. T. (1972). *Arch. Biochem. Biophys.* **149**, 388–401.
WINZLER, J. W. and BAUMBERGER, J. P. (1938). *J. cell. comp. Physiol.* **12**, 183–211.
ZOTIN, A. I. (1972). "Thermodynamic Aspects of Developmental Biology". Karger, Basel.
ZOTIN, A. I. and LAMPRECHT, I. (1976). "Thermodynamics of Biological Processes". Russian edition, Nauka Publisher, Moscow, 1976. English edition, de Gruyter, Berlin, 1978.
ZOTIN, A. I., SCHAARSCHMIDT, B. and LAMPRECHT, I. (1975). *Akad. Nauk SSSR, Skr. Biol.* pp. 593–598.

Microcalorimetric Studies of Tissue Cells *in vitro*

R. B. KEMP

Apart from investigations of heat production by blood (Levin, 1973; also see Levin, p. 131) and muscle (see Woledge, p. 145), there have been few studies of tissues (Boivinet, 1977) and eukaryotic cells employing calorimetry as an analytical tool (Kemp, 1975; Nikolic and Neskovic, 1976). It is understandable, perhaps, that the complex requirements and responses of excised tissues in culture have been an impediment to calorimetric study but this is not a valid reason for neglecting cell lines or even primary cultures. So, why? The reasons are complex and may partly stem from the fact that heat production to the average cell biologist is closely associated with that formal term of thermodynamics, enthalpy; a mysterious world from undergraduate classrooms. However, so long as the unsure do not blunder into the unknown, one need go no further than to measure Watts. After all, a calorimeter as an analytical instrument need hold no greater fears than spectrophotometers or electrodes: and how many users of these fully understand their theory?

It is undoubtedly true that, until recent years, calorimeters of the required sensitivity and capable of an acceptable work rate in producing data have not been commercially available. It is still a fact that a great deal of work needs to be done on the design of vessels as containers for the cells under study and part of this chapter will be devoted to this problem. However, perhaps the most difficult task is to convince the cell biologist and his sponsors that calorimetry will give a valuable return of important data for the considerable capital investment involved.

One of the greatest goals in this field is to study cellular interactions with molecules such as hormones, antigens and antibodies; particles, for instance viruses, mycoplasma and bacteria; and homo- and heterotypic cells. While the technical problems of monolayer cells and those in suspension are different, in neither case do we yet know the sources, magnitude and pattern of heat production. It would be a fair assumption to say that the greatest proportion of heat is derived from catabolic processes but, although defined media are available, there have been only a few studies of even glucose as a defined carbon source in the growth of cells lines (Kitos and Weymouth, 1964). This is because most cell types require essential amino acids in the medium for growth; and these acids can be employed by cells as a source of energy in catabolism (Dunn and Kemp, 1973). In fact, most researchers supplement the medium with a witches' brew called serum, rendering it impossible to state the carbon source in respiration at any given time. Thus, the first step in utilizing any given cell type for calorimetric study is to determine and control the parameters affecting heat production. This is the state of the game at present.

It is possible, indeed probable, that catabolic heat dwarfs all other sources, making it difficult to investigate most biosynthetic events, unless conditions are very strictly controlled. The problems in calorimeter design depend on whether the cells are grown in monolayer or in suspension. The former is more widely employed and thus better defined but has the greater design difficulties. Thus, not only will traditional considerations of growth media and gaseous phase influence confidence in the results, but so will vessel design.

If the above points are valid for anabolic processes, which may be 10 per cent of the total heat production, then they are certainly of critical importance for studies of cellular interactions. In an arbitrary fashion, the binding of a macromolecule or particle to a cell can be divided into the physical binding to the cell surface and the resultant cooperative effects. It is likely that the exceedingly large and, on the scale required, variable metabolic heat of the cell will prevent accurate measurements of the first phase for the foreseeable future, even when using a twin calorimeter. However, the metabolic and, possibly, organizational effects of binding, including the (undetected) heat of binding should be manifest. Indeed, Krakauer and Krakauer (1976) have already studied the metabolic response of equine lymphocytes under stimulation by the specific antigen, dinitrophenylated bovine serum albumen, and the mitogen, concanavalin A.

The biochemist interested in metabolism might ask why he should invest in the technology needed to measure heat production when traditional methods, such as oxygen uptake and assays for enzymic activities, substrates and products, have served him so well. Of course, calorimetry can only augment and not supplant these techniques; but it can provide strong indications of changes taking place and only detectable by a lengthy series of assays, while also giving possible explanations in energy terms for such events. Thus, changes in heat production, if carefully monitored, will always indicate an alteration in metabolism, though the absence of changes does not imply the reverse. Heat measurements, then, give the careful researcher broad and non-specific data pinpointing events and qualifying them. To the careless, the non-specificity of heat will give as many artifacts as facts and lead to many a blind alley. Elimination of such systematic errors is one of the objects of this chapter. Conversion of a few sceptics is the other: a feel for heat.

A. Vessel design

When studying the heat production of tissue cells, it is most important, of course, to design appropriate vessels for the calorimeter. In essence, there are two types of culture concerned, monolayer and suspension. The former is the more common but holds more calorimetric problems because the whole culture has to be placed in the calorimeter, rather than a proportion of it being pumped through the instrument from a culture vessel outside, the latter being the case for suspension culture.

Since changes in metabolism, whether inherent or induced, are comparatively slow, heat conduction calorimeters (Spink and Wadsö, 1975) have mainly been employed for studies of tissue cells and are well suited for this type of research. In their monolayer studies, Nikolic and Neskovic (1976) employed a commercial Calvet differential calorimeter (Setaram, France) developed from Tian's early design (Calvet, 1956). These workers shortened the characteristically long equilibration time of the instrument upon introduction of a sample from 8 h to 4 h using the Peltier effect (Spink and Wadsö, 1975). On the other hand, Krakauer and Krakauer (1976) flowed lymphocytes through the platinum tubing of the flow version (Sturtevant, 1969) of the Benzinger calorimeter (Benzinger and Kitzinger, 1963), marketed by Beckman Instruments (Benzinger, 1965). The volume of tubing in contact with the thermopiles of this machine is 2 ml and this, when associated with the fairly high rates of heat evolution by lymphocytes (initially 8 pW per cell), enabled Krakauer and Krakauer (1976) to use unmodified vessels.

In most other flow studies of tissue cells, including blood cells (see, for instance, Levin, 1973; Monti and Wadsö, 1973) and chick embryo fibroblasts (Kemp, 1975), the flow version of the LKB calorimeter has been employed. Based on a design by Wadsö (1968), this heat conduction instrument will be used in an illustration of the problems of vessel design.

In its standard form, the LKB flow calorimeter is equipped with a 24 K gold flow-through cell (Monk and Wadsö, 1968, 1969), 0·45 ml in volume. Thus, if the heat production is low, the average residence time can be critical and, in some cases, a limiting factor. Then again, in cases requiring a continuous gaseous phase, this vessel is not suitable owing to physical effects, including frictional heat differences and other zero effects. Nevertheless, this type of flow vessel has been used successfully in studies of blood (Levin, 1973) and of the interaction of chick embryo fibroblasts with adenosine triphosphate (Kemp, 1975).

In order to permit experiments using bacteria which require a high oxygen tension, Eriksson and Wadsö (1971) designed and constructed a stainless steel flow vessel of relatively large volume, 1·4 ml (Fig. 1). Thus, this removable vessel can be employed in studies of cells with a relatively low heat production (2 μW per 10^6 cells per hour) and a requirement for

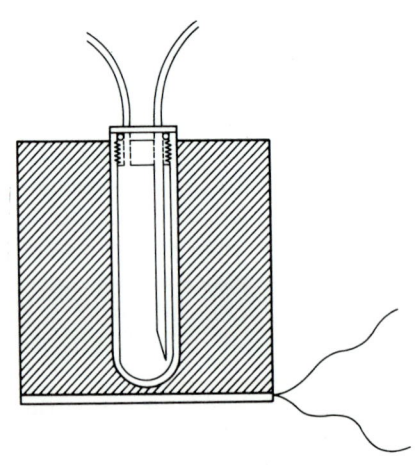

FIG. 1. Stainless steel aeration vessel of the LKB flow calorimeter. The steel vial has a screwdown lid with inlet and outlet ports and is removable from the block. (Kemp, 1975.)

aeration. In a series of experiments of this type using embryonic chick cells, Kemp (1975) showed that $2·5 \times 10^{-5}$ M malonic acid and 10^{-5} M iodoacetic acid inhibited heat production by 37 per cent and 22 per cent respectively after incubation of the cells for 4 h at 37 °C. The validity of calorimetry as a measure of respiratory metabolism can be gauged from results showing that oxygen uptake was inhibited by these two agents over the same time period, by 30 per cent and 17 per cent respectively (Dunn and Kemp, 1973). Nevertheless, in the search for an ideal design, the aeration vessel holds only part of the answer for tissue cells, that of larger volume. In terms of hydrodynamics, the vessel is less than adequate, there being a tendency for cells to be trapped at the bottom of the vessel in an unstirred layer. This accumulation of clumps of cells can lead to spurious peaks unless very high pumping speeds (> 100 ml h^{-1}) are employed, which will itself result in a very noisy recorder trace and possible damage to the cells, even when the pump is placed on the outlet tube of the instrument (Kemp, 1975). A small modification to the syringe needle-type of inlet to the aeration vessel, achieved by placing a short length of PVC as a sleeve over the tip of the needle, did alleviate the problem to a certain degree and afforded protection to the cells from the sharp cutting edge of the inlet. However, a better solution was sought.

In 1971, Kusano *et al.* described a batch sorption system suitable for the LKB calorimeter. At a slightly later date, a flow sorption vessel (Johansson, 1973) was introduced commercially and seemed to have certain hydrodynamic advantages over the aeration vessel. As can be seen from Fig. 2, a glass microcolumn contained in an aluminium block is part of a flow-through system. The microcolumn is removable and can be sterilized by

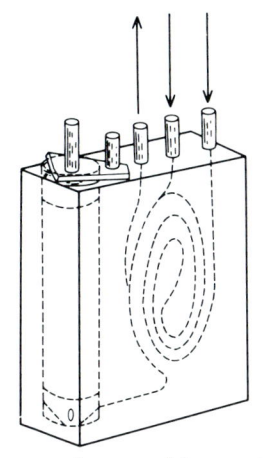

FIG. 2. Removable glass flow sorption vessel in a stainless steel block, which also contains a heat exchanger. Suitable choice of inlet and outlet allows upward or downward flow in the vessel. (Kemp, 1975.)

conventional means. The greatest disadvantage for studies of tissue cells in suspension seemed to be the comparatively small volume (0·48 ml) of the vessel. When using chick embryo cells, the calorimetric response was low, unless a slow flow rate (20 ml h^{-1}) was employed to increase the average residence time (product of flow rate and volume). At this flow rate, the hydrodynamics of the system are inadequate. The ratio of microcolumn diameter to that of the inlet and outlet is 10 : 1 and the transition region is poorly contoured. Thus, cells were trapped in unstirred cones at both the entrance and exit of the column, irrespective of the direction of flow. Therefore, for embryonic cells, the thermogram was little better than that produced using the aeration cell. However, the system did give useful data in respect of cells with a higher heat production, for instance L929 mouse fibroblasts in suspension (Fig. 3, also see Kemp, 1979). All in all, this little story emphasizes the old adage concerning the "right tools for the job". The flow sorption vessel is, of course, excellent for its design requirements and probably for bacterial studies, but only of limited value for tissue cells in suspension (see also p. 128). The quest continued.

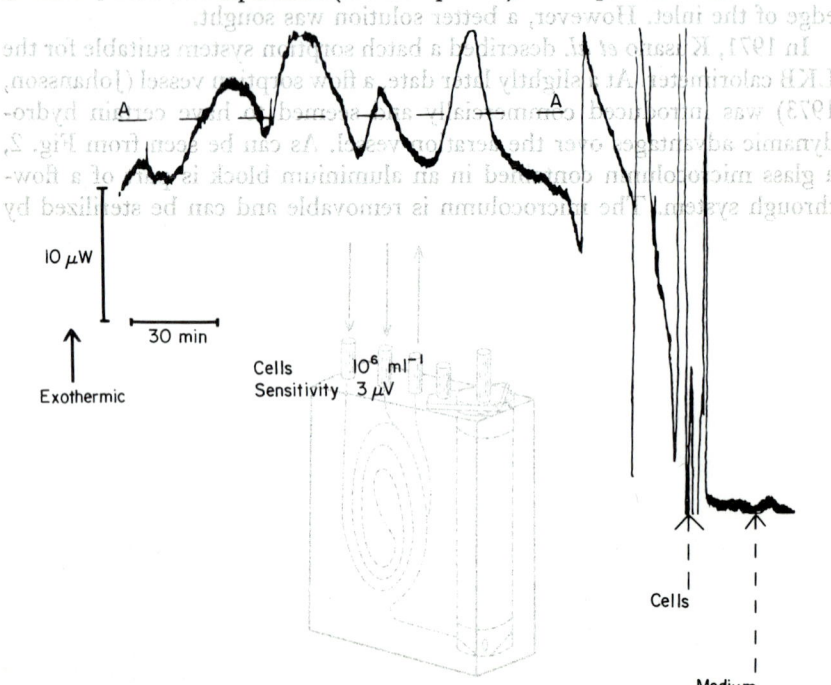

FIG. 3. Thermogram produced on the introduction of LS cells by upward flow (20 ml h^{-1}) into the sorption vessel of the calorimeter. The initial disturbance upon introduction of the cell suspension (10^6 ml^{-1}) is due to stopping the pump to transfer the line from medium to cell suspension. Note the highly irregular plateau with the arbitrarily drawn median line, A–A.

It was decided to stop trying to adapt commercially available vessels to a specific task, and design one to suit the requirements, hoping that it could be made. The physical constraint of the semiconducting thermopiles was one limiting factor. Whatever was designed had to fit into a space of about $4 \times 4 \times 1$ cm and needed to allow the maximum volume of cell suspension with the best possible hydrodynamics. This latter could be achieved using a tube of constant internal diameter. If one imagines a matchbox of the above dimensions then, to secure the maximum volume, the tubing should be wound in an orderly fashion in the box—on a small fixed spool. Simple enough! But what materials should one use? Cells seem satisfactorily to tolerate organic polymers, such as teflon, and these are easy for amateurs to wind. So, an aluminium box containing as much wound teflon tubing as possible and filled with heat sink compound became our standard vessel (Kemp, 1979). Using teflon transmission tubing of internal diameter 0·9 cm, the volume available within the vessel is 1·22 ml, which is close to that of the aeration vessel. In initial experiments, it was noted that clumps of cells still occurred and resulted in spikes on the recorder trace. Their origin was traced to joints at the primary and secondary heat exchangers within the air bath and block respectively (see Spink and Wadsö, 1975). These were replaced by coils of teflon tubing of the appropriate length. For studies in which oxygen tension is a rapid limiting factor, this modification would exacerbate one of the disadvantages of teflon; that it permits a significant amount of gaseous exchange.

The time constant for all the commercial flow vessels is close to 120 s (Spink and Wadsö, 1975). Using the calibration heater in a basal position, the time constant of the teflon vessel is slightly shorter, 115 ± 2 s. The relatively slow time constant of vessels in the LKB instrument has not proved a difficulty when investigating the heat production of tissue cells and, indeed, is much shorter than those of many other heat conduction calorimeters. However, it could prove troublesome in certain cases. Spink and Wadsö (1975) have stated that the position of the calibration heater may be critical in some cases and have stressed that, when testing modified designs, different heater positions should be tried, as this may reveal systemic errors. This procedure was undertaken when testing the teflon vessel but did not reveal any noticeable differences in time constant.

In the past, most biochemical studies of cells grown in culture have employed monolayer techniques, in which the cells are attached to a substratum, usually glass or some form of treated plastic. In order to remove cells from the surface, proteolytic enzymes, such as trypsin, or chelating agents, usually ethylene diamine tetra-acetic acid (EDTA), are used either singly or in combination. It is essential that the success of this process at the end of an experiment is monitored under the microscope. Therefore, ideally, a vessel containing monolayered cells within a calorimeter should be removable. This is how Nikolic and Neskovic (1976) conducted their studies, using an hermetically sealed glass dish (29 mm diameter, 15 mm high) in the Calvet calorimeter. However, the cells could only be observed at the beginning and end of the experiment, a restriction which would annoy many a fastidious tissue culturist. Like the hospital doctor, most tissue culture experts like to keep their "patients" under constant observation. Nikolic and Neskovic (1976) did the next best thing and set up parallel cultures outside the calorimeter.

Looking at the commercially available vessels for the LKB calorimeter and bearing in mind the criteria needed for monolayer studies, only three are easily and routinely removable, the aeration vessel and the two sorption vessels. In addition, the LKB ampoule-drop heat capacity calorimeter (Konicek et al., 1971) has vessels which, in the course of each experiment, are inserted into the measuring block in only a few minutes. Although the LKB batch calorimeter is available with optional glass vessels, these are fixed in position and, therefore, of little value, bearing in mind the required criteria. However, Cerretti et al. (1977) have used such vessels to investigate the thermal behaviour of HeLa S-3 and KB cells. Of the more suitable vessels, only the sorption types are made of glass. As earlier described, the vessel in the flow version is a vertical column of small volume. A means is required to increase the total surface area available for cells, such as a series of baffles or glass beads. This latter solution was adopted by Kemp (1975). The microcolumn was filled with glass beads (200 μm overall diameter) and injected with embryonic chick fibroblasts as a primary culture. The flow sorption vessel was adopted because there was the ability to perfuse the cells with fresh medium. Although some satisfactory results were obtained (Kemp, 1975; and p. 128), the overall potential of the systems seems limited. Once again, adaptation seems a poor alternative to a new design.

What is needed? Clearly, there is a need for a vessel with as large a volume as possible within the physical limits imposed by the thermopiles within the LKB calorimeter. It must be fabricated from materials which are compatible with cells and should be capable of allowing for perfusion. In order to give maximum surface area for cells to monolayer, a means must be sought satisfactorily to incorporate as many layers as possible in the vessel, possibly by adapting the Flow Capafusion system (Flow Laboratories Ltd, Irvine, Scotland) for perfusion of cells, to the designed vessel. Finally and ideally, the vessel should be removable from the instrument for visual examination, with only minimal temporal disturbance to the long-term thermogram. A difficult goal but one that is being attempted!

Before leaving the subject of vessel design, one should mention other minor constraints that the use of tissue cells implies for successes. Materials must not contain toxic substances which may elute into the culture medium. Many plastics contain measurable quantities of toxic cationic activators and plasticizers used in their manufacture. Metals and plastics may adsorb substances at one procedural step and these are then eluted over long periods, possibly with harmful effects. In this respect, bleach (10% v/v, domestic), so valuable in the sterilization of teflon tubing, adsorbs to the plastic and needs very careful removal by washing. Metals may corrode during long-term exposure to the relatively high salt content of tissue culture media. Glass is ideal, but difficult to work and easy to break!

B. Metabolic studies

It is not intended to give an exhaustive review of the literature in this section. In any case, an hour in the library would suffice for this purpose! Instead, data will be employed to illustrate the virtues and problems of calorimetry in studying the metabolism of cells. Krakauer and Krakauer (1976) emphasized that heat production is the parameter that most broadly reflects metabolic activity and employed this property when studying the antigenic stimulation of equine lymphocytes. It is important to note their finding that unstimulated lymphocytes initially evolved heat at a high rate (8 pW per viable cell), a result supporting that of Bandman et al. (1975) on human lymphocytes. The rate rapidly decayed over the first 24 h in culture (<1 pW per cell), stabilizing to approximately 0·4 pW per viable cell between 3 and 9 days in culture. Thus, the age of a culture is important and this was readily discovered by calorimetry. Selecting a standard preincubation of 24 h at 37 °C, Krakauer and Krakauer (1976) found that dinitrophenylated BSA-stimulated cells had an increased heat production after exposure for 2 days, to a maximum at 4–5 days. This time course paralleled an increase in DNA synthesis. However, inhibition by hydroxyurea of DNA synthesis alone did not significantly inhibit heat output, whereas inhibitors of RNA and protein synthesis, actinomycin D and cycloheximide respectively, had a profound effect. These findings were interpreted as signifying that transcription is the step at which selection of responses to stimulants might take place. The authors also showed that cytochalasin B inhibited heat production, believing that this agent would prevent the transmission of the mitogenic signal from the cell surface to the interior. However, the distinct likelihood that cytochalasin B inhibited sugar transport (Lin and Spudich, 1974) and, thereby, depressed heat output through the limitation of glucose for respiratory metabolism, could well have accounted for at least part of this finding. Nevertheless, the study clearly showed the potential of calorimetry in relating cause to effect in investigating the properties of tissue cells.

A similar type of study was undertaken by Ross and co-workers (Fletcher *et al.*, 1973; Ross *et al.*, 1973), who measured the activation of platelets from human blood (see also Levin, p. 131) by thrombin. Using a batch calorimeter, they showed that this agent caused a rapid and sustained increase in heat production of, at the maximum rate, 10^{-13} J platelet^{-1} min^{-1}. Using the metabolic inhibitors, 2-deoxy-D-glucose and antimycin, it was shown that the early part of the thrombin-induced heat response was attributable to antimycin-sensitive oxidative phosphorylation, whereas the later stages were due to glycolysis. These workers balanced out the endogenous heat production of the platelets by placing a second aliquot of platelet suspension in the other batch vessel of the calorimeter, the thermoelectric devices of the two vessels being wired in opposition. This method carries the implication that aliquots must be very carefully measured and also introduces a problem regarding a decreased heat production upon initiation of the reaction. While, on the face of it, such a decrease could be attributable to either less heat production or an endothermic reaction, depending on the nature of the experiment, this is only true if an increased heat output has not occurred in the reference vessel.

Kemp (1974, 1975) employed the fact that the interaction of rabbit antibodies with chick cells produced heat to demonstrate the presence or absence of certain antigens on the cell surface. In these studies, he was not concerned with the source of the heat produced, that is, binding *per se*, or the consequent conformational and metabolic changes, but with utilizing the changes in heat production as a screening test for cell–antibody interaction. This type of investigation, involving particle-to-cell interaction, has important medical implications, especially once the complete thermal behaviour of a given cell type has been worked out in detail.

For the reasons stated earlier (see p. 115), work in this laboratory has concentrated on evolving techniques for studying cells in suspension. Using the flow sorption vessel (Fig. 2) to estimate the heat output of LS cells (suspension-adapted L929 cells), large variations in the thermograms (see Fig. 3) occurred, probably owing to the trapping of cells in the vessel (see p. 118). However, a rough value for the heat production of these cells was obtained (Fig. 3; plateau A–A), but it was subject to an error of ± 33 per cent. At an approximate figure of 60 pW per viable cell, heat output of these cells was much higher than that of lymphocytes or embryonic chick fibroblasts (2 pW per cell; see Kemp, 1975), but closer to that of KB cells, 25 pW per cell (Cerretti *et al.*, 1977).

An immeasurably improved thermogram was obtained using the teflon vessel (see p. 119) as a flow chamber for LS cells (see Fig. 4). Using the same flow rate as for the sorption cell (20 ml h^{-1}), in general the heat output was 32.8 ± 2.3 pW per viable cell. It is believed that the lower value compared with that of the data from the flow sorption vessel was obtained because there was no accumulation of cells in the vessel. It is also possible that the greater exposure of the medium to teflon caused depletion of oxygen to a limiting value and this is currently under investigation. Despite this caution, it seems appropriate at this point to illustrate the potential of calorimetry in metabolic studies by reference to a series of experiments in which LS cells in logarithmic growth were treated with

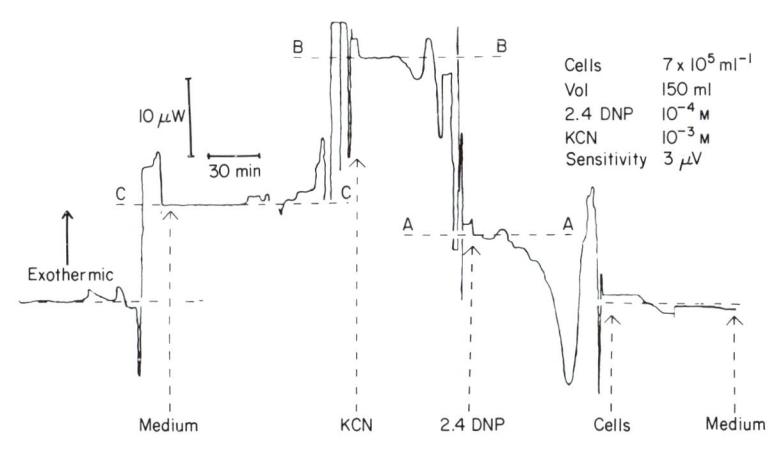

FIG. 4. Thermogram showing the effect of 2,4-dinitrophenol (DNP) and potassium cyanide (KCN) on a suspension of LS cells (7×10^5 cells ml^{-1}) pumped through the teflon flow vessel. Plateau A–A represents the heat production of untreated cells whereas the increase in heat production upon treatment with DNP is averaged as plateau B–B. Attenuation of DNP-stimulated heat production by KCN is given by plateau C–C.

an uncoupler of oxidative phosphorylation, 2,4-dinitrophenol (DNP), and an inhibitor, cyanide, of the respiratory chain at the step cytochrome a to a_3. The thermogram reproduced as Fig. 4 is qualitatively typical, but the heat production of the cells was lower than normal (10 pW per viable cell), owing to culture problems. It was chosen because the thermal response to agents could be seen without the normal base-line adjustments needed in order to remain on scale once DNP has been added. The difference between plateau A–A (Fig. 4) and the base line represents the heat production of LS cells pumped from a culture vessel containing cells suspended in a serum-rich medium (140 ml) at a concentration of $7 \cdot 0 \times 10^5$ ml^{-1}. The pump, situated between the calorimeter and waste, was stopped while DNP was added to the cell suspension. Hence, there was a disturbed trace until a new plateau, B–B, was attained at a far higher level (> 250 per cent) than for untreated cells. Addition of CN$^-$ ions, which alone generally reduced heat output by 35 per cent, caused an attenuation in heat production of DNP-treated cells (plateau C–C) to a value close to, but definitely above, the normal value for untreated cells (+45 per cent). The third plateau (C–C) would contain a component derived from the increasing number of cells during the time of the experiment (2·5 h) but, since the generation time of LS cells is 22·6 h, the difference is only partly accounted for in this way. The effects of the metabolic poisons were consistent throughout a series of 10 runs using cells metabolizing at the usual rate. The changes induced by DNP and CN^- ions were in agreement with measurements of oxygen uptake (Kemp, 1979), though the extent of the DNP effect was, perhaps, surprising. Further studies will be required before a full interpretation of the influence of CN^- ions on the DNP effect can be made and this will involve the use of other inhibitors of the respiratory chain such as rotenone and antimycin.

A potentially valuable rider to these experiments is the finding that the heat output of cells in decline decreases before other metabolic parameters or the length of generation time. The fact that this always occurs before a diminution of oxygen uptake requires further study. Nevertheless, it does indicate that the level of heat production by cells is an extremely sensitive indicator of their "state of health". This could be a boon in laboratories where large stocks of tissue culture cells are maintained.

A recent investigation by Cerretti *et al.* (1977) underlines the difficulties which can occur when using inappropriate vessels in calorimetric studies of tissue cells. They used an LKB Batch Microcalorimeter equipped with glass vessels which, if supplied with the instrument, are normally siliconized prior to delivery. Suspension-adapted KB cells, $1·2 \times 10^6$ per 6 ml of medium containing serum, were cultured in the glass vessel, which was rotated every 30 or 60 min over a 20-h period. Under these conditions, the cells grew at the same rate as cells in normal suspension culture. The heat production of cells at the start of an experiment in the batch calorimeter cannot be established because living cells are producing heat on introduction. Thus, no zero thermal base line can be achieved and, as in the work by Nikolic and Neskovic (1976), heat production is only seen as either a result of an increase in cell numbers or a change in output by existing cells; or possibly from both occurrences.

With each programmed rotation of the batch block, Cerretti *et al.* (1977) observed an endothermal deflection of 3–4 min in duration and equivalent to 1 mJ, if rotation was performed every 30 min, and 2·5 mJ at 60 min. This was immediately followed by an exothermic peak with a duration as much as 30–60 min, depending on the interval of rotation. This deflection was 6·2 mJ (30 min) to 12·3 mJ (60 min). It is likely that the exothermic and endothermic events, which were not produced by the medium alone, were simultaneous in occurrence. It was also noted that the size of the exothermic signal seemed to depend on the number of cells not attached to the glass vessel. This was confirmed using HeLa cells, a direct relationship being found between the settling of living cells and the decay of the exothermic plot. If the majority (>71 per cent) of the cells were attached, then only the endothermic deflection occurred on rotation of the block.

Cerretti et al. (1977) interpreted their findings on the basis that, prior to mixing the contents of the vessel, the cells were crowded at the bottom of the vessel. Release from this condition by rotation of the block would allow increased cell surface binding and transport of nutrients—the endothermic component. Increased availability of nutrients from the bulk phase would permit a raised metabolism and, thus, enhanced heat production—the exothermic phase. As the cells settled and a local environment of low nutrient and high waste product concentrations was set up, metabolism would decrease, as manifest by the declining heat output of the cells. This traumatic cycle of events may also explain the authors' unexpected finding that heat production did not increase with the gain in cell number over the 24-h period of observation. They cite a personal communication from Wadsö who also experienced a similar non-proportional thermal behaviour by HeLa cells growing in the batch vessel of an ampoule-drop calorimeter. Presumably, conditions of crowding existed in the ampoule.

As carefully researched and interesting as is the work of Cerretti et al. (1977), one is left with the impression of difficult problems in the calorimetry of tissue cells. This need not be the case if cells adapted to grow in suspension are passed through a suitable calorimetric flow vessel and cells accustomed to growing on a substratum are allowed to monolayer, after seeding at the correct concentration and under suitable conditions in the calorimeter. Nikolic and Neskovic (1976) have achieved the latter by inserting an hermetically sealed glass dish containing L929 cells into the thin-walled cylinder of a Calvet calorimeter. These cells form a monolayer on the glass surface and, at low inoculum in a medium containing calf serum, have a generation time of approximately 12 h. Over an 18-h period, an inoculum of 2.5×10^5 cells grew to 6.2×10^5 cells. As estimated from their graph (Nikolic and Neskovic, 1976), the additional 3.7×10^5 cells had a heat output of 25.5 µW, equivalent to 69 pW per cell. When a higher number of cells in medium (7.0×10^5) was introduced into a dish at the start of the experiment, less than half the cells divided, presumably owing to contact inhibition of growth, and the cell count was 10×10^5 at 18 h. Nevertheless, the additional 3.0×10^5 cells appear to have caused an increased heat production of 31 µW, or 103 pW per cell. This surprising result further illustrates a problem in the batch calorimetry of metabolizing cells, for the thermal base line is not that of the instrument containing nutrient medium, but that of the instrument containing the initial inoculum of cells in medium. The heat production of the initially seeded cells cannot be estimated accurately but the thermal behaviour of these cells may influence the thermogram, the form of which can strictly only be interpreted on the basis of the increase in cell numbers coincident with it. Thus, while the 31 μW in the experiment under discussion is only attributed to the growth in cells from 7 to 10×10^5, the undetermined heat production of the seeded cells may not be constant, thereby influencing the thermogram. The same consideration is also true for the thermogram from a low inoculum of cells ($2 \cdot 3 \times 10^5$). However, despite this problem, Nikolic and Neskovic (1976) calculated that the change in enthalpy for the growth of cells after low seeding was $8 \cdot 27 \times 10^{-7}$ cal per cell. From available published data, they calculated the total change in free energy for the biosynthesis of DNA, RNA, protein and lipid and the degradation of glucose in the Krebs cycle and in the Embden–Meyerhof–Parnas cycle as $-3 \cdot 67 \times 10^{-7}$ cal per cell. By adopting, perhaps slightly unwisely, the assumption that the entropic member (binding energy) in the Gibbs equation is much smaller than the enthalpic one at 37 °C, these authors arrived at a figure of $-4 \cdot 60 \times 10^{-7}$ cal per cell [$(8 \cdot 27 \times 10^{-7}) - (-3 \cdot 67 \times 10^{-7})$] for the bioenergetic processes in the growth, development and division of L929 cells. Despite the limited data in this report and reservations about the interpretations made, it is encouraging to find experiments of this nature being performed under suitable conditions.

There are available at this time several values for heat production of various cells in culture. Broadly speaking, the value per cell seems to reflect the generation time of the cells. This is, perhaps, not surprising since the overall metabolism of cells is a limiting factor in the cell cycle. The only conflicting figure, so far, is that of Nikolic and Neskovic (1976) for cells seeded at high density. The heat production of cells approaching contact inhibition of movement and growth is, thus, one of the areas for further investigation.

C. Cellular interactions

Few studies have been undertaken in which heat production has been observed for cells in interaction with other particles and, even then, only with molecules. Research on the thermal behaviour of cells in contact with antibodies and ligands has been mentioned previously (see p. 121), though the primary purpose of these investigations was not to study the interactions *per se*, but the metabolic consequences of them.

One potential field lies in investigating the properties of membrane-bound enzymes. The application of flow calorimetry to enzyme kinetics was pioneered by Beezer et al. (1973) and this would be an ideal method for studying the particulate enzymes of membranes, because the turbidity problems of conventional spectroscopy are not a limitation in calorimetry. This advantage is particularly true for cells, the plasma membranes of which are thought to contain many enzymes, some of which are vectorially orientated to the outer part of the membrane; for instance, a glycosyl-transferase system implicated in cell adhesion (see review by Kemp et al., 1973).

A crude attempt to demonstrate the presence of an ecto-adenosine triphosphatase (ATPase) at the surface of embryonic chick cells was made by Kemp (1972) using a batch calorimeter. The results seemed to indicate that the enzyme was present and this was later verified using a flow calorimeter (Kemp, 1975). Embryonic chick cells ($2 \cdot 0 \times 10^5$–$2 \cdot 0 \times 10^7$ cells ml^{-1}) were mixed with an excess of ATP (10^{-2} M—zero-order kinetics) in a culture medium at 37 °C. The mixture was pumped through the gold flow vessel of the calorimeter at a range of cell concentrations and a linear relationship between cell numbers (equivalent to enzyme concentration) and calorimetric base-line displacement (representing substrate hydrolysis) was obtained (see Fig. 5), equivalent to 8·5 μM of ATP hydrolysed by 10^6 cells in 1 h. The inhibitor, 4,4′-dithiodi(nicotinic acid),

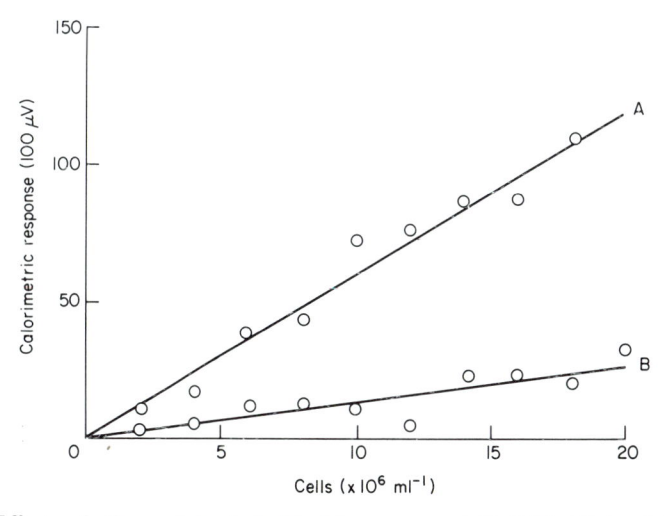

FIG. 5. Effect of the sulphydryl blocking agent 4,4′-dithiodi(nicotinic acid) (112 μM) (B) on the zeroth-order hydrolysis of 10 mM ATP by chick embryo fibroblasts (A) in terms of heat production (reproduced by permission from Kemp, 1975).

which does not enter cells but combines with sulphydryl groups at the cell surface (Mehrishi and Grassetti, 1969), arrested the hydrolase activity of the cells by 78·3 per cent, as indicated by the decreased calorimetric response (Fig. 5). This inhibition of a sulphydryl enzyme, ecto-ATPase, was of a similar degree to that determined by conventional spectrophotometric methods, at 80·6 per cent (Kemp, 1972).

Of studies of cellular interactions with viruses, bacteria and other cells, the fruit is ripe for picking but, perhaps, it will be a cactus fruit!

D. Conclusions

Vessel design holds the key to success in calorimetric studies of tissue cells. However, studies of cells in monolayer have different design problems to those utilizing cells in suspension. For most commercially available machines, suitable vessels are not available and, thus, artifacts are more easily discovered than true data. Cells in monolayer present the greatest problems owing to physical size limitations of the measuring space and, at least in experiments to the present time, the lack of a true thermal base line. The work of Nikolic and Neskovic (1976), Cerretti et al. (1977) and Kemp (1975) have all suffered from problems, the latter failing to maintain the growth of perfused cells on beads in the flow sorption vessel for more than 16 h. The solution to the base-line problem may be to adopt Wadsö's ampoule drop system (Konicek et al., 1971), incorporating a specialized vessel for cells growing in monolayer.

With the reservation concerning the teflon tubing, the matchbox design (see p. 119) seems to be appropriate for studying cells in suspension. Metabolic data can be achieved rapidly and the possibilities for on-line measurements of turbidity, oxygen tension and pH can be easily perceived by the ambitious experimenter. By locating the culture vessel in the thermostatic bath of the calorimeter as Ackland et al. (1976) have done for bacterial studies, the time for a given cell to reach the measuring vessel from the culture chamber can be reduced to about 60 s. Thus, one is in the position of being able to define the thermal behaviour of tissue cells in suspension and experimentally to alter metabolic parameters. Once this is achieved for given cell types, the biologically and medically important field of cellular interactions can be explored, with confidence, by calorimetry.

Acknowledgements

The personal research of the author was funded by the Science Research Council and the British Cancer Research Campaign and owes much to the skilled technical assistance of Mr J. Meredith and Mr P. Lloyd. LKB Produkter, AB, primarily in the person of Mr R. L. Taylor, has been an invaluable source of advice and expert knowledge.

References

ACKLAND, P. J., PRITCHARD, F. E. and JAMES, A. M. (1976). *Microbios Lett.* **3**, 21.
BANDMAN, V., MONTI, M. and WADSÖ, I. (1975). *J. clin. Lab. Invest.* **35**, 121.
BEEZER, A. E., STEENSON, T. I. and TYRRELL, H. J. G. (1973). "Protides of the Biological Fluids", vol. 20, p. 563. Pergamon Press, Oxford.
BENZINGER, T. H. (1965). *Fractions*, 1.
BENZINGER, T. H. and KITZINGER, C. (1963). *In* "Temperature—its Measurement and Control in Science and Industry" (J. D. Hardy, ed.), vol. 3, pt 3. Reinhold, New York.
BOIVINET, P. (1977). *In* "Application of Calorimetry in Life Sciences" (I. Lamprecht and B. Schaarschmidt, eds), p. 159. W. de Gruyter, Berlin and New York.
CALVET, E. (1956). *In* "Experimental Thermochemistry" (F. D. Rossini, ed.), p. 237. Interscience, New York.
CERRETTI, D. P., DORSEY, J. K. and BOLEN, D. W. (1977). *Biochim. biophys. Acta*, **462**, 748.
DUNN, M. J. and KEMP, R. B. (1973). *Cytobios*, **7**, 127.
ERIKSSON, R. and WADSÖ, I. (1971). "Proc. 1st Eur. Biophys. Congr.", vol. 4, p. 319.
FLETCHER, A. P., ROSS, P. D. and JAMIESON, G. A. (1973). "Protides of the Biological Fluids", vol. 20, p. 555. Pergamon Press, Oxford.
JOHANSSON, Å. (1973). "Protides of the Biological Fluids", vol. 20, p. 567. Pergamon Press, Oxford.
KEMP, R. B. (1972). *Science Tools*, **19**, 3.
KEMP, R. B. (1974). "Protides of the Biological Fluids", vol. 21, p. 141. Pergamon Press, Oxford.
KEMP, R. B. (1975). *Pestic. Sci.* **6**, 1.
KEMP, R. B. (1980). *Biochim. biophys. Acta.* In preparation.
KEMP, R. B., LLOYD, C. W. and COOK, G. M. W. (1973). *Progr. Surface Membrane Sci.* **7**, 273.
KITOS, P. A. and WAYMOUTH, C. (1964). *Exp. Cell Res.* **35**, 108.
KONICEK, K., SUURKUUSK, J. and WADSÖ, I. (1971). *Chemica Scripta*, **1**, 217.
KRAKAUER, T. and KRAKAUER, H. (1976). *Cell. Immunol.* **26**, 242.
KUSANO, K., NELANDER, B. and WADSÖ, I. (1971). *Chemica Scripta*, **1**, 211.
LEVIN, K. (1973). *Scand. J. clin. Lab. Invest.* Suppl. 135.
LIN, S. and SPUDICH, J. A. (1974). *J. biol. Chem.* **249**, 5778.
MEHRISHI, J. N. and GRASSETTI, D. R. (1969). *Nature, Lond.* **224**, 563.
MONK, P. and WADSÖ, I. (1968). *Acta chem. scand.* **22**, 1842.
MONK, P. and WADSÖ, I. (1969). *Acta chem. scand.* **23**, 29.

Monk, M. and Wadsö, I. (1973). *Scand. J. Lab. Clin. Invest.* **32**, 47.
Nikolic, D. and Neskovic, B. (1976). *Iugoslav. Physiol. Pharmacol. Acta*, **12**, 191.
Ross, P. D., Fletcher, A. P. and Jamieson, G. A. (1973). *Biochim. biophys. Acta*, **313**, 106.
Spink, C. and Wadsö, I. (1975). In "Methods of Biochemical Analysis" (D. Glick, ed.), vol. 23, p. 1. Wiley–Interscience, New York.
Sturtevant, J. M. (1969). *Fractions*, 1.
Wadsö, I. (1968). *Acta chem. scand.* **22**, 927.

Note added in proof:

To send the reader to the library for a further hour (see p. 121), one should mention some interesting recent studies. Following the personal communication from Wadsö (see p. 125) cited by Cerretti *et al.* (1977), Ljungholm *et al.* (1978) reported a value of 20–40 pW per cell for HeLa cells maintained in an ampoule microcalorimeter, the variation being ascribed to differences in "the state of the cell".

Interest has begun to turn towards relating heat production to other metabolic parameters. Using a batch microcalorimeter, Nedergaard *et al.* (1977) showed that noradrenaline-stimulated brown adipocytes from hamsters produced 8 pW per cell. By applying Ivlev's (1934) oxycalorific constant (226 kJ mol^{-1} oxygen), these authors demonstrated that heat production predicted from respirometric data was close to this experimental value. Isolated rat hepatocytes, in contrast to adipocytes and pancreatic β-cells (Gyfle and Hellman, 1975), have a relatively high rate of respiration (Jarrett *et al.*, 1979). The metabolic efficiency of these cells was relatively constant at 1·8 µmol of oxygen per J unless perturbed by conditions that markedly enhance substrate cycling, for instance extracellular 10 mM NH_4Cl and urease.

References

Gyfle, E. and Hellman, B. (1975). *Acta physiol. scand.* **93**, 179.
Ivlev, V. G. (1934). *Biochem. Z.* **275**, 49.
Jarrett, I. G., Clark, D. G., Filsell, O. H., Harvey, J. W. and Clark, M. G. (1979). *Biochem. J.* **180**, 631.
Ljungholm, K., Wadsö, I. and Kjellen, L. (1978). *Acta Path. Microbiol. Scand. Sect. B*, **86**, 121.
Nedergaard, J., Cannon, B. and Lindberg, O. (1977). *Nature, Lond.* **267**, 518.

Studies on Blood Cells

K. LEVIN

A. Introduction

Although calorimetry has been in use for quite a considerable time it is not until recently that it has been used for measurements of the heat production evolved by blood corpuscles. The reason for this is probably that calorimeters of adequate sensitivity have not been available to investigators who have had access to this kind of material. With the construction of modern heat-flow calorimeters with high sensitivity and with their industrial production and commercial availability, this situation has changed and a number of studies have recently been carried out showing that heat production from various kinds of formed elements in the blood can easily be performed on a routine basis. This field now seems to be under development and this trend can be expected to continue. The fundamentals of this development were laid down by Wadsö and his co-workers. In fact most of the work reviewed in this chapter was done with calorimeters built according to the original design of Wadsö (1968) and of Monk and Wadsö (1968, 1969). Some work has also been done with a modified version of the original calorimeter (Wadsö, 1974). The basic design of this calorimeter is a twin heat-flow calorimeter with two measuring positions interposed between thermoelements which are countercoupled. This design is advantageous since it compensates to a certain degree for heat disturbances in the surrounding heat sink. It also allows differential measurements to be made so that heat effects from side-reactions can be blanked out during one measuring cycle.

Another great advantage with the basic design is that the measuring cells can be of different types. By using the central part of the heat sink as a heat exchanger, rapid thermo-equilibrium between a liquid sample and the interior of the calorimeter can be achieved. Flow-through measuring cells can be used of single-channel type or two liquid streams can be mixed at the entrance of a flow cell. Alternatively, cuvettes for stationary liquid samples can be mounted in the measuring position and the design also allows the measuring unit to rotate so that two liquid samples can be mixed within the measuring cuvette after proper thermal equilibrium has been obtained. Due to the longer times necessary for each experiment the latter has only found limited use in the type of studies presently being reviewed. Besides the above-mentioned design with a simple flow-through design of the calorimeter, the ampoule drop calorimeter (Wadsö, 1974) has proved most useful when measuring the heat production from blood cells. This variant of the calorimeter is designed so that each sample is contained in a separate ampoule which can be prethermostated outside the calorimeter thereby shortening the calorimetric equilibration time. It can be said generally that the sensitivity of heat conduction calorimeters is such that heat production measurements can be made with reasonable precision on samples of blood corpuscles derived from 5 ml of blood or less.

The present review will cover reports of measurements made on erythrocytes (red blood corpuscles), leucocytes (white blood corpuscles), polymorphonucleated leucocytes (PMN) and lymphocytes (subtypes of leucocytes) and thrombocytes (platelets). It is not the purpose of the present review to cover the analytical work done on other constituents of blood; instead the reader is referred to a recent review (Levin, 1977).

B. General aspects

Here the general aspects of measurements on various types of blood cells will be covered and the present state of knowledge as regards the correlation of metabolic events to the heat production measured is presented.

1. LEUCOCYTES

Results of measurements of heat production from leucocytes and thrombocytes were the first to be published (Levin, 1971). The determinations were made with a commercially available calorimeter. A flow-through technique was used, the samples being drawn through the calorimeter by a peristaltic pump on the outlet tubing from the calorimeter. This was done in order to avoid mechanical damage to the leucocytes. Two other technical observations were pointed out in this paper. One was that the introduction of cell-free blood plasma from human subjects always indicated a heat production in the calorimeter. The other observation was that leucocytes in suspension and particularly PMNs tend to adhere to the walls of the tubing and to the walls of the measuring cell during their passage through the calorimeter so that the reading obtained was sometimes difficult to interpret.

The observation that a protein-containing solution or even an ultrafiltrate of blood plasma produces a base-line shift when introduced into a calorimeter has not yet been fully explained. The observation as such has been confirmed by others (e.g. P. D. Ross, personal communication) with different constructions and material in the measuring cells. In the original paper it was suggested that part of the phenomenon could be due to the oxidation of SH-groups but since the phenomenon is also obtained with teflon or other polymeric wall material this does not seem to be the only explanation. The phenomenon is characterized by a rapid attainment of steady state which does not shift for long periods of time during a flow type of experiment. A simple reaction between the proteins and the surface material cannot afford an explanation since the surface would be saturated after a certain time and the reaction would subside. It seems more likely that continuous oxidation of oxidizable substances takes place within and at the surface of the measuring cell (cf. Sand *et al.*, 1977).

The same type of reaction can be seen in a batch calorimeter when a protein-containing solution is allowed to wet previously dry parts of the cuvette. A slowly subsiding signal is then obtained.

In a later paper (Levin, 1973) I presented some experiments in which the flow calorimeter was modified to make it more suitable for leucocyte studies. The gold tubing in the heat exchanger and the measuring cells was replaced by teflon tubing. With this modification and with the introduction of air bubbles into the measuring stream during its passage through the heat exchanger, a functioning system was obtained. With this modified calorimeter it was possible to confirm the results obtained earlier. A value of $6 \cdot 2 \pm 1 \cdot 4$ pW per cell (mean value and standard deviation) was found for leucocytes isolated from blood when heparin was used as an anticoagulant. In contrast a value of $2 \cdot 2 \pm 1 \cdot 9$ pW per cell was found for leucocytes obtained after allowing the sample to coagulate with simultaneous defibrination. It was suggested that the latter value represented a truer basal level of unstimulated leucocyte heat production.

This system utilizing teflon tubing and air segmentation is, however, rather complicated to use and I hesitate to recommend it for routine work.

As far as the above-mentioned results are concerned, the level of basal heat production from leucocytes has been confirmed by Bandmann *et al.* (1975) who carried out similar measurements using an ampoule drop calorimeter. Their measurements seem to involve fewer problems than those encountered with a continuous flow technique. In general, a batch-type technique seems more suited for the handling of leucocytes, particularly since activated leucocytes are characterized by increased adhesiveness and hence will always be difficult to handle in a flow system. Compared to other techniques used for the study of leucocytes, calorimetry has certain advantages. The most frequently used technique for leucocyte studies is the measurement of oxygen consumption which can be done either with a Warburg-type equipment or with a gas electrode. In both cases a rather tedious preparation technique is necessary in order to separate the leucocytes from the oxygen-containing erythrocytes. This is not necessary with the calorimetric technique since the heat production from erythrocytes is very low in comparison with leucocytes and thus contamination of a leucocyte preparation with erythrocytes poses fewer problems.

Other techniques for leucocyte studies normally include some type of measurement of substrate turnover with or without a radioactive tracer. These types of studies will of course always be necessary in order to obtain a more precise picture of the function of different metabolic pathways within the leucocyte. With proper standardization of the experimental procedure it may, however, be possible to link a defined metabolic process to an observed heat production or a change in heat production.

In the present context the term "leucocytes" has been used for the total population. With the simple methods of preparation used here it is mainly PMNs that are responsible for the heat production. Depending on the particular suspension medium (Bandmann *et al.*, 1975) lymphocytes will produce as much heat as PMNs when suspended in plasma, whilst in buffer media lymphocytes produce less heat than the PMNs.

There is no direct evidence published as to which metabolic process gives rise to the heat production measured in leucocytes. It has been shown with other techniques (cf. Karnovsky, 1962) that glucose is metabolized in leucocytes by at least three different processes. These are anaerobic glycolysis, oxidative breakdown by the pentose pathway shunt and a conventional oxidative phosphorylation. From other observations it seems likely that the basal heat production mainly stems from the first and the third of these pathways while the pentose pathway is stimulated during phagocytosis and similar phenomena.

2. LYMPHOCYTES

As mentioned above, lymphocytes produce approximately the same amount of heat as do PMNs. Since 10^6 cells can normally be isolated from 1 ml of blood they have to be concentrated about twofold before measurements can be made. Higher concentrations of lymphocytes are not recommended since an effect known as the "crowding effect" then appears (Hedeskov and Esmann, 1966; Sand *et al.*, 1977). Some reports concerning the heat production of lymphocytes have been published (Bandmann *et al.*, 1975; Krakauer and Krakauer, 1976; Gorski and Levin, 1977). Basically lymphocytes appear to behave similarly to other leucocytes. No problems with adhesion have been reported and these cells do not of course participate in any phagocytic reactions. An increased metabolic activity in response to cell surface active substances was reported independently by Krakauer and Krakauer (1976) and Gorski and Levin (1977). This appeared after 24–48 h and continued for up to 5 days in lymphocyte cultures. Krakauer and Krakauer found that the stimulation of heat production preceded the onset of increased DNA production. The pattern of increased thermogenesis seems to be independent of whether a purified tuberculin factor or phythaemagglutinin (Levin and Gorski, 1977), concanavalin A or a specific antigen (Krakauer and Krakauer, 1976) is used as a stimulant.

With proper standardization, calorimetry may prove to be a suitable technique for the study of this type of reaction.

3. THROMBOCYTES AND COAGULATION

In addition to the heat production from lymphocytes the heat production from thrombocytes was also reported in my first paper (Levin, 1971). This was partly due to necessity since complete separation of leucocytes and thrombocytes is difficult to achieve by simple means of preparation and thrombocytes were hence present in the leucocyte preparations. A mean value of 68 fW per cell was found with a reference range of 36–110 fW per thrombocyte, indicating that the heat production from one thrombocyte is about 90 times less than that from one leucocyte. This seems to be a reasonable difference considering the smaller volume of the thrombocyte. There was considerable variation between samples obtained from the same individual measured on different occasions. This was interpreted to be an expression of great sensitivity on the part of the thrombocytes for different types of stimuli.

Ross *et al.* (1973) have studied the reactions of thrombocytes in a batch calorimeter. They investigated the heat production after adding different aggregating agents. They used ADP, collagen, Mn^{2+} and Cd^{2+} and found different patterns of heat production. Their investigation indicates that calorimetry could be used for the study of different types of reactions induced in thrombocytes.

Several reports have been presented on calorimetric measurements of the coagulation process in cell-free plasma (e.g. Laki and Kitzinger, 1956;

Bostick and Carr, 1973). The most interesting is that by Watt et al. (1974) who obtained very complicated thermograms after initiating coagulation by the addition of thrombin (Fig. 1).

By combining thrombocytes and plasma with suitable initiators of the coagulation process it should be possible to follow the entire coagulation

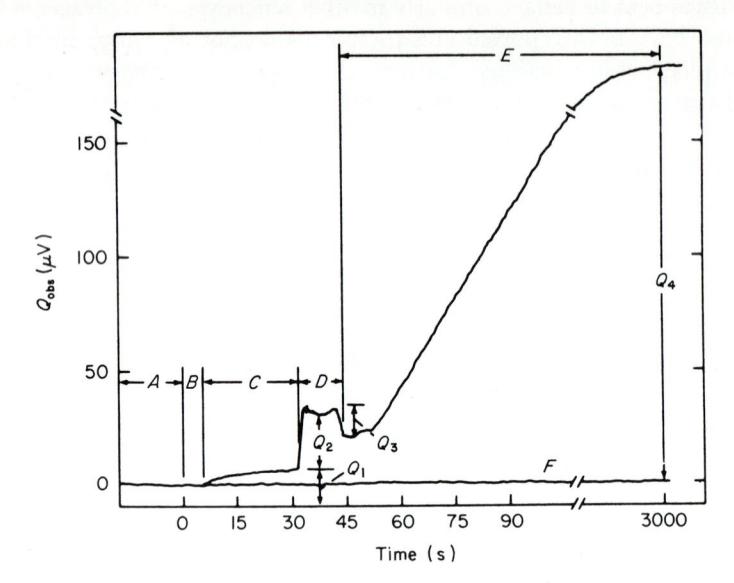

Fig. 1. Typical thermogram for the coagulation of plasma by thrombin at 20 °C. The sample contains 0·11 ml of plasma in a 4-ml volume composed of citrate, saline solution. After thermal equilibrium has been attained (segment A), 6 μl of thrombin solution (1000 units ml^{-1}) is added in 6 s. Coagulation occurs during the time segment, D. (Watt et al., 1974.)

process by measuring the heat production at different stages of the process. This should conceivably provide interesting information. However, there are no instruments available suitable for this kind of measurement since the flow principle can obviously not be used and it would be preferable to have a batch-type instrument with disposable cuvettes in order to ensure standardized measuring conditions as this process is very sensitive to various external influences.

4. ERYTHROCYTES

The blood cells most intensively studied with calorimetric techniques are the erythrocytes. Obviously the reason for this is that well-defined preparations of erythrocytes can easily be obtained in sufficient quantities.

Erythrocytes are less sensitive to variations in the preparative procedure than other types of blood cells and their behaviour can also be investigated in serum, plasma or artificial suspension media. Working with a 40 per cent suspension of erythrocytes there is no large discrepancy between measurements made in flow- or batch-type systems.

Several reports, including those of Boyo and Ikomi-Kumm (1972) and Levin (1973a), have arrived at the conclusion that the heat production of erythrocytes under standard conditions of pH 7·40 and 37 °C in media containing glucose is about 110 mW l^{-1} of packed cells from healthy humans (with 10–15 per cent coefficient of variation). The lowest coefficient of variation was obtained by Monti and Wadsö (1973, 1976a, b) who carefully standardized their preparative method. A constant finding in these experiments has been that only about half of the heat production obtained can be related to glycolysis and the breakdown of glucose to lactate. Figure 2 is reproduced from Levin et al. (1974) and demonstrates that during a prolonged incubation at 37 °C in artificial medium the heat production from the erythrocytes decreases as the glucose in the medium is depleted. However, even after complete glucose depletion substantial heat production still occurs. Amino acids in excess were added to one of the incubation media (marked with filled circles in the Figure). The general pattern of heat production remained unaltered and the conclusion was drawn that the prolonged heat production was not due to a degradation of amino acids. It was also shown that the erythrocytes were depleted of the energy-rich phosphate compounds soon after glucose depletion. Hence these compounds cannot be responsible for the prolonged heat production. At present it seems most likely that this heat production arises from lipid degradation in the erythrocyte membranes.

In a paper by Minakami and de Verdier (1976) several metabolic aspects of the heat production from the various steps of glycolysis were studied. They obtained a very good correlation between the heat production and the enthalpy changes of the different metabolic steps. Their studies confirmed other observations that during normal conditions most of the glucose is metabolized by glycolysis and very little goes via the pentose shunt. It is, however, easy to stimulate the penthose shunt by adding methylene blue (Monti and Wadsö, 1976c) resulting in a five- to ten-fold increase in the heat production.

C. Clinical aspects

A number of studies have been published where the heat production from different kinds of blood cells has been investigated in different patient groups. These studies have shown that a change in the heat production can often be observed as a result of a disease process. In some cases this can be measured with cells in their basal state while in other cases the differences only become apparent after the cells have been stimulated by some means.

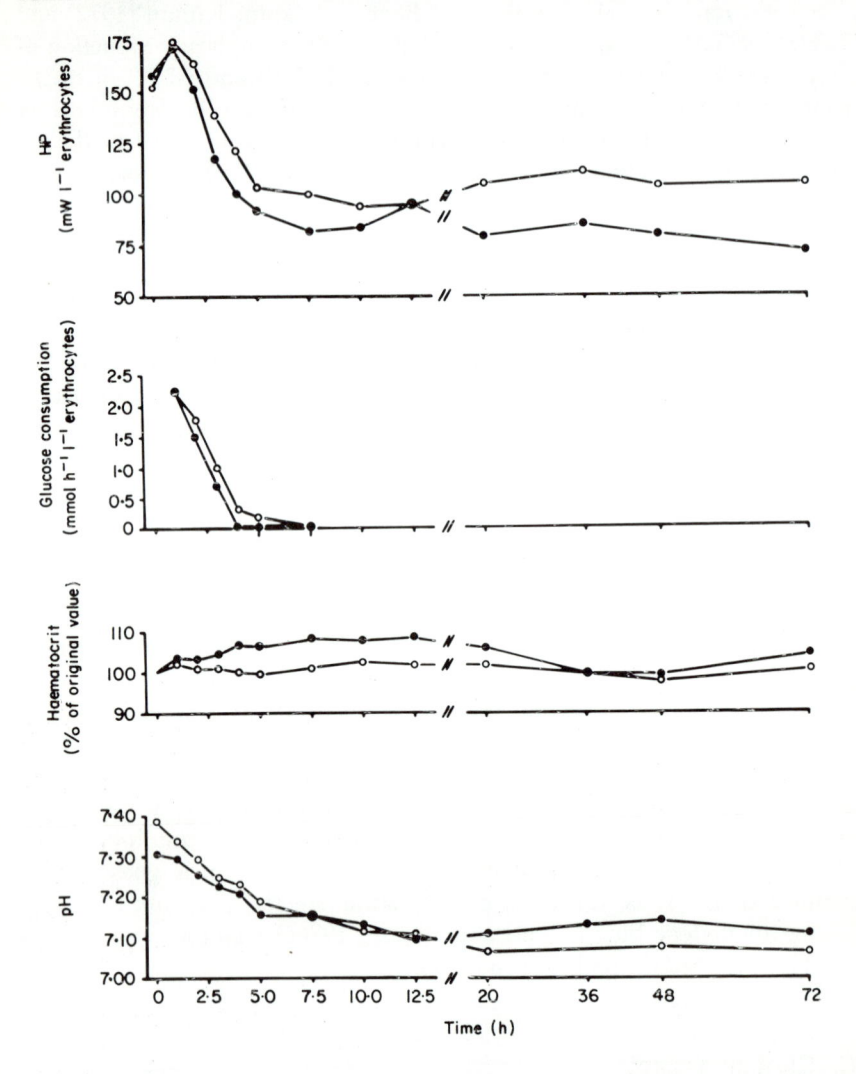

FIG. 2. Heat production (HP), glucose consumption, haematocrit and pH from erythrocytes during incubation. Mean values of three experiments (○) in a synthetic medium containing albumin, salts and glucose, and mean value of five experiments (●) in the same medium with amino acids added. (Levin et al., 1974.)

1. THYROID DISEASES

Thyroid hormones are known to stimulate enzyme production in various tissue cells. This has also been demonstrated to be true for blood cells. Thus it is not surprising that leucocytes from patients suffering from hyperthyreosis produce more heat than average leucocytes (Levin, 1971). A significantly lower heat production was also found in leucocytes from hypothyroid patients. Although the mean values of normals and patients differed significantly, the overlapping between the groups was too great to allow the results to be used as a diagnostic criterion. No such differences were found for the thrombocytes from the same patients. Monti and Wadsö (1976d) have found a similar relationship studying erythrocytes from hyperthyroid patients. They were also able to follow the effects of treatment and the subsequent decrease in heat production. Since the turnover time for erythrocytes is more than one month the decrease in erythrocyte heat production comes much later than the improvement in the patient's condition after proper treatment.

2. IMMUNOLOGICAL DISEASES

As mentioned above, phagocytosis causes a marked increase in the heat production from leucocytes. With this in mind an investigation was performed where leucocytes from patients suffering from the auto-immune disease Systemic Lupus Erythematosus (SLE) were exposed to an autohomogenate. This homogenate was prepared from the patient's own leucocytes. A very marked increase in the heat production was recorded, similar to that seen during phagocytosis. This finding was expected since the classical way of detecting this particular disease is the demonstration of material from other leucocyte nuclei within an otherwise normal leucocyte, the so-called LE-cell phenomenon. These patients are in fact known to produce antibodies directly against DNA. However, Levin and Thomasson in 1974 were not able to demonstrate that phagocytosis actually occurs during the test which we called the CAL-LE-test. Later unpublished experiments have shown that it is possible to produce a similar response from leucocytes when antigen–antibody reactions between proteins in solution take place in the presence of leucocytes and serum. The increased heat production during the CAL-LE test may thus not be caused by actual phagocytosis but can be regarded as an indication of antigen–antibody reactions taking place in the solution and triggering an increased metabolism in the leucocytes. In this way leucocytes can be used as biological amplifiers for the indication of the presence of immunological phenomena. As can be seen in Fig. 3, a clear separation was obtained between the patient groups and the normal subjects tested. The

CAL-LE test proved to give as good a diagnostic accuracy as any of the other tests it was compared with.

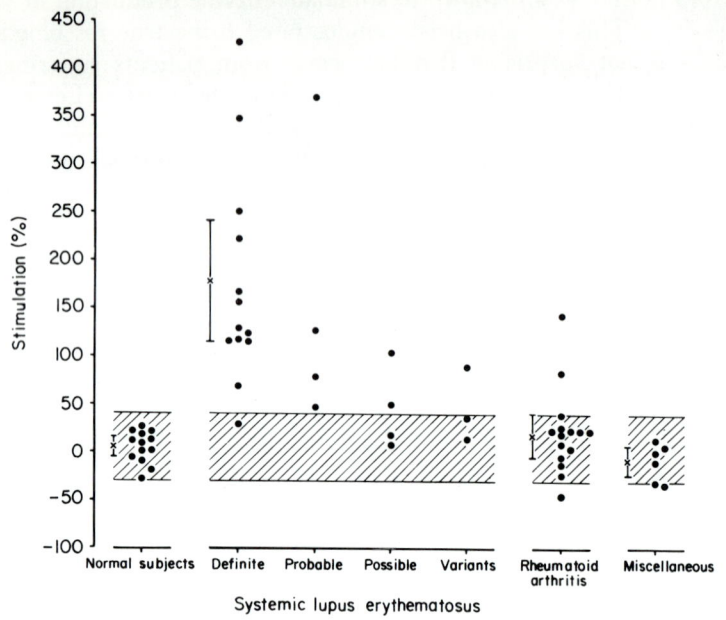

FIG. 3. Results obtained with CAL-LE test in normal subjects and patients. The crosses and vertical bars to the left of each column illustrate the mean value ±2 SEM for the groups that contain more than five subjects. The shaded area indicates the reference range calculated from the mean value ±2 SD of the normal subjects. (Levin and Thomasson, 1974.)

3. ANAEMIAS

The heat production from erythrocytes derived from anaemic patients can generally be said to be higher than that from normal erythrocytes (Levin and Boyo, 1971; Levin, 1973; Monti and Wadsö, 1973) irrespective of the cause of the anaemia. The reason for this is probably that the average erythrocyte population is younger in anaemic subjects and younger erythrocytes are metabolically more active. This phenomenon is clearly demonstrated in Fig. 4 which illustrates the increased heat production during treatment of two cases of pernicious anaemia. This increase is closely correlated to the number of immature erythrocytes, so-called reticulocytes. The increased heat production is particularly prominent in haemolytic types of anaemia (Monti and Wadsö, 1973).

Boyo and Ikomi-Kumm (1972, 1973) investigated the heat production by erythrocytes in sickle-cell disease. They found a markedly elevated heat production from these cells. They also found that carriers of this hereditary disease had a moderate increase in the heat production from their erythrocytes. Whether this is a specific phenomenon attributable to the abnormal haemoglobin configuration or a secondary effect has not been clarified. This increase in heat production appears to be greater than other reported metabolic changes in the sickle cells.

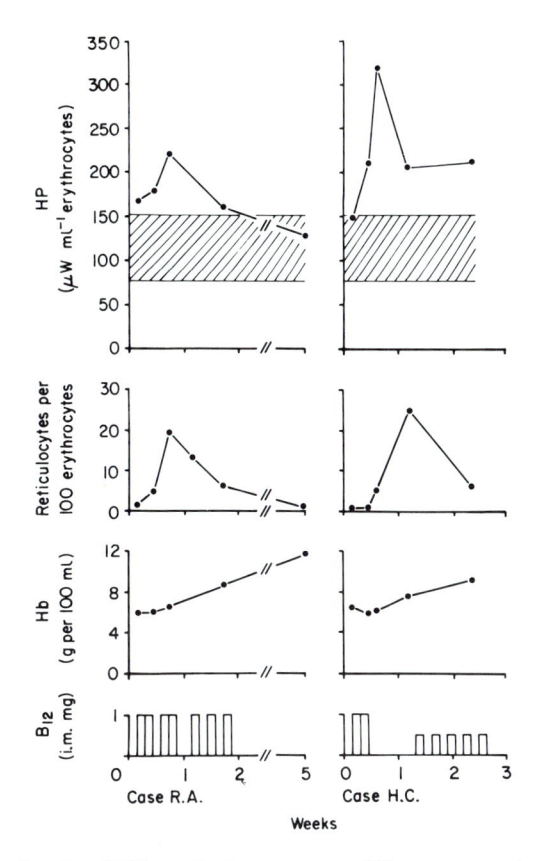

FIG. 4. Heat production (HP), reticulocyte count, Hb concentration in two cases of pernicious anaemia followed during the initial phase of treatment with cyanocobalamin by intramuscular injection (Levin, 1973b).

D. Concluding remarks

Calorimetric measuring techniques have been used successfully during the last decade for the study of blood cells. The technique appears to be convenient and allows measurements to be performed with ease on cell samples that can be prepared with simple methods. Although the measuring time is still rather long compared with other techniques, this is well compensated for by the simplicity of the method.

On the one hand, it can be an advantage to obtain information about the overall metabolic activity of one cell species while, on the other hand, it will of course seldom be possible to use this as a specific diagnostic indication. Specific stimulation or inhibition of one or several metabolic functions within the cell is, however, often easy to achieve. By such means it will be possible to obtain more specific and diagnostically more useful information from calorimetric measurements.

References

BANDMANN, U., MONTI, M. and WADSÖ, I. (1975). *Scand. J. clin. Lab. Invest.* **35**, 121–127.
BOSTICK, W. D. and CARR, P. W. (1973). *Am. J. clin. Pathol.* **60**, 330–336.
BOYO, A. E. and IKOMI-KUMM, J. A. (1972). *Lancet*, **1**, 1215.
BOYO, A. E. and IKOMI-KUMM, J. A. (1973). *In* "Protides of Biological Fluids" (H. Peters, ed.), pp. 559–562. Pergamon Press, Oxford.
GÓRSKI, A. and LEVIN, K. (1977). *Science Tools*, **24**, 27–29.
HEDESKOV, C. J. and ESMANN, V. (1966). *Blood*, **28**, 163–174.
KARNOVSKY, M. L. (1962). *Physiol. Rev.* **42**, 143–168.
KRAKAUER, T. and KRAKAUER, H. (1976). *Cellular Immunol.* **26**, 242–253.
LAKI, K. and KITZINGER, C. (1956). *Nature*, **178**, 985.
LEVIN, K. (1971). *Clin. chim. Acta*, **32**, 87–94.
LEVIN, K. (1973a). *Scand. J. clin. Lab. Invest.* **32**, 55–65.
LEVIN, K. (1973b). *Scand. J. clin. Lab. Invest.* **32**, 67–73.
LEVIN, K. (1977). *Clin. Chem.* **23**, 929–937.
LEVIN, K. and BOYO, A. E. (1971). *Scand. J. clin. Lab. Invest. Suppl.* **118**, 55.
LEVIN, K. and THOMASSON, B. (1974). *Acta med. Scand.* **195**, 191–200.
LEVIN, K., FÜRST, P., HARRIS, R. and HULTMAN, E. (1974). *Scand. J. clin. Lab. Invest.* **34**, 141–148.
MINAKAMI, S. and DE VERDIER, C.-H. (1976). *Eur. J. Biochem.* **65**, 451–460.
MONK, P. and WADSÖ, I. (1968). *Acta chem. scand.* **22**, 1842–1852.
MONK, P. and WADSÖ, I. (1969). *Acta chem. scand.* **23**, 29–36.
MONTI, M. and WADSÖ, I. (1973). *Scand. J. clin. Lab. Invest.* **32**, 47–54.
MONTI, M. and WADSÖ, I. (1976a). *Scand. J. clin. Lab. Invest.* **36**, 565–572.
MONTI, M. and WADSÖ, I. (1976b). *Scand. J. clin. Lab. Invest.* **36**, 573–580.
MONTI, M. and WADSÖ, I. (1976c). *Scand. J. clin. Lab. Invest.* **36**, 431–436.
MONTI, M. and WADSÖ, I. (1976d). *Acta med. scand.* **200**, 301–308.
ROSS, P. D., FLETCHER, A. P. and JAMIESON, G. A. (1973). *Biochim. biophys. Acta*, **313**, 106–118.

Sand, T., Condie, R. and Rosenberg, A. (1977). *Blood*, **50**, 337–346.
Wadsö, I. (1968). *Acta chem. scand.* **22**, 927–937.
Wadsö, I. (1974). *Science Tools*, **21**, 18–21.
Watt, D., Berger, R. L., Green, D. and Marini, M. A. (1974). *Clin. Chem.* **20**, 1013–1017.

The Heat Production of Intact Organs and Tissues

R. C. WOLEDGE

The heat produced by a number of organs has been measured, and in some cases studied in great detail. The measurement of heat output in itself allows few conclusions to be drawn about the behaviour of an organ. Only in the case where a function of the organ is thermogenesis, as may be the case for brown fat tissue, is the heat production of much intrinsic interest. In other cases interest centres on (a) how the heat rate varies with activity of the organ, (b) correlating the heat changes with biochemical changes, (c) studying the action of metabolic inhibitors, (d) comparing different tissues or comparing the tissues of different species. In all these applications both the strength and the weakness of heat production measurements is that they are non-specific; the heat observed is the sum of the contributions from all the chemical changes occurring in the tissue. This is a strength because heat production measurements can detect the effects of previously unsuspected processes; but it is also a weakness because it is impossible to draw conclusions about the extent of a specific process in the tissue from heat production methods alone, without the assumption, usually quite unjustified, that other processes are not contributing to the observed heat. Another potential error in the interpretation of heat production measurements is to assume that the absence of heat production implies an absence of biochemical reactions, or that the absence of alteration in the heat when an inhibitor is applied means that the inhibitor has no effect. These assumptions are false because certain biochemical processes can be almost thermally neutral, and because the tissue may have the ability to replace inhibited reactions with others that produce almost the same heat. These difficulties, although real (Woledge, 1975), should not be exaggerated. Provided that the heat measurements are carried out in parallel with suitable biochemical and, where appropriate, electrical and mechanical measurements, these problems of interpretation can be avoided. It is then possible to exploit the advantages of heat measurements, namely simplicity, speed of response, non-destructiveness and applicability to biological systems at every level of complexity.

A. Techniques

Five types of technique have been used to study the heat production of organs and intact tissues.

1. THERMOPILES

This technique, which A. V. Hill did so much to develop, is described in detail in his book (Hill, 1965). Briefly a thermopile consists of a number of metal thermocouples in series running between the tissue being investigated and a relatively massive metal frame. The couples are made very thin and are electrically insulated from, but in good thermal contact with, the tissue. The low thermal mass of the thermocouples and their insulation means that they are able to follow temperature changes in the tissue without much distortion. Thermopiles are mostly used to study rapidly changing rates of heat production, as during muscle contraction or nerve action potential, but they can also be used to measure low steady heat rates, such as the resting heat rate of muscle. The advantages of this technique are sensitivity and speed of response. Disadvantages are that errors can be caused by (a) non-uniformity of heat production in the tissue or (b) changes in osmotic pressure in the tissue. Wilkie (1963) describes a thermopile designed to avoid the first of these problems. Another limitation is that thermopiles cannot measure heat rates while the tissue is immersed in a solution; it must be held in (moist) gas while the actual measurements are made. Methods of constructing thermopiles are described by Hill (1965), Ricchiuti and Mommaerts (1965), Howarth (1970) and Mulieri *et al.* (1976, 1977).

2. HEAT-FLOW CALORIMETERS

The well-known designs of Calvet and Prat and of Wadsö are described elsewhere in this book and commercial versions of these instruments are available. The calorimeters constructed by Fales *et al.* (1967), by Smith and Woledge (1971) and by Chinet *et al.* (1977) should also be noted. This type of instrument differs from thermopiles principally in that the tissue is contained in a vessel of some kind and it is the temperature change of, or heat flow from, this vessel that is measured. This offers the possibility of immersing the biological preparations in solution, and provides immunity from osmotic errors and from errors associated with non-uniformity in the tissue. But the price of these advantages is a great loss in speed of response, and some loss of sensitivity.

3. PERFUSION CALORIMETERS

Dewar vessel perfusion calorimeters for measuring the heat production of intact hearts have been described by McDonald (1971) and by Coulson and Rusy (1973). In this method the heart is contained in a dewar vessel so that conductive heat loss is minimized. Most of the heat produced by the heart is carried away by the perfusion fluid and is measured by the difference in temperature between fluid entering and leaving the chamber. The method might be applicable to other perfused organs, but will only measure heat production while the organ is in a steady state.

4. *IN SITU* MEASUREMENTS

The heat production of certain organs *in situ* can be measured in intact animals by using thermocouples or thermistors to detect the rise in temperature which occurs when the circulation is occluded for a short time. In muscle this occlusion is brought about by the rise in pressure inside the muscle when it contracts. When a large mass of homogeneous tissue surrounds the thermocouple heat losses can be negligible. Flood and Aukland (1971) discuss the criteria for obtaining reliable results by this method in non-muscle tissue. The use of the technique in muscle is described by Edwards *et al.* (1974). Heat production by *in situ* tissues can also be estimated, in arbitrary units, without interrupting the circulation, using the method of "internal calorimetry". This technique (which uses heated thermocouples) was introduced by Grayson (Dosekun *et al.*, 1960) and was improved by Ohnhaus and Hunziker (1974). As no absolute calibration is possible with this method, the results will not be discussed in this review.

5. RADIOMETRY

Fraser and Carlson (1973) report the use of radiometry to measure muscle heat production. Although their method is not as sensitive as the thermopile technique, it does have outstanding time resolution. A limitation of their radiometer was that it could not be used at temperatures below 15 °C.

B. Calibration

Methods 2 and 3 may be calibrated by Joule heating. Method 1 may be calibrated by the method given by Hill and Woledge (1962) or by the ingenious Peltier heating method devised by Kretzschmar and Wilkie (1972, 1975). This method is probably also applicable to method 2. Methods 3, 4 and 5 are calibrated from the temperature sensitivity of the detector and the specific heat capacity of the tissue or perfusion fluid. Results are often expressed per unit weight of wet tissue and are given here in that form. As the wet weight is hard to determine reproducibly it might be better to use dried weight or some measure of protein content.

C. Results with different tissues

1. STRIATED MUSCLE

The heat produced by resting frog muscle under aerobic conditions at 20 °C is close to 0.17 mW g^{-1} (Hill, 1965, summarizing earlier work). Gore and Whalen (1968) report values of 0.24 mW g^{-1} at 22 °C and 0.4 mW g^{-1} at 27 °C. They show that the heat production is independent of oxygen tension provided this is sufficient to ensure diffusion of oxygen to the whole thickness of the tissue. Hill (1965) also summarizes data on O_2 consumption of resting frog muscle from which it is apparent that the aerobic resting heat can be accounted for by the oxidative metabolism which is occurring, suggesting that the tissue is otherwise in a steady state. Under anaerobic conditions the heat rate is about three times less than in aerobic conditions (Gore and Whalen, 1968). This is not what would be expected if the anaerobic muscle maintained a steady state with the same ATP turnover rate as aerobic muscle, and if the ATP was synthesized by glycolysis. If this were the case the anaerobic heat rate would be about two-thirds of the aerobic rate. So one at least of these assumptions about anaerobic resting muscle must be wrong.

In resting mammalian skeletal muscle (rat soleus) the resting heat rate is much higher than in frog muscle: 2.6 mW g^{-1} at 30 °C (Chinet et al., 1977). Treatment of the rat muscle with ouabain reduces the heat rate by about 5 per cent, but this fall is, with the muscle in the usual solution, followed by a rise to a higher than normal heat rate. By comparing this fall in the heat rate with the fall in sodium efflux produced by ouabain, Chinet et al. calculate the efficiency of the sodium pump as 34 per cent.

Several treatments of muscle other than the stimulation of contraction will increase the heat rate above the resting rate. These are treatment with solutions containing raised K^+ (Hill and Howarth, 1957), slight stretch (Clinch, 1968) and treatment with hypertonic solutions (Yamada, 1970). But the majority of the published work on muscle heat production concerns the effects during contraction and recovery from contraction. This is a large field which has been the subject of a number of comprehensive reviews in the last few years (Hill, 1965; Woledge, 1971; Abbot and Howarth, 1973; Homsher and Kean, 1977; Curtin and Woledge, 1978). However, for the benefit of the reader not wishing a detailed treatment, a brief summary of the main results will be given here, without references to the original literature, for which these reviews should be consulted. This work has mostly been done with frog muscle.

The rate of heat production during contractile activity is much higher than at rest. For instance during an isometric tetanus of frog muscle at 20 °C the steady heat rate is 200 mW g^{-1}, which is about 1000 times the resting rate. For rat muscle, however, the increase in heat rate on stimulation is less dramatic, about 100 times. The rapid heat production during contractions lasting a few seconds is followed by a much slower production of an about equal quantity of heat after contraction is over. This recovery heat may continue for 1 h (at 0 °C) before the heat rate returns to the resting rate. The heat produced during contraction (the initial heat) is the same in aerobic and anaerobic conditions, which suggests that it comes from non-oxidative reactions. This is confirmed by the fact that the O_2 consumption occurs mostly, or entirely, after contraction is over (Hill, 1940). It was at first supposed that the formation of lactic acid from glycogen was producing the initial heat. Following the discovery of ATP and phosphocreatine (PCr), it was realized that the breakdown of ATP and its resynthesis from PCr were responsible for much of the heat production. In the past few years it has become apparent that these are not the only sources of heat production important in contraction and that other, as yet unidentified, reactions contribute significantly.

The heat produced during contraction is dependent on the mechanical conditions. When muscles are stretched before contraction less tension is developed and less heat is produced. The heat production is not reduced as much as the tension development, and some heat (about one-fifth of the total) is still produced when no tension is developed. This suggests that some heat is produced in muscle by the system which switches it on and off. When muscle is allowed to shorten during contraction the heat rate is

TABLE 1

A comparison of heat rates during activity and rates of relaxation in some muscle tissues (these results are plotted in Fig. 1)

Animal	Muscle[a]	Temperature °C	Observed heat rate per unit tissue wet weight (in excess of resting rate) mW g^{-1}	Tension per unit cross-section area mN mm^{-2}	Heat rate "normalized" to a tension of 200 mN mm^{-2} mW g^{-1}	Relaxation time for tension to fall from 90% to 10% s	References
A. Skeletal muscles							
Frog (*Rana temporaria*)	Sartorius (twitch fibres)	17	149	256	116	0·087	Hill and Woledge (1962) Hill (1961)
		0	16	192	17	1·1	Hill and Woledge (1962) Hill (1961)
	Iliofibularis (slow fibres)	20	6	250	4·8	15	Floyd and Smith (1971)
Tortoise (*Testudo graeca*)	Rectus femoris (twitch fibres)	0	0·37	184	0·40	7·2	Woledge (1968)
Rat	EDL	27	136	290	94	0·12	Wendt and Gibbs (1973) Close (1969)
	SOL	27	21	240	17	0·42	Gibbs and Gibson (1972a) Close (1969)
Chicken	PLD	20	40	100	80	0·10	Canfield (1969, 1971)
	ALD	20	5·0	100	10	0·74	Canfield (1969, 1971)
Human	Quadriceps	37	70	—	—	0·27	Edwards *et al.* (1972, 1975a, b)

B. Smooth muscles

Snail (*Helix pomatia*)	Retractor pharynx	~20	—	—	9	4.5	Bozler (1930)
Rabbit	Rectococcygeus (anaerobic conditions)	27	2.9	158	3.7	12	Davey et al. (1975)
Guinea pig	Taenia coli	23	0.8	70[c]	2.3[b]	18	Mulvaney and Woledge (1972)
Mussel (*Mytilus edulis*)	ABRM phasic contractions	20	0.9	300	0.6	27	Baguet and Gillis (1967); Baguet et al. (1962); Baguet and Edmond (1972)
	ABRM tonic contractions	20	—	—	0.27[a,b]	200	Baguet and Gillis (1968); Hoyle and Lowy (1956)

C. Cardiac muscle

Rabbit	Papillary muscle	20	2.8	6.5[c]	86[b]	0.83	Gibbs (1969, 1974)

[a] Calculated from observations of oxygen consumption.
[b] These observations include oxidative "recovery" process. For comparison with the other results they should therefore be divided by 2, as has been done in plotting Fig. 1.
[c] Time-averaged values of tension. These tissues were intermittently active, not continuously active.
[d] SOL, soleus; EDL, extensor digitorem longus; PLD, posterior latissimus dorsi; ALD, anterior latissimus dorsi; ABRM, byssus retractor muscle.

higher than when the muscle is contracting isometrically (without shortening). During shortening external work is being done and the sum of the work rate and the heat rate therefore exceeds the isometric value even more than the heat rate does. The ratio of work rate to (heat rate + work rate) has a maximum value of about 0·45 in frog muscle (and 0·7 in tortoise muscle). This ratio has been called the efficiency but is not the same as the thermodynamic efficiency (Wilkie, 1961). However, if the same ratio is calculated including the recovery heat for oxidative recovery, the values, which are about half as great, do probably represent the efficiency for the overall process occurring.

The heat production and the chemical reactions occurring during relaxation (i.e. the fall of tension after the end of a period of stimulation) are not greatly different from those in contraction. After relaxation is over, other, much slower, heat productions occur which are associated with recovery, i.e. the aerobic or anaerobic resynthesis of the ATP which has been split, and the reversal of other changes which have occurred during contraction. There may be a small heat absorption at the start of the recovery period but during most of the recovery period heat is produced. The amount of heat produced is much greater in aerobic than in anaerobic conditions and its quantity can also be influenced by the nature of the buffers present. The quantitative relation of the recovery heat production to the recovery reactions which are occurring is unknown.

Besides the large volume of work on frog muscles, which consist mostly of twitch fibres, work has also been published on frog slow fibres (Floyd and Smith, 1971), on tortoise muscle (Woledge, 1968), on chicken fast and slow muscle (Canfield, 1969, 1971; Rall and Schottelius, 1973; Bridge, 1976), on rat muscle (Gibbs and Gibson, 1972; Wendt and Gibbs, 1973, 1974; Gower and Kretzschmar, 1976) and on human muscles *in situ* (Bolstad and Ersland, 1975; Edwards *et al.*, 1975a, b). The last study is noteworthy for the fact that the authors were able to account quantitatively for the heat produced during contraction by the amounts of ATP and PCr split and the amount of glycolysis which occurred. Some of the data from the experiments with different species are summarized in Table 1. As might be expected muscles which are faster in their mechanical properties are found to produce heat at a higher rate than mechanically slower muscles. Figure 1 illustrates this point by a comparison of the time required for relaxation with the heat rate during isometric contraction. The graph emphasizes the unusually low heat rate during tension maintenance in tortoise muscle, whereas frog slow muscle is remarkable for its rather high heat rate in comparison to its speed of relaxation.

Some other differences in the pattern of heat production between different skeletal muscles may also be noted. In frog twitch muscle and chicken fast muscle the heat rate is higher during the early part of a long isometric contraction and falls off during the first few seconds of the contraction. This effect is not seen in rat soleus, tortoise, frog slow muscle, or chicken slow muscle. Whereas in frog muscle and tortoise muscle the shortening heat referred to above is a conspicuous phenomenon, it is much smaller, and perhaps absent in rat and chicken muscles.

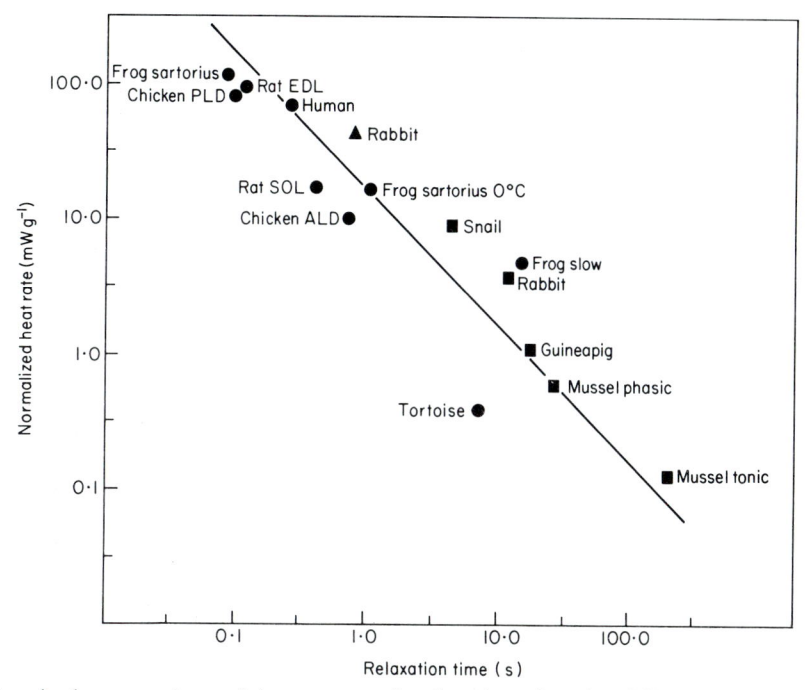

FIG. 1. A comparison of heat rate and relaxation time in different contractile tissues. The data plotted is tabulated in Table 1. ●, Skeletal muscle; ■, smooth muscle; ▲, cardiac muscle.

2. SMOOTH MUSCLE

Smooth muscles have been relatively little used for calorimetric investigation. The only comprehensive study on vertebrate smooth muscle is that by Davey et al. (1975). They used rabbit retrococcygeus muscle at 27 °C which is not spontaneously active. The resting heat rate could thus be determined and was 0·6 mW g^{-1}. When the muscle was stimulated the heat rate rose to 5 mW g^{-1}. In contrast to skeletal or heart muscle the heat production did not start abruptly on stimulation but increased slowly over a period of a few seconds. The heat produced during contraction is less in the absence of oxygen, suggesting that at least part of this heat is derived from oxidative processes. Thus, it is not possible to divide the heat into initial and recovery components. The ratio of work done to total heat and work had a maximum value in these experiments of 0·185. If, as seems possible, all the heat and work are derived from oxidative metabolism this figure would represent the efficiency of the tissue in converting the free energy from these processes into work (Wilkie, 1961). The paper by Davey et al. has an interesting discussion in which their results are compared with those from other types of muscles.

Some preliminary results with another mammalian smooth muscle, guinea pig taenia coli, are reported by Mulvany and Woledge (1972). At 23 °C this muscle is spontaneously active, producing a series of contractions during which tension is exerted for about half the time. The total heat rate (i.e. presumably including heat from oxidative reactions) is about 1·1 mW g^{-1}. When contractile activity was prevented by the absence of Ca or glucose the heat fell to 0·3–0·4 mW g^{-1}.

Two invertebrate smooth muscles have been investigated calorimetrically: the retractor pharynx of the snail by Bozler (1930, 1936) and the anterior byssus retractor (ABRM) of *Mytilus* by Baguet et al. (1962). In these muscles the oxidative recovery processes seem to occur largely after contraction is over (Bozler, 1930; Baguet and Gillis, 1967). In both cases the heat production is found to start slowly rather than abruptly, at the onset of stimulation; this seems therefore to be a general phenomenon in smooth muscle. In both cases also the heat rate was found to diminish considerably as an isometric contraction continued, a change which is accompanied, at least in the snail muscle, by a slowing of relaxation. Bozler also showed that the heat rate in the snail muscle could be greatly reduced, without reduction in tension development, by exposing the muscle to 1–10 per cent CO_2. This treatment also slows relaxation.

The heat rates observed for these smooth muscles are plotted in Fig. 1 where it can be seen that in comparison to the speed of relaxation these muscles produce heat at rates comparable with striated muscles, and are not remarkable for their low metabolic rate, as was claimed by Bozler.

3. HEART MUSCLE

Heat production has been studied using whole hearts of several species and in rabbit papillary muscles. The results obtained with the papillary muscle will be described first. They are reviewed in greater depth by Gibbs (1974). Papillary muscles remain quiescent if not stimulated, so the resting heat rate can be measured. The value given by Chapman and Gibbs (1974; see also Chapman et al., 1976) is 2·6 mW g^{-1} at 20 °C with pyruvate as substrate and 1·8 mW g^{-1} with glucose as substrate. The reason for this unexpected effect of changing to a different substrate is unknown. As in skeletal muscle, the resting heat rate is increased by stretching (Gibbs et al., 1967). The resting heat rate of rabbit papillary muscle at 20 °C is about the same as that of rat skeletal muscle at 30 °C, suggesting that if skeletal and cardiac muscle of the same species were compared at the same temperature the resting heat rate of cardiac muscle would be several fold greater. This is perhaps associated with the greater surface to volume ratio of the cardiac muscle cells.

Stimulation increases the heat rate of cardiac muscle considerably but not by such a large factor as in skeletal muscle. In cardiac muscle the heat due to oxidative "recovery" processes follows the "initial" heat more closely than in skeletal muscle, and the two components are not easily separated (Chapman and Gibbs, 1974). However, Gibbs has shown (1969) that the heat during contraction itself is little changed by blocking oxidation and glycolysis by treating the muscle with N_2 and IAA. Nevertheless, most work has concentrated on measuring the total heat produced (initial and recovery) which is about 15 mJ g^{-1} for each contraction. If the muscle is stimulated at a very short length no tension is developed but some heat, about 2 mJ g^{-1}, is still produced (e.g. Gibbs and Gibson, 1969, 1970a). This heat is referred to as "tension-independent" and the extra heat produced at longer lengths is referred to as "tension dependent". Part of the tension independent heat may be due to the internal tension development and internal shortening which probably occur under these conditions, part is also likely to be due to the Na$^+$ pumping and Ca^{2+} pumping activities occurring during and after contraction (Chapman et al., 1970). Another condition under which heat is produced without tension development is on stimulation in hypertonic solution. Again about 2 mJ g^{-1} is produced (Alpert and Mulieri, 1977). (This is a measurement of "initial heat" and therefore not easily comparable with Gibbs' measurements of total heats.) As the mode of action of hypertonic solution is not entirely clear, it is again hard to interpret this result. The size of the tension independent heat is increased by changes which increase the contractility of heart muscle: increase in external calcium concentration (Gibbs and Vaughan, 1968; Chapman et al., 1970), increase in frequency of stimulation (Gibbs and Gibson, 1970a), decrease in temperature (Gibbs and Vaughan, 1968; Gibbs and Gibson, 1969), and treatment with ouabain (Gibbs and Gibson, 1969) and isoprenaline (Gibbs and Gibson, 1972b). These treatments also increase the tension and the tension-dependent heat, but not the ratio of these two. Whether the tension-independent part of the heat is length dependent, as contractility is (Jewell et al., 1975), is unknown. Heart muscle is damaged when stretched to lengths much beyond the optimum for tension development, so tension cannot be reversibly suppressed by stretching as it can be in skeletal muscle. It has been noted, however, by Mulieri and Alpert (personal communication) that heat production does not fall with tension at lengths slightly beyond l_0. This suggests that the tension-independent heat may well be length dependent.

Energy production in contractions of papillary muscle during which shortening occurs has been observed by Gibbs and Gibson (1970b). In contrast to frog skeletal muscle the energy output (or the rate of energy output) (during shortening) is not greater than in an isometric contraction at the same muscle length. It *is* greater, however, than in an isometric contraction at a shorter length which produces the same force. The ratio of work/work+heat has a maximum value of about 0·21 (Chapman et al., 1976). This probably represents the thermodynamic efficiency (Wilkie, 1961) as the heat is probably derived from oxidative processes. The value is similar to those for smooth and for most skeletal muscles when the oxidative recovery processes are included.

The heat rate during activity of papillary muscle is compared to that of skeletal muscles in Table 1 and Fig. 1. When due allowance is made for (a) the low time-averaged tension exerted by papillary muscle, (b) the fact that oxidative processes are included in the heat measurements, the heat rate is seen to be fairly typical for the relaxation rate observed.

Using the intact rabbit heart at 25 °C in a perfusion calorimeter, Coulson (1976) measured heat output and oxygen uptake as a function of the pressure developed which was varied by varying the diastolic size of the heart. Extrapolation to a zero tension gives a "tension-independent heat" of about 5 mJ g^{-1} for each beat which is rather larger than the value for papillary muscle which under similar conditions would probably be about 3 mJ g^{-1} (Gibbs, 1974). The difference is perhaps due to the greater "internal" tension development and shortening which are possible in the intact heart. The maximum heat output per beat was about 10 mJ g^{-1} which is rather *less* than that observed for papillary muscle. Coulson also shows that the heat produced can be accounted for by the oxidation occurring. This suggests that the heart is in a steady state and that little anaerobic metabolism is occurring. A similar conclusion was reached for the dog left ventricle *in situ* by Kjekshus and Mjos (1971) and for the dog heart in a perfusion calorimeter by Neill and Huckabee (1966). The latter authors also showed that when oxygen consumption was artificially restricted the heat output was greater than could be explained by either oxidation alone or the combined effects of oxidation and glycolysis.

Boivinnet and Rybak (1969a, b), using a Calvet calorimeter, measured the heat production of frog hearts which had been opened and pinned out under a known tension. The power output at 27 °C was found to be

2·95 mW g^{-1}, equivalent to 2·46 mJ g^{-1} per beat. The energy output is thus small compared to that of mammalian heart. Herold (1975; see also Herold and Cudey, 1972a, b) has used a Calvet calorimeter to study the heat produced by beating snail hearts. The hearts contracted isometrically under a tension that was varied. The rate of heat output increased with tension as did the rate of beating. The heat output per beat also increased with tension. In a separate series of experiments the work output was measured with perfused hearts. The work rate and rate of beating increased with the filling pressure. Comparing these two series of experiments, Herold calculates the ratio of work rate (perfused) to heat rate (isometric) for various rates of beating. This ratio has a maximum of 0·26. It is doubtful if this represents the efficiency of the heart as the heat rate might well be different when the heart is isometric rather than perfused and working. The weights of the hearts are not given so that the results cannot be compared with the other heart muscle experiments discussed here.

4. NERVE

The older work on the heat produced by nerves is reviewed by Feng (1936). At that time it was known that a heat production of about 8 μJ g^{-1} nerve occurred at about the time of transmission of an impulse and was followed by a much larger and slower heat production during recovery. It was discovered in 1958 (Abbott et al., 1958) that the initial heat in fact consisted of two phases, first a heat production, followed after about 0·1 s by a heat absorption almost as large. Later work (Abbott et al., 1965; Howarth et al., 1968, 1975) has shown that these two phases probably occur respectively during the rising and falling phases of the action potential. The size of each phase is about 100 μJ g^{-1} in rabbit vagus nerve and about 300 μJ g^{-1} in pike olfactory nerve, which contains a greater area of nerve membrane per gram. The most probable explanation of these heat changes is that they arise from changes in the dielectric medium of the cell membrane as the voltage gradient to which it is subjected changes. Studies of the temperature coefficient of the membrane capacity support this idea. The question is further discussed by Abbott and Howarth (1973) in reviewing work on nerve heat and by Howarth (1975). It was shown by Howarth et al. (1968) that the recovery heat after a short burst of impulses was stopped by ouabain or by substitution of Li$^+$ for Na$^+$. These facts support the idea that the recovery heat is caused by the accelerated action of the sodium pump necessary to remove the sodium that entered the fibres during the impulses. The heat produced by nerves is discussed very thoroughly in relation to electrical, optical and metabolic phenomena by Ritchie (1974).

The small size and rapid time course of the nerve heat are such that its detection and measurement is a remarkable achievement. The study of this work is therefore recommended to all calorimetrists interested in the limitations of biological microcalorimetry.

5. ELECTRIC ORGANS

The heat production of various electric organs has been studied by inserting thermistors into them. The electric discharge was found to be accompanied by heat absorption (Keynes and Aubert, 1964; Aubert and Keynes, 1968; Keynes, 1968) and followed by a slower and larger heat production. The heat absorption is greatest when the organ is performing external electrical work during the discharge, when it amounts to as much as 600 μJ g^{-1}. Under open circuit electrical conditions, when no external work is done, the cooling is about 250 μJ g^{-1}. Aubert and Keynes (1968) and also Abbott and Howarth (1973), who review this work, consider and reject a number of explanations for the open circuit cooling. They speculate that the heat absorption is due to synthesis of ATP by the reversal of the sodium potassium pump. This is feasible and invites a biochemical test.

6. BROWN ADIPOSE TISSUE

In spite of the large amount of work on the oxygen consumption of this tissue, which is often described as measurement of thermogenesis, there are few published calorimetric observations. Chinet et al. (1977) have found the heat rate of unstimulated brown fat from rats to be 3·6 mW g^{-1} at 30 °C. This is 50 per cent greater than the rate observed at the same temperature in rat soleus muscle. Ouabain treatment produced a small reduction (about 5 per cent) in the heat rate, followed by a much larger increase. Giradier et al. (1976) observed the effect on the heat rate of this tissue of stimulation with norepinephrine. The maximum heat rate they observe is greater than the basal rate by 15 mW g^{-1}. Nedergaard et al. (1977) have observed the heat produced by isolated cells from hamster brown adipose tissue at 37 °C, both in the resting state and on stimulation by noradrenaline. The resting heat rate was 110 μW per 10^6 cells and the maximally stimulated rate 820 μW per 10^6 cells. If, as they suggest, 10^8 cells are equivalent to 1 g of tissue, these values are equivalent to 11 mW g^{-1} (resting) and 82 mW g^{-1} (stimulated). The authors suggest, on the basis of these figures, the heat production of this tissue is not of importance in arousal of the torpid hamster. In both of these studies of brown fat it was shown that the heat produced could be accounted for by the oxidation of fat that was occurring. Evidently heat production can be correctly inferred from the oxygen consumption, at least for these tissues in these conditions.

7. BRAIN

Flood and Aukland (1971) have measured the heat production in goat brain *in situ* during temporary circulatory arrest. They find that the rate of heat production is 7·7 mW g^{-1} for grey matter and 4·2 mW g^{-1} for white matter. The rate remains constant for a surprisingly long time after the circulatory arrest, beyond the time when the oxygen supply would be exhausted. A switch from aerobic metabolism to glycolysis would be expected to *increase* the heat rate in a tissue using glucose as substrate, if the rate of ATP usage remained constant. This increase may mask the early effects of a fall in the rate of ATP usage. Flood and Aukland's paper also contains a useful discussion of the criteria for assessing the reliability of *in situ* heat production measurements.

8. KIDNEY

Aukland *et al.* (1969) measured heat rate in the cortex and outer medulla of dog kidney *in situ*. Following circulatory arrest a linear temperature rise was observed lasting 3–5 s. From the rate of this rise the metabolic rate was found to be 30 mW g^{-1}. This is about the amount expected from the oxygen consumption. Aukland *et al.* and also Sejersted *et al.* (1971) and Lie *et al.* (1975) have used this technique to study the action of various inhibitors and diuretics on the metabolic rates.

9. LIVER

Although liver has been studied by "internal calorimetry" (Dosekun *et al.*, 1960; Ohnaus and Tilvis, 1976), the heat production has not been observed in absolute units. Baisch (1977) describes experimental arrangements designed to study perfused liver in a Calvet microcalorimeter, but no experimental results from this approach have yet been published.

References

ABBOTT, B. C. and HOWARTH, J. V. (1973). *Physiol. Rev.* **53**, 120–158.
ABBOTT, B. C., HILL, A. V. and HOWARTH, J. V. (1958). *Proc. R. Soc.* **B148**, 149–187.

ABBOTT, B. C., HOWARTH, J. V. and RITCHIE, J. M. (1965). *J. Physiol.* **178**, 368–383.
ALPERT, N. R. and MULIERI, L. A. (1977). *Basic Res. Cardiol.* **72**, 153–159.
AUBERT, X. and KEYNES, R. D. (1968). *Proc. R. Soc. London, Ser. B*, **169**, 241–263.
AUKLAND, K., JOHANNESEN, J. and KIIL, F. (1969). *Scand. J. clin. Lab. Invest.* **23**, 317–330.
BAGUET, F. and EDMOND, J. L. (1972). *Archs int. Physiol. Biochim.* **80**, 831–833.
BAGUET, F. and GILLIS, J.-M. (1967). *J. Physiol.* **188**, 67–82.
BAGUET, F. and GILLIS, J.-M. (1968). *J. Physiol.* **198**, 127–143.
BAGUET, F., MARECHAL, G. and AUBERT, X. (1962). *Archs int. Physiol. Biochim.* **70**, 416–417.
BAISCH, F. (1977). *In* "Applications of Calorimetry in Life Sciences" (I. Lamprecht and S. Schaarsshmidt, eds). Walter de Gruyter, Berlin and New York.
BOIVINET, P. and RYBAK, B. (1969a). *J. Physiol., Paris*, **61**, S1–92.
BOIVINET, P. and RYBAK, B. (1969b). *Life Sci.* **8**, II, 11–20.
BOLSTAD, G. and ERSLAND, A. (1975). *Acta physiol. scand.* **95**, 73A.
BOZLER, E. (1930). *J. Physiol.* **69**, 442–462.
BOZLER, E. (1936). *J. cell. comp. Physiol.* **8**, 419–438.
BRIDGE, J. H. B. (1976). Ph.D. Thesis, University of California at Los Angeles.
CANFIELD, S. P. (1969). Ph.D. Thesis, University of London.
CANFIELD, S. P. (1971). *J. Physiol.* **219**, 281–302.
CHAPMAN, J. B. and GIBBS, C. L. (1974). *Cardiovasc. Res.* **8**, 656–667.
CHAPMAN, J. B., GIBBS, C. L. and GIBSON, W. R. (1970). *Circulation Res.* **27**, 601–610.
CHAPMAN, J. B., GIBBS, C. L. and GIBSON, W. R. (1976). *J. molec. Cell Cardiol.* **8**, 545–558.
CHINET, A., CLAUSEN, T. and GIRARDIER, L. (1977). *J. Physiol.* **265**, 43–61.
CLINCH, N. F. (1968). *J. Physiol.* **196**, 397–414.
CLOSE, R. (1969). *J. Physiol.* **204**, 331–346.
COULSON, R. L. (1976). *J. Physiol.* **260**, 45–63.
COULSON, R. L. and RUSEY, B. F. (1973). *Cardiovasc. Res.* **7**, 859–869.
CURTIN, N. A. and WOLEDGE, R. C. (1978). *Physiol. Rev.* **58**, 690–761.
DAVEY, D. F., GIBBS, C. L. and MCKIRDY, H. C. (1975). *J. Physiol.* **248**, 207.
DOSEKUN, F. O., GRAYSON, J. and MENDEL, D. (1960). *J. Physiol.* **150**, 581–606.
EDWARDS, R. H. T., HILL, D. K. and JONES, D. A. (1972). *J. Physiol.* **227**, 26–27P.
EDWARDS, R. H. T., HILL, D. K. and MCDONELL, M. (1974). *J. appl. Physiol.* **36**, 511–513.
EDWARDS, R. H. T., HILL, D. K., JONES, D. A. and HOSKING, G. P. (1975a). *Clin. Sci. molec. Med.* **49** (3), P24.
EDWARDS, R. H. T., HILL, D. K. and JONES, D. A. (1975b). *J. Physiol.* **251**, 303–315.
FALES, J. T., CRAWFORD, W. J. and ZIERLER, K. L. (1967). *Am. J. Physiol.* **213**, 1427–1432.
FENG, T. P. (1936). *Ergebn. Physiol.* **38**, 73–132.

FLOOD, S. and AUKLAND, K. (1971). *J. appl. Physiol.* **30**, 238–000.
FLOYD, K. and SMITH, I. C. H. (1971). *J. Physiol.* **213**, 617–631.
FRASER, A. and CARLSON, F. D. (1973). *J. gen. Physiol.* **62**, 271–285.
GIBBS, C. L. (1969). *In* "Comparative Physiology of the Heart" (F. V. McCann, ed.). Birkhauser, Basel.
GIBBS, C. L. (1974). *In* "The mammalian Myocardium" (G. A. Langer and A. J. Brady, eds), pp. 105–134. John Wiley, New York.
GIBBS, C. L. and GIBSON, W. R. (1969). *Circulation Res.* **24**, 951–967.
GIBBS, C. L. and GIBSON, W. R. (1970a). *Circulation Res.* **27**, 611–618.
GIBBS, C. L. and GIBSON, W. R. (1970b). *J. gen. Physiol.* **56**, 732–750.
GIBBS, C. L. and GIBSON, W. R. (1972a). *Am. J. Physiol.* **223**, 864–871.
GIBBS, C. L. and GIBSON, W. R. (1972b). *Cardiovasc. Res.* **6**, 508–515.
GIBBS, C. L. and VAUGHAN, P. (1968). *J. gen. Physiol.* **52**, 532–549.
GIBBS, C. L., MOMMAERTS, W. F. H. M. and RICCHIUTI, N. V. (1967). *J. Physiol.* **191**, 25–46.
GIRARDIER, L., SEYDOUX, J., GIACOBINO, J. P. and CHINET, A. (1976). *In* "Regulation of Depressed Metabolism and Thermogenesis" (L. Jansky and X. J. Musacchia, eds), pp. 196–212. Thomas, Springfield, Illinois.
GORE, R. W. and WHALEN, W. J. (1968). *Am. J. Physiol.* **214**, 277–286.
GOWER, D. and KRETZSCHMAR, K. M. (1976). *J. Physiol.* **258**, 659–671.
HEROLD, J. P. (1975). *Comp. Biochem. Physiol.* **52A**, 435–440.
HEROLD, J. P. and CUDEY, G. (1972a). *J. Physiol. Paris*, **65** (3), 433A.
HEROLD, J. P. and CUDEY, G. (1972b). *C.r. Soc. Biol.* **166**, 561–564.
HILL, D. K. (1940). *J. Physiol.* **98**, 207–227.
HILL, A. V. (1961). *J. Physiol.* **159**, 518–546.
HILL, A. V. (1965). "Trails and Trials in Physiology". Edward Arnold, London.
HILL, A. V. and HOWARTH, J. V. (1957). *Proc. R. Soc. London, Series B*, **147**, 21–43.
HILL, A. V. and WOLEDGE, R. C. (1962). *J. Physiol.* **162**, 311–333.
HOMSHER, E. and KEAN, C. J. (1978). *Ann. Rev. Physiol.* **40**, 93–131.
HOWARTH, J. V. (1970). *Q. Rev. Biophys.* **3**, 429–458.
HOWARTH, J. V. (1975). *Phil. Trans. R. Soc.* **270**, 425–429.
HOWARTH, J. V., KEYNES, R. D. and RITCHIE, J. M. (1968). *J. Physiol.* **194**, 745–793.
HOWARTH, J. V., KEYNES, R. D., RITCHIE, J. M. and VON MURALT, A. (1975). *J. Physiol.* **249**, 349–368.
HOYLE, G. and LOWY, J. (1956). *J. exp. Biol.* **33**, 295–310.
JEWELL, B. R. (1977). *Circulation Res.* **40**, 221–230.
KEYNES, R. D. (1968). *Proc. R. Soc. London, Series B*, **169**, 265–274.
KEYNES, R. D. and AUBERT, X. (1964). *Nature, Lond.* **203**, 261–264.
KJEKSHUS, J. K. and MJOS, O. D. (1971). *Scand. J. clin. lab. Invest.* **28**, 379–388.
KRETZSCHMAR, K. M. and WILKIE, D. R. (1972). *J. Physiol.* **202**, 66–67P.
KRETZSCHMAR, K. M. and WILKIE, D. R. (1975). *Proc. R. Soc. London, Series B*, **190**, 315–321.
LIE, M., JOHANNESEN, J. and KIIL, F. (1975). *Am. J. Physiol.* **229**, 55–59.
MCDONALD JR, R. H. (1971). *Am. J. Physiol.* **220**, 894–900.

MULIERI, L. A., LUHR, G., TREFY, J. and ALPERT, N. R. (1976). *The Physiologist*, **19**, 307.
MULIERI, L. A., LUHR, G., FREFRY, J. and ALPERT, N. R. (1977). *Am. J. Physiol. Cell Phys.* **2**, 146–157.
MULVANY, M. J. and WOLEDGE, R. C. (1972). *J. Physiol.* **229**, 20–21P.
NEDERGAARD, J., CANNON, B. and LINDBERG, O. (1977). *Nature, Lond.* **267**, 518–520.
NEILL, W. A. and HUCKABEE, W. E. (1966). *J. clin. Invest.* **45**, 1412–1417.
OHNAUS, E. E. and HUNZIKER. (1974). *Pflügers Arch.* **347**, 255–260.
OHNAUS, E. E. and TILVIS, R. (1976). *Acta hepato-gastroento.* **23**, 404–408.
RALL, J. A. and SCHOTTELIUS, B. A. (1973). *J. gen. Physiol.* **62**, 303–323.
RITCHIE, J. M. (1974). *Prog. Biophys. molec. Biol.* **26**, 147–188.
SEJERSTED, O. M., LIE, M. and KIIL, F. (1971). *Am. J. Physiol.* **220**, 1488–1493.
SMITH, I. C. H. and WOLEDGE, R. C. (1972). *J. Physiol.* **226**, 9–11P.
WENDT, I. R. and GIBBS, C. L. (1973). *Am. J. Physiol.* **224**, 1081–1086.
WENDT, I. R. and GIBBS, C. L. (1974). *Am. J. Physiol.* **226**, 642–647.
WILKIE, D. R. (1961). *Prog. Biophys. biophys. Chem.* **10**, 260–298.
WILKIE, D. R. (1963). *J. Physiol.* **167**, 39P.
WOLEDGE, R. C. (1968). *J. Physiol.* **197**, 685–707.
WOLEDGE, R. C. (1971). *Prog. Biophys. molec. Biol.* **22**, 39–74.
WOLEDGE, R. C. (1975). *Pestic. Sci.* **6**, 305–310.
YAMADA, K. (1970). *J. Physiol.* **208**, 49–64.

The Identification and Characterization of Microorganisms by Microcalorimetry

R. D. NEWELL

A. Introduction

The application of microcalorimetry to microbiological problems is just one example of the extension of physical techniques to biological systems. It has been emphasized many times that the conventional microbiological methodology has not basically changed since Pasteur's original experiments. The increased attention paid to microbiological analysis has been fuelled by the ever-increasing numbers of specimens routinely analysed in the food industry, public health (environment) and clinical laboratories. The opportunities seized have involved automated procedures of conventional methods and new conceptual approaches adapting new principles and technology. The emphasis is obviously toward rapid results and automation; and is evidenced by the recent interest in these new techniques and procedures (e.g. Hedén and Illéni, 1975; Johnston and Newsom, 1977).

An identification procedure must demonstrate good discrimination and absolute resolution between different microorganisms. Conventional examination procedures are found in all standard microbiology texts; these involve morphological examination, cultural characteristics, biochemical tests (e.g. fermentation ability, proteolysis, lipolysis and reducing ability) and serological tests. The detail to which these examinations are performed is governed by the demands of the specific type of laboratory. Apart from facilities, time is the determining factor.

Ideally the identification scheme requires the organism to be obtained in pure culture on an unselective medium; a single colony is selected and inoculated on to a secondary plate which is used to provide the inoculum for the subsequent examinations. Often in the clinical diagnostic laboratory this time delay cannot be tolerated and therapy begins before the infecting organism is identified; only when antibiotic insensitivity is encountered does the identity of the organism become important. Tests based purely on biochemical criteria are sometimes insufficient and require confirmation with morphological examination. For example, a micro yeast identification scheme can give results in 24–48 h, but to identify *Candida albicans* it is necessary to observe germ tube formation in serum (2–3 h) and chlamydospore production in corn meal agar (several days). Because of the serious increase in opportunistic mycotic infections in debilitated patients (e.g. cardiac surgery, kidney transplants) improved methods for identification have been developed (e.g. Huppert *et al.*, 1975; Land *et al.*, 1975). The problem with these infections is that successful therapy requires rapid antimycotic sensitivity testing.

The difficulties facing routine microbiological laboratories are the low numbers as well as the types of organisms encountered; and elsewhere in this volume are accounts of the application of microcalorimetry to enumeration (see also Cliffe *et al.*, 1973; Beezer *et al.*, 1974). It is a general criticism of all methods that all equally fail to detect the low numbers of organisms present in contaminations. However, each particular field of routine microbiology has its own specific problems; at the risk of stating the obvious, these fields divide into two principal areas, the industrial and the clinical.

a. In the food industry microbiological assay is a question of screening products to detect significant levels of contamination (usually few positives) and the subsequent identification of the organisms. In this context calorimetry has been successfully applied to the non-destructive analysis of vacuum-packed food (Lampi *et al.*, 1974).
b. Clinical investigations require in addition to the question of numbers and identity, the testing for antibiotic sensitivity. Antibiotic therapy is often the prime consideration and it is quite usual for the growth of the infecting organism to be countered before it is absolutely typed. Calorimetry has also been successfully used in antibiotic sensitivity testing (e.g. Binford *et al.*, 1973) where the calorimetric results correlated well with the results of plate diffusion assay. Using this calorimetric method it was possible to distinguish between bacteriocidal and bacteriostatic agents. It is, however, statistically most likely that infecting organisms from similar types of infection can be provisionally identified, and antibiotic therapy begun before the microorganism is typed.

The fermentation industries also require the rapid screening of microorganisms, particularly in contaminations (e.g. *Lactobacilli* in beer). They also require the screening of microorganisms for the selection of particular characteristics, either of the microbial mass or of the fermentation product. Such cultures are often produced from uniform backgrounds before being subjected to analysis.

It is to meet some of these requirements that conventional identification tests have been miniaturized so that a large number of identification kits are now commercially available. Automated systems involving complex and sophisticated procedures have also been developed (e.g. Hedén and Illéni, 1975) and the current emphasis is towards computer analysis of an automated system (Hedén *et al.*, 1977). New techniques that have been used include impedance measurements of growing cultures, pyrolysis gas chromatography of isolates and microcalorimetry.

This contribution is therefore concerned with the application of microcalorimetry to the questions of identification and characterization of microorganisms, particularly since it has been demonstrated that over 200 clinically significant bacteria can be identified by means of their specific thermal patterns (Russell *et al.*, 1975). The deployment of microcalorimetry in organism identification is a specific example of the general application of microcalorimetry in analysis; this topic has been thoroughly reviewed recently by Spink and Wadsö (1975), and in special areas elsewhere in this volume. Many of the illustrations are drawn from our own studies of yeast growth which have been made over the past five years.

B. Microbiological microcalorimetry

It will be apparent from the preceding contributions concerned with microbial growth (cf. Lamprecht, p. 43, and Belaich, p. 1) that the calorimetrically derived thermogram represents the net effect of all the physical and chemical processes occurring in the reaction vessel of the calorimeter. These processes include all metabolic reactions, heats of binding, mechanical effects (e.g. viscous flow) and sedimentation. To emphasize the obvious, the technique is non-specific and the interpretation of any result requires extreme caution.

The microcalorimetric investigations reporting microbiological identification have all been growth-dependent processes and are therefore subject to all the constraints operating upon all growth processes (e.g. Dean and Hinshelwood, 1966). Indeed these considerations are the critical area of microbiological microcalorimetry and unfortunately the constraints and limitations acting on the technique are only now being slowly and painfully learnt. It appears that the classical thermochemist's caution and fundamental microbiological principles have been temporarily forgotten.

Forrest (1969) demonstrated that in calorimetrically observed microbial growth studies only the catabolic reactions are detected and that the observation of anabolic processes requires an improved instrument sensitivity. Consequently the application of calorimetry to microbial identification can only be achieved by studying catabolic processes. If a small inoculum is used (initial culture density of $\sim 10^4$ cells ml^{-1}) the heat production is initially almost undetected in even the most sensitive instruments presently available and this, in part, coincides with the lag phase of growth. During exponential growth the rate of heat evolution begins to increase exponentially concurrent with the synthesis of new

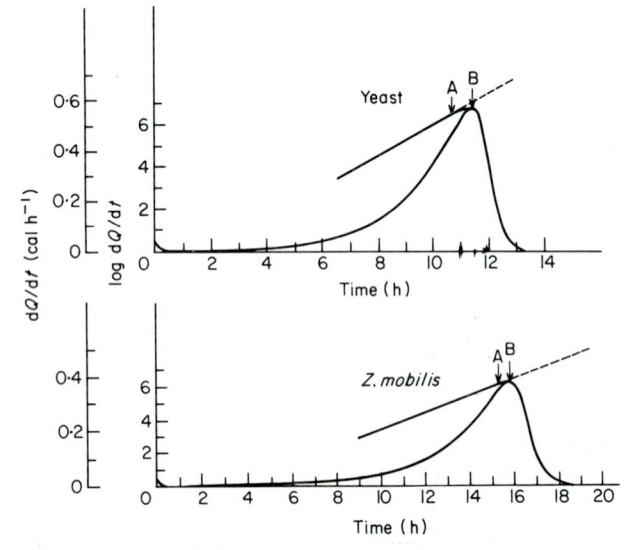

Fig. 1. Classic thermograms of *Saccharomyces cerevisiae* and *Zymomonas mobilis* grown anaerobically in glucose-limited medium batch calorimeter. (Belaich *et al.*, 1968.)

cell material. The rate of heat production per unit cell mass remains constant at the highest level reached by the cells. The rate of synthesis of new cell material, degradation of energy source, appearance of catabolic products and heat evolution are described by the same exponential function. When exponential growth is limited by energy source the heat evolution rate declines from the exponential rate as the substrate is exhausted and returns to a base-line value. This classic thermogram is illustrated in Fig. 1, and can provide valuable thermochemical data (e.g. Belaich *et al.*, 1968; see also contributions by Belaich, p. 1, and Lamprecht,

p. 43). Stationary phase cells so produced have been observed to have no calorimetrically detectable endogenous metabolism (Murgier et al., 1968). However, the heat output rate does not necessarily return to the initial base line before the growth of the microorganism, but in batch cultures heat evolution can remain at a small value above the initial base line over several days (Lamprecht, unpublished observations). The same phenomenon has been noted in this laboratory for a period of several hours after the exhaustion of glucose (Fig. 2) in growing cultures of *Candida albicans*.

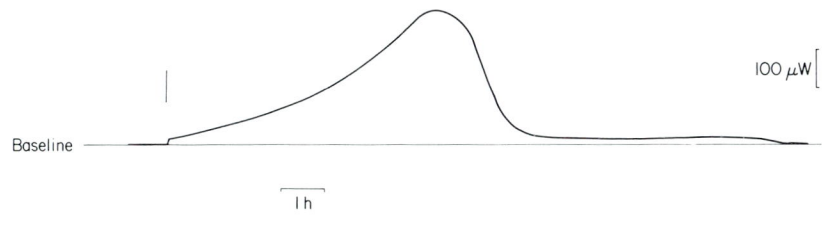

FIG. 2. Thermogram of *Candida albicans* demonstrating long plateau after exhaustion of glucose before return to base line, flow calorimeter.

Endogenous metabolism could be responsible for this heat output since washed cells resuspended in buffer produce a significant heat output (Newell, unpublished observations; Hoogerheide, 1976).

The application of calorimetry to the identification of microorganisms has been a recent development made possible by the availability of sensitive and, in some cases, commercial instruments. The possibility of this application is implicit in the work of Rubner (1906a, b), who used a Beckman thermometer to record the temperature changes in growing cultures (Fig. 3). The shape of these thermograms varied with the differences in growth medium and with different bacteria. That identification of microorganisms could be achieved by the pattern of thermal events as a function of time was also implied by Prat et al. (1946). Recent studies using a number of different sensitive microcalorimeters have resulted in the demonstration that identification by calorimetry may indeed be possible (Boling et al., 1973; Staples et al., 1973; Monk and Wadsö, 1975; Russell et al., 1975; Ljungholm et al., 1976) (see Figs 4 and 5).

It has further been claimed that at least 200 clinically significant bacteria can be identified from their thermograms grown on a single medium (Russell et al., 1975).

To date the traditional caution of the thermochemist in respect of these claims and their interpretation has been voiced rarely (e.g. Schaarschmidt and Lamprecht, 1976; Beezer et al., 1978). The use of the term "identification" implies an absolute property. If we consider a simple chemical example, the ultraviolet spectrum of benzene is always the same. In comparison calorimetric derivations of microorganisms are not absolute. The thermograms will vary considerably with calorimeter design and operation. The types of microcalorimeter currently in use for biological work have been described by Spink and Wadsö (1976) and it is unnecessary

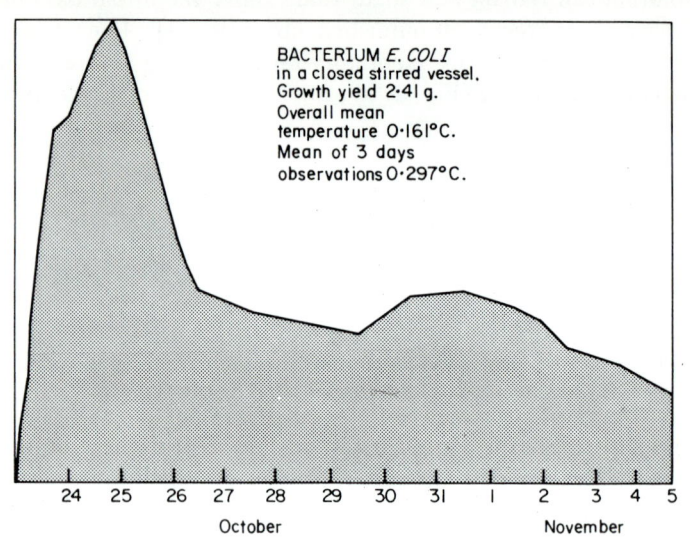

FIG. 3. Record of temperature changes in a growing culture (Rubner, 1906b).

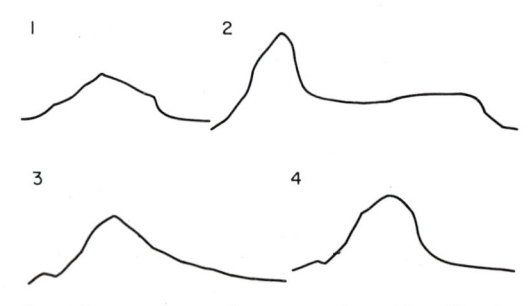

FIG. 4. Representative thermograms demonstrating identification. 1, *Proteus morganti*; 2, *Proteus mirabilis*; 3, *Proteus rettgeri*; 4, *Proteus vulgaris*. (Russell et al., 1975.)

to go into great detail in this review. The types of cellule readily applicable to microbiological studies are ampoule drop (batch) and the flow-through cellule. This latter design can be operated in a segmental flow or stopped flow fashion; the flow aeration cellule (Eriksson and Wadsö, 1971) has also been used to a limited extent in identification.

A brief examination of the different types of cellule and their operation in relation to identification has been made by Monk and Wadsö (1975). The thermograms for *Streptococcus faecalis* var. *liquefaciens* varied with the different calorimeter type (flow-through and ampoule) and with the

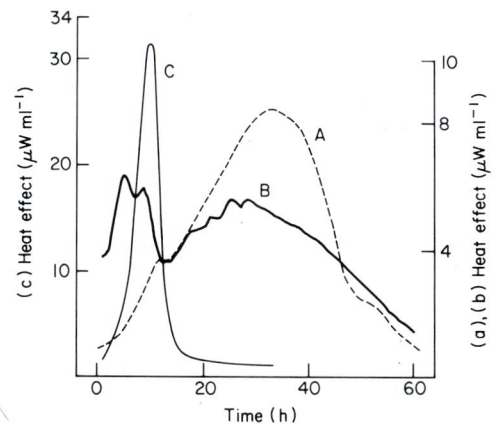

FIG. 5. Representative thermograms demonstrating identification. (A) *Acholeplasma laidlawii* (B) *Acholeplasma granularum* and (C) *Ureaplasma urealyticum*. (Ljungholm et al., 1976.)

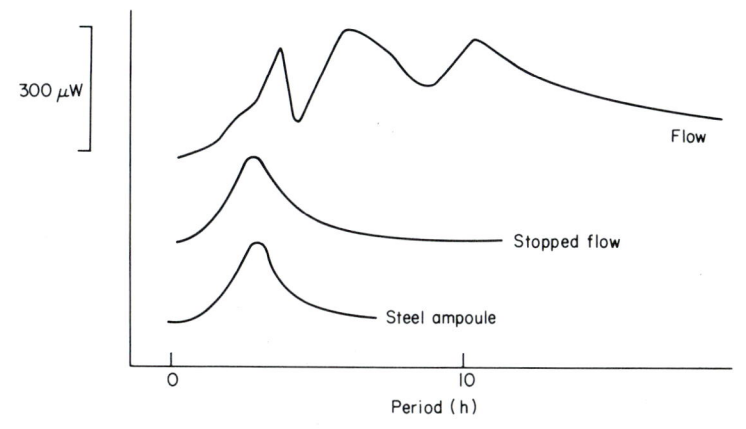

FIG. 6. Variation of thermogram with different calorimeter type for *Streptococcus faecalis* var. *liquefaciens* (Monk and Wadsö, 1975).

operation of the flow-through cellule (Fig. 6). Lamprecht and Meggers (1969) demonstrated that stirring a batch culture influenced the magnitude and time dependency of the thermogram detail compared with the thermogram recorded in the absence of stirring (Fig. 7).

It is the wide number of different calorimeter types which produces the divergence in derived thermograms, even when strictly controlled inocula are used. Again, the cause of the differences is that the method is dependent on the growth of microorganisms which is itself influenced by the physical parameters of the growth chamber (e.g. Solomons, 1969).

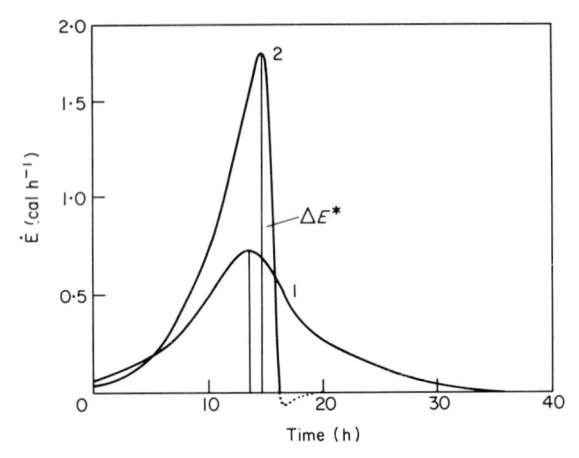

Fig. 7. Thermograms for *Sacch. cerevisiae* grown in anaerobic batch calorimeter. 1, unstirred; 2, stirred. (Lamprecht and Meggers, 1969.)

C. Growth considerations

At the risk of oversimplifying this contribution, it is necessary to discuss the fundamental aspects of growth; although they are interrelated, they can be discussed under the following headings.

1. pH

The bulk of the work published on bacterial identification has employed batch microcalorimetry. The reaction chambers of these instruments are usually of small capacity (~ 4 ml) and they are operated without stirring. The growths achieved are, therefore, essentially anaerobic, and the cell population is heterogeneously dispersed throughout the growth medium. Consequently localized variation of pH and gas concentration is developed. Fluctuation in yeast cell density, due to fermented carbon dioxide in a batch calorimeter, has been shown to cause fluctuation in the thermal pattern (Schaarschmidt et al., 1973). Fluctuation of pH will have two effects, metabolism may vary as a function of pH and the heats associated with the removal of substrates and the appearance of metabolites will likewise depend upon their respective degrees of protonation.

2. AEROBIC–ANAEROBIC

The major disadvantage of batch systems is that no adequate control over the atmosphere of the incubation is possible in these incubations. Apart from the carbon or energy source, the gas phase is the second most crucial factor in the growth of facultative anaerobes, as has been demonstrated in calorimetric examinations of yeasts and bacteria, for glucose-limited growth (Figs 6 and 7).

The oxygen available for the growth of microorganisms is of course dependent on the rate of oxygen uptake (proportional to the cell number) and the rate of replenishment (flushing rate of oxygen and solubility in the medium). In the batch system used by Russell et al. (1975) the incubations were shown to be essentially anaerobic. Increasing the inoculum volume resulted in a small initial "hump" in the thermogram (Fig. 8); this was caused by the presence of oxygen, and so permitted limited aerobic growth. This small peak was not found to be a specific part of the organisms profile by Russell et al. but it could be intensified by prior flushing of the medium with oxygen, or, eliminated by nitrogen. Since organisms have different K_ms for oxygen (Wimpenny, 1968), it would be expected that the aerobic peak will vary for different microorganisms. The finding of Russell et al. merely emphasizes the disadvantage of a sealed ampoule growth chamber.

Inoculum volume and the efficiency of aeration are intimately related. Using the same aeration rate (fixed-rate stirred culture vessel linked to a flow microcalorimeter) an increase of around an order of magnitude in the inoculum volume resulted in a linear increase in heat output rate rather than the expected exponential increase (see Figs 9 and 10). The aeration rate was therefore limiting growth at moderate cell densities. The limitation of the growth of facultative anaerobes by oxygen cannot be overemphasized, especially in calorimetric work. It is commonly disregarded in flow incubations (Poole et al., 1973), particularly when using the segmental flow aeration vessel designed by Eriksson and Wadsö (1971). The limitations of this vessel's performance are, unfortunately, not widely accepted and its use has led to the erroneous interpretation of phenomena (e.g. Poole et al., 1973). Using synchronous cultures of *Saccharomyces pombe* these authors found a linear increase heat output rate accompanied an exponential increase in cell mass. A crucial error in this work was the use of simulated incubations to determine cell mass and the biochemical analysis. The calorimetric growth was oxygen limited and hence a linear heat output rate obtained. The aeration cell has therefore imparted a false sense of security to many investigations. The fact that there is a maximum solubility of oxygen in water (Wimpenny, 1968), which is reduced by the presence of salts in growth media, and a maximum cell density that can be maintained aerobically using mild aeration has not been appreciated—even classically.

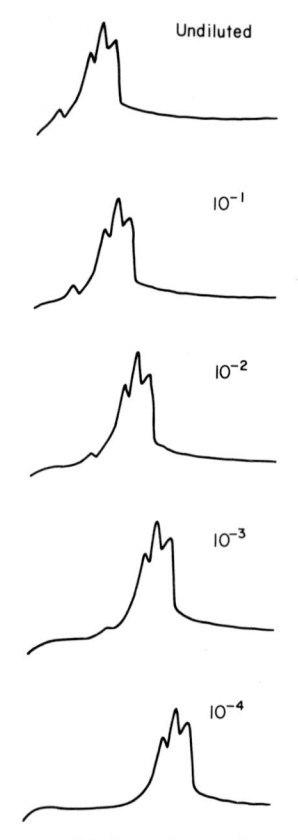

FIG. 8. Variation of thermogram with inoculum volume, *Proteus vulgaris* (Russell et al., 1975).

Calorimetric studies on typical laboratory-scale cultures are not common, but a large reaction batch calorimeter (300 ml medium) has been used with aeration (1 l h^{-1}) for studies on bacterial growth and antibiotic action (Jaworski et al., 1968). This calorimeter is unfortunately of limited sensitivity and linear heat output rates were observed for L-forms of *Proteus mirabilis*. This again could be the result of insufficient oxygenation to ensure total aerobic growth rather than a characteristic of the L-form growth.

The forced aeration rate used in laboratory fermentors is not possible with calorimeters, principally because of the small culture volume, evaporation problems and the high noise-to-signal ratio which would be encountered. For flow microcalorimetric cultures vigorous aeration rates

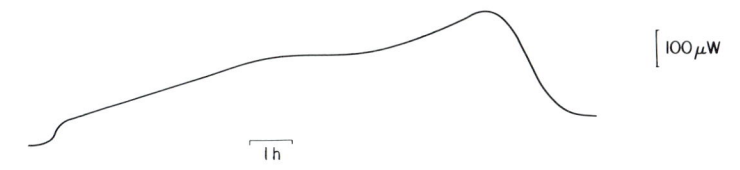

FIG. 9. Linear increase in heat output rate *Sacch. cereviasiae* (Newell, 1975).

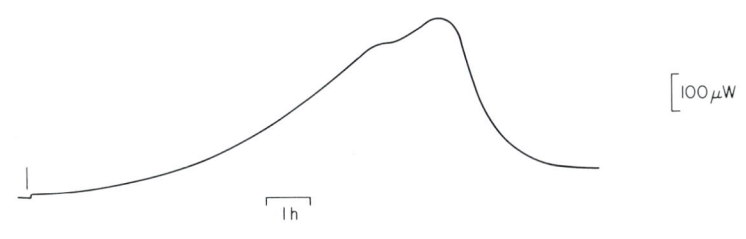

FIG. 10. Exponential increase in heat output rate, *Sacch. cerevisiae*. Culture grown under identical conditions to Fig. 9, except inoculum volume was reduced by one tenth. (Newell, 1975.)

can be used, but the time delay in pumping from the fermentor to the calorimeter vessel and the return to the fermentor, if used in cycle, will again lead to oxygen-limited growth.

Total aerobic growth will, like anaerobic growth, be evidenced by a simple exponential increase in heat output rate followed by a decay of the signal (Fig. 2). When aeration becomes limiting the resulting thermograms increase in complexity (e.g. Figs 11 and 12). This increase in detail of the thermogram is perhaps required for demonstrating possible differences between different organisms. Partial aerobic growth (biphasic growth with respect to oxygen) is a necessity in these incubations. Total anaerobic growth conditions may not yield much thermogram detail unless a number of different fermentation pathways are utilized by the microorganisms under investigation. The employment of the perfusion-ampoule drop chamber (Wadsö, 1974) would be a considerable advantage over the ampoule drop chamber in achieving a certain amount of aerobic growth.

These same problems of ill-defined aeration conditions, encountered in batch and ampoule calorimeter systems, are also found in the application of impedance measurements (e.g. Hadley and Senyk, 1975) to identification. The impedance measurement is again growth dependent. The culture is sealed in an ampoule (0·1 ml) and no aeration is made; the time course of the growth is the same order as for calorimetry and 10^3–10^4 bacteria ml^{-1} can be detected within 2 h.

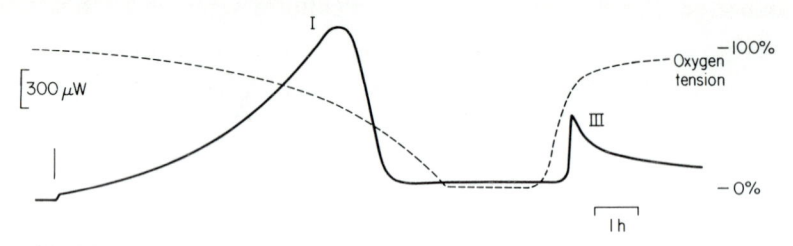

FIG. 11. Thermogram of oxygenated culture of *Kluyv. fragilis* (Newell, 1975).

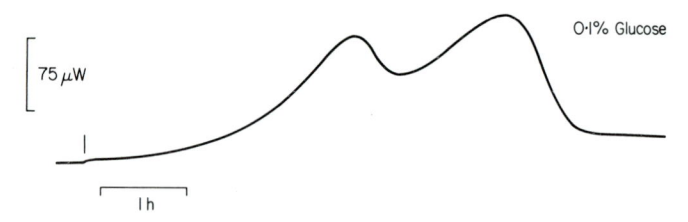

FIG. 12. Thermogram of aerated culture of *Kluyv. fragilis* (Newell, 1975).

At this point two simple statements illustrated by the influence of oxygen should be emphasized.

1. Calorimetric observations require the reaction being measured, be it growth or a chelation reaction, to be strictly defined if meaningful thermochemical and hence metabolic data are to be determined.

2. The application of calorimetry in identification places less emphasis on the strict control of growth (i.e. aerobic versus anaerobic) except that the growth conditions be exactly reproducible. The desired conditions are a "defined uncertainty" in which it is possible to produce a "structured" thermogram. This therefore gives more detail with which it may be possible to discriminate between different species.

These statements are contradictory and it is obvious that only one purpose can be satisfied, they are mutually exclusive. They also equally well apply to considerations of the growth medium, as will be demonstrated below (section 4). The use of calorimetry in identification adopts a pragmatic approach. The extrapolation of the calorimetric data to specific reactions is a mammoth task and the association of specific metabolic reactions to "heat peaks" can only be certain when analysis is performed on identical test samples of cultures which have been treated identically to the calorimetric suspensions in all respects.

3. INOCULUM VOLUME

The effect of inoculum volume on microcalorimetric monitored growth has been dismissed in some accounts (e.g. Kallings, 1976), since it is claimed only to modify the time of appearance of the thermogram structure (Russell et al., 1975, Fig. 8). It is however evident that time is the important parameter in this method; the dynamic series of events that is metabolism is presented on a time basis and any shift in maxima or minima along the abscissa will complicate any analysis to identify the organism. This has been underlined by Russell's group, who used a standardized inoculum of 500 organisms; although the structures of the thermograms of some organisms were similar, they could be differentiated on the basis of timing and of amplitude in the thermal effect. The influence of inoculum volume and its relationship to aeration were demonstrated for a yeast culture monitored in a flow microcalorimeter (see section C.2).

4. MEDIUM COMPOSITION

Organisms when supplied with a simple substrate (e.g. limited glucose) and defined growth conditions will produce the simple classic thermogram (Fig. 1). The growth rate of a particular microorganism and substrate binding data (K_s) could be determined from these thermograms (e.g. Belaich and Belaich, 1975). However, to maximize differences between strains, a large number of substrates would have to be investigated; this again is the volume problem. To maximize the discrimination complex media have been adopted; this is most likely based upon a traditional microbiological decision to use a rich general-purpose medium (e.g. brain heart infusion). It is now recognized that such a general medium may not be the optimal choice for the proposed calorimetric procedure. The thermogram of a particular microorganism has been shown to be dependent upon the specific medium used in the calorimeter (Russell et al., 1975). These findings are not surprising, since the adaptation period to different substrates will vary with the microorganism and, importantly, its history (see section C.5).

For a *Kluyv. fragilis* glucose-limited culture grown under incomplete aeration and monitored in a flow microcalorimeter two peaks were obtained in the thermogram, an additional peak was obtained for the diauxic growth on glucose plus maltose (see Figs 13 and 14). The presentation of *Kluyv. fragilis* to a mixture of five substrates again increased the structure of the thermogram (compare Figs 15 and 16; these cannot be compared with Figs 13 and 14, since the experimental conditions in the microfermentor were different). Returning to the aerobic–anaerobic argument, if the growth conditions for the mixture of five sugars was made anaerobic a less structured thermogram (Fig. 17) was obtained.

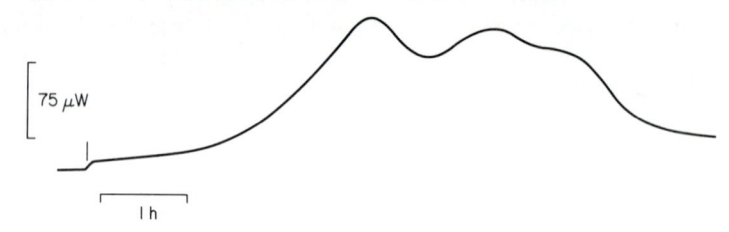

Fig. 13. *Kluyv. fragilis*, biphasic growth on glucose (Newell, 1975).

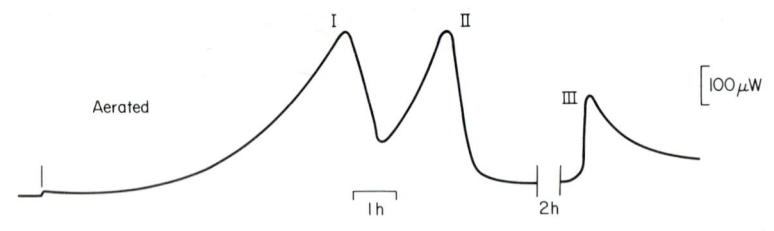

Fig. 14. *Kluyv. fragilis*, biphasic growth on mixed sugar substrate (glucose + malt extract) (Newell, 1975).

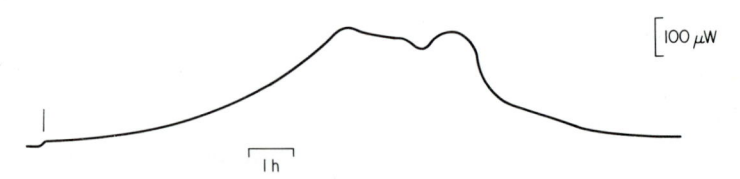

Fig. 15. Structured thermogram for *Kluyv. fragilis* grown on mixture of sugars (0·15 per cent glucose and 0·05 per cent each of sucrose, maltose, galactose and lactose) (Newell, 1975).

In these incubations with mixed substrates the calorimetric procedure is combining several of the traditional assimilation–fermentation tests in a single growth experiment. It is completely dependent upon the fundamentals of growth and detects the rates of substrate utilization. However, the number of substrates is restricted to catabolites. This is not to say that the effect of vitamins cannot be examined by microcalorimetry, although statements have been made in several accounts (e.g. Forrest,

1969) that the effect of vitamins on growth cannot be studied microcalorimetrically. In the absence of biotin no exponential growth of *Candida albicans* was obtained in an aerated strictly defined medium (Fig. 18).

A number of laboratories are currently engaged in a detailed examination of the effects of different growth constituents. It is therefore imperative to standardize all calorimetric media so that the slightest variation in batches of medium are eliminated. This excellent lead has been set by

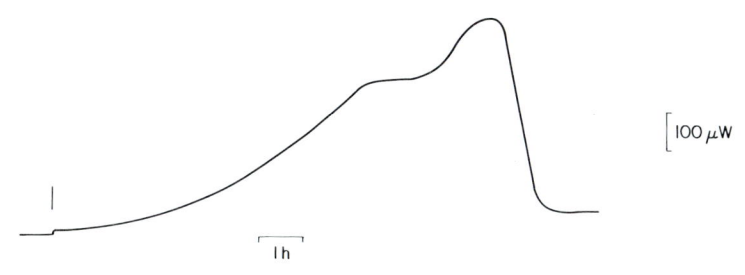

FIG. 16. *Kluyv. fragilis* grown on 0·35 per cent glucose, under identical conditions to those described by Fig. 15 (Newell, 1975).

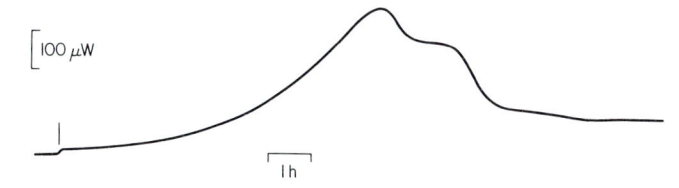

FIG. 17. *Kluyv. fragilis* grown anaerobically on mixture of five sugars (see Fig. 15) (Newell, 1975).

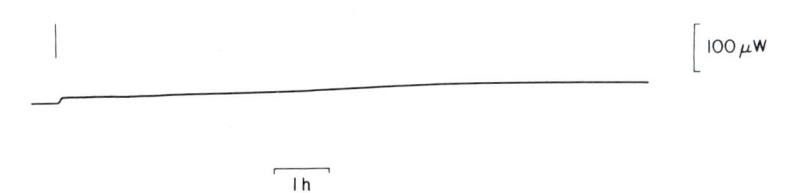

FIG. 18. *Candida albicans*, aerobic growth limited by biotin deficiency.

Boling et al. (1973). However, ideally the medium should be strictly defined.

5. INOCULUM HISTORY

The most fundamental consideration with regard to identification is the origin of the inoculum. Although no differences in the calorimetric thermogram were shown for bacteria previously cultured on different complex media and presented to a common medium in an unstirred batch calorimeter (Russell *et al.*, 1975), recent evidence (Beezer *et al.*, 1977) has shown that *Escherichia coli* D2102 grown in three different media, washed and resuspended to the same cell density gave substantially different rates of heat output for glucose respiration. The growth rates of *C. albicans* prepared from different media and inoculated into a common medium also show differences in flow microcalorimetry (Cawson *et al.*, unpublished observations).

However, an ampoule calorimeter has been used in the diagnosis of septicaemia and meningitis (Kallings and Hoffer, 1977, unpublished communications) using direct inoculation of blood or cerebrospinal fluid samples; the final concentration of blood in the sample was 10 per cent. Blood donated by different individuals gave essentially identical thermograms for one isolate of *E. coli*. But this work begs the real question. Will the same strain of organism give identical thermograms when isolated from the blood of different individuals, or any other common source (e.g. urine, cerebrospinal fluid)? The answer is not yet known; the calorimetric technique may give diagnostic results when material from common sources is tested. However, a more sensible calorimetric application which has clinical relevance is that of antibiotic sensitivity testing.

D. Characterization

1. CHARACTERIZATION OF MICROORGANISMS

From the previous considerations of the factors influencing growth it is not possible to claim universal identification of microorganisms by calorimetry. It is in the area of characterization that calorimetry has a major role to play, rather than the preserve of identification. A readily foreseen application is that of continuity of strain performance, particularly in the industrial field of fermentation, for microbial crop or fermentation product (e.g. antibiotic, alcoholic beverages). The standardized performance of a strain could be checked in a calorimetric system enabling continuous growth monitoring, in the flow cellule, rather than discontinuous sampling for biochemical tests or end-point analysis.

A simple example of the characterization of microbial activity was demonstrated by a flow microcalorimetric investigation of yeast respiration; the same order of activity was obtained as far as the respirometric method for different batches of commercial baking yeast (Newell, unpublished observations). It has also been demonstrated that different types of yeast when subject to the same preparation history give different thermograms when grown under identical conditions in a flow microcalorimeter (Beezer et al., 1978). It was possible to discriminate between baker's, brewer's and distiller's yeast (see Figs 19 to 21). These thermograms

FIG. 19. Baker's yeast (Newell, 1975).

FIG. 20. Brewer's yeast, *Sacch. cerevisiae* NCYC 239 (Newell, 1975).

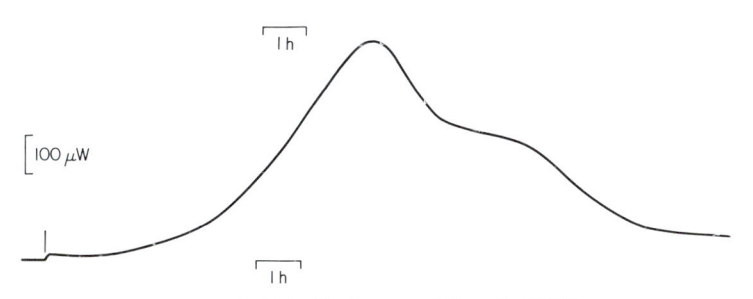

FIG. 21. Distiller's yeast (Newell, 1975).

contain important kinetic data for the sequential utilization of sugars by the yeast. Sequential fermentation of sugars has been demonstrated for baker's yeast in panary fermentation (Larmour and Bergsteinsson, 1936; Koch et al., 1954), and for brewery yeast fermentation of wort (Phillips, 1955). The thermogram for *Kluyv. fragilis* grown on the mixed sugar substrates (Fig. 15) shows considerably more fine detail than growth on glucose alone (Fig. 16).

The calorimetrically monitored growth is therefore detecting the sequencing of substrate oxidation and more importantly the rates of substrate utilization, which provides valuable data for recording the progress of fermentation. Since it is an economic decision to use complex medium for many industrial fermentations, the calorimeter could valuably be used to investigate the progression of a fermentation and be compared to biochemical tests, and eventually substitute for conventional analysis.

The performance of a standard strain could therefore be investigated during its routine maintenance and the selection of strains for a particular efficiency of fermentation could be made.

2. GROWTH MEDIUM STANDARDIZATION AND ANTIBIOTIC SENSITIVITY SCREENING

The reverse of organism standardization, that of medium standardization, has recently been reported (Ripa *et al.*, 1977). Standardized inocula of *E. coli*, *Staphylococcus aureus* and *Neisseria meningitidis* have been used to

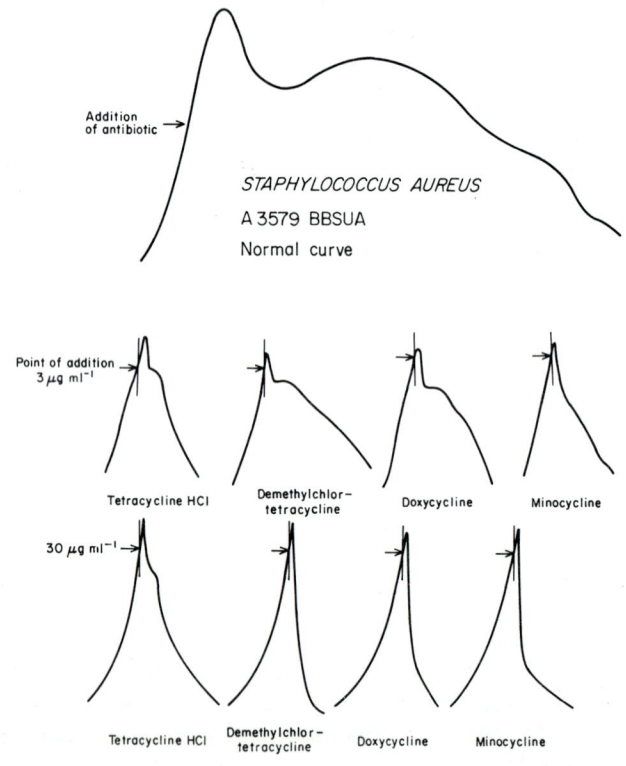

FIG. 22. Effect of tetracyclines on the heat output of *Staphylococcus aureus* (Semenitz and Tiefenbrunner, 1977).

study the effect of the addition of sodium polyanethosulphonate, heparin, sucrose and Isovitalex® to blood culture media. The results indicated that the application of the calorimetric technique had considerable advantages, significantly of speed, over conventional analytical approaches.

Other recent excursions into characterization have been applied to antibiotic action. The work of Binford et al. (1973) in distinguishing bacteriocidal and bacteriostatic antibiotics on 15 clinically important bacteria demonstrates the application of microcalorimetry to antibiotic sensitivity testing. Batch calorimetric procedures are limited when the kinetics of an inhibitory process are studied. The inhibition of growth by

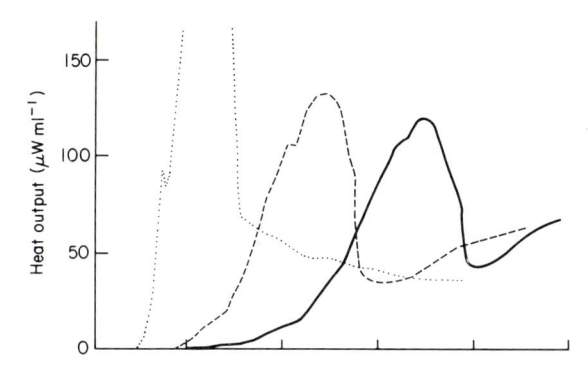

FIG. 23. Heat output of *E. coli* grown in presence of tetracycline and doxycycline. —, Tetracycline; - - -, doxycycline; . . . ., control. (Mårdh et al., 1977.)

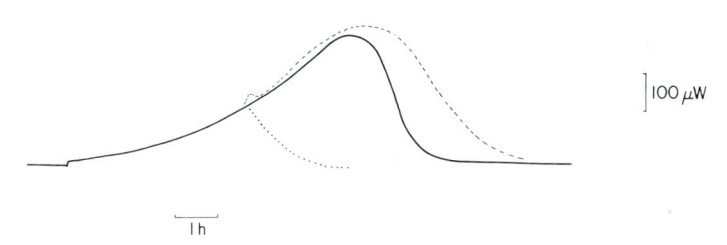

FIG. 24. Effect on the heat output of nystatin addition to growing cultures of *C. albicans*.

antibiotics, namely tetracyclines, Fig. 22 (Semenitz and Tiefenbrunner, 1977) and Fig. 23 (Mårdh et al., 1977), monitored in ampoule-drop chambers, demonstrates that the calorimetric method can provide a wealth of profiles. However, these thermograms are not necessarily easy to interpret, particularly if components in the medium rescue the treated cells or directly react with the antibiotic (e.g. metal chelation). In addition, if glycolytic processes are not directly inhibited the oxidation of the remaining substrate (e.g. glucose) will continue to contribute to the heat output rate.

Studies of antimycotic agents on *C. albicans*, grown under completely aerobic conditions and monitored in a flow microcalorimeter, can be used to determine sensitivities of growing cultures to a drug and also to distinguish fungicidal and fungistatic doses. If we take the polyene antibiotic nystatin, an exponentially growing culture of *C. albicans* is rapidly killed at high doses, whilst at low doses an increased heat output rate above a control thermogram is obtained (Fig. 24). This demonstrates the non-specific nature of calorimetry, which is both its strength and its weakness.

E. Conclusion

It must be apparent that the calorimetric method has severe restrictions when applied to the identification of microorganisms. Thermograms that are distinct under strictly defined conditions can be varied at ease by changes in culturing conditions—medium composition, calorimeter type, viable inoculum volume and inoculum history. Therefore, for an identification procedure it is extremely inflexible and makes rigid demands on all operations. For universal application the requirements for a standardized inoculum history and viable cell inoculum are the most serious disadvantages, since all the other factors are capable of control. Inoculum history immediately restricts the calorimetric technique to isolates from similar

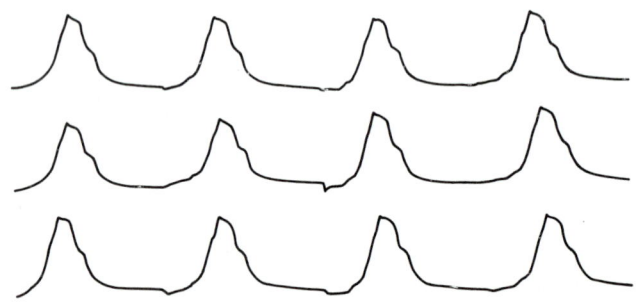

FIG. 25. Reproducibility of the same isolate of *E. coli* (Russell *et al.*, 1975).

types of background, e.g. blood, urine. Such biological problems are inevitable, but in the sphere of calorimetry it has only been emphasized that calorimetric investigations have specific shortcomings. Different calorimetric studies are not necessarily comparable, principally because of variation in instrumentation and calorimetric procedures (Monti and Wadsö, 1976).

The comparison of thermograms requires more than a subjective decision. Computer analyses of bacterial growth patterns have been necessary to provide objective assessment of thermograms (Johansson *et al.*, 1975). Two methods were used by these workers, correlation analysis of the thermograms and the first derivative of the thermogram (i.e. d_2P/dt^2 versus t). The work by Russell et al. (1975a) demonstrates remarkable reproducibility (Fig. 25). However, on close examination (Russell et al., 1975b) slight variations in the peak amplitude and decay of the heat output rate were detected. Since the calorimetric method is so sensitive, slight errors in the preparation of medium, inoculum volume and non-standardization of the whole calorimetric procedure (e.g. timing from inoculation to calorimeter presentation) are the most likely cause of the divergences in the thermogram for the same strain of microorganism. It is the precision with which the experiment can be reproduced which may be a serious limiting factor.

At least 200 cultures of clinically significant organisms, predominantly *Enterobacteriaciae*, representing 24 genera and 47 species, have been investigated (Russell et al., 1975b). In general virtually all organisms were distinguished; thermogram diversity within a species was not found to be prevalent. However, exceptions were found. To distinguish between *E. coli* and *Salmonella paratyphi B* growth was repeated in a glucose medium. No identifiable thermal pattern was demonstrated among the serological groups of *Salmonella*.

An interesting proposition is whether a particular group of organisms has similar features in the thermogram. The evidence suggests that species can be identified by virtue of their thermograms, but large groups (e.g. genera) may be too varied for a generalized outline thermogram to be made. A wide diversity was obtained with a single species, namely *E. coli*. For 13 different cultures no uniformity of shape with established biochemical variation was apparent (Russell et al., 1975a). It is again emphasized that the calorimetric incubation does not record a mere end-point analysis, as with conventional biochemical tests, but the kinetics of the process. *Escherichia coli* was the most thoroughly investigated species. Only a general profile was obtained for all strains; three major distinctions (or groupings) comprising 72, 17 and 6 per cent of the total cultures examined were made. Profile diversity was not so pronounced among other species (e.g. 92 per cent of *Proteus mirabilis* gave a single profile, whilst *Klebsiella pneumoniae* gave only two profiles of 98 and 2 per cent frequency).

The fundamental question to be answered is whether calorimetry has any advantage over conventional techniques, including miniaturized methods, for identification? Only multichannel calorimeters would provide the necessary throughput required by a routine laboratory. Significant disadvantages to the exploitation of the calorimetric technique are the expense of current commercial calorimeters which can only be regarded as research instruments, since their throughput load is poor. A commercial multichannel instrument still has to be produced. The multichannel batch calorimeters have the important advantage of simplicity, but suffer from the inherent disadvantages of all batch systems compared to flow or even perfusion systems. For these latter cellule designs would be desirable. A further prospect is the development of more sensitive instruments which would reduce the amount of biological material required.

It is inappropriate to claim that the calorimetric technique can achieve anything more than academic identification. However, it is in the narrower field of characterization that the calorimeter has immediate application, in assessing the performance of microorganisms from a standard origin, evaluating media batches (quality control), or quantifying antibiotic sensitivity.

Acknowledgements

I wish to thank Professor R. A. Cawson and Dr A. E. Beezer for their advice in preparing this work. The award of a research fellowship by the Wellcome Trust is gratefully acknowledged.

References

BEEZER, A. E. (1977). *In* "Applications of Calorimetry in the Life Sciences" (I. Lamprecht and B. Schaarschmidt, eds). Walter de Gruyter, Berlin.

BEEZER, A. E. BETTELHEIM, K. A., O'FARRELL, S. M., AL-SALIHI, S. and SHAW, E. J. (1977). *In* "2nd International Symposium on Rapid Methods and Automation in Microbiology" (H. H. Johnston and S. W. B. Newsom, eds). Learned Information Ltd, Oxford and New York.

BEEZER, A. E., NEWELL, R. D. and TYRRELL, H. J. V. (1979). *Antonie van Leeuwenhoek*, **45**, 55–63.

BELAICH, A. and BELAICH, J.-P. (1976). *J. Bact.* **125**, 19–24.

BELAICH, J.-P., SENEZ, C. and MURGIER, M. (1968). *J. Bact.* **95**, 1750–1757.

BINFORD, J. S., BINFORD, L. F. and ADLER, P. (1973). *Am. J. clin. Path.* **59**, 86–94.

BOLING, E. A., BLANCHARD, G. C. and RUSSELL, W. J. (1973). *Nature, Lond.* **241**, 472–473.

CLIFFE, A. J., McKINNON, C. M. and BERRIDGE, N. J. (1973). *J. Soc. dairy Technol.* **26**, 209–210.

DEAN, A. R. C. and HINSHELWOOD, C. (1966). "Growth, Function and Regulation in Bacterial Cells." Oxford University Press, London.

ERIKSSON, R. and WADSÖ, I. (1971). *In* "Proceedings First European Biophysical Congress, Baden" (E. Broda, A. Locker and H. Springer-Lederer, eds), vol. 4. Verlag der Wiener Medizinischen Akademie, Austria.

FORREST, W. W. (1969). *In* "Biochemical Microcalorimetry" (H. D. Brown, ed.), pp. 165–180. Academic Press, New York and London.
HADLEY, W. K. and SENYK, G. (1975). *In* "Microbiology—1975" (D. Schlessinger, ed.). American Society for Microbiology, Washington, D.C.
HEDÉN, C.-G. and ILLÉNI, T. (1975). "New Approaches to the Identification of Microorganisms". John Wiley, New York and London.
HEDÉN, C.-G., ILLÉNI, T. and KUHN, I. (1976). *In* "Methods in Microbiology" (J. R. Norris, ed.), vol. 9, pp. 15–50. Academic Press, London, New York and San Francisco.
HOOGERHEIDE, J. C. (1975). *Rad. environ. Biophys.* **11**, 295–307.
HUPPERT, M., HARPER, G., SUN, S. H. and DELANEROLLE, V. (1975). *J. clin. Microbiol.* **2**, 21–34.
JAWORSKI, A., SEDLACZEK, L., CZERNIAWSKI, E. and ZABLOCKI, B. (1968). *Acta Microbiol. Polonica*, **17**, 219–230.
JOHANSSON, A., NORD, C.-E. and WADSTRÖM, T. (1975). *Science Tools*, **22**, 19–21.
JOHNSTON, H. H. and NEWSOM, S. W. B. (eds) (1977). "2nd International Symposium on Rapid Methods and Automation in Microbiology", Cambridge, 1976. Learned Information (Europe) Ltd, Oxford and New York.
KALLINGS, L. O. (1977). Unpublished observations (LKB Appl. Note No. 309).
KALLINGS, L. O. and HOFFNER, S. (1976). *In* "2nd International Symposium on Rapid Methods and Automation in Microbiology" (H. H. Johnston and S. W. B. Newsom, eds). Learned Information (Europe) Ltd, Oxford and New York.
LAMPI, R. A., MIKELSON, D. A., ROWLEY, D. R., PREVITE, J. J. and WELLS, R. E. (1974). *Fd Technol., Champaign*, **28**, 52–60.
LAMPRECHT, I. and MEGGERS, C. (1969). *Z. Naturforsch.* **24b**, 1205–1209.
LAND, G. A., VINTON, E. C., ADCOCK, G. B. and HOPKINS, J. M. (1975). *J. clin. Microbiol.* **2**, 206–217.
LUNGHOLM, K., WADSÖ, I. and MÅRDH, P.-A. (1976). *J. gen. Microbiol.* **96**, 283–288.
MÅRDH, P.-A., RIPA, T., ANDERSON, K.-E. and WADSÖ, I. (1976). *Antimicrob. Ag. Chemother.* **10**, 604–609.
MONK, P. and WADSÖ, I. (1975). *J. appl. Bacteriol.* **38**, 71–74.
MONTI, M. and WADSÖ, I. (1976). *Scand. J. clin. Lab. Invest.* **36**, 573–580.
NEWELL, R. D. (1975). Ph.D. Thesis, University of London.
PHILLIPS, A. W. (1955). *J. Inst. Brew.* **61**, 122–126.
POOLE, R. K., LLOYD, D. and KEMP, R. B. (1973). *J. gen. Microbiol.* **77**, 209–220.
PRAT, H., CALVET, E. and FRICKER, J. (1946). *Rev. canad. Biol.* **5**, 247–250.
RIPA, K. T., MÅRDH, P.-A., HOVELIUS, B. and LJUNGHOLM, K. (1977). *J. clin. Microbiol.* **5**, 393–396.
RUBNER, M. (1906a). *Arch. Hyg.* **57**, 193–243.
RUBNER, M. (1906b). *Arch. Hyg.* **57**, 244–268.
RUSSELL, W. J., FARLING, S. R., BLANCHARD, G. C. and BOLING, E. A. (1975a). *In* "Microbiology—1975" (D. Schlessinger, ed.), pp. 22–31. American Society for Microbiology, Washington, D.C.

Russell, W. J., Zettler, J. R., Blanchard, G. C. and Boling, E. A. (1975b). *In* "New Approaches to the Identification of Microorganisms" (C.-G. Hedén and T. Illéni, eds), pp. 101–134. John Wiley, New York and London.
Schaarschmidt, B. and Lamprecht, I. (1976). *Experientia*, **32**, 1230–1234.
Schaarschmidt, B., Lamprecht, I. and Stein, W. (1973). *Biophysik*, **9**, 349–355.
Semenitz, B. and Tiefenbrunner, F. (1976). *In* "Applications of Calorimetry in Life Sciences" (I. Lamprecht and B. Schaarschmidt, eds). Walter de Gruyter, Berlin.
Solomons, G. L. (1969). "Materials and Methods in Fermentation". Academic Press, New York and London.
Spink, C. and Wadsö, I. (1975). *In* "Methods of Biochemical Analysis", vol. 23. (D. Glick, ed.), pp. 1–159. Wiley–Interscience, New York.
Staples, B. R., Prosen, E. J. and Goldberg, R. N. (1973). National Bureau of Standards NBSIR 73-181, Washington, D.C.
Wadsö, I. (1974). *Science Tools*, **21**, 18–21.
Wimpenny, J. M. T. (1969). *In* "Microbial Growth", 19th Symp. Soc. Gen. Microbiol. (P. Meadows and S. J. Pirt, eds), pp. 161–205. Cambridge University Press, London.

Microcalorimetry in Diagnostic Medical Microbiology

K. A. BETTELHEIM
and
E. J. SHAW

A. Introduction

Heat changes are phenomena which are common to all chemical reactions whether they occur in the inanimate environment of the test tube or within the living bacterial cell. It is only with the development of sophisticated, sensitive instruments such as the adiabatic calorimeter of Forrest (1961, 1972), the Calvet microcalorimeter (Calvet, 1962) and the LKB flow microcalorimeters (Beezer and Tyrrell, 1972) that the small heat production of bacterial metabolism has become measurable and has enabled the application of this useful technique to diagnostic microbiology.

The diagnostic microbiology laboratory is concerned as to whether (i) any bacteria are present in a specimen from a site which is normally sterile, (ii) pathogenic bacteria are present in a site which contains its own indigenous flora, (iii) the identification of the organism causing the disease and finally (iv) the antibiotic sensitivity of the infecting microorganism. Occasionally the concentration of the infecting bacteria is also of importance.

B. Microcalorimeters

For these criteria microcalorimeters of two basic designs have been used. The first is a multichamber instrument. Here the culture is allowed to grow in a static tube and the heat generated regularly monitored by the calorimeter. These data are then stored in a computer and a chart of heat production versus time constructed. This then gives the characteristic heat profile of the organism. This method has been used extensively by Dr Boling and his co-workers (Boling et al., 1973; Russel et al., 1975) to identify bacteria. Between 10^5 and 10^6 organisms per ml are usually required to generate detectable heat.

The second form of instrument is a flow calorimeter. Here the bacteria have already multiplied outside the chamber and the culture is presented to the calorimeter for a short duration (usually 3–5 min). The amount of heat produced during this time is proportional to the number of bacteria present.

C. Bacterial identification by microcalorimetry

Boling et al. (1973) and Monk and Wadsö (1975) have shown that the pattern of heat production is reproducibly characteristic for certain strains of bacteria pathogenic for man. The former group of workers used a batch calorimeter and allowed the organism to grow up for 3 h yielding a heat output of about 50 μcal s^{-1} ml^{-1}. This application of microcalorimetry is extensively discussed by Newell (p. 163).

Members of the order Mycoplasmatales are currently considered more important in medical microbiological studies but most of the present methods of determining mycoplasma growth are not as reliable as one would desire. Ljungholm et al. (1976) showed that calorimetric measurements of mycoplasma growth in teflon-coated stainless-steel ampoules in a batch instrument were more precise than conventional methods. They note that a particular advantage is being able to record growth from 10^6 organisms ml^{-1} upwards instantly and continuously. This suggests that calorimetry may be particularly suitable for the examination of clinical materials for the presence of some of the more fastidious organisms.

D. Examination of clinical specimens

1. URINARY INFECTION

Urinary infection is caused chiefly by a limited number of organisms, *Escherichia coli*, *Proteus* sp., *Klebsiella* sp., *Pseudomonas* sp., *Streptococcus faecalis* and *Staphylococcus albus*. These organisms are part of the normal bowel and skin flora and therefore also frequently contaminate urine specimens. To differentiate between contaminating and infecting microorganisms, the concept of "significant bacteriuria" was developed by Kass (1957). Urinary infection is said to be present if, in a specimen of urine taken with special precautions (mid-stream specimen), there are at least 10^5 organisms per ml of urine of one strain of bacteria (i.e. "pure growth") present in the specimen. This has been modified somewhat in that urinary infection is also said to be present if a pure growth of between 10^4 and 10^5 organisms per ml of the same strain are present in serial specimens from the same patient (Kunin, 1974). This therefore suggests that the presence or absence of significant bacteriuria could be correlated with heat production.

For an evaluation of the heat output of infected urine the LKB flow microcalorimeter is particularly suitable (Beezer and Tyrrell, 1972) and lends itself easily to further automation because the test procedure depends purely on passing a specimen through a tube by means of a peristaltic pump for a given period of time. After replacing the specimen by a wash solution, the probe is placed into the next specimen.

A number of preliminary studies (Beezer *et al.*, 1974) were undertaken with solutions of nutrient broth containing *E. coli*, *Proteus morganii*, *S. faecalis* and *Pseudomonas aeruginosa* at concentrations greater than 10^4 cells per ml. These studies, which demonstrated that it was possible to enumerate these organisms, have been extended (Beezer *et al.*, 1974, 1977) and a procedure developed which correlates bacterial cell numbers for several taxonomic groups with the observed heat effect as described microcalorimetrically (Fig. 1). In this procedure a base line is established for the microcalorimeter with a solution of glucose (0·1 M) in phosphate buffer at pH 6·5, passing through the flow cell at a rate of 55 ml h^{-1}. The bacterial suspension to be studied is added to the glucose-buffer reservoir to give a dilution of 1 : 10. The response of the instrument is given in terms of percentage deflection which is proportional to the heat effect.

As the numbers of organisms which need to be enumerated in urine specimens vary between 10^3 and 10^5 ml^{-1} it was considered that some form of preincubation was necessary because the instrument does not significantly detect numbers of bacteria below 10^5 ml^{-1}. Thus for a study (Bettelheim *et al.*, 1976; Beezer *et al.*, 1977) on enumeration of bacteria in clinical specimens of urine 0·6 g of Oxoid No. 3 nutrient broth powder was added to 25 ml of urine which was then incubated at 37 °C for 2 h before presentation to the calorimeter. Two hundred and ninety-nine randomly selected specimens were examined by this procedure and by standard cultural methods. The results are given in Table 1; for the purposes of interpreting these results a significant response by the calorimeter was considered to be a deflection greater than 5 per cent. These results suggest that this is an effective screening procedure for detecting more than 10^5 organisms ml^{-1}.

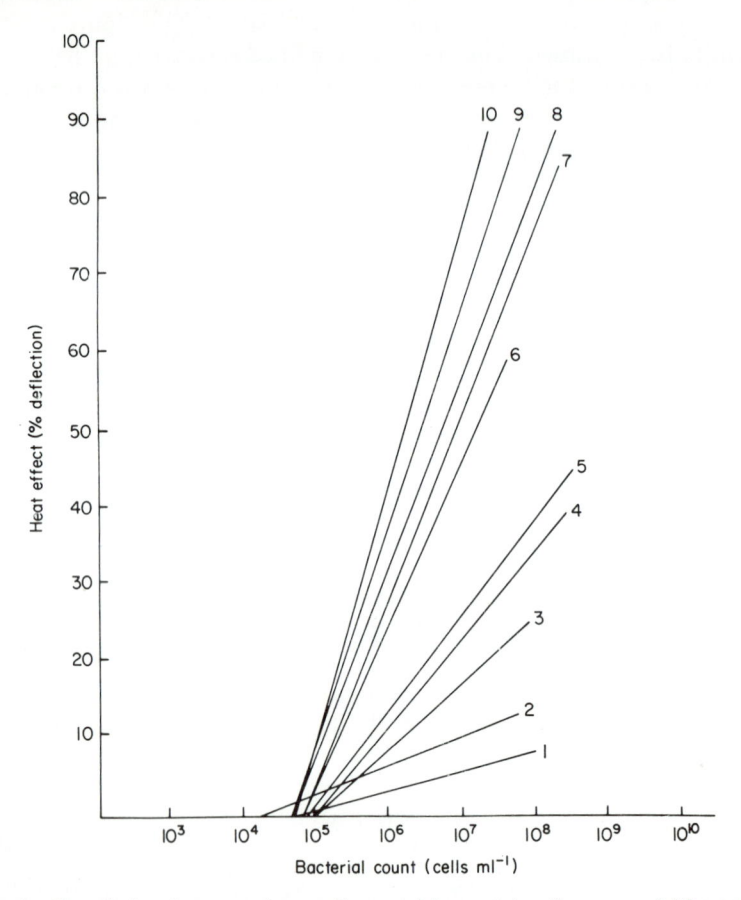

Fig. 1. Correlation between heat effect and bacterial cell count of 15 strains of bacteria recently isolated from infected urines. 1, *Staph. albus* (1 strain); 2, *Ps. aeruginosa* (1 strain); 3, *Proteus* sp. (1 strain); 4, *Strep. faecalis* (3 strains); 5, *Ps. aeruginosa* (1 strain); 6, *Proteus* sp. (1 strain); 7, *Klebsiella* sp. (1 strain); 8, *Proteus* sp. (1 strain); 9, *Staph. albus* (1 strain); 10, *E. coli* (4 strains). (Reproduced by permission of the Editors of the *Journal of Medical Microbiology*.)

In order to increase the sensitivity of the procedure in the range of 10^3–10^5 organisms ml^{-1} a preincubation of 3 h was attempted. Preliminary experiments suggest that this is feasible.

TABLE 1

Comparison of microbiological and microcalorimetric assay of urine specimens
(Bettelheim et al., 1976)

Microbiological assay organisms (ml^{-1})	Number of specimens	0–5%	6–10%	11–20%	>21%
<10^3	102	80	18	4	0
10^3–10^5	47	25	10	11	1
>10^5					
Mixed	56	4	5	4	43
Pure	94	0	2	4	88
TOTAL	299	109	35	23	132

Number of urine specimens giving % deflection indicated

2. SEPTICAEMIA

Blood is normally sterile and the presence of microorganisms is abnormal. The detection of bacteraemia is therefore an important clinical problem. Organisms may be present in very small numbers (1–10 ml^{-1}) and still cause disease. Current methods for the detection of bacteraemia depend almost entirely on the multiplication of bacteria until detectable visually. This usually takes a minimum of 18 h and up to 6 weeks. There have been several methods introduced in an endeavour to detect bacteraemia more rapidly. Examples of these are measurements of the changes in electrical impedance and radiometric assay of metabolized ^{14}C-labelled glucose (Hadley and Senyk, 1975; Randall, 1975).

Microcalorimetry as a technique is eminently suitable for the detection of bacteraemia, as heat production is a universal characteristic of all bacteria.

Evaluation of microcalorimetry as a potential method for detecting bacteraemia has been limited so far to measurement of the heat produced by strains previously isolated from infected individuals by conventional techniques. Strains of *E. coli* and *S. aureus* when grown in the presence of blood produced similar heat profiles to those produced after growth in nutrient broth (Mårdh et al., 1976). However, heat production was dependent on constituents present in the blood as it was shown that much greater heat was produced in the presence of the anticoagulant sodium polyanethosulfonate (0·5 per cent) than heparin (600 iu ml^{-1}). Moreover the same substance could have different effects on different bacterial species. Isovitalex® was found to stimulate the growth of

S. aureus and *E. coli* but depress the growth of *Neisseria gonorrhoea*. From these studies these workers also concluded that microcalorimetry may be a useful tool in the evaluation of culture media (Ripa *et al*., 1977).

F. Antibiotic sensitivity testing

The studies of Binford *et al*. (1973) on the use of microcalorimetry for antibiotic sensitivity testing were mainly designed to develop a method of screening antibiotics. They note that those antibiotics which cause the metabolic heat of a given organism to decrease rapidly and thus act bacteriocidally can be distinguished by this method from bacteriostatic antibiotics which would cause only a gradual decrease in metabolic heat.

For their studies they used a Beckman Model 190 B Microcalorimeter and tested *E. coli, Streptococcus* sp. and *Pseudomonas* sp. isolated from clinical specimens against eight different antibiotics. The basis of their test system was to present a suspension of the organism to be tested to the calorimeter followed by testing the same suspension with one of the antibiotics. This was repeated every 15 min till all the antibiotics were tested. Finally, a sample of culture was introduced alone again as control.

They found an over-all correlation of 225 tests out of 291 agreeing with the agar diffusion methods of studying antibiotic sensitivities. A problem they noted was the development of anaerobiosis in the sensitizing cell during the test and alteration of the culture during the period of testing. The authors consider that these problems can be easily overcome and that a very suitable method of rapid antibiotic resistance determination could be developed.

Other workers (Mårdh *et al*., 1976b, c) have noted that different but closely related antibiotics (minocycline, doxycycline, oxytetracycline and tetracycline) produce different changes in the heat effect of one strain of *E. coli*. Immediately on addition of the antibiotics heat production decreased sharply and remained at a low level for about 1 h with tetracycline, for 9 h with minocycline and 16 h with doxycycline and oxytetracycline. This work suggests that microcalorimetry may supplement pharmacokinetic studies in establishing optimum doses and dose intervals in antibiotic therapy.

F. Virological studies

Although it is to be expected that viruses would effect the metabolic heat of tissue culture cells, no formal studies on these aspects appear to have been undertaken.

Voitsekhovsky and Polyak (1972) used microcalorimetry to study the effect of various influenza viruses on the respiratory enzymes of primary cultures of chick embryo cells. They noted a decrease in glycolysis, pyruvate utilization in the Krebs cycle and cytochrome C utilization. Simultaneously the heat released during the utilization of $NADH_2$ and ATP increased. This type of experiment suggests that heat changes could be used to monitor viral infections of cells and possibly diagnose the presence of virus particles.

G. Conclusion

The production of heat by living matter is a universal phenomenon. In recent years calorimeters have been developed of sufficient sensitivity to permit studies to be carried out on the relatively small amount of heat produced by microorganisms. To date it has been shown that heat output is a regular and characteristic feature of bacterial cells. Although it is clear that with pure cultures of bacteria it is possible to identify, enumerate and determine antibiotic sensitivity patterns on the basis of heat production, the possibility of achieving the same with clinical materials is still being investigated. At present most work has been directed towards the examination of urine; other areas such as the examination of CSF blood and dialysis fluid have not been reported. As continuous-flow methods are currently available, to which this technique could be adapted, it has potential for the automation of some clinical microbiological determination.

Acknowledgement

This work has been supported in part by a grant from the Department of Health and Social Security.

References

BEEZER, A. E. and TYRRELL, H. J. V. (1972). *Science Tools*, **19**, 13–16.
BEEZER, A. E., BETTELHEIM, K. A., NEWELL, R. D. and STEVENS, J. (1974). *Science Tools*, **21**, 13–16.
BEEZER, A. E., BETTELHEIM, K. A., AL-SALIHI, SABRIA and SHAW, ELIZABETH J. (1977). *Science Tools*, **25**, 6–8.
BETTELHEIM, K. A., O'FARRELL, S. M., AL-SALIHI, S., SHAW, E. J. and BEEZER, A. E. (1976). *In* "2nd International Symposium on Rapid Methods and Automation in Microbiology" (H. H. Johnston and S. W. B. Newsom, eds), p. 8. Learned Information (Europe) Ltd, Oxford and New York.

BINFORD, J. S., BINFORD, L. F. and ADLER, P. (1973). *Am. J. clin. Path.* **59**, 86–94.
BOLING, E. A., BLANCHARD, G. C. and RUSSEL, W. J. (1973). *Nature, Lond.* **241**, 272–273.
CALVET, E. (1962). *In* "Experimental Thermochemistry" (H. A. Skinner, ed.), vol. 2, pp. 385–410. John Wiley, New York.
FORREST, W. W. (1961). *J. scient. Instrum.* **38**, 143–145.
FORREST, W. W. (1972). *In* "Methods in Microbiology" (J. R. Norris and D. W. Ribbons, eds), vol. 6B, pp. 285–318. Academic Press, New York and London.
HADLEY, W. K. and SENYK, G. (1975). *In* "Microbiology—1975" (D. Schlessinger, ed.), pp. 12–21. American Society for Microbiology, Washington, D.C.
KASS, E. H. (1957). *Arch. Int. Med.* **100**, 709–714.
KUNIN, C. M. (1974). "Detection, Prevention and Management of Urinary Tract Infections", 2nd ed., p. 26. Lea and Febiger, Philadelphia.
LJUNGHOLM, K., WADSÖ, I. and MÅRDH, P.-A. (1976). *J. gen. Microbiol.* **96**, 283–288.
MÅRDH, P.-A., RIPA, T., LJUNGHOLM, K., WADSÖ, I. and ANDERSSON, K. E. (1976a). *In* "Rapid Methods and Automation in Microbiology" (H. H. Johnston and S. W. B. Newsom, eds), p. 7. Learned Information (Europe) Ltd, Oxford and New York.
MÅRDH, P.-A., ANDERSSON, K. E., RIPA, T. and WADSÖ, I. (1976b). *Scand. J. infect. Dis. Suppl.* **9**, 12–16.
MARDH, P.-A., RIPA, T., ANDERSSON, K. E. and WADSÖ, I. (1976c). *Antimicrobial Ag. Chemotherapy*, **10**, 604–609.
MONK, P. and WADSÖ, I. (1975). *J. appl. Bact.* **38**, 71–74.
RANDALL, E. L. (1975). *In* "Microbiology—1975" (D. Schlessinger, ed.), pp. 39–44. American Society for Microbiology, Washington, D.C.
RIPA, K. T., MÅRDH, P.-A., HOVELIUS, B. and LJUNGHOLM, K. (1977). *J. clin. Microbiol.* **5**, 393–396.
RUSSELL, W. J., FARLING, S. R., BLANCHARD, G. C. and BOLING, E. A. (1975). *In* "Microbiology—1975" (D. Schlessinger, ed.), pp. 22–31. American Society for Microbiology, Washington, D.C.
VOITSEKHOVSKY, B. L. and POLYAK, R. YA. (1972). *Acta Virol.* **16**, 313–322.

Microcalorimetric Investigations of Drugs

A. E. BEEZER
and
B. Z. CHOWDHRY

A. Introduction

The study of the analysis of drugs has, essentially, three objectives. Firstly, the quantitative and qualitative assay of pure and impure drugs in diverse matrices. Secondly, to investigate the mode of action of the drug at the molecular level. Thirdly, the design and clinical evaluation of new drugs. Both quantitative and qualitative analyses of drugs are important in these fields.

Clinical medicine requires high precision, reproducibility and accuracy in quantitative drug measurements. The clinician expects rapid results (1–2 h or sooner) and this often demands that the laboratory be able to deal with large numbers of samples. Metabolic studies require detection of very low levels of drugs: between microgram and nanogram levels. The clinician often requires to know the effective concentration of drugs (i.e. bound and unbound) in a variety of body fluids (serum, urine and cerebrospinal fluid)—sometimes at close intervals of time. The sensitivity of microorganisms to different drugs may also be required to be known when there is a choice of therapy. Moreover, many drugs are administered in combination. More than one drug may therefore require to be monitored at the same time in body fluids.

The above comments also apply to veterinary medicine. The requirements for pharmaceutical analysis are similar to those given above; however, the limits of detection are not so low (Hughes and Wilson, 1973; Fairbrother, 1977).

Mode of action studies of drugs are required so that the molecular action of the drug can be made selective, to help in the design of new drugs and in order that drugs can be used as tools in scientific experiments.

Combined drug therapy may be necessary in the initial therapy of severe undiagnosed infection, and for treatment of mixed infections where the pathogens do not share an antibiotic susceptibility. Combined therapy may prevent the development of resistance to certain drugs. Secondly, two antibacterial drugs may be given in the hope of achieving synergy—a greater antimicrobial activity than the sum of the single agents. The clinical relevance of synergy is often difficult to prove and results of classical microbiological *in vitro* tests may not always predict the activity of antibiotics in therapy where pharmokinetic behaviour is also important (Anonymous, 1978).

The uses of microcalorimetry in the analysis of drugs will be examined in this contribution. An evaluation of the extent to which microcalorimetric techniques have fulfilled the objectives given above will then be made.

From the earliest days of the antibiotic era the quantitative and bioactivity analytical methods of choice were biological (Hughes and Wilson, 1973). Nevertheless with the advent of purified antibiotic drugs there has been a growing tendency towards physical and chemical methods of assay (Quinn, 1974; Fairbrother, 1977). This is in spite of advances made in methods of microbiological analysis of antibiotics (see, for example, Cosgrove, 1978). The advantages and disadvantages of both microbiological and physico-chemical methods of analysis for drugs have been examined by Hughes *et al.* (1973) and Chowdhry (1975). Certain antibiotics have to be assayed (quantitatively and for bioactivity) by biological (microbiological) methods because of the legal requirements of official bodies (Hughes *et al.*, 1973). Also, drugs often require to be assayed by biological methods because it is necessary to examine them in the mode in which they will be used, e.g. antimicrobial drugs against micro-organisms. Classical microbiological methods such as serial dilution, agar diffusion and disc methods (Kavanagh, 1963) for antibiotic assays each have disadvantages (Newell, 1975). Such tests are (1) time consuming (4–18 h), (2) of low reproducibility (cf. Cosgrove, 1978) and precision, and (3) open to subjective errors. Furthermore, they are subject to vagaries and variations inherent in classical microbiological assays, for instance the use of unstandardized innocula, leading to conflicting results between or within laboratories (see, for example, Lightbown *et al.*, 1963). Microbiological methods require large numbers of observations for satisfactory results, are labour intensive and little advantage is gained in automation.

Each of the classical microbiological methods show disadvantages particular to themselves. Photometric methods require that the antibiotic does not interfere in the measurement at the wavelength at which turbidity is measured. However, in theory, this method could be used to measure the kinetics of antibiotic action although it may be considered somewhat crude. Respirometric methods rely upon the fact that any indicators used must not interfere with the assay by, for instance, complexing with antibiotic. The theory of plate agar diffusion methods is ill understood from a molecular point of view (the methods rely upon the diffusion of antibiotic in a gel) (Kavanagh, 1963). If an antibiotic preparation is not homogeneous then difficulties may be encountered in assessing the results of the assays. Synergistic or antagonistic effects of drug molecules on microorganisms are also difficult to assess by plate agar diffusion methods. Photometric methods require that the antibiotic does not interfere in the measurement at the wavelength at which turbidity is measured. On the other hand, respirometric methods rely upon the fact that any indicators used must not interfere with the assay by, for example, complexing with the antibiotic. Tube dilution technique results depend on the growth of microorganisms in a control tube (Kavanagh, 1963) and in test tubes. However, in the latter growth may or may not have occurred depending on time of assay. Usually such assay results are read (by optical density measurements) before a non-growth condition is attained.

In view of the disadvantages of classical microbiological techniques, new methods for the analysis of drugs acting against microorganisms have been investigated. Isothermal heat conduction microcalorimetry is one such technique. Microcalorimetric methods have been utilized for many years in microbiological research (Forrest, 1972). The emphasis has usually been to correlate observed heat effects with thermodynamic data. In such work both macro- and micro-systems have proved to be useful. The use of microcalorimetry as an analytical tool in microorganism drug interactions is, as pointed out above, a relatively recent innovation.

B. Calorimetric methods

The use of isothermal heat conduction microcalorimetry in the examination of microorganism–drug interactions is based upon the fact that the complex sequence and conjunction of chemical events that is metabolism results in heat evolution and to the recording of a complex thermogram in growth medium.

In growth medium typical thermograms show, initially, an exponential increase in heat output rate followed by peaks and troughs. If a drug is added to the incubation vessel containing microorganisms then there will be a difference in the thermogram according to the concentration and timing of the drug addition. Elsewhere in this book appears discussion of the metabolism of bacteria and yeast (Belaich, p. 1, and Lamprecht, p. 43) as well as "identification" of microorganisms by microcalorimetric methods (Newell, p. 163). Drugs can also be added to microorganisms which are metabolizing a single substrate, e.g. glucose. A "simpler" thermogram is obtained than for microorganisms in growth medium.

Both batch and flow microcalorimeters (Spink and Wadsö, 1976) can be used for the study of microorganisms and microorganism–drug interactions. Batch techniques, however, have certain disadvantages for such studies. Good (exact) control of gaseous conditions, particularly aerobic atmospheres, are difficult to maintain in batch instruments. Microbial manipulations, such as addition of antibiotics to microorganisms, cannot be carried out easily without upsetting the thermal equilibrium of the system. Temperature equilibration of microcalorimetric cell contents in batch systems is time consuming. Homogeneous mixing of antibiotics and microorganisms cannot always be ensured. Physical effects such as absorption of materials to the walls of microcalorimetric cells may also pose problems in certain instances (cf. Wadsö, p. 247). Batch techniques do not allow sampling during the experimental period. This contributes to difficulties in interpretation of the thermograms obtained.

Some batch-type instruments have disadvantages particular to themselves. In isoperibol instruments, for example, the uncertainty introduced in slow processes is usually inconveniently large. Most available isoperibol calorimeters are not true microcalorimeters. Adiabatic shield calorimeters cannot be used to study microorganisms since they are based on adiabatic principles and as with isoperibol instruments are unsuitable for studies of isothermal metabolic processes. Thermometric techniques also suffer from the disadvantages highlighted above. Flow-type microcalorimetric instruments such as the thermoelectric heat microcalorimeter and the labyrinth-flow microcalorimeter (Reid, 1976) have the disadvantage of slow response times.

There are certain experimental requirements which have to be met in studies of microorganisms and microorganism–modifier interactions. The culture vessel should be repeatedly sterilizable; a sampling device or an apparatus for parallel culture is desirable; provisions for agitation (cellular suspensions must be homogeneous) and retention of a controlled gaseous atmosphere are also required. Continuous monitoring of physico-chemical parameters of the cell suspension is necessary.

It is often essential to analyse the cell suspension, for instance in terms of medium composition. An incubation medium of defined chemical composition is therefore necessary. This makes for better reproducibility in the systems studied and ease of analysis of medium constituents.

The ability to add cellular modifiers to the incubation vessel speedily, conveniently and without changing the properties of the system under study (e.g. thermal equilibrium) at the time of addition is also advantageous.

Microbiological incubation vessels should be able to accommodate and not disturb any devices used for monitoring of microbiological suspensions within it. Exact temperature control and prevention of contamination by microorganisms other than those being studied are also required. Any microcalorimetric techniques which are used to measure the properties of living microorganisms in the presence and absence of antibiotics must also fulfil certain criteria. These include:

1. Analytical criteria; good reproducibility, high precision and heat sensitivity. The latter allows (a) both large and small cell numbers to be studied, (b) subtle changes in the properties of the system being measured, e.g. metabolism of cells to be detected.

2. The ability to operate isothermally at different temperatures. Different microbiological inocula, for instance bacteria and yeast, may be required to be monitored at different temperatures (for optimal growth for example). Secondly, the nature of microorganism–drug interactions may also change with temperature and therefore require to be monitored.

3. The ability to detect kinetic phenomena (i.e. the microcalorimeter should have low thermal inertia).

4. The technique should ideally be non-invasive, i.e. not alter the properties of the system under study during measurements or introduce artifacts.

5. Physical effects in the microcalorimetric cells (condensation, evaporation and surface adsorption phenomena) must be absent.

6. It is helpful to be able to examine different types of microbial cultures (chemostat, turbidostat, continuous, batch, liquid or semisolid cultures) under aerobic or anaerobic conditions.

7. Monitoring of cell suspensions over long periods of time (i.e. retention of operational stability). Microbiological phenomena often occur over a wide span of time periods.

8. Convenience and ease of use.

9. Be commercially available.

10. Negligible heat effects due to stirring of cultures.

11. Relatively short equilibration time (biological materials are often labile).

12. Be usable both continuously and intermittently.

13. Data acquisition and treatment should be capable of automation.

14. It is advantageous if other analytical techniques can be used simultaneously with the proposed microcalorimetric techniques for parallel measurements.

In sum the technique should offer advantages over alternative or preceding techniques used for similar purposes in the past.

Flow microcalorimeters of the type designed by Wadsö (1976) satisfy many of the demands made by the requirements for the study of microbial processes. They can also fulfil, to a large extent, the criteria required of the microcalorimetric apparatus.

Available flow microcalorimeters (including the Wadsö design) do have certain disadvantages when used for microbial studies including cell-modifier interactions:

1. Sensitivity of heat measurements may not be sufficient (e.g. the heat output of 10^2–10^4 cells ml^{-1} cannot be detected). Most of the studies conducted with microorganisms to date have examined the catabolic events in cells because the heat sensitivity required for the study of anabolic processes cannot be achieved at present.

2. The time resolution of the instruments is often low. Batch instruments monitor whole reactions. In flow microcalorimeters there is usually a lag time between the initiation of a response and its detection, that is, the initial response is not monitored.

3. Resulting from (2) above changes in microbial metabolism during transport from the incubation vessel to the microcalorimetric cells may occur, for instance, in dissolved gas concentration (anaerobic experiments) and temperature along the flow lines.

4. Clogging or sticking of cells along flow lines and in the microcalorimetric cell, as well as the presence of gas bubbles, can occur and should be avoided. Otherwise erroneous instrument signal output will be observed. Wall growth of microorganisms in the microcalorimetric cell during prolonged growth experiments has to be avoided. It is essential therefore to clean out the microcalorimetric cells thoroughly in between experiments.

5. The physical nature of the material under study must be such that it is amenable to pumping through the microcalorimetric tubing. This is not a very severe problem in microbiological studies unless gel-like materials are used.

C. Effect of cell modifiers on cellular metabolism

1. MICROCALORIMETRIC MECHANISTIC STUDIES OF CELL-MODIFIER INTERACTIONS

The pioneering studies of the effect of antibiotics on growing microorganisms by microcalorimetry were carried out by Prat (1953) and Zablocki et al. (1962). Prat used a batch Calvet instrument whilst Zablocki used a modified isoperibol microcalorimeter. Prat demonstrated that the addition of 30·0 mg of streptomycin (1 ml) to a culture of *Escherichia coli* in broth (4·0 ml) at 297·4 K resulted in a marked fall in the heat output, which was reduced to one-tenth the original value after 6 h (Fig. 1).

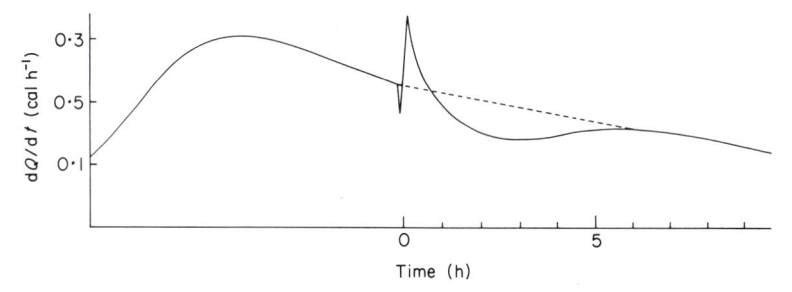

FIG. 1. Interaction of streptomycin with *E. coli* using a batch Calvet instrument (Prat, 1953).

Using a stirred culture, the depression of the thermogram was found to be very much more rapid than if a static culture was used. The effectiveness of the antibiotic was limited by its rate of diffusion through the medium in the latter case. Prat (1953) also studied the interaction of rimafon with *E. coli* (BCG) by microcalorimetry. He realized the technique could be used to obtain information concerning the mechanism of action of drugs (Fig. 2). Zablocki et al. (1962) found that penicillin (in terms of the latent period of action) affected the character of the thermogenesis curves of *E. coli* (6×10^6 cells ml^{-1}) according to the phase of growth at which the antibiotic was added to the culture. At 298 K the latent period decreased the later the antibiotic was added to the cells in the logarithmic phase. Effects of streptomycin, chlortetracycline and oxytetracycline on *E. coli* were also studied. Streptomycin added to a culture in the logarithmic growth phase had a latent period of 15 min. The effects of oxytetracycline and chlortetracycline differed from those of streptomycin, the latent period being shorter. Penicillin and oxytetracycline added during the logarithmic growth phase of *S. typhi* ($1·4 \times 10^6$ cells ml^{-1}) produced a similar change in the case of oxytetracycline as in the presence of *E. coli*. The effect of penicillin differed in the period of latency from that with *E. coli* (Fig. 3). The concentration of antibiotics used was not quoted.

Binford et al. (1973), using a stopped-flow version of the Beckmann microcalorimeter, tested the sensitivity of various bacterial species (*E. coli* Y, *Streptococcus* and *Pseudomonas*) to eight antibiotics (erythromycin, ampicillin, carbenicillin, cephalothin, gentamycin, kanamycin, colistin

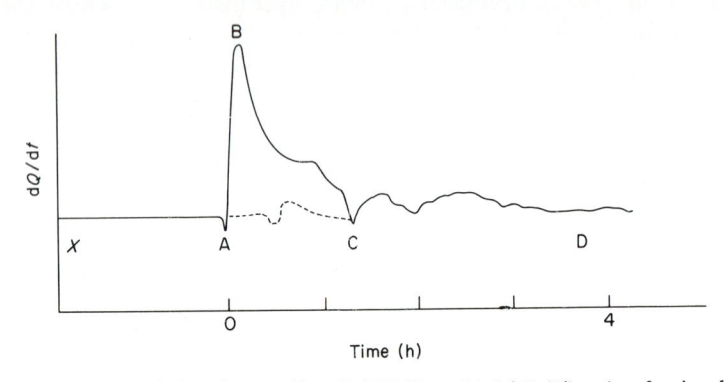

FIG. 2. The effect of rimafon on *E. coli* (BCG) at 34·6 °C. The rimafon is added at point A. The dotted line represents the thermogram for *E. coli* (BCG) in the absence of the drug. (Prat, 1953.)

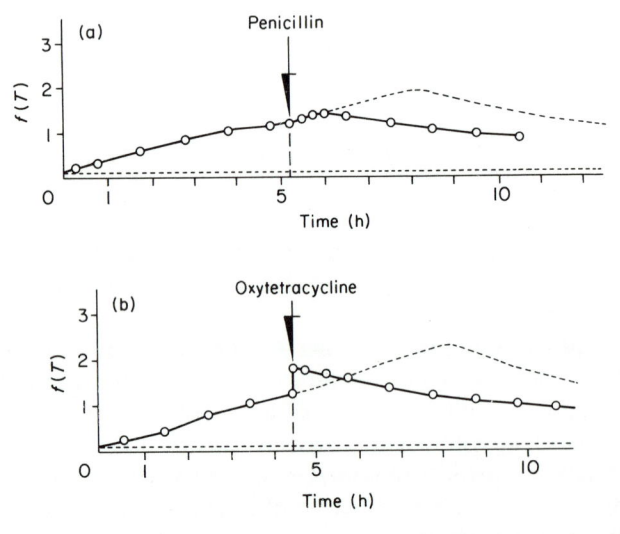

FIG. 3. (a) Interaction of penicillin with *S. typhi*. (b) Interaction of oxytetracycline with *S. typhi*. (Dashed lines indicate nature of thermogram in absence of antibiotics.) (Zablocki et al., 1962.)

and gantrisin). The experiments were carried out at 310 K; both anaerobic and aerobic atmospheres were used. The method used allowed distinction between bactericidal and bacteriostatic antibiotics. Results were attained 12–24 h sooner than with the disc agar diffusion method. The percentage correlations with results obtained by disc agar diffusion method were 77 and 87 per cent under anaerobic and aerobic conditions respectively. The bacterial antibiotic sensitivity method described by Binford et al. is not easily automatable and is still time-consuming in relation to the requirements of the clinician.

Wadsö (1969) has shown that the addition of ampicillin to growing cultures of *E. coli* at 310 K results in a different thermogram from that for *E. coli* cultures alone. An LKB flow microcalorimeter was used. The experiments indicated that the sensitivity of *E. coli* to ampicillin could be determined by microcalorimetry.

The effect of tetracyclines (minocycline, doxycycline, oxytetracycline and tetracycline hydrochlorides) on *E. coli* (10^4 cells ml^{-1}) in the logarithmic growth phase has been tested by microcalorimetry (Mårdh et al., 1976). Both flow and stainless steel ampoule batch microcalorimeters were used, the experiments being carried out at 310 K. Experiments in the flow microcalorimeter were conducted under aerated conditions. Clear differences in the capacity of the tetracyclines to suppress the metabolism of the test organism were obtained above and below the minimum inhibitory concentration (MIC) of the antibiotics. The order of activity of the antibiotics to suppress the metabolism of the test organism was minocycline > doxycycline > oxytetracycline > tetracycline. There was a variation in the extent and duration of the inhibitory effect on the metabolism of *E. coli* as judged from the decrease in heat production (Fig. 4). It is thought, by the authors, that this information may be of use in conjunction with pharmokinetic data in establishing times of dose intervals in antibiotic therapy.

Similar experiments to those of Wadsö using tetracycline antibiotics (tetracycline, demethyl tetracycline, doxycycline and minocycline) have been conducted by Semenitz et al. (1977a). A flow microcalorimeter with a 1·0 ml steel ampoule measuring cell was used. Concentrations of 10^6 cells ml^{-1} of medium were obtained at the beginning of the experiments. All four tetracyclines, at concentrations of 3–40 μg per 10^6 cells, caused similar effects on growing cultures of *E. coli* H3579, antibiotics being added 1·5 h after the beginning of measurable heat production. These antibiotics in the presence of *Staphylococcus aureus haemolyticus* A3579 also gave similar thermograms for all antibiotics at 3·0 and 30·0 μg ml^{-1} concentrations. Almost immediate inhibition of heat production occurred upon addition of the antibiotics to the cultures.

Staphylococcus aureus haemolyticus A2953 in the presence of tetracycline gave different thermograms from those obtained with minocycline and doxycycline, both the latter giving similar responses. The antibiotics were used at levels of 30 μg ml^{-1}.

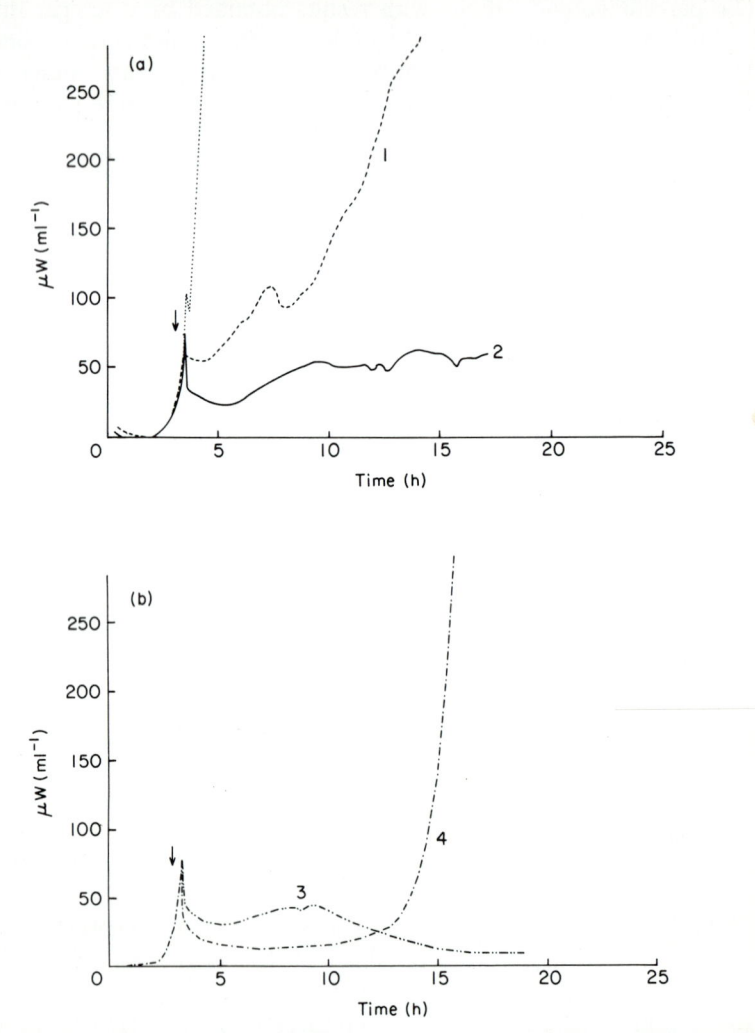

FIG. 4. (a) Heat effects produced by *E. coli* cultured in the presence of tetracycline (1) and doxycycline (2). The thermogram of a non-antibiotic-containing culture is included. (b) Heat effects produced by *E. coli* cultured in the presence of oxytetracycline (3) and minocycline (4). The antibiotics were added (indicated by arrows) in the log growth phase at a heat production of 40 μW ml^{-1}. The concentration of the antibiotics was 1·60 μg ml^{-1}. (Mårdh et al., 1976.)

Using a flow microcalorimeter, the effect of tetracycline (a bacteriostatic antibiotic) on *E. coli* (H3579, BBSU) during the exponential growth phase has been tested (Semenitz et al., 1977b). The experiments were carried out at 307 K using undefined complex growth medium. The addition of 1 or 3 µg of tetracycline caused immediate inhibition of the heat production of the test bacterial population. With the addition of 3, 6 or 12 µg of ampicillin the onset of inhibition was directly dependent upon the concentration of the antibiotic. The combined addition of 1 µg of tetracycline and 3 µg of ampicillin effected immediate inhibition as in the case of tetracycline and a strong decrease of residual activity with practically unchanged bacterial density. The study of antibiotics, such as those above, in the presence and absence of each other may be of significance in therapeutic studies of the effectiveness of antibiotics in combination.

Beezer et al. (1978a) have studied the effect of antifungal drugs in combination on *Saccharomyces cerevisiae* SQ1600. They used an LKB flow microcalorimeter. Combinations of clotrimazole with amphotericin B and amphotericin B with 5-fluorocytosine (5-FC) were studied in both growth and respiration media. Combinations of 5-FC with amphotericin did not show synergistic or antagonistic effects compared to amphotericin B alone. Combinations of clotrimazole with amphotericin B showed antagonistic effects in both respiration and growth medium. Experiments on the adsorption of clotrimazole and amphotericin B to yeast cells indicated that competition for sites on the cell membrane between the two drugs explained this antagonism.

The effects of the antibiotics ampicillin, streptomycin, chloramphenicol, tetracycline and sulphadimidine on two strains of *E. coli* have been studied by flow microcalorimetry (Beezer et al., 1978b). The experiments were carried out using a semi-defined growth medium and conducted at 298 K. Antibiotics were added 50 min after the addition of $2 \cdot 0 \times 10^7$ cells ml^{-1} to the incubation vessel. The exponential heat output of the bacteria changed upon the addition of antibiotics, the exact change depending on the type and concentration of antibiotic (see, for example, Table 1). Similar results were obtained for the other antibiotics. Half MIC amounts of the antibiotics produced none or only marginal effects. MIC quantities of antibiotics prevented any further increase or caused a decline in heat output. Sulphadimidine was an exception, however, and produced no effect even at MIC levels. This could be due to the high numbers of cells used—2×10^7 cells ml^{-1} in the microcalorimetric cells compared to 2×10^5 cells ml^{-1} in MIC experiments.

The antibacterial activity of oleandomycin and erythromycin (macrolide antibiotics) against *Staphylococci* and *Enterococci* has been compared

TABLE 1

The effect of ampicillin on *E. coli* 1 (sensitive) and *E. coli* 2 (resistant) bacterial cells

Antibiotic	Concentration μg ml^{-1}	E. coli	Microcalorimetric effect
Ampicillin	1·0	1	No effect
	2·0 (MIC)	1	Decline in heat output (20 min)
	5·0	1	Decline in heat output (8 min)
	5 000·0	2	No effect
	10 100·0	2	Decline in heat output (30 min)

by microdilution MICs and microcalorimetry (Semenitz, 1978). An LKB 2107 microcalorimeter was used at 35 °C and with initial cell concentrations of 5×10^7 cells ml^{-1}. The MIC of erythromycin for *S. aureus* strains was one-tenth and for *Enterococci* one-half that of oleandomycin. The thermograms showed that both antibiotics interfered with bacterial metabolism at concentrations significantly lower than the MIC (Fig. 5). Amounts of 0·08–0·02 μg ml^{-1} of the antibiotics altered the thermograms of *S. aureus* in an identical way. These two antibiotics show similar clinical effectiveness. Semenitz suggested that differing *in vitro* activity by MIC tests might be caused either by different pharmokinetics or by interference with bacterial metabolism at sub-MIC concentrations.

Examples of the interaction of cellular inhibitors with microorganisms not given above include the work of Mårdh (1978) who has examined the effect of cinoxacin against *E. coli*. Other studies not quoted in the foregoing examples have been carried out on the interaction of drugs with plant cells (Tillbert *et al.*, 1971) and cell organelles (Spink and Wadsö, 1976) using isothermal heat-leak microcalorimetry.

D. Quantitative analysis of drugs

Jensen *et al.* (1976) have used a Tronac isothermal titration microcalorimeter to examine the interaction of *Streptococcus faecalis* growing in complex medium with penicillin G (potassium salt) and tetracycline hydrochloride at 310 K (Fig. 6). Levels of tetracycline down to 2·3 ng ml^{-1} could be determined within 40 min of addition of the culture to the microcalorimeter. It was suggested that very low levels of tetracycline could be determined in human serum by this method. Data on the mode of action of tetracycline were also obtained indicating, according to the authors, that the action of tetracycline involved at least two distinct steps both of which involved zero-order kinetics, and related to transport and/or protein inhibitor binding. Perhaps further experiments are required, however, before such conclusions can be accepted. The experiments of Jensen et al. (1976) can be criticized on the following grounds. The

Oleandomycin 50 μg ml^{-1}

Oleandomycin 10 μg ml^{-1}

Oleandomycin 2 μg ml^{-1}

Oleandomycin 1 μg ml^{-1}

Oleandomycin 0.4 μg ml^{-1}

Oleandomycin 0.2 μg ml^{-1}

Oleandomycin 0.08 μg ml^{-1}

Oleandomycin 0.02 μg ml^{-1}

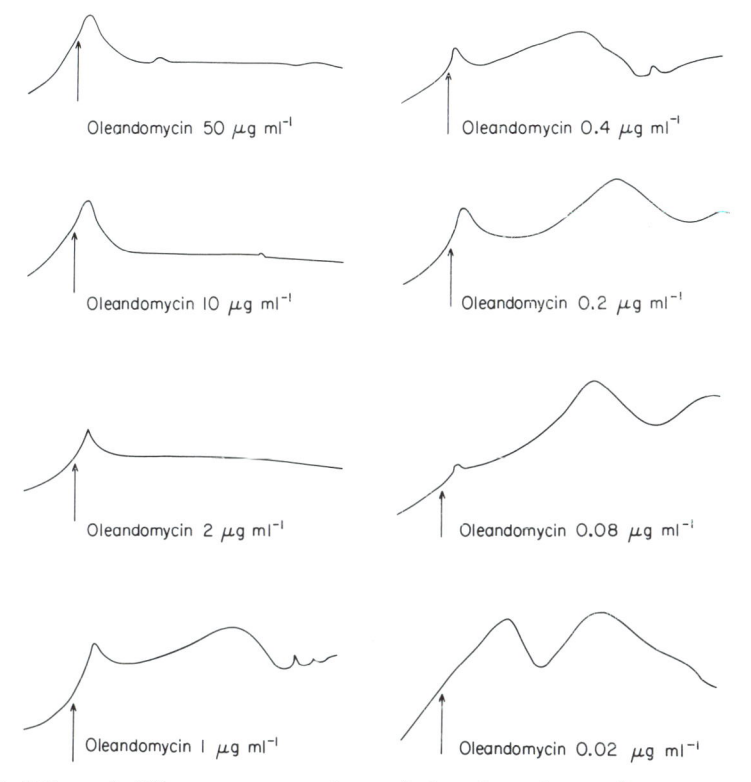

FIG. 5. Effect of different concentrations of oleandomycin on thermograms of *S. aureus* A 13665 (Semenitz, 1978).

composition of the atmosphere in the microcalorimetric cell was not controlled and a reproducible inoculum was not used (no measurements of cell numbers were given). The large volume of DMSO used (4 per cent) may have had an effect on the results obtained, i.e. affected the action of the antibiotics. A large microcalorimetric cell volume was also used. No relationship between dose of antibiotic and microcalorimetric response was given. The final disadvantage is that after introduction of the growth culture into the microcalorimetric cell, time is required to achieve temperature equilibration of the reaction vessel.

The thermal effects of addition of doxycline (0–0·25 μg ml^{-1}) to growing cultures of *Mycoplasma hominis* in a steel ampoule heat-leak isothermal microcalorimeter can be used for the quantitive analysis of the antibiotic.

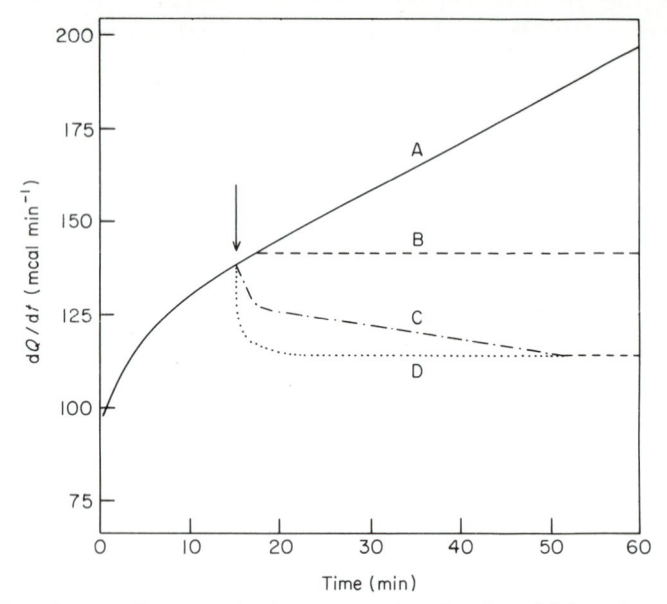

FIG. 6. Plot of rate of heat production versus time for the addition of samples to *S. faecalis* (25 ml of 4 per cent DMSO broth). A, the standard curve. The arrow indicates the injection of (B) 100 μl of 12·6 μg penicillin "G" per μl, (C) 100 μl of 1·2 μg tetracycline HCl per μl and (D) 100 μl 11·8 μg of tetracycline HCl per μl. (Jensen *et al.*, 1976.)

The experiments carried out at 310 K indicated that microcalorimetry may be a valuable technique for studying the antibiotic susceptibility of organisms belonging to the order *Mycoplasmatales*. These experiments were carried out under essentially anaerobic conditions. In such experiments the fact that the antibiotic may not be homogeneously mixed with the cell culture is of concern. However, this does not detract from the fact that the method can be used for quantitative analysis of the antibiotic. The dose–response relationship was not given (Ljungholm *et al.*, 1976).

The interaction of polyene antibiotics (nystatin, filipin, pimaricin, amphotercin B, candicidin and lucensomycin) and synthetic antifungal drugs (clotrimazole, 5-fluorocytosine) with the yeast *S. cerevisiae* NCYC 239 has been examined in a series of studies by Beezer *et al.* (1977a, b).

This microorganism was used because it did not aggregate under the experimental conditions used. This ensured that maximum surface area of cells was exposed to the drugs. An LKB flow microcalorimeter (10700-1) was employed in all the studies.

This series of studies is unique in that they are the only examples of microcalorimetric experiments which have used liquid nitrogen stored inocula. A main factor in microbiological analysis is the critical dependence of the assay system on the inoculum. The metabolic performance of inocula is one of the major sources of variation in all microbiological incubations and in particular is a source of inconsistency in microbiological assays. Variations can be attributed to inoculum size, inoculum history and phase of growth (Beezer et al., 1976).

It has been found that the adaption of a defined cell growth procedure together with rigorous standardization of freezing procedure and storage of cells using liquid nitrogen results in a highly reproducible inoculum (Beezer et al., 1976). The advantages of liquid nitrogen stored inocula include the following: they can be stored in large quantities and over a lengthy time span; the freezing and unfreezing procedures are simple. The experiments are also unique because the interaction of drugs with microorganisms was examined under both growing and non-growing (respiration only) conditions. All experiments were conducted at 303 K (unless stated otherwise) and under anaerobic conditions. The addition of drugs to yeast cells suspended in phthalate buffer (pH 4·5) containing 10 mM glucose resulted in thermograms of the type shown in Fig. 7. 5-Fluorocytosine did not give a response under the conditions of experimentation. This is expected since this antimycotic interferes with DNA synthesis and would therefore only demonstrate its effects in growth medium (Beezer et al., 1977b).

Flow microcalorimetric experiments at 298 K for nystatin raw material with yeast cells gave a linear relationship between a response parameter and the logarithm of the applied dose (Beezer et al., 1977a). Heat-treated nystatin was also analysed. At 303 K, however, the relationship between dose and response found for the polyenes and clotrimazole was linear. The response parameter depended upon the nature of the drug and/or its concentration (Beezer et al., 1977b). Flow microcalorimetry can thus be used for the quantitative analysis of these drugs between 10^{-5} M and 10^{-6} M levels. The quantitative analysis of polyene antibiotics by flow microcalorimetry has advantages compared to the use of the classical disc agar diffusion method (Table 2).

Lucensomycin occupies a border-line position in the microcalorimetric response of yeast cells to polyene antibiotics; it gives two types of responses according to its concentration (Fig. 7). At concentrations below 5×10^{-6}

M per 10^7 cells ml^{-1} it gives a sigmoidal response and is acting as a fungiostatic antibiotic (Beezer et al., 1977b). Above this concentration the thermogram is similar to that of nystatin, amphotericin B and candicidin. At high concentrations, i.e. above 5×10^6 M per 10^7 cells ml^{-1} the antibiotic is fungicidal. Lucensomycin differs from the polyene antibiotic pimaricin structurally only in the alkyl side-chain ($CH_3(CH_2)_3-$ for the former; CH_3- for the latter). Pimaricin, however, gives a sigmoidal

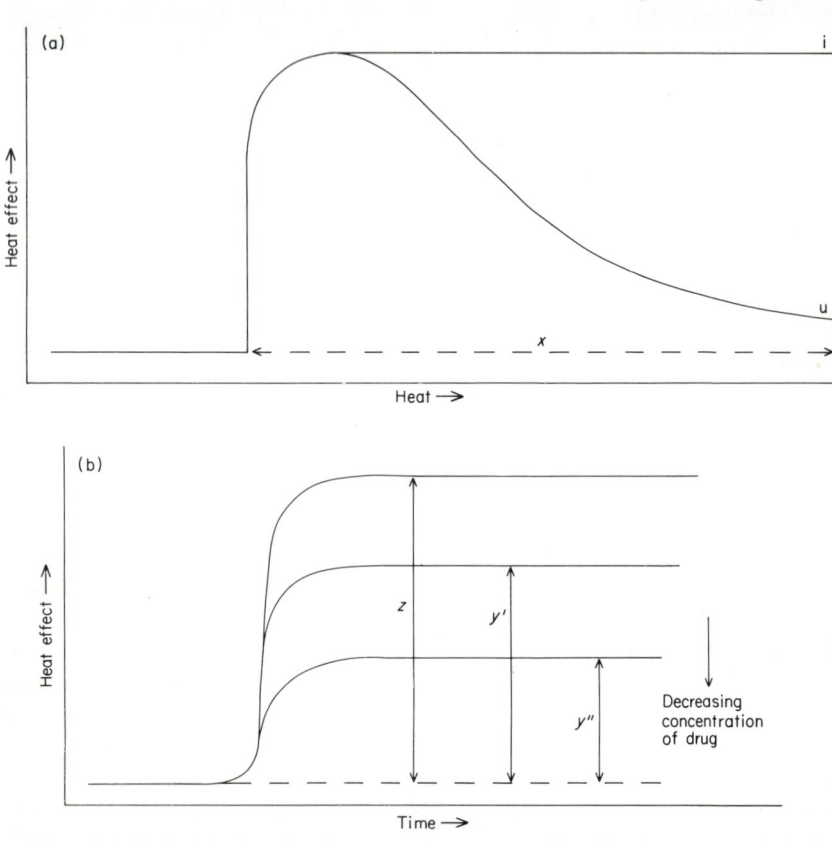

FIG. 7. Typical thermograms of the interaction of polyene antibiotics with *S. cerevisiae* NCYC 239 respiring in phthalate buffer (pH 4·5) containing 10 mM glucose. (a) Yeast cells alone (i); yeast cells in the presence of the polyene antibiotic nystatin (ii). (b) Yeast cells in the presence of the polyene antibiotics pimaricin, lucensomycin and the antimycotic drug clotrimazole. x is the response parameter, time in minutes for nystatin, amphotercin B, candicidin, filipin and lucensomycin at high concentration. z is the percentage deflection or dQ/dt for yeast alone. y' and y is the percentage deflection of dQ/dt in the presence of pimaricin, lucensomycin at low concentration and clotrimazole. (B. Chowdhry, Ph.D. Thesis, London University, 1979.)

TABLE 2

Comparison of microcalorimetric and plate agar diffusion methods for polyene antibiotic assay

	Microcalorimetry	Agar plate diffusion test
Reproducibility (%)	3	5–10 (cf. Cosgrove, 1979)
Lowest determinable concentration (unit ml^{-1})	0·1	20
Range (unit ml^{-1})	0·1–100	20–100
Time per assay (h)	1	16
Capacity for automation	High	Automation not very successful
Convenience and Simplicity	High	High
DMF in experiments (%)	0·3	100

microcalorimetric response at all concentrations. This is an example of how small changes in structure can affect the biological activity of drugs, the change in activity being reflected by changes in thermograms.

The microcalorimetric experiments allow some measure of the relative rates of the reactions of the drugs with yeast cells. At equimolar concentrations the drugs appear to have the potency order described in Table 3. The potency judgement used here is that of "time to kill", the most potent drug being the one which at equimolar concentrations inhibits the respiration of the standardized liquid nitrogen stored inoculum in the shortest time. The order for equimolar concentrations of the antibiotics is different from that obtained when half the MIC concentrations of the drugs were applied (Table 3). The "normal" ranking order of MICs is as shown in Beezer (1977b). There is therefore here a problem of the definition of potency. This problem of inconsistency of results of the activity of antibiotics when flow microcalorimetric experiments are compared with MICs obtained by classical methods has also been found for other antibiotics by Wadsö (see Mårdh et al., 1976; Semenitz et al., 1977a). The MIC is currently regarded as the criterion for antibiotic evaluation. Mårdh et al. (1976), for example, found that the MIC for tetracyclines (minocycline, doxycycline, oxytetracycline and tetracycline) by the broth dilution technique was the same. However, their bioactivity as measured by microcalorimetry differed. Greenwood (1976) has discussed the disadvantages of the use of MIC values. Microcalorimetry has provided information which adds further to the need to question the

TABLE 3

Comparison of potency rankings for half MIC and equimolar solution of polyene antibiotics

Antibiotic	Minimum inhibitory concentration (μg ml^{-1})	Assumed MW	$\frac{1}{2}$ MIC[a]	Equimolar solutions
Candicidin	0·03	1200	1	2
Amphotericin B	0·09–0·5	924	2	1
Nystatin	0·8–3·1	926	5	5
Filipin	5·0	654	3	4
Pimaricin	0·9–15·0	665·7	4	3
Lucensomycin	—	707·7	—	Between 4 and 3

Microcalorimetric potency ranking (concentration)

[a] Half MIC concentration taken was half the lowest value in the quoted range.
Note the wide variations in MICs. This is characteristic of interlaboratory results for the assay of polyenes—see, for example, the WHO report (Lightbawn, 1963) establishing the international standard for nystatin.

validity and uses of parameters for antibiotic assays obtained by classical microbiological methods.

Beezer also examined the effect of pH, metal ions (calcium and magnesium) and sterols (ergosterol and cholesterol) on the respiration of yeast cells (Beezer et al., 1980b).

Pharmaceutical formulation excipients had a negligible effect on the action of nystatin as measured by microcalorimetry. Tetracycline hydrochloride which is present in Mysteclin tablets shows only a small synergistic effect when studied in combination with bulk nystatin raw material alone. The magnitude of its "potentiating" effect was only just outside the error limits (3 per cent) attached to the experiments. Nystatin complexed with polyvinylpyrrolidone showed less activity against yeast cells (Beezer et al., 1977a) as measured by flow microcalorimetry than it did as measured by the agar plate diffusion test. The bioactivity and quantitative assay of N-acetyl-nystatin has also been carried out using flow microcalorimetry (Beezer et al., 1980a).

For the polyene antibiotics the form of the thermograms as a function of temperature (Beezer et al., 1977b) and comparison with those reported previously (Beezer et al., 1977a) suggest that the kinetics and hence the mechanism of the interaction process is exceedingly complex. However, the processes to which these thermograms relate include, at least, adsorption, diffusion and enzyme inhibition which are not, as yet, very clearly understood.

The order of activity of polyene–antibiotic–yeast cell interactions differ between yeast cells which are not growing (respiring only) and those actively growing (Beezer et al., 1980c). The effect of polyene antibiotics on S. cerevisiae NCYC 239 at 303 K under anaerobic conditions and growing in rich medium is shown in Fig. 8. The antibiotics were added

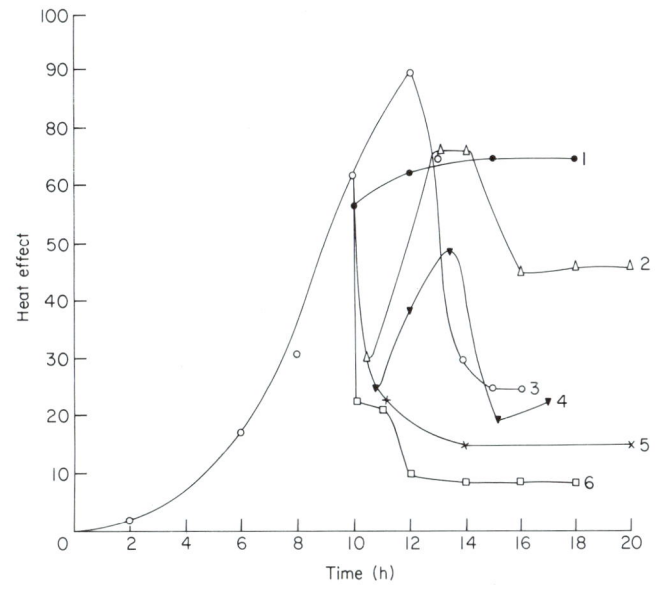

FIG. 8. Effect of polyene antibiotics on the growth thermograms of S. cerevisiae NCYC 239 growing in a complex medium under anaerobic conditions. 1, Pimaricin; 2, amphotercin B; 3, no antibiotic; 4, candicidin; 5, nystatin; 6, lucensomycin. (B. Chowdhry, Ph.D. Thesis, London University, 1980.)

during the exponential growth phase of the yeasts. The order of activity of the antibiotics obtained is lucensomycin, nystatin, candicidin, amphotericin B and pimaricin. It is noteworthy that lucensomycin at high concentration (10^5 M per 10^7 cells ml^{-1}) shows a biphasic activity. The inhibition of growth of yeast cells by polyene antibiotics can also be used for the quantitative assay of the antibiotics (Beezer et al., 1980c).

The antibacterial actions of cephalexin and cephaloridin against E. coli (0:119, H, 19) and S. aureus (483) have been studied by flow and ampoule microcalorimetry (Arhammer et al., 1978). The experiments were carried out at 37 °C using undefined growth media, the antibiotics being added to the cells in the logarithmic growth phase. MIC tests were also conducted to study the microorganism–antibiotic reactions. The addition of cephalexin 2·5 μg ml^{-1} (5 × MIC) to *S. aureus* caused a decrease in heat production. Similarly when 9·0 μg ml^{-1} (2 × MIC) of this drug was added to cultures of *E. coli* heat production decreased. Two to three hours after the drugs had been added no heat effects could be registered for the following 6–8 h after which an increase in the heat production occurred. A direct relation between drug concentration and response, that is heat effects produced, was found when increasing concentrations of cephalexin, i.e. 1–50 μg ml^{-1} (2–100 × MIC) were added in the logarithmic growth phase to cultures of *S. aureus*. The effects of heat production by microorganisms were related to the influence on viable counts, pH and optical density of the medium.

E. Model studies of cell-modifier interactions

Isothermal heat conduction batch microcalorimetry has been used to examine how the hydrolysis of exogenous adenosine triphosphate (ATP) by tissue cells is modified by inhibitors of the action of ATP (Kemp, 1972). Inhibitors including ouabain, oligomycin, mersalyl- and sulphydryl-blocking agents were used in microgram quantities. Chick fibroblastic cells suspended in Eagles MEM at 310 K were used as test cells. The results of the experiments indicated that the hydrolysis of exogenous ATP by cells can be arrested by the inhibitors. The study is of interest *per se* because it indicates that the activity of an ouabain and oligomycin-sensitive ATPase usually considered to be orientated on the inner border of the plasma membrane might be affected by exogenous ATP. Because of the complexity of biological systems, model studies have been carried out to give some insight into ill-understood biological phenomena. Many such studies have been carried out by microcalorimetry. The binding of molecules such as 5-methoxy-tryptamine (an acridine dye) to deoxyribonucleic acid (DNA) (calf thymus) and poly A is an example. The thermodynamic parameters for the binding of 5-methoxy tryptamine to DNA and poly A have been examined by mixing DNA or poly A (in sodium cacodylate 1 mM EDTA, pH 7·0) with the dye. An LKB batch microcalorimeter was employed. The results of the experiments are shown in Table 4. It can be seen that there is virtually no change between the thermodynamic parameters determined at 298 K and 313 K for DNA. Thus the variation in C_p ($C_p = d\Delta H/dT$) is zero, indicating that the structure of the complex does not change between these two temperatures. The authors indicated that aromatic amino acids could play a special role

TABLE 4

Thermodynamic parameters for the interactions of 5-methoxy-tryptamine with DNA and poly A

	K	$K \times 10^{-4}$ M^{-1}	ΔH kJ mol^{-1}	ΔS J mol^{-1} K^{-1}
DNA	298	2	−10·2508	46·024
	313	1·6	10·46	46·024
Poly A	298	0·5	8·368	92·048

in the formation of protein acid complexes due to the direct interaction between bases and the aromatic ring (Durand, 1976).

The ion selectivity of neutral carrier antibiotics is mainly due to selective complex formation of these antibiotics with metal cations (Urry, 1975). Various methods have been used to determine the complex formation constants of carrier antibiotic–metal complexes. The advantage of microcalorimetry is that it permits the simultaneous determination of the enthalpy ($\Delta H°$), the free enthalpy ($\Delta G°$) and the entropy ($\Delta S°$) of complex formation. The technique requires only very small quantities of the antibiotics. Früh et al. (1971, 1972) have investigated the influence of solvents and ligand structure on the thermodynamics of complexation reactions between metal ions (e.g. K^+ and Na^+) and ionophores (valinomycin, nonactin, monensin, dibenzo-18-crown-6 and pentaglycine) by batch microcalorimetry. The reactions were carried out by mixing the ionophores with metal thiocyanates. The thermodynamic parameters for the interaction of nigericin and monensin with sodium and potassium ions by batch microcalorimetry (Table 5) were found to be in good agreement with results obtained by e.m.f. and relaxation techniques. The

TABLE 5

Thermodynamic parameters for the interactions of metal ions with nigericin and monensin

Antibiotic	Metal iron	ΔH^0 kJ mol^{-1}	ΔG^0 kJ mol^{-1}	ΔS^0 J mol^{-1}	Log K	K M^{-1}
Nigericin	NA$^+$	+6·9 ±11%	−22·2	+98	3·9	8 × 10^{-3}
	K$^+$	−4·7 ±7%	−32·0	+93	5·6	4 × 10^{-5}
Monensin	Na$^+$	−16·2 ±2%	−34·3	+61	6·0	1 × 10^6
	K$^+$	−15·6 ±2%	−26·0	+35	4·6	4 × 10^{-4}

selectivity order $K^+ > Na^+$ (nigericin) and $Na^+ > K^+$ (monensin) was found. This ionic selectivity is the same as that observed in biological systems. It was found that complex formation constants varied with the dielectric constant of the solvent. Trends in $\Delta S°$ were discussed in terms of translation entropies relating to solvent molecules around cations. The influence of macrocyclic structure on ion selectivity was also discussed. Studies such as those cited above have given insight into the mechanism of action of synthetic and natural ionophores. Christensen *et al.* (1977) have carried out similar experiments to those of Simon *et al.* (1972) on crown ethers. The thermodynamics of lipid–protein interactions have been studied by isothermal heat leak batch microcalorimetry (Rosseneu *et al.*, 1976). Such experiments may be of value in gaining further understanding of the membrane which in turn could help in the understanding of drug–membrane interactions.

F. Advantages of isothermal microcalorimetry in analysis of drug activity

Microcalorimetry, particularly flow microcalorimetry, is of general application to the analysis of the bioactivity of both antibacterial and antifungal drugs. Organisms belonging to the order *Mycoplasmatales* can also be examined for drug sensitivity by, for instance, ampoule microcalorimetry. Besides microcalorimetry, other methods for recording growth and metabolism in microbial cultures have certain disadvantages as outlined by Eriksson (1977). Microcalorimetry, because of the universality of its application and its ability to detect subtle changes in metabolism of microorganisms, may give information not indicated or difficult to obtain by other methods that use one specific reaction property for observation in the study of microorganism drug interactions.

The microcalorimetric observations of cell–modifier interaction makes no demand upon the mechanism of action of the drug save that there be an action of the drug that leads to a change in the heat effects observed. Moreover, differences seen in qualitative thermograms may be strongly indicative of a differing mode of action or show that the overall process differs quantitatively between, for example, closely related drugs. Information on the rate of interaction of the drug with microorganisms is contained within thermograms. Such information is not often easily attained by conventional microbiological methods. To date it has not been possible, however, to deduce any simple mathematical relationship between applied drugs and responding cells. This is largely due to the fact that component reactions of drug–cell interactions, e.g. adsorption, diffusion, enzyme inhibition etc., are not well understood at present.

The potentially high reproducibility, sensitivity, speed, simplicity and capability of automation as well as convenience are other advantages of microcalorimetric techniques compared to classical microbiological techniques. Microcalorimetry can distinguish between those drugs which inhibit respiration and those which inhibit growth processes. One microcalorimetric technique, flow microcalorimetry, has the advantage that it mimics to some extent the conditions found in the body in that a flowing stream of substances are involved. The physico-chemical basis of some microbiological techniques such as the plate-diffusion test are ill understood. Microcalorimetric experiments are, however, potentially easier to understand from a physico-chemical basis. Microcalorimetric techniques also have advantages compared to other physico-chemical techniques such as the use of ion-selective electrodes. This is that microcalorimetry is of universal application whereas other techniques are limited. For example, ion selective electrodes can only be used for drugs that cause an efflux of ions from the cells. Moreover, microcalorimetry is one of the few techniques that measures the overall metabolism of the cell in a non-invasive manner. Other techniques, however, often measure a component reaction only. Microcalorimetric methods do not require transparent solutions as do many analytical procedures and this is of particular importance in such fields as microbiology.

G. Future studies

Many more studies of the interaction of microorganisms with natural and synthetic drugs by microcalorimetry are to be expected in the future. Further model system studies, for example, the effect of drugs on cell-free protein systems using batch microcalorimeters (Berthe-Corti, 1977) and the use of flow microcalorimeters as titration apparatus, are also expected. Thermodynamic parameters for the interaction of drugs such as duanomycin with DNA and poly A etc. would be useful. The study of the interaction of drugs with cellular systems other than microorganisms is also envisaged, using isothermal heat leak microcalorimetry. Wadsö has conducted exploratory experiments of the growth of HeLa cells using microcalorimetry (Wadsö, 1978). In the future it may be possible to assay potential anticarcinogenic molecules quantitatively by this technique.

The examination of the interaction of erythrocytes with drugs by isothermal microcalorimetry is also an interesting possibility (e.g. polyene antibiotics). Information concerning possible toxicity (or side-effects) of drugs may be obtainable. Many microcalorimetric studies of erythrocytes have been conducted in recent years (Levin, p. 131). Drugs are being increasingly used in combination during clinical therapy (Anonymous, 1978). Microcalorimetric techniques could be significant in giving information as to the effectiveness of drugs used in this mode. As yet few examples of such experiments exist.

Further improvements in the technological aspects of microcalorimeters will probably occur in the future. Microcalorimeters which are multichannel require to be available commercially, particularly for use in clinical medicine. A microcalorimeter in which the microcalorimetric cell functions as an incubation (fermentation) vessel may be very useful for microorganisms and microorganism–drug interactions. Such a calorimeter has been designed and tested (Fujita *et al.*, 1976). However, further improvements in the design of microcalorimeters of this type are envisaged. Such a microcalorimeter would have many advantages over presently used flow microcalorimeters. Since there would be no lag time from the addition of the antibiotics to the incubation vessel and the response, the whole reaction between microorganisms and drugs would be observed. The above development could be combined with a microcalorimeter in which the heat sensors are sensitive enough to detect the heat output of 10^2–10^3 cells ml^{-1}. This would decrease the time required to examine clinical specimens of microorganisms from body fluids in relation to, for example, the sensitivity of microorganisms to antibiotics which are to be used for clinical therapy. More sensitive microcalorimeters would also enable smaller concentrations of antibiotics to be examined than is presently possible. Finally, very sensitive microcalorimeters are required in order to study antibiotic kinetic phenomena in microorganisms.

In the future more defined growth medium will probably be used during studies of microorganisms. There is a need for more uniformity in the experimental parameters used during cell–drug interaction studies. This includes factors such as number of cells per millilitre, temperature, incubation medium constituents, prehistory of microorganisms, e.g. cultivation conditions, etc. This would make comparisons between antibiotics which have similar modes of action and the antibiotic sensitivity of different microorganisms easier. If an increase in heat sensitivity of microcalorimeters were combined with ability to have multichannel instruments then such microcalorimeters could be of use to the clinician. The resistance of microorganisms to drugs could be examined by microcalorimetry too. Future studies of drug–microorganism interactions by microcalorimetry could be carried out in the presence of plasma or serum components. Such studies may reveal more about the pharmokinetics of drugs.

H. Differential scanning calorimetry (DSC)

The precise biological and pharmacological actions of drug molecules are largely undetermined (Triton et al., 1977). In spite of widespread interest little direct attention has been paid to the problem of interaction of molecules with cell biomembranes, their components or related model systems. Evidence to support the membrane as the site of action of various types of molecules (including drugs) was, in the past, largely derived from the observation that these agents partition into or bind to and exert their effect on isolated cell plasma membranes. For several drugs their biological activity has been correlated to their lipid solubility and/or to their lipid (n-octanol–water–buffer) partition coefficients. There are, however, limitations to this approach (Jain et al., 1977). Evidence from techniques such as radioactive labelling and radio-immunofluorescence have helped to establish the presence of, and in some cases the position of, exogenous molecules in biomembranes. Such techniques do not necessarily establish that interaction of the exogenous molecules with membrane components occurs and that it leads subsequently to the biological action of the drugs.

There is a widening belief that interaction with the lipid constituents of the membranes (resulting in an alteration of the physico-chemical characteristics of lipid) may play an important role, direct or indirect, in the mode of action of many drugs (Chapman, 1975). The drug–lipid interaction may in turn elicit or modulate a variety of biological and physiological phenomena such as catalytic and transport functions of the cell. More specifically, drug molecules can affect the thermotropic phase transition of lipid–water (liposome) systems which are used as models of biological membranes (Bangham, 1971). Such drugs include tranquillizers, antidepressants, anaesthetics and antibiotics.

The cooperativity of interaction of lipid acyl chains in cell membranes shows significant long-range (over several molecular diameters) order–disorder (gel (solid)→liquid crystalline phase) transitions. These transitions have now been well characterized both in isolated lipids, lipids in liposomes and lipids in isolated cell membranes (cf. Chapman and Bach, p. 275). Lipid phase transitions are endothermic and can be detected by a variety of physico-chemical techniques including DSC, e.s.r., n.m.r., X-ray diffraction, etc. (see Lee, 1975) as well as biological techniques (Fourcans and Jain, 1975). It has been argued that the overall course of lipid phase transitions is best described by thermodynamic methods (Melchoir et al., 1977). In order to obtain an understanding of the molecular events which occur during such transitions, however, it is best to use as many of the above techniques as possible. Studies of lipid transitions in the presence or absence of drug molecules has been carried out using either commercially available DSC apparatus (e.g. Perkin–Elmer DSC 1B and DSC 2B; Perkin–Elmer, Norwalk, Connecticut, USA) or modified commercial instruments (Melchoir et al., 1977). Melchoir et al. (1977) have dealt with the reasons for, and the requirements of DSC apparatus suitable for biological studies. Such instruments require to be very sensitive since the enthalpy of the order–disorder transition is between 1 and 10 cal g^{-1} of membrane lipid. Amounts of 10–100 μg of lipid are usually used in experimental studies. Order–disorder transitions in biomembranes can occur over a temperature range as broad as 303 K to 313 K, starting and ending very gradually. Often these transitions occur at temperatures as low as −30 °C or −40 °C (243–233 K). DSC allows the heat of transition of phospholipids to be readily measured and a peak is obtained the shape of which reflects the extent of the transition as a function of temperature. The technique is very sensitive to minor changes in the position and shapes of peaks.

In interpreting phase transition data a large number of parameters are derived from phase transition profiles; half-height width, area, shift at the half height, temperature at the beginning (T_c) and the end of the transition (T_e) for each peak. Other parameters may also be derived from the thermograms (Jain et al., 1977). Observations are made therefore concerning whether or not drugs remove the (pre)transition endotherm, decrease or increase the temperature at which the phase transition occurs, the range over which transition occurs, the heats of transition involved and the number of peaks in the presence and absence of drugs. The width of transition at half-height is taken as a qualitative measure of the extent to which the added drug can widen the temperature range over which both fluid and gel phase can coexist. It has been claimed that DSC data can give information as to whether the whole bilayer or only a part of it undergoes a change in fluidity (Eliasz et al., 1976). Thus from the type of transition profile one can not only determine whether or not an additive would "fluidize" (Chapman, 1975) or "solidify" the bilayer, but a modified profile also provides the information about the thermodynamic properties of the new phase (Melchoir et al., 1977). Such experimental parameters combined with information obtained from other techniques has been used to give an indication of:

1. how the presence of the drug may alter the physico-chemical properties of the lipid;
2. the nature of the lipid–drug interaction;
3. the position of drugs in lipid bilayers;
4. the biochemical and physiological implications of drug–lipid interactions in cell membranes.

1,2-Dipalmitoyl phosphatidylcholine (1,2-DPPC), a common phospholipid component of eucaryotic (and some procaryotic) cell membranes, undergoes two endothermic phase transitions (Eliasz, 1976) which occur at 34·5 °C (307·5 K)—heat of transition 7·1128 J mol^{-1}, pretransition endotherm—and a main transition at 41·5 °C (314·5 K)—heat of transition 34·31 J mol^{-1}. 1,2-DPPC has been used in many drug–lipid phase transition studies. Other phospholipids or their derivatives either synthetic or extracted from cell membranes have also been used. Besides drug–phospholipid interactions, tertiary mixtures of drug and phospholipid and another component, for instance, sterols or metal ions, have also been investigated (Chapman, 1975). Occasionally whole cell membrane–drug interactions have been studied too (Melchoir et al., 1977). The importance of phase transitions in lipids to the mode of action of drugs and antibiotics has been shown most successfully in the case of the ionophores, gramicidin A, nonactin and valinomycin. Gramicidin A, a polypeptide ionophore, is thought to transport ions by means of a pore mechanism (Chapman, 1975). With this molecule, the pretransitional peak is affected at low polypeptide concentrations, suggesting that the packing of the lecithin polar group has been affected. In addition to this the heat involved in the main lipid endothermic transition is markedly reduced. It has been concluded (Chapman, 1975) that the molecule is interdigitated among the lipid chains preventing chain crystallization from occurring. This result is also interesting in view of the results on black lipid membrane systems. Krasne et al. (1971) have shown that gramicidin A was able to mediate potassium ion transport above and below the transition temperature of the lipid, forming a black lipid film. It is possible that, whilst the bulk of the lipid below the transition temperature was rigid, the lipid immediately adjacent to the gramicidin was fluid.

Nonactin and valinomycin are observed to act as ion carriers (Stark, 1972) and so must be able to diffuse freely throughout a lipid bilayer. The effectiveness of the ion carriers is found to decrease drmatically below the temperature of the lipid systems studied (Krasne et al., 1971). In black lipid membranes (constituted of glyceryl dipalmitate and glyceryl distearate in decane), valinomycin was not capable of increasing the conductivity of the films at temperatures below the transition temperature of the lipids. Experiments on *Acholeplasma laidlawii* also confirmed this conclusion (in the region of the gel-to-liquid crystalline phase transition). Valinomycin-induced efflux of K$^+$ is (nearly) zero below the temperature of the gel–liquid crystalline phase transition. In the temperature range above the phase transition the valinomycin inducible K$^+$ leakage gradually increased with increasing temperature (Fig. 9) (Van Deenen et al., 1976–77). Recent studies have shown that the permeability properties of bilayers of phospholipids to small molecules start to increase just below the main transition and reach a maximum at the transition temperature. Passive permeation of non-electrolytes through *A. laidlawii* membranes at temperatures in the range of lipid phase transition appears to occur predominantly in the regions of the membrane that are still in the liquid

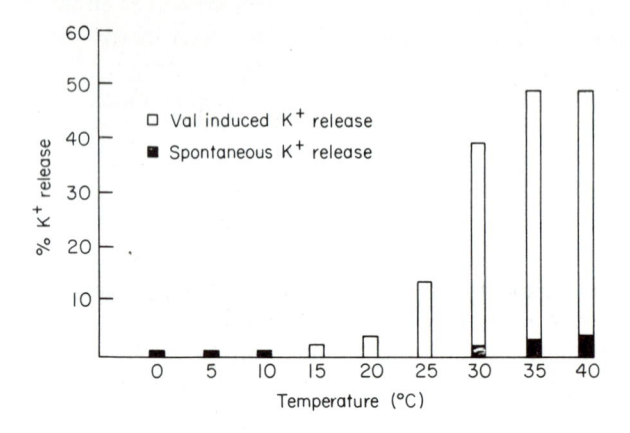

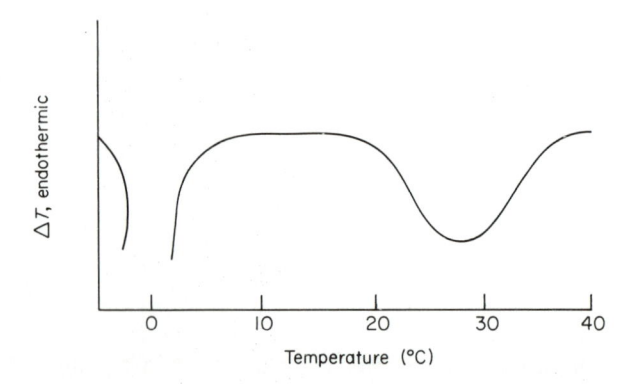

FIG. 9. Leakage of K^+ from cells of *A. laidlawii* in relation to the lipid phase transition (Van Deenen *et al.*, 1976–77).

crystalline state (Van Deenen *et al.*, 1976–77). The lipid fluidity of cell membranes has also been shown to be an important parameter in various other functions of biological membranes, beside permeability properties (see, for example, Trauble, 1975).

I. Examples of studies of model–membrane–drug interactions by DSC

Jain et al. (1976) have pointed out that the relative activity sequences obtained for a large number of adamantane, protoadamantane and homoadamantane derivatives (Fig. 10) cannot be accounted for by simple

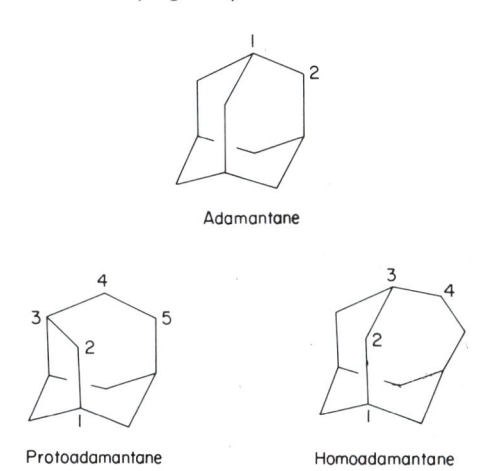

FIG. 10. Structures of adamantane derivatives used by Jain et al. (1976). Typical substituent groups included oxygen, hydroxyl, halogen and aliphatic functions. The derivatives used included: 1, Substituted adamanatanes; 2, substituted adamantanes; 2, substituted protoadamantanes; 5, substituted protoadamantanes; 2,5,-disubstituted protoadamantanes; 2, substituted homoadamantanes; 4, substituted homoadamantanes. (Jain et al., 1976.)

considerations of n-octanol–water partition coefficients, substitution constants based on free energy relationships or the relative polarities and sizes of substituent groups. The interaction of adamantane derivatives with DL-1,2-dipalmitoyl phosphatidylcholine–water system measured by DSC shows, however, that the cage compound derivatives modify the phase properties and under some conditions may induce a phase separation in the doped bilayer. For these derivatives the temperature at the beginning of the lipid phase transition decreases and the half-width increases—as the concentration of additive increases. The broadening of the phase transition profile depended on the concentration and molecular structure of the additive. It was concluded that the position and orientation of a solute within the lipid bilayer is a critical factor in determining its relative potency. Moreover, it was found that relatively minor modifications in the structure of a solute significantly altered its impact on the phase transition behaviour of a bilayer.

Jain and Wu (1977) have also studied the effects of a large number of other drugs on the phase transition behaviour of DL-dipalmatoyl–lecithin bilayers (Fig. 11; Table 6). The effect of the drugs ("additives") on the phase transition behaviour of the lipid could be categorized into certain groups. It was concluded that the type of effect on the phase transition profile depends on the nature of the additive, whereas the extent of the effect depends on the concentration of additive. Their observations were

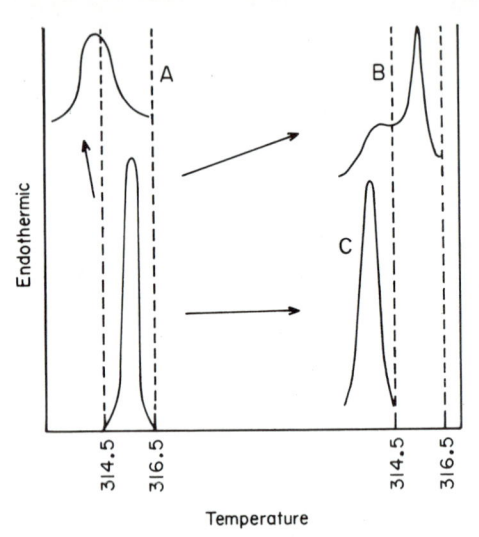

FIG. 11. Differential scanning calorimetric transition profiles for dipalmitoyl lecithin liposomes doped with various drugs (Table 6). Three types of profiles have been observed for the drugs. In type A, an asymmetrical broadening of the main peak is observed. This broadening is accompanied by an increase in the half-height width and a decrease in T_c. In type B profiles, the area of the main peak decreases, and the area of the shoulder at lower temperatures increases with increasing drug concentration. T_c for the main peak seems to remain unchanged, whereas T_c for the shoulder decreases at higher drug concentrations. In type C profiles, the main peak shifts to lower temperatures at successively higher drug concentrations. No broadening of the peak is observed. The total area under the curves (enthalpy of transition) seems to remain constant in all these profiles. In the temperature range for transition, several phases may coexist within the plane of the bilayer. The modified transition profiles are thought to be related to changes in bilayer fluidity. (Jain et al., 1975.)

consistent with the hypothesis that the type of effect induced by additives that are structurally related is qualitatively the same, but the magnitude of the effects is different for different compounds in the same series at the same concentration. The type of effect induced by an additive on the phase transition profile of the bilayer is thought by these workers to be

TABLE 6

Localization of drugs in lipid bilayers as indicated by DSC studies

Drug	Localization of drug in lipid bilayers
1. Alkanols	C_1–C_8 methylene (Fig. 12) C_9–C_{16} methylene
2. Uncouplers: 2,4-Dinitrophenol, picric acid, carbonyl, cyanide, phenylhydrazone, m-CLCPP, p-trifluoromethoxy-CCP, tetrachloro-trifloromethyl-benzimidazole	Glycerol backbone
3. Ionophores: i. Valinomycin, tetraphenylboron, dicyclohexyl-18-crown-6, Di-t-butyl-18-crown-6	C_1–C_8 methylene
ii. Monensin, nigericin, dianemycin, A 23187	Glycerol backbone
4. Local anaesthetics: procaine, nesacaine, benzocaine, butacaine, tetracaine, dibucaine, xylocaine, marcaine	C_1–C_8 methylene
5. Tranquillizers: prochloroperazine, trifluoroperazine, trimeprazine, chlorpromazine	C_1–C_8 methylene
6. Inhalation anaesthetics: diethyl ether, chloroform, halothane, fluorane, enflurane	C_9–C_{16} methylene

related to the position of localization (Table 6) and orientation of the additive along the thickness of the bilayer, localization in turn being determined by, amongst other factors, the presence of the polar and apolar groups in the compound—and by their geometric arrangements within the molecule.

Chapman and co-workers (see Frenzel et al., 1978) have put forward an empirical rule that compounds that penetrate the interior of the lipid bilayer and disrupt the chain packing result in a lowering of the heat of the lipid phase transition. Those drugs that remain at the surface of the bilayer and interact electrostatically with the polar headgroups of the lipid primarily affect the transition temperature.

Other examples of studies of model–membrane–drug interactions are given in Table 7.

Local anaesthetics are one of the best examples of a group of drugs which have been shown to have an effect on the fluidity of phospholipids by a large number of techniques. Studies on the interaction of local anaesthetics with artificial lipid membranes, including monolayers, black lipid films and phospholipid vesicles, have established that there is a good correlation between the ability of local anaesthetics to interact with phospholipids and their ability to modify membrane properties. Such techniques include

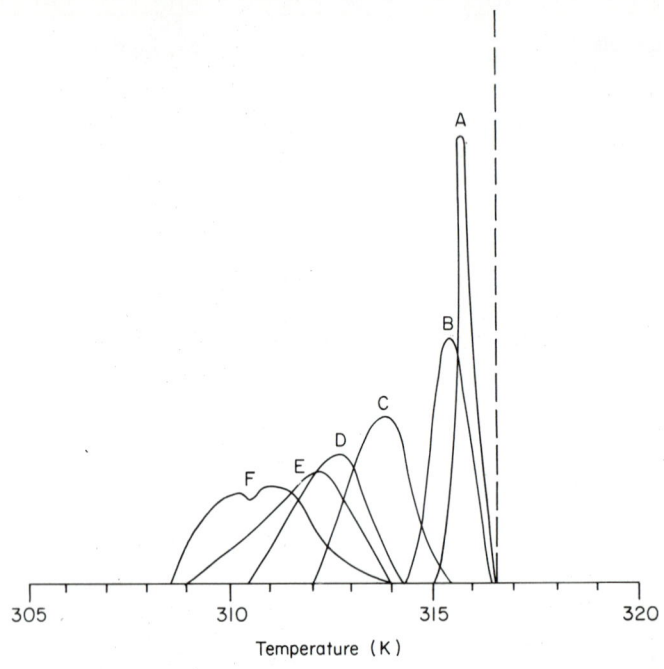

FIG. 12. Phase transition profiles for dipalmitoyl lecithin liposomes modified with successively higher concentrations of n-hexanol. The profiles are broadened and both T_c and T_e shift towards lower temperature. The behaviour illustrated by this set of profiles is qualitatively typical of those induced by C_5 through C_{10} alcohols. Hexanol: A, 0 mM; B, 2·5 mM; C, 7·5 mM; D, 12·0 mM; E, 15·0 mM; F, 19·5 mM. (Jain et al., 1977.)

DSC, n.m.r. and e.s.r. However, this fluidizing effect was only observed at anaesthetic concentrations above the concentration sufficient to induce anaesthesia. It has been found that anaesthetics such as cannibinol in the presence of dipalmitoylphosphatidylcholine-cholesterol may have a condensing instead of a fluidizing effect on lipids. Other anaesthetics such as pentobarbital condense or fluidize the lipid according to the cholesterol content (Pang et al., 1978).

DSC is also widely used for the qualitative (purity) analysis of drugs (Zynger, 1975; Liptay, 1973; Ferrari, 1974; Karoly, 1976; Nikolics, 1976;

TABLE 7

Examples of lipid drug studies using DSC

Drug	Lipid or cell membrane	Effect of drug	Conclusion drawn	Reference
n-Alkanols (e.g. butanol, ethanol, hexanol, decanol, octanol, tetradecanol, and dodecanol)	1,2-Titetradecyl-sn-glycero-3-phosphoryl-choline in excess water	From C_8 to C_{11} increasing molar proportions up to 1 : 1 raised T_c, magnitude of this effect being greater for the longer chain alkanols. At higher molar proportions than 1 : 1 T_c decreased, this effect becoming more pronounced with shorter chain lengths. Butanol, ethanol have no effect on T_c	Effect of n-alkanols dependent on chain length can be interpreted, in part, on the basis of their distribution coefficients between water and n-alkane	Hui (1973)
Dibucaine	DPPG and DPPC residues	Dibucaine produces a significant reduction in gel–liquid transition temperature. Effect more marked in presence of calcium. Neutral lipid (DPPC) only affected at high concentrations of dibucaine (Fig. 13(a), (b))		Papadhadjopoules et al. (1975)
Aspirin, desipramine, chlordiazepoxide, procaine, chlorpromazine	1,2-DPPC	At concentrations as low as 2% mol mol^{-1} chlorpromazine removes pretransitional endotherm.	Chlorpromazine causes lipid to alter tilt of hydrocarbon chains; affects the long-range	Cater et al. (1974a)

TABLE 7 (*cont.*)

Drug	Lipid or cell membrane	Effect of drug	Conclusion drawn	Reference
		Increasing concentration of chlorpromazine progressively lowers the temperature of the main endotherm until at 50% mol mol^{-1} multiple endotherms occur. All drugs lower transition temperature of lipid (Fig. 13). Even at molar/molar ratio heat of transition is not markedly decreased (Fig. 13(c))	organization of both lipid chains and polar head groups	
Trimethylalkyl ammonium iodides ($C_6C_9C_{12}C_{18}$)	1,2-DPPC	Shift of melting temperature towards lower values. Maximum effect by C_9 and C_{12}	—	—
Anthracycline antibiotic derivatives: (1) daunomycin derivatives; (2) adriamycin derivatives	1,2-DPPC and phosphatidyl serine (PS)	Adriamycin derivatives effect 1,2-DPPC thermotropic behaviour. Daunomycin produces decrease in melting temperature of 1,2-DPPC. All derivatives decrease enthalpy of melting between 15% and 25%. Daunomycin causes shift in melting	Suggested that drugs fluidize membranes according to structure	—

		temperature of PS. Other antibiotics cause decrease in enthalpy of melting of PS, not melting temperature	
Inhalation anaesthetics (cyclopropane)	Phosphatidyl-serine bilayers	Two new peaks with lower enthalpy of melting and at lower temperature. In presence of Ca^{2+} main peak not shifted; additional peak appears at lower temperature 66·5 °C→63·0 °C and 60·5 °C (pH 3·1), in presence of Ca^{2+}, 66·5 °C and 58·0 °C in presence of nitrogen	In the presence of an inert gas the ionic environment of a lipid can moderate the effect a particular inert gas may have. Presence of inert gas can cause calcium to dissociate apparently from the bilayer to form a separate calcium phase	
DMF, DMSO, pyridine-N-oxide, tetramethylurea (inducers of cell differentiation)	Dimyristoyl phosphatidyl glycerol liposomes	All drugs cause appearance of new transition at higher temperature. Divalent cations enhance effect. Local anaesthetics inhibit effect	Tendency to promote less fluid or more ordered membranes. Parallel studies with Friend leukemia cells suggest that induction of differentiation by cryoprotective agents may be result of the interaction of these agents with cell membranes	Lyman *et al.* (1976)
(a) Structurally related morphine-like compounds; (b) anti-depressant molecules similar to desipramine	1,2-dipalmitoyl- and 1,2-dimyristoyl-phosphatidylcholine–water mixtures	Lipid pretransition endotherm removed. Both sets of drugs affect main phase transition temperature of lipid—	Drugs alter the tilt of hydrocarbon chains at low concentration and then fluidize membranes at higher concentrations.	Cater *et al.* (1974b)

TABLE 7 (*cont.*)

Drug	Lipid or cell membrane	Effect of drug	Conclusion drawn	Reference
		some drugs more than others. Shifts of T_c very sensitive to structure of drugs. Morphine itself no effect	At low concentrations drugs decrease the area per lipid polar group at the lipid–water interface. *No obvious correlation between structure of derivative and the extent of the decrease in transition temperature*	
Hashish components: (a) 1-tetrahydrocannabinol; (b) cannabidiol	1,2-DPPC	Decrease melting temperature and enthalpy of melting. At ratio of 1 : 5 of drug to lipid two peaks appear in transition profile "suggesting a phase separation in the drug and 1,2-DPPC mixture"	No specific action on phospholipid. Drugs have fluidizing effect on 1,2-DPPC liposomes. Might damage permeability properties of biological membranes	Bach *et al.* (1976)
Polyene antibiotics (filipin, pimaracin, lucensomycin, nystatin, amphotericin B)	1-Oleoyl-2-stearoyl-sn-glycero-3-phosphoryl-choline and 1,2-dielaidoyl-sn-glycero-3-phosphorylcholine in presence and absence of cholesterol	—	Polyene antibiotics only cause effect in presence of sterol	Norman *et al.* (1972)

Polyene antibiotics	*A. laidlawii* membrane grown on elaidic acid	Addition of polyene antibiotics causes an increase in the energy content of the phase transition	Polyene antibiotics form complexes with cholesterol in the membrane	Kruyff *et al.* (1974)
(a) Diethazine; (c) chlorpromazine (Fig. 14)	1,2-DPPC	Lower the main phase transition temperature of lipid. At low concentrations the drugs affect the pretransition endotherm. Negligible change in heat of transition. Effect of (a) and (b) (Fig. 15)	Drugs cause a higher fluidity in the fatty acid chain region and a mobility reduction of the polar head group. Dialkyl amino-alkyl chains located near polar head groups. Ring system does not penetrate far beyond glycerol backbone into hydrocarbon phase	Frenzel *et al.* (1978)
Polymixin B (a) Gramicidin S (b)	1,2-DPPC	(a) In increasing concentration reduces main endothermic transition and removes it completely at ratios of 1 : 1 of polymixin B : lipid. (b) Does not remove the transition but shifts the transition to lower temperatures	(b) Interacts with polar region. (a) Interacts with both polar and apolar region of lipid	Pache *et al.* (1972a)
Amphotericin B	Dimyristoyllecithin plus cholesterol	Position of peak and heat absorbed under the very broad main transition did not change significantly with incorporation of amphotericin B (Fig. 16)	Amphotericin B is shielded from lipid by cholesterol	Bunow and Levin (1978)

TABLE 7 (*cont.*)

Drug	Lipid or cell membrane	Effect of drug	Conclusion drawn	Reference
Chlorothricin	1,2-DPPC	Increasing amounts remove endothermic transition step by step until it completely disappears at drug: lipid ratio of 1 : 1. At lower ratios no shift of transition occurs. (Fig. 17)	Reduction in membrane fluidity. Lytic action of antibiotic due to non-polar association with lipid	Pache and Chapman (1972b)

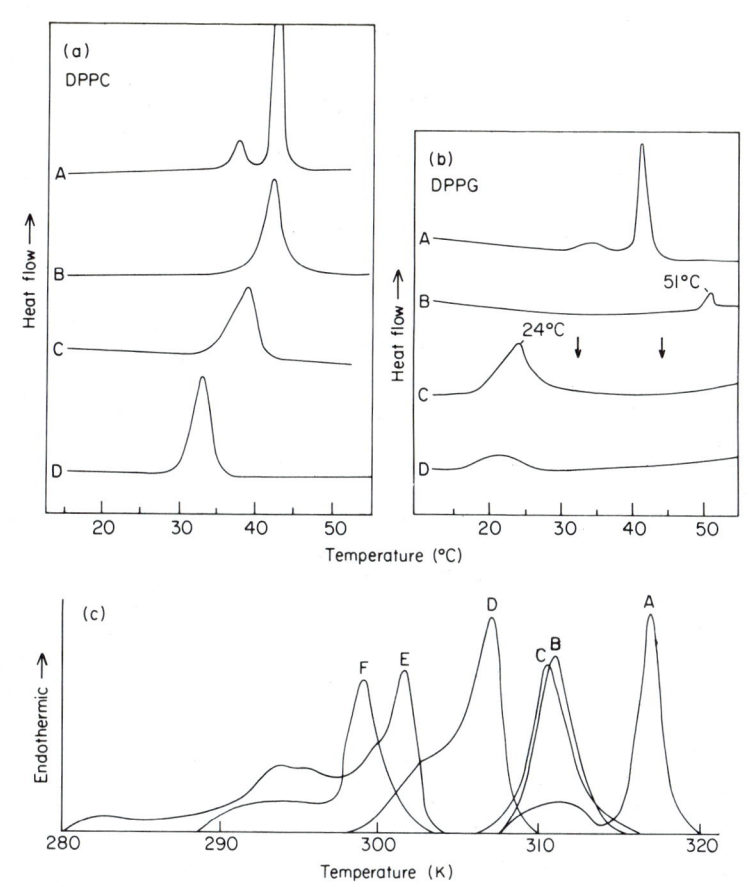

FIG. 13. (a) Differential scanning calorimetry of dipalmitoylphosphatidylcholine (DPPC) vesicles in the presence of different concentrations of dibucaine hydrochloride: A, no dibucaine; B, 4×10^{-4} M dibucaine; C, 2×10^{-3} M dibucaine; and D, 1×10^{-2} M dibucaine. (b) Differential scanning calorimetry of dipalmitoylphosphatidylglycerol (DPPG) vesicles in the presence of different concentrations of dibucaine and Ca^{2+}: A, vesicles in 100 mM NaCl buffer. The remaining curves were obtained with vesicle populations under the same conditions except for the presence of: B, 6×10^{-4} M $CaCl_2$; C, 4×10^{-4} M dibucaine. The arrows indicate the mid-point of the endothermic peak in the presence of 2×10^{-4} and 1×10^{-4} M dibucaine; D, 6×10^{-4} M Ca^{2+} and 4×10^{-4} M dibucaine. (Papahadjopoules et al., 1975.) (c) Differential scanning calorimetric heating curves of the 1,2-dipalmitoylphosphatidylcholine water system at molar/molar ratios with no additions (A), aspirin (B), procaine (C), chlordiazepoxide (D), chlorpromazine (E), desipramine (F). All samples were first mixed in a chloroform solution. The chloroform was then removed under a flow of N_2, the last traces being expelled by placing the samples under vacuum for at least 3 h. The samples were then made up in excess of water (2 : 1) and dispersed by heating the samples above the transition temperature of the lipid and mixing on a bench vibrator. The samples were examined on a Perkin–Elmer DSC 2B microcalorimeter with a heating rate of 5 K min^{-1} and a range setting of 9·37 mJ s^{-1} (2 mcal s^{-1}). (Cater et al., 1974.)

Rejholic, 1977). Ancillary information, e.g. the thermal stability and crystal form of drugs, can also be obtained from the thermograms when the method is used in conjunction with thermogravimetric and X-ray diffraction techniques respectively. The method is, however, restricted to 2–4 mg quantities of sample.

FIG. 14. Structures of phenothiazine derivatives (1). (a) Diethazine hydrochloride, 10-(2-diethylminoethyl)phenothiazine. (b) Chlorpromazine hydrochloride, 2-chloro-10-(3-diemethylaminopropyl)phenothiazine hydrochloride.

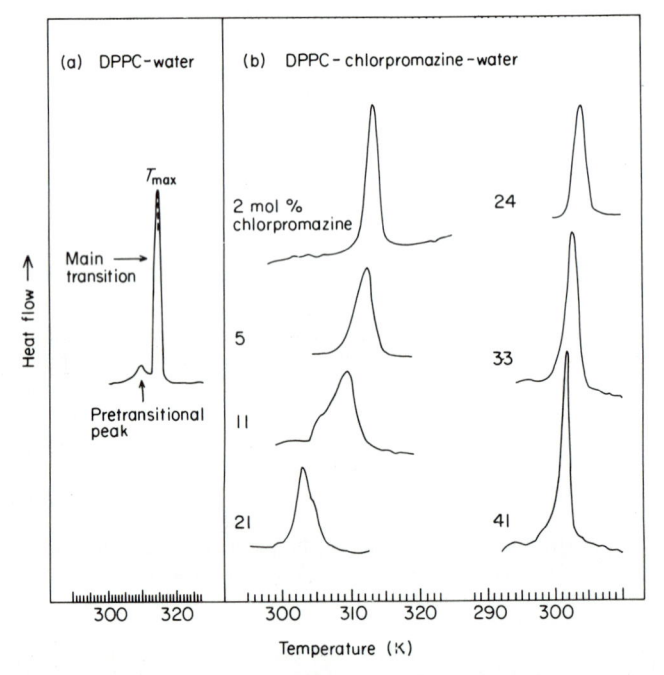

FIG. 15. Differential scanning calorimetry for dipalmitoylphosphatidylcholine (DPPTC)–chlorpromazine–water mixtures. Heating rate, 4 K min^{-1}. (a) Transition profile for a pure simple dipalmitoylphosphatidylcholine–water dispersion without any drug. (b) Some thermograms of the dipalmitoylphosphatidylcholine–water system at different drug concentrations (downward: 2, 5, 11, 21, 24, 33 and 41 mol% of chlorpromazine). (Frenzel et al., 1978.)

J. Drawbacks of DSC studies on model membrane–drug studies

The experiments reported on the effect of drugs on the thermotropic behaviour of lipids have certain drawbacks. Most of the effects of drugs on phase transitions of lipids have only been observed at relatively high concentrations of the drugs compared to the concentrations at which they

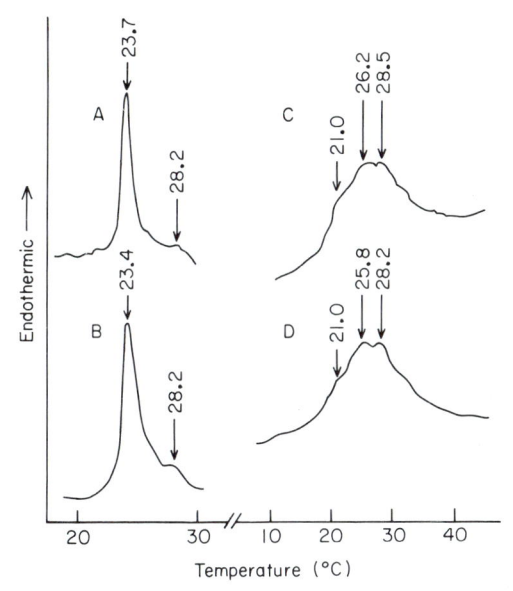

FIG. 16. Effect of amphotericin B on heating curves for dimyristoyllecithin–cholesterol multilayers. A, dimyristoyllecithin–cholesterol (8 : 1). B, dimyristoyllecithin–cholesterol (8 : 1) plus amphotericin B (cholesterol : amphotericin B- 3 : 1). C, dimyristoyllecithin–cholesterol (4 : 1). D, dimyristoyllecithin–cholesterol (4 : 1) plus amphotericin B (cholesterol : amphotericin, 6 : 1). The ordinate in scans A and B is, on the basis of dry weight of dimyristoyllecithin, about four times that in C and D. For reference, the main transition for pure dimyristoyllecithin occurs at 24·0 °C and the pretransition occurs at 16·0 °C. The main transition peak half-widths were 1·5 and 1·2 °C for A and B, respectively; 12 °C for C and D; and 0·9 °C for pure dimyristoyllecithin. (Transition temperatures are averages of two or more scans.) (Bunow et al., 1978.)

induce biological (or clinical) effects. Biological membranes contain a wide variety of lipids and other structural components such as proteins and metal ions (Trauble, 1975). However, many phase transition studies to date have been limited to single phospholipid species—and mixtures have not been widely used. This is mainly due to difficulties relating to interpretation of results. Another limitation is that only the bulk or average properties of the entire (phospholipid) system is observed in the experiments. Consequently localized effects which may be of critical importance in an organized membrane system are not readily discernible. The solubility properties of additives such as drugs in cell membranes, which are anisotropic, may differ from those in isotropic bulk solvents

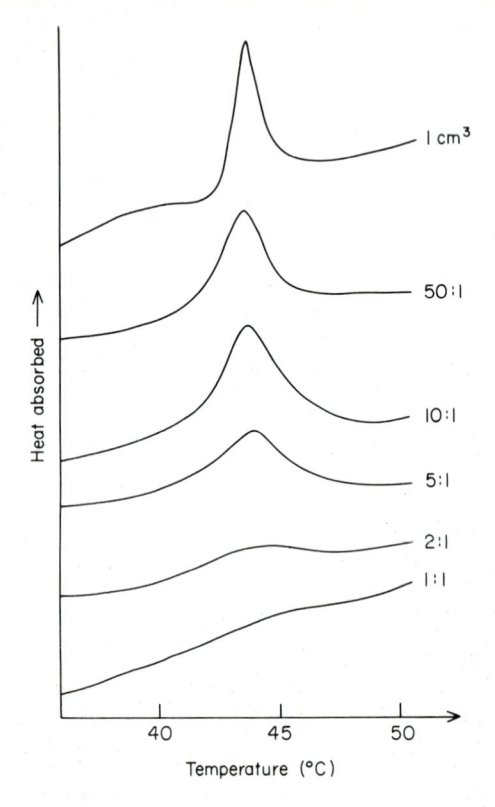

Fig. 17. Endothermic transition of dipalmitoyllecithin and dipalmytoyllecithin–chlorothricin mixtures. Ratios indicate the proportion of lecithin to antibiotic. (Pache *et al.*, 1972.)

such as phospholipids, which may produce different effects on the phase transition of membrane phospholipids compared to bulk phospholipids. If the only way of producing a change in the fluidity state (Trauble, 1975) of lipids in a membrane were by varying the temperature, then biologically important functions associated with such changes would be unlikely. However, fluidity changes can be produced by other factors apart from temperature changes (Jacobsen, 1975; Trauble, 1975). Furthermore, changes in fluidity of lipids have been detected by many other physicochemical techniques apart from DSC such as n.m.r. (Lee, 1972; Levine, 1973) and e.s.r. (Hubbell and McConnell, 1971).

Lastly, most of the lipids of living cell membranes are thought to be in the fluid state (Chapman, 1975). In order, therefore, to examine the effect of drugs on lipid fluidity, tests should be carried out when the lipid is in the fluid state before the addition of the drug to the lipid. This is not generally the case. However, it has been found that even when the membrane lipids are all in a completely fluid state, i.e. above their transition temperature, drugs will shift the phase boundary of the lipids and thus still affect the fluidity of the membrane (Chapman, 1975).

DSC can be used to study liquid–crystalline lyotropic systems (Sackmann, 1974). Liquid–crystalline lyotropic systems play some role in the chemistry of colloids and interfaces. They are also of importance in the study of membranes, particularly in connection with the structure and function of biological membranes. The crystal structure of drugs can also be studied by DSC.

K. Future studies

The ability of certain drugs to change the fluidity of a lipid region in a membrane could also affect the metabolism of certain drugs. It has been suggested that the cytochrome P450 system which is the drug-metabolizing system situated in the microsomal fraction of the liver may be enclosed in a "rather rigid phospholipid halo" (Chapman, 1975a). The rate of metabolism of a drug could be determined not only by its lipophilic nature but also by its ability to fluidize the phospholipid halo surrounding the cytochrome P450 complex (Chapman, 1975a).

The possibility of transporting drugs in the circulatory system of man and animals using liposomes has been recently suggested (Tyrrell, 1976). It is important to understand the effect of the drugs on the phase transition behaviour of the lipids used for the liposomes since this could be important in affecting the stability and diffusional characteristics of the drug-carrier system. It may be that studies of drug interactions of the sort indicated by DSC experiments will be useful as a method of indicating broad ranges of pharmacological action and help in the design of new drug molecules. Further studies of this nature are therefore to be expected. The range of different effects obtained with diverse compounds on lipid phase transition properties also offers a means for introducing various degrees and types of perturbation into membrane systems.

L. Adiabatic flow microcalorimetry

Recently several flow microcalorimeters based on the adiabatic principle have been successfully constructed (Reid, 1976). When combined with immobilized enzymes or microorganisms it may be possible to use these techniques for the sensitive assay of drugs.

M. Thermometric titrimetry and thermogravimetric analysis

Thermometric titrimetry (Tyrrell and Beezer, 1968) and thermogravimetric techniques (Wendlandt, 1974), although applicable to the quantitative and thermal stability analysis of drugs respectively, have not, in general, been used for analysis of microgram quantities of drugs.

N. Catalytic thermometric titrimetry (CCT)

Greenhow (1977) has applied catalytic thermometric titrimetry to the quantitative analysis of drugs at microgram levels. In CTT a monomer which can undergo ionic polymerization is added to a non-aqueous solution of the drug under test prior to titration with a suitable titrant. The end-point is indicated by an exothermic reaction (involving temperature rises of 6 °C or greater, depending upon the nature and amount of sample) which results from the ionic polymerization of the monomer and which is initiated by the titrant following neutralization of the sample.

In general acid–base and non-aqueous iodometric reactions (Greenhow, 1977) have been used for the analysis of pure drugs, pharmaceutical formulations and drug extracts from natural materials. Examples of drugs determined by CTT include sulphonamides, sulphanilamide derivatives, barbiturates, thioamides, thiols, alkyl aralkyl and heterocyclic tertiary amines (e.g. antipyrine and cinchonine), alkaloids and related compounds, catecholamines and metal dithiocarbamates (Table 8). Typical CTT curves for alkaloids are shown in Fig. 18. It has been suggested (Greenhow, 1977) that CTT could be used as an alternative to existing non-aqueous potentiometric and visual indicator titrimetric methods for the assay of acidic and basic active constituents in pharmaceutical formulations, e.g. amine hydrochlorides. The technique is rapid (capable of giving results within minutes), simple to perform and requires inexpensive apparatus. Manual, semi-automatic and automatic methods may be employed. The relative standard deviation of results obtained by CTT vary from 0·07 to 1·82. The precision that can be obtained in the determination of the

TABLE 8

Examples of the quantitative analysis of drugs by catalytic thermometric titrimetry

Reference	Compounds	Type of titration	Solvents	Titrant	End-point indicator	Remarks
Greenhow et al. (1976)	(a) Metal dithiocarbamates $\left[\begin{array}{c}R\\ \diagdown\\ R\end{array}\!\!N\!-\!C\!\!\begin{array}{c}S\\ \diagup\\ S\end{array}\right]_n^{M^{n+}}$	Iodometric $\Delta T = 3\text{--}8\ °C$ $0\!\cdot\!002\text{--}0\!\cdot\!5$ mEq	DMF or acrylonitrile	Iodine reagent	Cationic polymerization of ethyl vinyl ether	0·002 mmol detectable. Method compared to aqueous iodimetry. Results agree to within 1%
Greenhow et al. (1973a)	(b) Catecholamines, e.g. (−)-adrenaline, L-noradrenaline, adrenaline hydrogen tartrate, dopamine hydrochloride, L-dopa, DL-dopa, L-α-methyl-dopa, D-α-methyl-dopa, (+)-corbasil	CTT of acidic and basic functions	Formic acid, acetic acid, DMF, NN-diethyl-3-amino-propionitrile, hexamethylphosphoramide, N-methylmorpholine	Perchloric (basic functions), tetra-n-butyl hydroxide (acidic functions)	Methyl styrene (basic functions), acrylonitrile (acidic functions)	L-dopa contact of tablets and capsules determined by CTT compared with British Pharmacopoeia. Non-aqueous titration and u.v. Results differed by 1·6%. 0·0001 mEq catecholamines detectable

TABLE 8 (*cont.*)

Reference	Compounds	Type of titration	Solvents	Titrant	End-point indicator	Remarks
	HO–⬡–C(H)(R₁R₂)–N(R₃R₄)H $\Delta T(°C)$ 5–15 °C (0·0001–1 mEq)					
Greenhow et al. (1973c)	(c) Alkaloids, alkaloidal salts and purine bases. (Strychnine, nicotine, atropine, quinine, papaverine, caffeine, theophylline, quinine-hydrochloride ephedrine hydrochloride, ephedrine sulphate codeine phosphate, atropine methonitrate)	Titration of bases $\Delta T(°C)$: 5–15 °C (0·001–0·1 mEq)	Acetic acid, 1,4-nitroethanol-1,2-dichloroethane or chloroform		Methyl styrene cationic polymerization	0·0001 mEq detectable, e.g. 33 µg strychnine, 8·5 µg nicotine. Technique applicable to raw materials, crude drugs, formulations or natural materials. Results compare well with previously available titrimetric methods

drug content of formulations is similar in some cases to that achieved with pure drugs, e.g. for L-dopa (Greenhow, 1973a). CTT may therefore be of value for the industrial analysis of pharmaceutical formulations.

CTT is limited in its applications by the different combinations of indicator reagents, titrants, sample solvents and intrinsic lack of selectivity

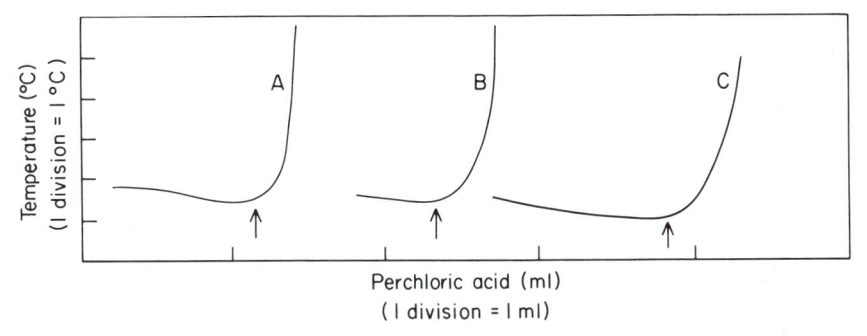

FIG. 18. Catalytic thermometric curves for atropine (A), theophylline (B) and nicotine (C) using ionic polymerization of methylstyrene to indicate end-points. Arrows indicate theoretical end-points (allowing for blank titrations). (Greenhow et al., 1973.)

of the method of end-point determination. Qualitative analysis of drugs is not possible using CTT. In order to obtain useful and accurate results the use of pure solvents is recommended. A further problem is that reaction stoichiometries may not always be exact, e.g. for many compounds the iodometric method gives results which indicate that iodine combines with the titrand in fractional molar amounts. This does not detract from its value as an analytical method. The requirements of reproducibility of results and linearity of calibration graphs (drug dose versus titrant amount) can be met. Mixtures of drugs, especially closely related ones, cannot be readily and easily analysed by this method.

CTT has not been used for the analysis of drugs in body fluids. Drugs, such as catecholamines which occur at submicrogram levels in body fluids and tissue extracts might be analysed if a concentration step could be introduced prior to analysis by CTT. Depending on the titrant if the end-point is very sharp, if weak titrants are effective and by taking extreme care to avoid contamination from the atmosphere as well as using well-insulated automatic equipment, determinations at the submicrogram level might also be possible for certain catecholamines. CTT can be used to study the basic properties and indirectly the structures of organic bases. Greenhow et al. (1973b) have titrated organic bases with 0·1 M perchloric acid in acetic acid and boron trifluoride diethyl etherate in dioxan. It was found that p-toluidine was more basic than p-nitroaniline, the ratios being 0·7 and 0·0 respectively with boron trifluoride titrant, under the conditions used for the titrations.

O. Summary

Microcalorimetric methods can be used for a variety of different purposes in the study of drug molecules (Table 9). Isothermal heat leak microcalorimetry allows examination of the bioactivity and mechanism of action

TABLE 9

Uses of microcalorimetric methods in analysis of drugs

I. ISOTHERMAL MICROCALORIMETRIC METHODS
  A. *Flow microcalorimetry*
   1. Distinguish between bacteria (fungio)static and bacterio(fungio)cidal drugs
   2. Measure the bioactivity of drugs either singly or in combination. This includes the relative biopotency of closely related drugs, both in pure (raw) form and in pharmaceutical formulations. Synergism and antagonism of combinations of drugs
   3. Effect of pharmaceutical formulation excipients on the activity of drugs
   4. Kinetics of drug action (including pharmokinetics)
   5. Model studies, e.g. using microcalorimeter as a titration apparatus
   6. Effect of environmental factors, e.g. pH, incubation temperature, medium constituents, on microorganism–drug interactions
  B. *Batch microcalorimetry*
   1. Complexation studies of (model) compounds with drugs
   2. Bioactivity and quantitative analysis of certain drugs
II. ADIABATIC MICROCALORIMETRIC METHODS
  A. *Catalytic thermometric titrimetry*
   1. Quantitative analysis of drugs
  B. *Differential scanning calorimetry*
   1. Mode of action studies of drugs
   2. Crystal structures of molecules

(at the macro level) of antibiotics and synthetic antimycotics with living microorganisms. Batch and flow microcalorimetry employing live organisms can also be used for the quantitative assay of drugs. Flow microcalorimetry is particularly suitable for the study of microorganism–drug interactions. The technique shows advantages over both batch microcalorimetric techniques and non-microcalorimetric methods for the study of antimicrobial activity.

Catalytic thermometric titrimetry can also be used for the quantitative analysis of drugs. The quantitative analysis of drugs by microcalorimetric techniques is sensitive; microgram levels of drugs can be determined. Few examples to date exist of the application of microcalorimetric techniques to the analysis of mixtures of natural and/or synthetic drugs.

Qualitative analysis of drugs is not possible by microcalorimetric techniques. Very few examples exist of the application of microcalorimetric techniques to the assay of drugs in body fluids at present. The extension of the use of microcalorimeters to study other cell types besides microorganisms is expected.

Adiabatic microcalorimetry in the form of the differential scanning calorimeter can be used for studies of model reactions of drugs on model membrane systems, thereby contributing to the mechanism of action studies of drugs.

Both isothermal heat leak microcalorimetry and adiabatic calorimetry (DSC) can be used to help in the evaluation and design of new drugs. The use of flow microcalorimetry for analysis of bioactivity of N-acetyl nystatin is an example (Beezer et al., 1979b).

References

ANONYMOUS (1978). *Lancet* (8080), 80–82.
ARHAMMER, M., MÅRDH, P.-A., RIPA, T. and ANDERSSON, K. E. (1978). *Acta path. microbiol. scand., sect. B*, **86**, 59–65.
BACH, D., RAZ, A. and GOLDMAN, R. (1976). *Biochim. biophys. Acta*, **436**, 889–894.
BANGHAM, A. (1971). *New Scientist*, 63–65.
BEEZER, A. E. (1973). "M.T.P. International Reviews of Science", Physical Chemistry Series 1 (T. S. West, ed.), vol. 13, pp. 71–93. Butterworths, London.
BEEZER, A. E. (1977). *In* "Applications of Calorimetry in Life Sciences" (I. Lamprecht and B. Schaarschmidt, eds). Walter de Gruyter, Berlin and New York.
BEEZER, A. E. and CHOWDHRY, B. Z. (1980a). *Talanta*, **27**, 1–5.
BEEZER, A. E. and CHOWDHRY, B. Z. (1980b): To be published.
BEEZER, A. E. and CHOWDHRY, B. Z. (1980c). To be published.
BEEZER, A. E. and COSGROVE, R. F. (1978). Unpublished observations.
BEEZER, A. E., NEWELL, R. D. and TYRRELL, H. J. V. (1976). *J. appl. Bact.* **41**, 197–220.
BEEZER, A. E., NEWELL, R. D. and TYRRELL, H. J. V. (1977a). *Analyt. Chem.* **49**, 34–37.
BEEZER, A. E., CHOWDHRY, B. Z., NEWELL, R. D. and TYRRELL, H. J. V. (1977b). *Analyt. Chem.* **49**, 1781–1784.
BEEZER, A. E., MILES, R. J., SHAW, E. J. and WILLIS, P. (1979). *Experientia*, **35**, 795–796.

BERTHE-CORTI, L. (1977). *In* "Applications of Calorimetry in Life Sciences" (I. Lamprecht and B. Schaarschmidt, eds), pp. 85–96. Walter de Gruyter, Berlin and New York.
BINFORD, J. S., BINFORD, L. F. and ADLER, P. (1973). *Am. J. clin. Path.* **59**, no. 1, 86–94.
BURROW, M. R. and LEVIN, I. W. (1978). *Biochim. biophys. Acta*, **464**, 202–216.
CATER, B. J., CHAPMAN, D., HAWES, S. M. and SAVILLE, J. (1974a). *Biochem. Soc. Trans.* **2**, 971–973.
CATER, B. J., CHAPMAN, D., HAWES, S. M. and SAVILLE, J. (1974b). *Biochim. Biophys. Acta*, **363**, 54–69.
CHAPMAN, D. (1975a). *Q. Rev. Biophys.* **8**, 185–236.
CHAPMAN, D. (1975b). *Chemy. Ind.*, 98–100.
CHOWDHRY, B. Z. (1975). M.Sc. Thesis, Analytical Chemistry, London University.
CHOWDHRY, B. Z. (1980). Ph.D. Thesis, London University.
CHRISTENSEN, J. (1977). Calorimetry Workshop, Isle of Ischia.
COSGROVE, R. F. (1978). Ph.D. Thesis, London University.
COSGROVE, R. F., BEEZER, A. E. and MILES, R. J. (1979a). *J. Pharm. Pharmac.* **31**, 83.
COSGROVE, R. F., BEEZER, A. E. and MILES, R. J. (1979b). *J. Pharm. Pharmac.* **31**, 171.
DURAND, M., MAURIZOT, T. C. and HELENE, C. (1976). *FEBS Lett.* **71**, 9–12.
EHASZ, A. W., CHAPMAN, D. and EWING, D. F. (1976). *Biochim. biophys. Acta*, **448**, 220–230.
ERIKSSON, R. (1977). L.K.B. Application Note 267.
FAIRBROTHER, J. E. (1977). *Pharm. J.* 509–513.
FERRARI, H. J. and PASSARELLO, N. J. (1974). *In* "Analytical Calorimetry" (R. S. Porter and J. F. Johnson, eds), vol. 3, p. 321. Plenum, New York and London.
FORREST, W. W. (1972). *In* "Methods in Microbiology" (J. R. Norris and D. W. Ribbons, eds), vol. 6B, pp. 285–318. Academic Press, London and New York.
FOURCANS, B. and JAIN, M. K. (1975). *Adv. Lipid Res.* **12**, 147–182.
FRENZEL, J., ARNOLD, K. and NUHN, P. (1978). *Biochim. biophys. Acta*, **507**, 185–195.
FRISCHLEDER, H. and GLEICHMAN, S. (1977). *Stud. Biophys.* **64**, 95–100.
FRUH, P. U. and SIMON, W. (1972). *In* "Protides of the Biological Fluids" (H. Peeters, ed.), vol. 20. Pergamon Press, Oxford.
FRUH, P. U., CLERE, J. T. and SIMON, W. (1971). *Helv. chim. Acta*, **54**, 1445–1450.
FUJITA, T., NUNOMURA, K., KAGAMI, I. and NISHIKAWA, Y. (1976). *J. gen. appl. Microbiol.* **22**, 43–50.
GOLDMAN, R., FACCHINETTI, T., TAX, A. and BACH, D. (1978). *Biochim. biophys. Acta*, **512**, 254–269.
GREENHOW, E. J. (1977). *Chem. Rev.* **77**, 835–854.
GREENHOW, E. J. and SPENCER, L. E. (1973a). *Analyst.* **98**, 485–492.
GREENHOW, E. J. and SPENCER, L. E. (1973b). *Analyst.* **98**, 81–89.

Greenhow, E. J. and Spencer, L. E. (1973c). *Analyst,* **98**, 98–102.
Greenhow, E. J. and Spencer, L. E. (1976). *Analyst,* **101**, 777–785.
Greenwood, D. (1976). *J. antimicrob. Chemother.* **2**, 312–313.
Hubbell, W. L. and McConnell, H. M. (1971). *J. Am. Chem. Soc.* **93**, 314–319.
Hughes, D. W. and Wilson, W. L. (1973). *Can. J. pharm. Sci.* **8**, 67–74.
Hui, F. K. and Barton, P. G. (1973). *Biochim. biophys. Acta,* **296**, 510–517.
Jacobsen, K. and Papahjopoulos, D. (1975). *Biochemistry,* **14**, 152–158.
Jain, M. K. and Wu, N. M. (1977). *J. Memb. Biol.* **34**, 157–201.
Jain, M. K., Wu, N. M. and Wray, L. V. (1975). *Nature, Lond.* **255**, 494–495.
Jain, M. K., Wu, N. M., Morgan, T. K., Briggs, M. S. and Murray, R. K. (1976). *Chem. Phys. Lipids,* **17**, 71–78.
Jensen, T. E., Hansen, L. D., Eatough, D. J., Sagers, R. D. and Izatt, R. M. (1976). *Thermochemica Acta,* **17**, 65–71.
Karoly, N. (1976). *Acta Pharm. Hung.* **46**, 205–212.
Kavanagh, F. (1963). "Analytical Microbiology". Academic Press, New York.
Kemp, R. B. (1972). *Sci. Tools,* **19**, 3–5.
Krasne, S., Eisenman, G. and Szabo, G. (1971). *Science,* **174**, 412–415.
Kruyff, B. de, Greef, W. J. de, Eyk, R. V. M., Van Demel, R. A. and Van Deenen, L. L. M. (1973). *Biochim. biophys. Acta,* **298**, 479–499.
Kruyff, B. de, Gerritsen, W. J., Oerlemans, A., Demel, R. A. and Van Deenen, L. L. M. (1974). *Biochim. biophys. Acta,* **339**, 44–56.
Lee, A. G. (1975). *Prog. Biophys. Molec. Biol.* **29**, 3–56.
Levin, K. (1977). *Clin. Chem.* **23**, 929–937.
Levine, Y. K. (1972). In "Progress in Biophysics and Molecular Biology" (J. A. V. Butler and D. Noble, eds). Pergamon Press, Oxford.
Lightbown, J. W., Kogurt, M. and Uemeera, K. (1963). *Bull. Wld Hlth. Org.* **29**, 87–00.
Liptay, G. (1973). "Atlas of Thermoanalytical Curves". Heydon and Son, London.
Ljungholm, K., Wadsö, I. and Mardh, P. A. (1976). *J. gen. Mircob.* **96**, 283–285.
Lyman, G. H., Papadjopoulos, D. and Preisler, H. D. (1976). *Biochim. biophys. Acta,* **448**, 460–473.
Mårdh, P.-A. (1978). *J. antimicrobial Chem.* **4**, 73–78.
Mårdh, P.-A., Ripa, T., Andersson, K. E. and Wadsö, I. (1976). *Antimicrobial Ag. Chemother.* **10**, 604–609.
Melchoir, D. L., Scavitto, F. J., Walsh, M. T. and Stein, J. M. (1977). *Thermochemica Acta,* **18**, 43–71.
Möschler, H. J., Weder, H. G. and Schwyzer, R. (1971). *Helv. chim. Acta,* **54**, 1437–1440.
Newell, R. D. (1975). Ph.D. Thesis, London University.
Nikolics, K., Gal, S., Sztatisz, J. and Nikolics, K. (1976). *Acta Pharm. Hung.* **46**, 205–212.
Norman, A. W., Demel, R. A., Kruyff, B. de, Guerts, van Kessel, W. S. M. and Van Deenen, L. L. M. (1972). *Biochim. biophys. Acta,* **290**, 1–14.
Overath, P. and Trauble, H. (1973). *Biochemistry,* **12**, 2625–2634.

PACHE, W. and CHAPMAN, D. (1972). *Biochim. biophys. Acta*, **255**, 348–357.
PACHE, W., CHAPMAN, D. and HILLABY, R. (1972). *Biochim. biophys. Acta*, **255**, 358–364.
PANG, K. Y. and MILLER, K. W. (1978). *Biochim. biophys. Acta*, **511**, 6–9.
PAPAHADJOPOULOS, D. (1976). *Nature, Lond.* **262**, 360–363.
PAPAHADJOPOULOS, D., JACOBSEN, K., POSTE, G. and SHEPHERD, G. (1975). *Biochim. biophys. Acta*, **394**, 504–519.
PRAT, H. (1953). *Rev. can. Biol.* **12**, 19–34.
QUINN, P. A. (1974). *J. Cell Biol.* **12**, no. 12, 73–76.
REID, D. S. (1976). *J. Physiol. (E), Scient. Inst.* **9**, 601–609.
REJHOLEC, V. (1977). *Čslká Farm.* **26**, no. 3, 101–103.
ROSSENEU, M., SOETEWEY, F., BLATON, V., LEIVENS, J. and PEETERS, H. (1976). *Chem. Phys. Lipids*, **17**, 38–56.
SACKMANN, E. (1974). *Ber. Bunsenges. Phys. Chem.* **78**, 929–932.
SEMENITZ, E. (1978). *J. antimicrob. Chemother.* **4**, 455–459.
SEMENITZ, V. E. and TIEFENBRUNNER, F. (1977a). In "Applications of Calorimetry in Life Sciences" (I. Lamprecht and B. Schaarschmidt, eds), pp. 251–260. Walter de Gruyter, Berlin and New York.
SEMENITZ, V. E. and TIEFENBRUNNER, F. (1977b). *Arzneimittel-Forsch. 1 Drug Research*, **27**, no. 11, 2247–2251.
SPINK, C. H. and WADSÖ, I. (1976). *Meth. biochem. Analysis*, **23**, 1–159.
STARK, G., BENZ, R., POHL, G. W. and JANKO, K. (1972). *Biochim. biophys. Acta*, **266**, 603–618.
TILLBERG, E., KYLIN, A. and SUNDBERG, I. (1971). *Pl. Physiol.* **48**, 779–782.
TRAUBLE, H. and EBL, H. (1975). In "Functional Linkage in Bimolecular Systems" (F. O. Schmitt, D. M. Schneider and D. M. Crothers, eds), pp. 59–101. Rowen Press, New York.
TRITTON, T. R., MURPHEE, S. A. and SARTORELLI, A. C. (1977). *Biochem. Pharmac.* **26**, 2319–2325.
TYRRELL, D. A. HEATH, T. D., COLLEY, C. M. and RYMAN, B. E. (1976). *Biochim. biophys. Acta*, **457**, 259–264.
TYRRELL, H. J. V. and BEEZER, A. E. (1968). "Thermometric Titrimetry". Chapman and Hall, London.
URRY, D. W. (1975). *Int. J. quantum Chem. Quantum Biol. Symp.* **2**, 221–235.
VAN DEENEN, L. L. M., GIER, J. DE, DEMEL, R. A., KRUYFF, B. DE, BLOK, M. C., VAN DER NEUTKOK, E. C. M., HAEST, C. W. M., VERVERGAERT, P. H. J. TH. and VERKLEY, A. J. (1975–76). *Ann. N.Y. Acad. Sci.* **264**, 124–141.
WADSÖ, I. (1969). *Svensk. Kenvsk. Tidskrift*, **81**, 28–32.
WADSÖ, I. (1976). *Biochem. Soc. Trans.* **4**, 561–565.
WADSÖ, I. and LJUNGHOLM, K. (1978). Unpublished observations.
WENDLANDT, W. W. (1974). "Thermal Methods of Analysis" (second edition). John Wiley, New York.
ZABLOCKI, B. and CZERNIASKI, E. (1962). *Bull. Acad. pol. Sci., Sér. Cl. II*, **10**, no. 6, 209–213.
ZYNGER, J. (1975). *Analyt. Chem.* **47**, 1380–1384.

Some Problems in Calorimetric Measurements on Cellular Systems

I. WADSÖ

A. Introduction

In all calorimetric investigations certain instrumental requirements must be fulfilled: the sensitivity and the base-line stability must be adequate, and the accuracy—or sometimes merely the reproducibility—must be compatible with the problem chosen. Further, it is often a requirement that the time resolution of the instrument be appropriate for the kinetics of the process studies. In calorimetric experiments on cellular systems it is, in addition, necessary to choose well-defined physiological conditions. The availability of oxygen or the concentration of carbon dioxide are often critical parameters in this connection. Cell sedimentation and adhesion can cause problems and different kinds of artifacts can easily be recorded in calorimetric work on cellular materials. In this contribution some microcalorimetric designs and working procedures related to problems of this nature are examined. Techniques suitable for studies of microorganisms, blood cells and tissue cell systems are discussed. For those readers who are not very familiar with calorimetry, a brief review of calorimetric principles and of a few microcalorimetric designs suitable for cell studies will first be given. A more detailed treatise of this latter subject has recently been given elsewhere (Spink and Wadsö, 1976).

B. Some calorimetric design principles

Several different calorimetric principles and a wide variety of practical designs have found use in biological calorimetry. It is convenient to group them under two major headings: essentially (i) adiabatic calorimeters, and (ii) heat conduction calorimeters.

1. ADIABATIC CALORIMETERS

In an ideal adiabatic calorimeter there is no heat exchange between the calorimetric vessel and the surroundings. The power P is proportional to the measured rate of temperature increase dT/dt, and

$$P = \varepsilon \frac{dT}{dt} \tag{1}$$

where ε is a calibration constant. The heat quantity evolved q will thus be proportional to the temperature change ΔT:

$$q = \varepsilon \Delta T \tag{2}$$

Reaction and solution calorimeters, as well as combustion calorimeters, are often of the "isoperibolic" type (also called "isothermal jacket calorimeters"). In this type the calorimetric vessel is separated by an efficient insulation from a surrounding thermostat, but some heat exchange will take place and the calorimetric type can be characterized as "nearly adiabatic". In accurate work corrections must be applied for the heat exchange with the surroundings. Even for moderately long reaction periods (of the order of 1 h) the heat exchange term will often cause significant uncertainties.

If the temperature of the surroundings follows that of the calorimetric vessel no net heat exchange will take place. For endothermic processes this can, in principle, easily be arranged by simultaneous electrical compensation in the reaction vessel. For exothermic processes an "adiabatic shield" can be inserted between the vessel and the thermostat. The temperature difference between the shield, usually consisting of a metal envelope, and the calorimetric vessel is kept at zero during the entire measurement by evolution of a suitable electrical heat effect. Adiabatic shield calorimeters are important in studies of slow processes and for heat capacity calorimeters.

2. HEAT CONDUCTION CALORIMETERS

In the ideal heat conduction calorimeter heat released is quantitatively transferred from the reaction vessel to a surrounding heat sink, normally consisting of a metal block. With such calorimeters it is a property proportional to the total heat flow from the calorimeter which is measured. Heat conduction calorimeters are also called "heat leakage calorimeters" or "heat flow calorimeters".

Most frequently the heat flow is recorded by positioning a "thermopile wall" between the calorimetric vessel and the heat sink. The temperature difference over the thermopile will give rise to a voltage signal V which is proportional to the heat flow dq/dt:

$$\frac{dq}{dt} = \varepsilon V \qquad (3)$$

where ε is a calibration constant. Under steady-state conditions the heat flow is equal to the power evolved,

$$P = \varepsilon V \qquad (4)$$

However, in practice the simple relationship (4) often holds, to a good approximation, for a non-steady-state process, provided that the changes in the rate of heat production are slow. When the rate change is fast, there will be a significant distortion of the voltage–time curve relative to the true heat effect curve. In order to obtain precise kinetic information by use of heat conduction calorimeters it is then necessary to reconstruct the thermogram by use of time-constant data for the calorimeter. Such calculations are discussed in some detail by Randzio and Suurkuusk (p. 311).

Thermopile heat conduction calorimeters have proved to be very suitable for studies of small heat effects or heat quantities and for processes of long duration. Most microcalorimeters currently used in biological work are of the thermopile heat conduction type. It is, as yet, very rare to find that the signals from thermopile conduction calorimeters are corrected to the true kinetic curves. Often the biological processes are so slow that the distortion is insignificant. However, for comparatively rapid rate changes such as those observed in bacterial identification experiments (see Newell, p. 163), it is likely that the corrected and uncorrected curves will be significantly different.

By use of a thermopile it is also possible to employ the Peltier effect principle for an active transport of heat released in the process. The cooling power produced, $-P$, is proportional to the current through the thermopile, I. In practice, this cooling power will be superimposed on the Joule heating power, produced throughout the circuit. This heat power will partly be transmitted to the thermopile junctions (corresponding to an effective resistance r). The cooling power is thus given by

$$-P = k(I - rI^2)$$

Normally the temperature of the "thermoelectric heat pump calorimeter" is kept constant, and the power produced and P will then be identical. With the well-known Calvet microcalorimeters it is possible to use the thermopile heat conduction principle and the thermoelectric heat pump method in combination.

3. MICROCALORIMETERS

There exists no well-defined difference between "ordinary" calorimeters or "macrocalorimeters" and "microcalorimeters", but the micro-prefix usually indicates a very sensitive instrument, often requiring only small sample quantities. With microcalorimeters used for biological work, the power sensitivity is usually in the order of $0{\cdot}1$–1 μW. Sample volumes for instruments used in batch experiments are often of the order of one to a few millilitres, but sensitive instruments with vessels in the order of 100 ml are frequently also called microcalorimeters. In biological work flow microcalorimeters are often used. A wide variety of flow vessels have been described (volumes are normally in the order of $0{\cdot}5$–2 ml). Vessels can be of the flow-through type or they can be intended for measurements of mixing processes. Flow microcalorimeters are normally used in continuous-flow experiments, but flow-through vessels are also employed in stopped-flow measurements.

Microcalorimeters used for power measurements on biological systems are in most cases arranged as twin instruments. Normally, the signals from two calorimetric units are connected in opposition and it is thus the differential signal which will be recorded. A twin calorimeter can also be operated as a "thermal balance", where the process in one of

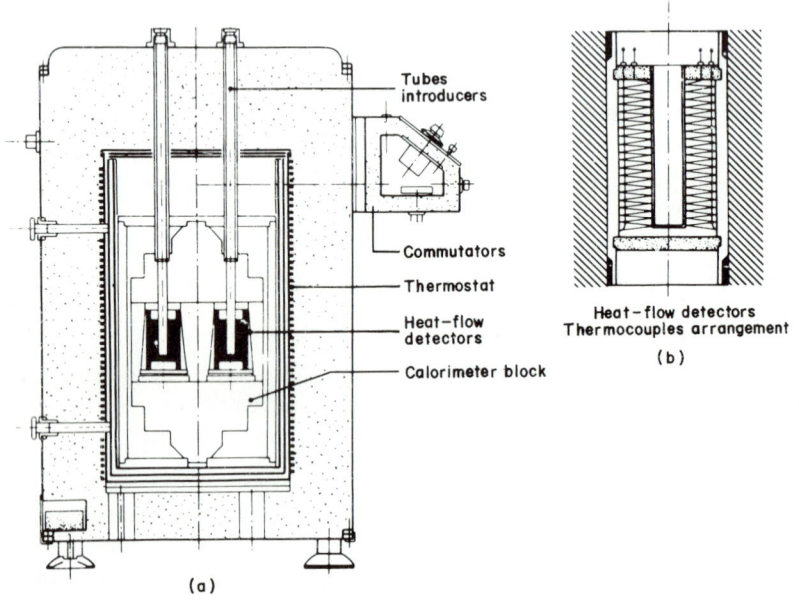

Fig. 1. (a) Schematic view of a Calvet microcalorimeter. (b) Detector arrangement. (Reprinted by permission of Setaram, Lyon, France.)

the reaction vessels is simulated electrically, so that the temperatures of both vessels are kept equal. The main advantage with twin calorimeters is that thermal disturbances from the surroundings, which ideally influence both vessels identically, cancel out. This property is of particular importance in studies of low heat effects during long experimental periods. Another useful property is that the heat effect produced by one sample can be directly compared with that from a reference sample.

In Figs 1 to 3 typical heat-conduction microcalorimeters currently used in biological work are shown. Figure 1 shows schematically the design of a standard Calvet microcalorimeter marketed by Setaram, Lyon, France. A large aluminium block serves as the heat sink for two calorimetric units. In each unit a thin-walled metal cylinder defines the space for the calorimetric vessel and is surrounded by a wire-wound thermopile which is in

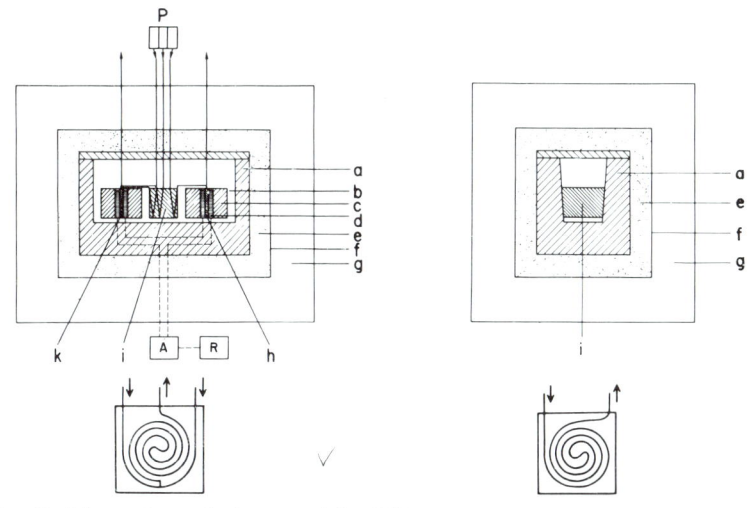

FIG. 2. Flow microcalorimeter of the LKB type (Monk and Wadsö, 1968). *Upper left*: a, main heat sink; b, air space; c, aluminium block; d, semiconductor thermopiles; e, styrofoam insulation; f, stainless steel container; g, thermostatted water bath; h, flow-through cell; i, heat-exchanger unit; k, mixing reaction cell. A is an amplifier and R is a recorder. *Upper right*: transverse section through the flow microcalorimeter. *Lower left*: mixing cell. *Lower right*: flow-through cell. (Spink and Wadsö, 1976.)

contact with the heat sink. The reaction vessel often consists of a simple tube with a stopper but a range of different reaction vessels have been designed; see, for example, the stirred mixing vessel by Belaich and Sari (1969) and the vessel described by Lamprecht and Schaarschmidt (1973) where the liquid is agitated by a piston (Fig. 4). In vessels described by Schaarschmidt and Lamprecht (1973) the liquid content can be stirred;

in addition quartz light guides are used for irradiation of the sample in the vessel or for measurements of its optical density. Volumes of reaction vessels used with Calvet instruments are frequently of the order of 15 ml but substantially larger vessels are also used.

Figures 2 and 3 show two twin microcalorimeters of the type marketed by LKB-Producter, Bromma, Sweden. Each calorimetric unit has a

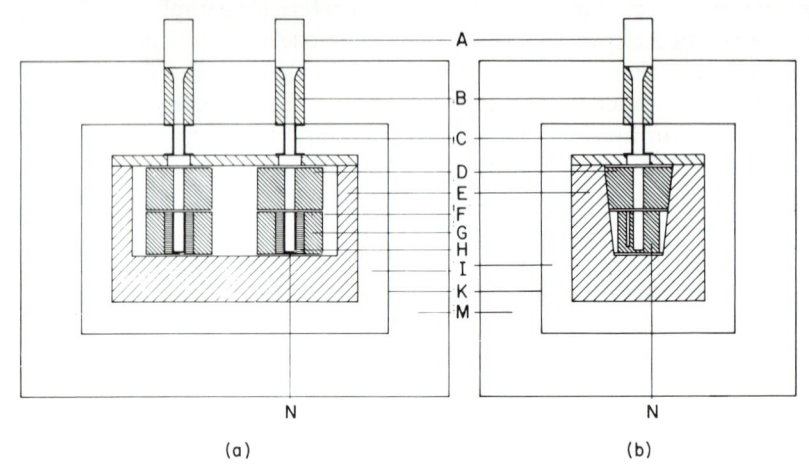

FIG. 3. Microcalorimeter for insertion of ampoules (LKB 2107–123/124). Schematic view of the calorimeter. (a) Longitudinal section; (b) transverse section. A, steel tube; B, copper constriction; C, steel tube; D, aluminium block; E, main heat sink; F, air space; G, aluminium block; H, thermocouple plate; I, air space; K, steel container; M, water thermostat; N, aluminium plate with hole for the ampoule. (Wadsö, 1974.)

sandwich-like construction with the calorimetric vessel in its centre. On each side of the vessel there are semiconducting thermocouple plates. These are surrounded by small aluminium blocks, which are in thermal contact with the main calorimetric block. Figure 2 shows schematically the LKB flow calorimeter (cf. Monk and Wadsö, 1968). Between the calorimetric units there is a heat exchange unit consisting of gold or teflon tubes which are in good thermal contact with the main block. Usually the two calorimetric units are equipped with different flow cells, in the standard LKB instrument a mixing vessel and a flow-through vessel, respectively, made from gold tubes held between thin copper plates. Gold tube diameter (i.d.) is 1 mm and the volume of the flow vessel is about 0·5 ml. Other flow-through cells used with the LKB instrument are shown in Figs 9 and 11.

The instrument shown in Fig. 3 (Wadsö, 1974) is designed for use with simple O-ring sealed ampoules (1–10 ml) or, for example, perfusion vessels of the type shown in Fig. 6. The ampoules are normally not introduced directly into the thermopile zone but are allowed to be thermostatted in the two consecutive heat exchange positions (B, D).

Other LKB calorimeters which are used in cellular work include a batch-mixing calorimeter (where the block can be rotated) and a sorption calorimeter (which can be used with a micro-column, within which cells can be attached). This latter instrument is very similar to the flow instrument shown in Fig. 2.

Although less commonly used in biological work than the Calvet and the LKB instruments, several other modern microcalorimeters, useful for heat effect measurements on cells, have been described (see Spink and Wadsö, 1976).

C. Some notes about errors and artifacts

Heat evolution or absorption takes place for all kinds of processes and systematic errors can, for this reason, easily be introduced into the results of calorimetric measurements. The risk of significant systematic errors is naturally greater in microcalorimetry than in work where comparatively large quantities of heat are dealt with. It is important to be aware of the possible sources of such errors: mechanical effects (friction etc.), evaporation or condensation processes, adsorption processes etc. (cf. Spink and Wadsö, 1976), but one must also watch for different kinds of artifacts. Fortunately, in practice many errors will cancel out by the procedure used for standardization (calibration) of the method.

1. CALIBRATION AND TEST EXPERIMENTS

Calorimetric experiments on cellular systems usually have the character of analytical experiments rather than thermodynamic measurements. It may then seem unnecessary to carry out calibrations of the instruments in terms of well-defined and correct energy units. However, it is believed that there is a general and a lasting value in well-documented energy data for biological systems. Further, in order to make a precise comparison between results from different studies, it is essential that the results are expressed in well-defined units. Thus, arbitrary units like "mm recorder deflection" should be avoided. Heat effects are correctly expressed in terms of watts (J s^{-1}).

Calorimeters are normally calibrated electrically, which is very convenient, but the experimenter should be aware of the risks associated with systematic errors. The actual measurement of electrical energy, or effect, is today a trivial procedure, which can easily be made with an accuracy far exceeding the needs in biological experiments. The problem is to make sure that the electrical energy is released in a manner that is closely comparable with that of the process studied. For practical reasons microcalorimeters used in biological work are not always well suited for strict comparisons; for instance, it is sometimes difficult to produce a calibration heater of an ideal design, or to place it in the best position. However, it is suggested that, when a new calorimetric design is tested or when modifications have been made on an existing instrument, calibration experiments should be performed with different types of heaters and/or with heaters placed in different positions. It is usually possible to incorporate a heater which can be judged to be nearly ideal, although it may not be realistic to use it in routine work. Comparison of calibration values from such heaters with those obtained with the regular heater can give the experimenter a realistic feeling for the magnitude of possible errors in the calibration value.

Problems that arise in the calibration of flow-through vessels used for power measurements of cell suspensions will now be discussed. Normally, methods used for their calibration are far from ideal and it is rare that the results are checked by a suitable test reaction. Therefore, results obtained with flow-through vessels in current use should be judged with some caution. Provided that kinetic corrections are not needed, the shape of the thermograms are presumably correct, but the power values may be seriously in error.

Let us first consider a vessel of the type indicated in Fig. 2, which consists of a thin gold tube spiral placed between copper plates. It has been shown that the heater problem is not very critical in this case; the copper plates will distribute the heat evenly along the flow path. However, with flow-through vessels of the type shown in Figs 9(a) and 11, rather serious errors can be introduced if the heater is placed in the aluminium block and not spread out inside the flow cell (which would be inconvenient). The errors can be reasonably large if the thermal contact between the flow vessel and the surrounding block is poor and if high flow rates are used (cf. Gustafsson and Lindman, 1977). With such vessels there is also a risk that errors will be caused by the entrapment of gas in the vessel so that the volume of the heat-producing suspension contained in the vessel is less than expected.

There is another, more general, problem with the assessment of the "practical" value for the volume of flow-through vessels in heat conduction calorimeters. In an experiment with a cell suspension some heat will be produced in the flow line between the heat exchanger and the flow vessel. A part of this heat will be recorded by the instrument, but in an electrical calibration experiment there will be no corresponding power produced.

In order to control electrical calibration values, and to assess a value for the "practical volume", it is sometimes rewarding to use a steady-state reaction mixture and compare calibration values for a flow-through vessel operated at different flow rates, including stopped flow. Alternatively, a calibration value can be transferred from a well-calibrated calorimeter to one for which an electrical calibration procedure is not very suitable. We have found it convenient to use the following enzyme–substrate mixture for this purpose. A strong solution of acetylcholine (c. 10 per cent) in 0·5 M Hepes buffer is adjusted to pH 7·4 and is mixed with a suitable amount of choline esterase. With a reaction rate at 37 °C corresponding to a heat production of 30 μW the power will decrease linearly by about 2 per cent per hour. The decrease in power stays linear for at least 15 h.

As yet, a test reaction for which the power can be precisely predicted, or can be calculated after performing a simple non-calorimetric rate determination, does not seem to have been developed.

In flow calorimetric experiments it is necessary to account for the power produced by the viscous heating in the flow vessel. With a heat conduction instrument this is incorporated in the "baseline value". For a cell suspension the position of the base line is usually established by pumping pure suspension medium (e.g. a buffer, nutrient broth, plasma) through the calorimeter. The base-line value thus accounts for the viscous heating, but in addition other effects, possibly due to insufficient temperature equilibration of the incoming flow, pressure effects on the thermopile and on the liquid flow, will also be included. Often the base line is close to the value at zero flow rate but for viscous liquids, and at high flow rates, the difference can be substantial. It is then very important that the liquid used in the base-line experiments has flow properties which are closely related to those of the cell suspension. At low cell concentrations there is normally no problem in this respect but for dense suspensions it is important to watch out for errors. As an example, Ikomi-Kumm (1977) observed in experiments with a 40 per cent suspension of erythrocytes a linear decrease in the power value with decreasing flow rates. The power value extrapolated to zero flow rate agreed within uncertainty limits with that obtained by use of a static method (Monti and Wadsö, 1973) where the same sample preparation method was used. The variation in (apparent) power value with the flow rate could be due to frictional effects which were not compensated for by the base-line value. But it could also be due to actual differences in metabolic rates.

In cases where the viscosity changes during the experiment, such as in microbial growth experiments, it is obviously possible to obtain a false calorimetric growth pattern for this reason. There appears to be no good working procedure, by which such effects can be overcome, except by employing low flow rates and using flow vessels where the frictional effects are minimized. Such conditions are in practice frequently difficult to arrange because of problems with supply of oxygen or due to sedimentation effects.

2. UNEXPECTED PROCESSES

It is often an advantage that calorimetric methods are so unspecific—one feature is that unexpected phenomena are likely to be detected. But there is also a risk that such effects can be misinterpreted. Two examples from studies with blood platelets (thrombocytes) will illustrate this point. Ross et al. (1973) have reported an investigation using an LKB batch microcalorimeter where the cells were mixed with different reagents. In control experiments they observed that large bursts of heat were evolved even when the cells, suspended in saline, were mixed with the same medium. Different sample preparations gave different heat effects. If the gentle agitation and mixing process was repeated, a similar but less prolonged heat production was observed. From results of other experiments it was concluded that the "agitation effect" was not due to any changes in the oxygen tension or due to the accumulation of any inhibiting metabolites. It may first be noted that if the control mixing experiments had not been performed, results of the mixing process between cell suspension and reagents dissolved in saline could have been gravely misinterpreted. We should also be aware that such "agitation effects" could remain undetected in a continuous-flow experiment where they would give rise to a constant power value. Probably such effects could be discovered if different flow rates were tested or if stopped-flow experiments were performed.

Another example is provided by the work on platelets by Bandmann et al. (1975). Using a static ampoule calorimeter method (cf. Fig. 3), they observed, during some preliminary experiments, very high power values which rapidly declined and levelled out. It was subsequently found that the initial high power value was due to aggregation of the platelets which occurred at the measurement temperature (37 °C) when the cells had been kept in the cold for a short time (0·5 h at 4 °C) before the calorimetric experiment. If these measurements had been performed as continuous-flow experiments it is possible that the results could have been misinterpreted. If a process in the cellular material can be initiated by the temperature change occurring during the transport to the flow-through vessel, the corresponding power will appear as a constant, regardless of the kinetics of the process. It may then easily be mistaken as part of a basal steady-state power value for the cells.

D. Respiration

In many types of calorimetric experiments with aerobic cells it is a difficult problem to arrange a sufficient supply of oxygen. A direct and efficient aeration of the medium in the calorimetric vessel is usually not possible as it is difficult to avoid disturbing evaporation effects.

1. STATIC CONDITIONS

Let us first consider the amount of oxygen which may be contained in a given volume of an aqueous medium. Pure water in equilibrium with air dissolves about 0.2 μmol oxygen ml^{-1} at 37 °C. If saturated with air, 1 ml of water, or aqueous solution, will thus contain enough oxygen for the complete oxidation of about 33 nmol of glucose, corresponding to an enthalpy change of about 100 mJ. For a steady-state process this would correspond, for example, to a power of the order of 13 μW during 2 h. If the aqueous medium is equilibrated with pure oxygen instead of air the amount of oxygen dissolved increases by approximately a factor of 5. Thus, with dilute cell suspensions and with very sensitive calorimeters, it is possible to make rather long experiments (hours) with aerobic cells without other provisions than air or oxygen saturation of the medium prior to the experiment.

In cases where the liquid medium is shallow or where the cellular material is floating, significant amounts of oxygen can be transferred to the biological material from a gas phase above the liquid medium even in a static system (cf., for instance, the experiments with fat tissue reported by Boivinet et al., 1968).

2. STIRRING OR AGITATION OF THE MEDIUM

If the liquid medium is stirred or agitated the rate of oxygen uptake from a gas phase will increase. Further, for a static system where the cells are not uniformly suspended in the medium, e.g. when the cells are attached to a solid support or when sedimentation takes place, only a fraction of the oxygen dissolved in the medium may be utilized before anaerobic conditions will develop at some local points.

Figure 4 shows, as an example, a reaction vessel for a Calvet microcalorimeter where a cell suspension can be agitated by a piston fitted with horizontal discs. The gas phase above the suspension is sealed from the atmosphere and it is possible to work with gas pressures in the vessel as high as 10 kPa cm^{-2}. The vessel, which has a total volume of 100 ml, is normally half filled with liquid.

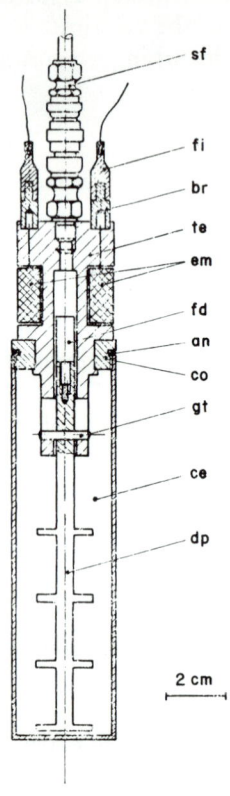

FIG. 4. Agitation cell with controlled pressure capability for use with a Calvet microcalorimeter. Agitation is accomplished by pulse-charging the electromagnet (em), which causes the agitator (dp) to move up and down in time with the pulses. The cell (ce) has a volume of 100 ml (50 ml liquid and 50 ml gas phase) and is sealed through a series of swagelock fittings (sf). (Lamprecht and Schaarschmidt, 1973.)

In the LKB batch microcalorimeter (Wadsö, 1968) the liquid contents of the two compartments of the reaction vessels are mixed and brought in intimate contact with the gas phase by rotation of the calorimeter by one complete turn (360°) and back. Total volume of the reaction vessel is about 10 ml; usually it is charged with about 6 ml of liquid reagents. This instrument was primarily designed for studies of biochemical reactions but has also been used for studies of aerobic cellular systems like tissue cells. When used for this purpose usually repeated mixing cycles are performed—see works by Cerretti *et al.* (1971) on HeLa cells and by Nedergaard *et al.* (1977) on isolated cells of brown adipose tissue.

A somewhat similar procedure for agitation of a liquid–gas system has recently been described by Fujita *et al.* (1976). A twin calorimeter, being a modified version of the design by Amaya *et al.* (1966) (cf. also Takahashi, 1973), was fitted with reaction vessels specially designed for aerobic microbial experiments. Part of the system is shown schematically in Fig. 5(a), (b). The twin calorimeter is operated as a thermal balance; the heat evolved in the active vessel is compensated by electrical heating in the reference vessel (in case of endothermic processes by Peltier effect cooling).

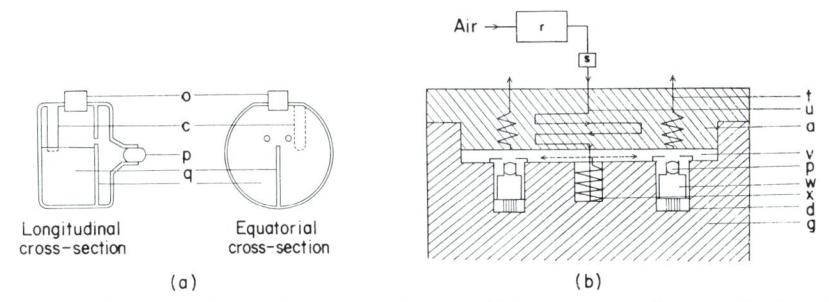

FIG. 5. Calorimeter for studies of aerobic microbial growth. (a) Longitudinal and equatorial cross-sections of a culture vessel. o, Silicone rubber cap; c, calibration heater; p, cotton-wool plug; q, partition walls. (b) Aeration assembly. r, Peristaltic pump; s, sterilizing filter; t, outlet copper tube; u, inlet copper tube; a, lid; v, air reservoir; p, cotton-wool plug; w, culture vessel; x, heat exchange unit; d, thermopile plate; g, heat sink. (Fujita *et al.*, 1976.)

The calorimetric vessel (Fig. 5(a)) consists of a short-necked cylindrical flask made from glass. The total volume is 40 ml, but the amount of liquid contained in the vessels is usually only 5 ml in order to secure efficient agitation and aeration. The agitation is achieved by continuous rotation of the calorimetric block, one turn and back. The opening at the neck of the vessel is stopped by a cotton-wool plug, which is prevented from being wetted by the liquid medium by a perforated partition wall, parallel to the bottom plane of the vessel. Another partition wall, vertical to the bottom of the vessel, separates medium and inoculum prior to the start of the experiment and serves also as a baffle plate during the agitation.

During a measurement moist air is pumped at a rate of 15 ml min^{-1} into an air reservoir in the calorimetric block above the vessels (Fig. 5(b)). Before reaching the reservoir the air has to pass a sterilizing filter outside the calorimeter and a heat exchange tube embedded in the lid of the calorimetric block. From the reservoir the air can diffuse through the cotton plugs of the reaction vessels. Excess air and exhaust gas diffusing from the vessels leave the calorimeter through outlet tubes. After being charged with medium and inoculum the equilibration time is 3 h. The sensitivity of the instrument is moderate; the highest sensitivity range used corresponds to a recorder deflection of about 40 μW mm^{-1}. However, for aerobic microbial growth experiments this sensitivity is usually adequate.

3. GAS PERFUSION

The enthalpy of vaporization of water is very high, 43·9 kJ mol^{-1} at 25 °C, which means that more than 2 mJ of heat is absorbed per microgram of water evaporated. If measurements are made at a high sensitivity it is therefore difficult to avoid significant heat effects from evaporation or condensation processes when a flow of gas is bubbled through an aqueous medium in a calorimetric vessel. Before being introduced to the calorimetric vessel the gas must be equilibrated very closely to the temperature

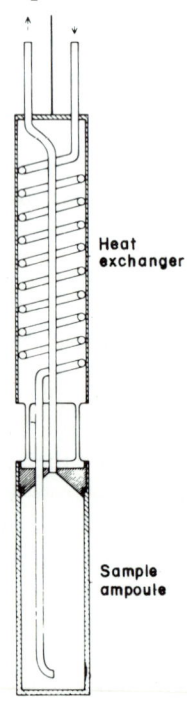

FIG. 6. Reaction vessel for gas or liquid perfusion used with calorimeter shown in Fig. 3 (Wadsö, 1974).

and vapour pressure of the medium. The gas perfusion vessel, shown in Fig. 6, was designed with these requirements in mind. This vessel is used with the heat conduction calorimeter shown in Fig. 3. The vessel consists of two parts: a sample ampoule, volume 0·5 or 5 ml, and an equilibration coil. The coil is made from stainless steel tubing, which is pressed to the inner surface of a thin-walled steel tube. The equilibration unit is permanently connected to the screw lid of the ampoule by means of thin steel strips. When this calorimetric vessel is in the measurement position within the calorimeter, the equilibration coil is in thermal contact with the heat exchange block D, Fig. 3. Before the gas is introduced to the equilibration coil it is saturated with moisture at a temperature close to that of the calorimeter. In order to further equilibrate the gas, a cotton thread wetted with the same medium as used in the experiment is placed inside the inlet tube and in the upper part of the equilibration coil. By this arrangement there will be established a well-defined gas–liquid equilibrium before the gas enters the sample container, positioned in the "heat-effect-sensitive" part of the calorimeter. The temperature difference between the equilibration block and the sample is low (c. 10^{-4} °C at a heat evolution of 10 μW) and evaporation or condensation effects are hardly significant at gas flow rates $\leqslant 10$ ml h^{-1}. Results of some test experiments are shown in Fig. 7. The variation in

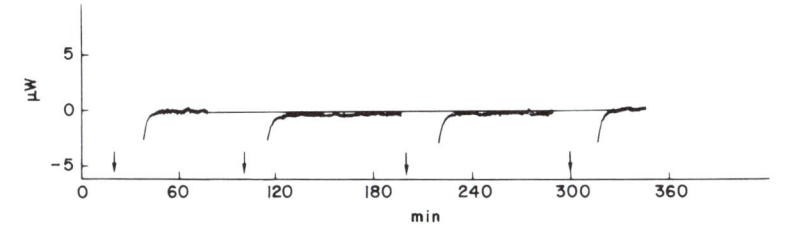

FIG. 7. Results from a series of test experiments with the perfusion vessel shown in Fig. 6. The ampoule contained 0·6 ml of water and the air flow rate was 10 ml h^{-1}. Arrows indicate the time when the vessel was inserted into tube B, shown in Fig. 3. (Wadsö, 1974.)

base-line position between experiments was about 0·5 μW. With gaseous flow (10 ml h^{-1}, 37 °C) the base-line stability over a period of 24 h typically corresponds to 2 μW. The vessel cannot be used to perfuse a liquid which tends to form foam, e.g. blood plasma.

Several years ago Poe et al. (1967) described a differential Dewar calorimeter, primarily designed for subcellular particles such as mitochondria. The sample vessel, volume 50 ml, was fitted with an aeration tube and a vibrating oxygen electrode, which also functioned as a stirrer.

More recently Nakamura and Matsuoka (1978) described a magnetically stirred dewar flask calorimeter, volume 5 ml, which was fitted with an oxygen electrode.

4. FLOW CALORIMETRY

For many types of cells the most convenient calorimetric method is flow calorimetry, where the cell suspension is continuously pumped from a reservoir outside the calorimeter through a flow-through vessel of the calorimeter. The method has been applied in many growth experiments involving yeast and bacteria, as well as in studies of blood cells and tissue cells. The cell suspension can be efficiently aerated in the reservoir without causing any thermal disturbances of the calorimetric system. However, for concentrated and fast-growing aerobic cell suspensions the dissolved oxygen may be rapidly consumed in the flow line during passage to the calorimetric vessel. The transport time for the cells between the reservoir outside the calorimeter and the calorimeter vessel is often of the order of 3–5 min and a normal residence time in the flow cell is about 1–2 min.

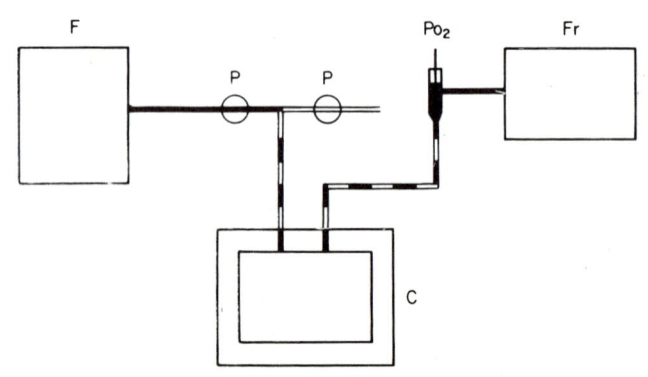

FIG. 8. Schematic diagram showing arrangement for segmented gas-suspension flow in aerobic microbial growth experiments. F, fermentor; P, pump; C, calorimeter; Po_2, oxygen electrode; Fr, fraction collector kept at low temperature. (Eriksson and Wadsö, 1971.)

Considerably higher oxygen consumption can be tolerated if a mixed flow of liquid suspension and gas is used. An experimental assembly used for this purpose is shown in Fig. 8 (cf. Eriksson and Wadsö, 1971). A cell suspension, e.g. a growing bacterial culture, is kept under controlled conditions in a fermentor from which it is pumped to a T-piece where it is met by a constant flow of air or another gas mixture. A segmented flow of suspension and gas will pass through the heat exchanger where gas–liquid phase equilibrium is established at the proper temperature, and then to the calorimetric vessel. With 1 mm (i.d.) teflon tubes and equal flow rates of liquid and gas the length of the alternating gas–liquid segments is of the order of 1 cm. At the exit of the flow calorimeter the oxygen pressure of the suspension can be tested. The suspension leaving the calorimeter can be returned to the fermentor or can be pumped to a fraction collector, which is kept at low enough temperature essentially to stop the microbial activity.

If a segmented flow of liquid and gas is pumped through the standard flow-through vessel of the LKB flow calorimeter there will usually be significant disturbances of the calorimetric signal (often a more or less regular noise in the order of 10 µW). Presumably this is due to evaporation or condensation effects resulting from small pressure variations in the system. Flow cells of the type shown in Fig. 9(a), (b) have been found to be more suitable for this purpose. Figure 9(a) shows a cylindrical

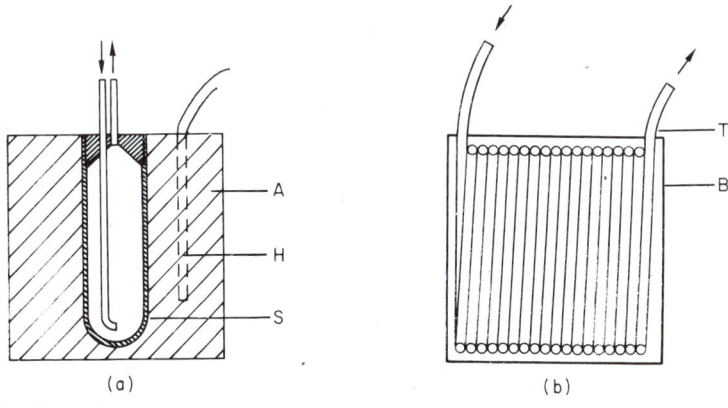

Fig. 9. Flow-through vessel used for segmented gas-liquid flow (cf. Fig. 11). (a) Steel insertion ampoule contained in an aluminium block (Eriksson and Wadsö, 1971). S, steel vessel with O-ring seal; H, calibration heater; A, aluminium block. (b) Teflon tube embedded in tin in a thin-walled metal box (Wadsö, 1973). T, teflon tube; B, metal box.

vessel made from stainless steel and closed with an O-ring sealed screw lid. Through the lid there are inlet and exit tubes. The cylinder is positioned in the hole of an aluminium plate which is surrounded by the thermocouple plates in the calorimeter block.

The performance of this flow vessel was tested in several growth experiments with *Escherichia coli*, grown aerobically on glucose at 27 °C (Eriksson and Wadsö, 1971; Eriksson and Holme, 1973). Bacterial suspension and pure oxygen were both pumped at a rate of 40 ml h^{-1}. It was found that the oxygen concentration tended to be limited during the logarithmic growth if the cell density exceeded 3 g l^{-1}. Without use of mixed liquid–gas flow, oxygen starvation occurred at a cell density lower by an order of magnitude.

When used with a mixed liquid–gas flow, the flow vessel shown in Fig. 9(a) has a somewhat poorly defined liquid volume, as a small air volume can be trapped below the exit tube. The true mean liquid volume should therefore be determined empirically for different flow rates and gas/liquid ratios. This difficulty is avoided with the flow vessel shown in Fig. 9(b) (cf. also Kemp, p. 113). It consists of a teflon tube embedded in tin in a thin-walled steel or brass box positioned between the thermocouple plates. The inner diameter of the teflon tube should preferably not be larger than 1 mm. If wider tubes are used the flow of the gas segments tends to be unpredictable. A flow-through cell designed for aeration of yeast cultures having particularly fast sedimentation rates is discussed below (Fig. 11).

For cells attached to a solid support, it seems as if liquid perfusion techniques, where fresh and oxygenated medium is pumped continuously through the system, are the most suitable. The LKB sorption microcalorimeter has been used by Kemp for such experiments with chick fibroblasts (1975, cf. Kemp, p. 113). The calorimetric vessel contained a small column (0·5 ml) filled with glass beads, average diameter 0·2 mm. During an incubation period outside the calorimeter, the cells were allowed to attach to the glass beads forming a monolayer. The column was then placed in the calorimetric vessel and was continuously perfused at a rate of 5 ml h^{-1} with culture medium. With the estimates made above for oxygen solubility, it is clear that this perfusion rate is more than sufficient for an adequate supply of oxygen in experiments like this, where the heat effect level normally is of the order 10 μW.

The air perfusion vessel shown in Fig. 6 is also useful for liquid perfusion experiments, maximum liquid rate being about 20 cm^3 h^{-1}.

5. BIOTECHNICAL PROCESSES

For systems with a very high oxygen consumption, such as many aerobic microbial growth processes of biotechnical interest, there appears, as yet, to be no suitable calorimeter available. In addition to problems with oxygenation, such cultures are often very viscous, which makes the conventional flow or agitation procedures, used with existing calorimeters, unsuitable. In such cases where the heat effects produced are large, typically of the order of a few watts per litre, it may be preferable to use a simple thermometric technique such as that described by Mou and Cooney (1976). These workers used a thermistor to measure the rate of temperature increase of the fermentation broth contained in a 7 l fermentor.

During the temperature measurement the temperature control system of the fermentor was turned off.

An alternative method would be to measure the flow rate of the cooling water and its temperature increase, preferably together with estimates of possible variations in power drawn into the fermentor by the stirrer and any cooling effect due to a non-perfect equilibration of the air introduced. This latter approach should be particularly suitable for large fermentors where the surface-to-volume ratio is small and thus most of the heat evolved will be removed by the controlled cooling effect.

6. MEASUREMENTS ON SOIL

In soil microorganisms occur in large numbers, although generally the majority of them are either dormant or resting because of lack of suitable energy sources or growth factors. For a soil sample the heat effect can vary within wide limits, but typically it is in the range of 5–50 $\mu W\ g^{-1}$ (25 °C).

Normally, the microbial metabolism in the top layer of soil is aerobic. Oxygen is thus consumed and carbon dioxide is produced. During the course of a recent methodological study it was shown that the microbial heat production in soil can be drastically influenced by the gas phase composition in the calorimetric vessel (Ljungholm *et al.*, 1979; cf. Mortensen *et al.*, 1973). Calorimetric experiments were performed with the type of heat conduction calorimeter shown in Fig. 3. During the calorimetric measurements, soil samples were hermetically enclosed in 10-ml stainless steel ampoules. The time for the calorimeter to reach thermal equilibrium after introduction of the ampoule was about 30 min; baseline stability of the instrument was about 1·5 μW per 24 h.

There often occurred a rapid reduction in the heat effect, particularly if the soil had been enriched with glucose. Results indicated that the decreased heat effect was due to high levels of carbon dioxide, rather than to depletion of oxygen. In order to avoid significant changes in the gas phase during long-term experiments the following ampoule technique was therefore developed. The samples were enclosed in polythene ampoules with top and bottom consisting of membranes made from 1 mm silicone rubber (Fig. 10). This latter material has a high permeability both for oxygen and for carbon dioxide. The plastic ampoule fits snugly into the calorimetric ampoules. During an extended experimental period (weeks or months), the plastic ampoule can be exposed to a controlled atmosphere outside the calorimeter, except for the brief calorimetric observation periods during which the insert ampoule is enclosed by the steel ampoule.

Gas chromatographic analysis verified that there is no significant depletion of oxygen or increased concentration of carbon dioxide in the insert ampoules charged with soil, provided that the ampoules are freely exposed to air. However, if they were enclosed in the calorimetric ampoule inhibitory concentrations of carbon dioxide were observed within a few hours, in particular for samples enriched with glucose. It was shown that carbon dioxide which has accumulated in the insert ampoule leaks out within 1 h if it is freely exposed to air.

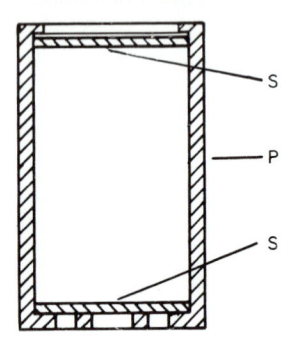

Fig. 10. Insertion plastic ampoule used for measurements of microbial activity in soil (Ljungholm *et al.*, 1979). S, silicone rubber membrane; P, polyethene plastic ampoule.

An alternative method for long-term calorimetric experiments with soil is to store a large sample of soil under suitable conditions and to charge the calorimetric ampoule for each calorimetric observation. However, random differences between samples then tend to obscure small systematic changes occurring in the soil. Measurements performed separately on each individual sample are therefore more informative.

E. Sedimentation

Partial or complete sedimentation of cellular material during a calorimetric experiment can in some cases lead to serious errors and in any case to poorly defined power values. For some types of cells like granulocytes and lymphocytes very pronounced "crowding effects", i.e. inhibition of the metabolic rates with increasing cell concentration, have been demonstrated. For lymphocytes this inhibitory effect increases exponentially with cell concentration (Hedeskov and Essman, 1966). In such cases it is important to know not only the total quantity of cells or the average cell concentration in the reaction vessel; for a well-defined power value the concentration must also be uniform. In other cases, where one cannot talk about "crowding effects" the metabolic rate can be affected by local differences in concentrations of, for example, oxygen or of pH, which may develop as a result of sedimentation.

If the liquid medium is not agitated it is likely that temperature gradients will develop in the calorimetric vessel if sedimentation takes place. In the case of a well-designed thermopile heat conduction calorimeter, such temperature gradients *per se* should not lead to any error in the power determinations as it is a representative part of the total heat flow which is measured. However, with, for instance, an isoperibolic calorimeter using one thermistor as temperature sensor, an adequate agitation of the liquid is always important.

In a flow calorimetric experiment sedimentation of cells prior to the flow reaction vessel can lead to systematically low power values. Probably more common, the cells may preferentially sediment in the flow vessel and power values recorded can then be increasingly larger than the value representative for the nominal cell concentration.

For many types of cells like bacteria, blood platelets, or mycoplasmas there are normally no problems with sedimentation, even in static vessels, during short calorimetric experiments. For long experiments sedimentation can easily be prevented by a very gentle stirring or agitation or by using a flow method. However, for large and heavy cells, and with aggregates of cells, it can be very difficult to prevent sedimentation during a calorimetric experiment. Below some problems noted for yeast cells and for blood cells are discussed.

1. YEAST CELLS

Yeast cells normally have a very fast sedimentation rate. Several types of batch calorimetric vessels have been designed to meet this problem and at the same time provide for an aeration of the suspension—see examples discussed in the earlier paragraph. Here some experiences with flow calorimetric procedures will be summarized. In a flow microcalorimeter used for measurements on cellular material the flow lines normally have an i.d. of 1–1·5 mm and flow rates used are commonly of the order 50 ml h^{-1} or less. Under such conditions it is likely that yeast cells will sediment to an appreciable extent in the flow lines or in the calorimetric flow vessel. Experiences in our laboratory have indicated that with a mixed flow of yeast suspension and gas there will normally be no sedimentation in the teflon tubes leading to the calorimeter (cf. also Brettel *et al.*, 1972; Poole *et al.*, 1973). However, with the type of flow vessel shown in Fig. 9(a) there will still be a risk of sedimentation of heavy particles. For this reason the narrow tube system used with the vessel shown in Fig. 9(b) is normally preferable. For some yeast cultures there appear to be particularly difficult sedimentation problems. Gustafsson and Lindman (1977) have recently reported a flow calorimetric study using an LKB flow microcalorimeter involving the halotolerant yeast *Debaryomyces hansenii*. They found both flow vessels shown in Fig. 9 to be unsuitable because of sedimentation. After testing several modifications of the vessel shown in Fig. 9(a) these workers obtained good results with the vessel shown in Fig. 11. It consists

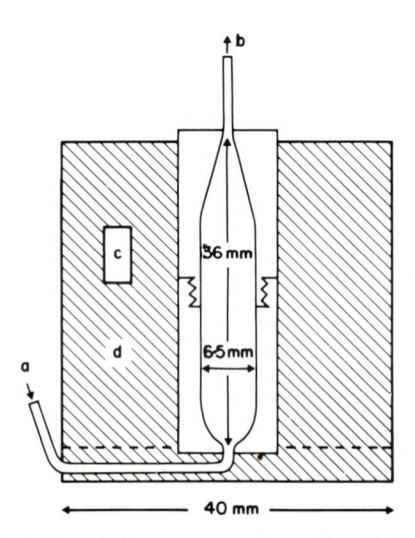

FIG. 11. Flow vessel designed for segmented gas–liquid flow. a, Inlet tube; b, outlet tube; c, calibration heater; d, aluminium block. (Gustafsson and Lindman, 1977.)

of a cylindrical tube with a conical upper part, volume 0·9 ml. The calorimetric experiments were normally performed with very high liquid flow rates, up to 170 ml suspension h^{-1}, segmented with air with a flow rate of 50 ml h^{-1}. The LKB flow calorimeter was for this reason equipped with extra heat exchangers. Also with lower flow rates, about 50 ml h^{-1} each of suspension and of air, good results were obtained and no tendency to sedimentation was observed even in long experiments (20 h). Results suggested, however, that the electrical calibration method used probably led to significant systematic errors (cf. p. 253). Electrical calibration values should thus preferably be checked by use of a test reaction.

2. BLOOD CELLS

Calorimetric measurements on blood cells, and other human cells, are of general physiological interest, but such studies have so far mainly been motivated by their potential use in clinical analysis. In order to make a calorimetric technique interesting for clinical work it must be characterized by a high sensitivity and small sample requirements. Further, precision must be high and working procedures should be simple, reliable and suitable for automatization. For this reason we have in our group used a static ampoule method in most of our work on blood cells. Calorimeters of the type shown in Fig. 3 have been used, normally together with simple sample ampoules. (In routine analytical work ampoules could be of a disposable type.) One millilitre or less of cell suspension, corresponding to *in vivo* concentration conditions, is adequate and the precision of the power measurements is more precise than that of other methods tested, e.g. colorimetric procedures. A major disadvantage with the method is, however, that significant sedimentation of erythrocytes and leucocytes can take place during a measurement.

For erythrocytes, which sediment rapidly, there is no evidence of any significant crowding effect and there is no problem with supply of oxygen. Normally excellent steady-state values are recorded which supports the view that variation in cell concentration (due to sedimentation) will not significantly influence the heat production. It has been noted, however, that if the suspension is kept statically in the ampoule for about 1 h, the pH will be slightly lower in the dense bottom part of the suspension compared to the upper part. The difference is of the order of 0·03 pH units corresponding to a power difference of about 4 per cent (Monti and Wadsö, 1976).

More serious problems occur with granulocytes and lymphocytes (for which fractions the measured power values are very low, typically of the order 5 μW ml^{-1}). With these cells, values for the heat production per cell decrease significantly with increasing cell concentration. For granulocytes suspended in phosphate buffer the decrease is about 1 per cent, when the cell concentration increased by 1 per cent, within the concentration range $2-10 \times 10^6$ cells ml^{-1} (Bandmann and Wadsö, 1977). Such results can be interpreted as a crowding effect, but it should be noted that the experiments were not performed under well-defined conditions of cell concentration, as sedimentation occurred during the measurement. The experimental conditions may still be acceptable in clinical analytical experiments, however, as long as the conditions are strictly defined. Similar results were obtained in experiments where the suspension was gently agitated by air perfusion using the vessel shown in Fig. 6, but in this case the conditions were less well defined because of an extensive cell adhesion to the vessel (see the following section).

F. Adhesion

Many types of cells have a tendency to adhere to the walls of the calorimetric vessel or to the tubes in a flow calorimeter. As for sedimentation there are two different problems which can be caused by such effects. First, the metabolic activity for a cell which is free in suspension may be different to one which has been attached to a wall. For flow systems, we have the additional problem of poorly defined quantities of cells contained in the calorimetric vessel.

1. BACTERIA

For experiments of short duration wall growth by bacteria can usually be neglected. But it is a common experience that in extended bacterial growth experiments, such as with continuous cultures, the performance of analytical sensors can be seriously impaired by cell attachments. Eriksson and Holme (1973) have reported studies on continuous culture experiments using an LKB flow calorimeter which was equipped with a flow vessel of the type shown in Fig. 9(a). Experiments were performed aerobically using a mixed flow of oxygen and bacterial suspension, (Fig. 8). The experimental set-up operated satisfactorily for some time in the monitoring of the metabolic rate of the continuous culture contained in the fermentor. However, after 10–15 h there were signs of wall growth in the calorimetric vessel as shown by drifts in the recorded power values for the system which supposedly was at steady state. Possibly a teflon tube vessel of the type shown in Fig. 9(b) would have been less susceptible to wall growth. In any case, it ought to be practically possible to clean such vessels, at suitable intervals, by brief flushings with a cleaning liquid.

2. GRANULOCYTES

Among the major fractions of blood cells it is the granulocytes which are most difficult to handle in a calorimetric experiment. Earlier, problems with their sedimentation and crowding effects were mentioned. However, their pronounced tendency for adhesion to most surfaces seems to be even more problematic. In a typical calorimetric experiment run under static conditions 5×10^6 cells were contained in the 1 ml calorimetric ampoule. This was made from stainless steel but was coated with teflon (Bandmann and Wadsö, 1977). After about 1 h at 37 °C the average cell count in the suspension had decreased by 10 per cent if a phosphate buffer was used as suspension medium. In plasma suspension up to 50 per cent of the granulocytes had become attached to the wall of the ampoule. If the perfusion ampoule (Fig. 6) was used with an air flow of 20 ml h^{-1}, about 50 per cent of the cells disappeared from a phosphate suspension during 1 h. If no air flow was used the decrease was the same as for the standard ampoules, about 10 per cent. For cells with these properties the most suitable conditions for measurements would probably be obtained if the cells were allowed to adhere quantitatively to the walls of the reaction vessel or to some support contained therein (see section F.3).

Levin (1970; cf. Levin, p. 131) has reported flow calorimetric experiments on mixtures with platelets and leucocytes suspended in plasma. An LKB calorimeter equipped with a standard flow-through vessel made from 18 or 24 carat gold tubing was used. As is normally done in the case of delicate cell material, and with cells which are sticky, the peristaltic pump was placed after the calorimeter and the suspension was thus sucked through the instrument. Cell counts performed on the suspension leaving the calorimeter showed a substantial reduction in concentration of granulocytes, at times 50 per cent of the original count. On no occasion did Levin observe any significant reduction in numbers of platelets or monomorphonuclear leucocytes (i.e. mainly lymphocytes). Attempts to overcome the retention of the granulocytes by siliconizing the tube system were unsuccessful. The instrument base line was established with plasma before and after passage of the cell suspension. Sometimes very substantial base-line shifts were noted, typically of the order 5 μW, which approximately corresponds to 25 per cent of the total power recorded for the leucocyte mixture. The results clearly show that the granulocytes adhered to a significant extent to the flow vessel and that such effects can lead to very large errors for which it is difficult to correct.

In a subsequent study on leucocyte mixtures Levin (1973) used a modified design of the LKB calorimeter where all flow lines, including heat exchanger and the flow vessel, were made from teflon tubing, i.d. 1·2 mm. In order to get an even flow of the leucocytes air bubbles were introduced into the liquid stream before the calorimeter. Levin noted that when a leucocyte suspension was passed through the teflon tubing uninterrupted by air bubbles, there was a pronounced tendency for the cell concentration to increase in the first part of the sample. The air bubbles were removed before reaching the calorimetric vessels. With this method Levin obtained a considerably better return to the base line after the leucocytes had been sucked through the system. However, from experiences in our group we are left with the impression that flow arrangements of the type described by Levin (1973) are difficult to get into a dependable performance.

3. TISSUE CELLS

Studies with tissue cells are frequently intentionally made with the cells attached as monolayers to a solid support, such as the walls of the reaction vessel or to glass beads contained in the vessel. A calorimetric technique using this latter method was employed by Kemp (1975) in a study on chick fibroblasts (see also Kemp, p. 113).

In the recent study by Ljungholm *et al.* (1978) heat effect measurements were made on HeLa cells, which had adhered to the bottom of a static calorimetric ampoule. The ampoule, containing 3 ml of cell suspension (*c.* 1.5×10^8 cells) and about 4 ml of air, was incubated statically for 16–18 h at 37 °C. During this time 90–95 per cent of the cells adhered to the bottom of the ampoule (area 2·5 cm²). The adhered cells were washed and fresh medium was added to the ampoule before the ampoule was presented to the calorimeter. Measurements of oxygen concentration, performed in experiments run in parallel to the calorimetric measurements, indicated that the oxygen concentration remained at an adequate level.

Another example of calorimetric measurements on tissue cells in monolayer under static conditions has been reported by Nicolić and Nešcović (1975) (see Kemp, p. 113).

G. Conclusions

Calorimetric investigations on living cells represent a vast experimental area, where many specialized instrumental properties and working procedures are needed. Usually, properties like instrument sensitivity, stability or precision are not the most problematic properties in current microcalorimetric experiments with cells. They rather deal with the characterization of the conditions for the cells in the calorimetric vessels, during the calorimetric measurements, which are not always satisfactory. It is felt that further attention must be given to the design of specialized reaction vessels where the following conditions for the cells can be verified during a calorimetric measurement: supply of oxygen, sedimentation, adhesion, cell concentration in a flow vessel, pH, etc. More attention should also be given to the development of processes suitable for tests and calibrations of microcalorimeters for heat effect measurements.

Finally, for many cellular systems it is important that values will be determined for the influences of various experimental parameters on the heat effect values. In cases where such relationships are known—in particular for non-growing cells—it will be possible to recalculate well-documented results to a chosen set of standard conditions. The use of such "standard" power values is believed to be essential when results of different investigations are to be compared.

References

AMAYA, K., TAKAGI, S. and HAGIWARA, S. (1966). Abstract of papers, 2nd Japanese Calorimetry Conference, Tokyo.
BANDMANN, U. and WADSÖ, I. (1977). Unpublished.
BANDMANN, U., MONTI, M. and WADSÖ, I. (1975). *Scand. J. clin. Lab. Invest.* **35**, 121–127.
BELAICH, J. P. and SARI, J. C. (1969). *Proc. natn. Acad. Sci. U.S.A.* **64**, 763–770.
BOIVINET, P., GARRIGUES, J. C. and GRANGRETTO, A. (1968). *C.r. Séanc. Soc. Biol.* **162**, 1770–1774.
BRETTEL, B., CORTI, I., LAMPRECHT, I. and SCHAARSCHMIDT, B. (1972). *Studia Biophysica (Berlin)*, **34**, 71–76.
CERRETTI, D. P., DORSEY, J. K. and BOLEN, D. W. (1977). *Biochim. biophys. Acta*, **462**, 748–758.
ERIKSSON, R. and HOLME, T. (1973). *Biotechnol. Bioeng. Symp.* No. 4, 581–590.
ERIKSSON, R. and WADSO, I. (1971). In "First European Biophysical Congress", Baden (E. Broda, A. Locker and H. Springer-Lederer, eds), Part IV, pp. 319–327. Verlag der Wiener Medizinischen Akademie.
FUJITA, T., NUNOMURA, K., KAGAMI, I. and NISHIKAWA, Y. (1976). *J. gen. Microbiol.* **22**, 43–50.
GUSTAFSSON, L. and LINDMAN, B. (1977). *FEMS Microbiol. Lett.* **1**, 227–230.
HEDESKOV, C. J. and ESMANN, V. (1966). *Blood*, **28**, 163–174.
IKOMI-KUMM, J. A. (1977). *IRCS Med. Sci.* **5**, 320.
KEMP, R. B. (1975). *Pestic. Sci.* **6**, 311–325.
LAMPRECHT, I. and SCHAARSCHMIDT, B. (1973). *Bull. Soc. Chim., France*, 1200–1201.
LEVIN, K. (1970). *Clinical chim. Acta*, **32**, 87–94.
LEVIN, K. (1973). *Scand. J. clin. Lab. Invest.* **32**, 67–73.
LJUNGHOLM, K., KJELLÉN, L. and WADSÖ, I. (1978). *Acta Path.* **86**, 121–124.
LJUNGHOLM, K., NORÉN, B., SKÖLD, R. and WADSÖ, I. (1978). Unpublished.
MONK, P. and WADSÖ, I. (1968). *Acta chem. scand.* **22**, 1842–1852.
MONTI, M. and WADSÖ, I. (1976). *Scand. J. clin. Lab. Invest.* **36**, 565–572.
MORTENSEN, U., NORÉN, B. and WADSÖ, I. (1973). *Bull. Ecol. Res. Commun. (Stockholm)*, **17**, 189–197.
NAKAMURA, T. and MATSUOKA, I. (1978). *J. Biochem.* **84**, 39.
MOU, D.-G. and COONEY, C. L. (1976). *Biotechnol. Bioeng.* **18**, 1371–1392.
NEDERGAARD, J., CANNON, B. and LINDBERG, O. (1977). *Nature, Lond.* **267**, 518–520.
NICOLIĆ, D. and NEŠKOVIĆ, B. (1976). *Jugoslav. Physiol. Pharmacol. Acta*, **12**, 191–197.

Poe, M., Gutfreund, H. and Estabrook, R. W. (1968). *Acta Biochem. Biophys.* **122**, 204–211.
Poole, R. K., Lloyd, D. and Kemp, R. B. (1973). *J. gen. Microbiol.* **77**, 209–220.
Ross, P. D., Fletcher, A. P. and Jamieson, G. A. (1973). *Biochim. biophys. Acta*, **313**, 106–118.
Schaarschmidt, B. and Lamprecht, I. (1973). *Experientia*, **29**, 505–506.
Spink, C. and Wadsö, I. (1976). *In* "Methods of Biochemical Analysis" (D. Glick, ed.), pp. 1–159. John Wiley, New York.
Takahashi, K. (1973). *Agric. biol. Chem.* **37**, 2743–2747.
Wadsö, I. (1968). *Acta chem. scand.* **22**, 927–937.
Wadsö, I. (1973). Unpublished.
Wadsö, I. (1974). *Sci. Tools*, **21**, 18–21.

Calorimetric Studies of Biomembranes and their Molecular Components

D. BACH
and
D. CHAPMAN

A. Introduction

In recent years there has been considerable interest in the structure and function of biological membranes. The electron microscope revealed that all the various organelles of the cell are compartmentalized by their own membrane system. Thus, in addition to the outer or plasma cell membrane, there are also membranes associated with the nucleus, the mitochondria, the endoplasmic reticulum, etc. It has also been realized that biological membranes act not only as semipermeable barriers to ions and various molecules, but also act as organizing and transduction systems. They organize the enzymes of mitochondrial membranes and the chlorophyll for the quantum conversion process of the chloroplast.

Calorimetric studies of biological membranes (biomembranes) have become popular in recent years following the observation of marked endothermic phase transitions of some of the important components making up their structure. This itself followed from the separation by biochemists of various biomembrane systems from cell structures and the analysis of their molecular components. It was shown that phospholipid molecules (as well as proteins) were important components of biomembrane structures. In certain biomembranes another important component, cholesterol, is also present. Carbohydrate groups are also found but usually attached to the lipids or the protein molecules.

Phosphatidic acid

$$R'-\overset{O}{\underset{\|}{C}}-O-\underset{\underset{\underset{H_2}{C}-O-\underset{\underset{OH}{\|}}{\overset{O}{\underset{\|}{P}}}-OH}{|}}{\overset{\overset{H_2}{C}-O-\overset{O}{\underset{\|}{C}}-R''}{|}}{CH}$$

Phosphatidylcholine

$$R'-\overset{O}{\underset{\|}{C}}-O-CH\text{ (with sn-2 acyl and sn-3 phosphocholine)}$$

Phosphatidylethanolamine

Ethanolamine plasmalogen (phosphatidalethanolamine)

Phosphatidylserine

Phosphoglyceride structures

Continued...

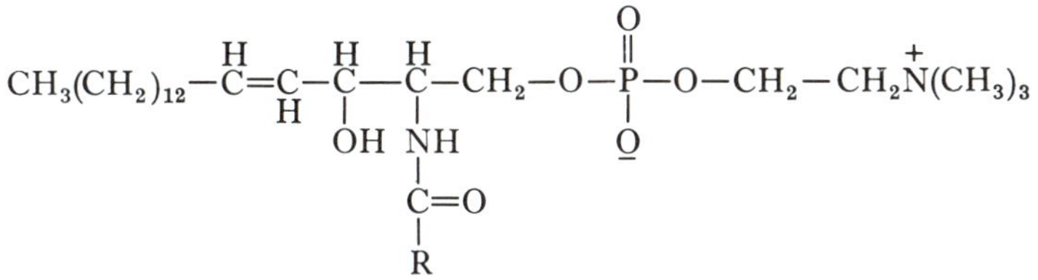

DIFFERENTIAL SCANNING CALORIMETRY

$$CH_3(CH_2)_{12}-\overset{H}{\underset{H}{C}}=\overset{H}{\underset{OH}{C}}-\overset{H}{\underset{NH}{C}}-\overset{H}{\underset{|}{C}}-CH_2-O-\overset{O}{\underset{\underline{O}}{\overset{\|}{P}}}-O-CH_2-CH_2\overset{+}{N}(CH_3)_3$$
$$\underset{R}{\overset{|}{C=O}}$$

Sphingomyelin

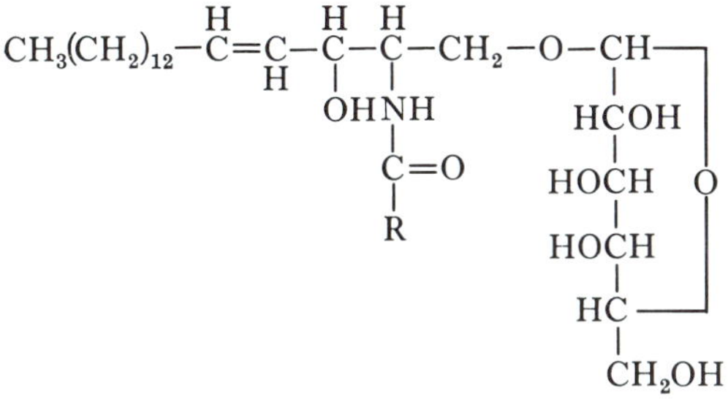

(Galacto-) cerebroside

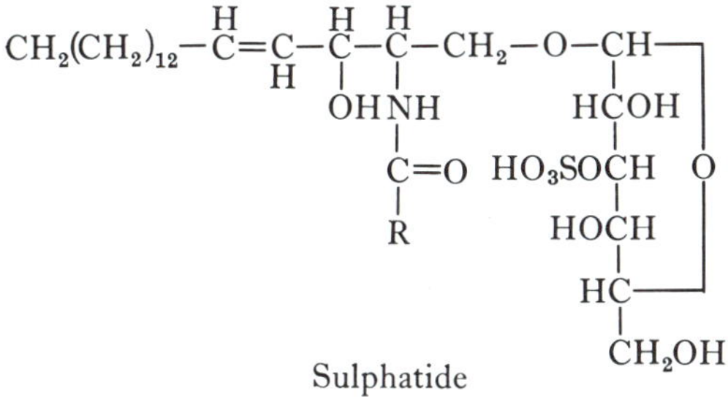

Sulphatide

Sphingolipid structures

FIG. 1. The structures of phosphoglycerides and sphingolipids.

The lipid molecules making up the matrix of biomembranes are often phospholipid molecules and any given biomembrane may contain various phospholipid classes, e.g. lecithins, phosphatidyl ethanolamines, phosphatidyl serines and sphingomyelins. In natural biomembranes these various lipid classes are arranged in an asymmetric fashion. The structures of some of these molecules are shown in Fig. 1. Associated with each of these phospholipid classes in natural biomembranes is a range of fatty acids of varying chain length and degrees of unsaturation. The fatty acid distribution is usually characteristic for a given membrane system. Thus excitable membranes contain as many as six double bonds whilst myelin membranes contain fatty acids which are much more saturated.

After some uncertainties it was realized that natural biomembranes are usually arranged in a bilayer arrangement comprised of a two-molecule-thick layer of lipid molecules. The phospholipid molecules therefore form the basic matrix into which the other components fit (e.g. the cholesterol molecules and the protein molecules). Some proteins penetrate into the lipid bilayer matrix (intrinsic or integral proteins) whilst other proteins remain attached to the outside.

The lipid molecules themselves (e.g. the lecithins) when in water spontaneously form this bilayer structure and studies of these systems have been most valuable for providing insight into the behaviour and the mutual interactions which the components have in the final composite biomembrane structures. Calorimetric studies of the phospholipid molecules in the anhydrous state, in water, with cholesterol and with intrinsic proteins have all been made and give important information. In this contribution we shall describe studies which have been made of the thermal properties of these molecular components and natural biomembranes using calorimetric methods.

Two main calorimetric techniques have been applied to the studies of phospholipids and biomembranes. These are differential thermal analysis and differential scanning calorimetry.

Scanning calorimetry, as contrasted with isothermal calorimetry, is carried out with the material under study heated at a programmed rate and the enthalpy accompanying any thermally induced change evaluated. The sensitivity of the method is much increased if a differential instrument is employed in which the enthalpy requirement of the sample under test is compared with that of reference material. This technique, which is closely related to differential thermal analysis, has been employed for many years in the characterization of a wide variety of materials, and highly developed commercial instruments are available.

In the commercially available instruments (Perkin Elmer Corporation DSC 1B, DSC 2, Du Pont 990) the sample and reference are heated independently so that their temperatures are at all times equal or in a constant relation to one another. During a thermal transition the differential power required to maintain the sample temperature is measured. This differential power is directly related to the heat absorbed or evolved by the sample during the transition. The temperature of the sample and reference holders is recorded. In the case of the first two instruments calibration of the temperature scale is required using substances of known melting point. Solid samples are examined as compressed discs encapsulated in aluminium or gold pans. Volatile samples are sealed into aluminium pans of 20 μl or 80 μl capacity. The sample atmosphere is dynamic with a purge gas of nitrogen or helium flowing through the sample chamber continuously. For sharp well-defined transitions the transition temperature corresponds to the point of departure from the baseline. The displacement of the DSC base line from the isothermal is related to the heat capacity of the sample. The heats of transition are obtained from the areas of the transition peaks. Absolute values for specific heat are obtained by comparison with an external standard.

B. Thermodynamic data from DSC

The primary importance of DSC lies in its ability to give detailed thermodynamic information for transitions. If we consider the two-state process $A \rightleftharpoons B$, the direct information obtained $\Delta H = M\Delta q$ usually evaluated at T_c and $\Delta C_p = M\Delta c_p$, where Δc_p is the change in specific heat. It is evident that the equilibrium constant is equal to unity at T_m so that at this temperature $\Delta G^0 = 0$ and

$$\Delta S^0 = \Delta H_{T,m}/T_m \qquad (1)$$

For a two-state process $A \rightleftharpoons B$ the enthalpy of phase change is given by the van't Hoff equation:

$$\Delta H_{VH} = 4RT_m^2/(T_2 - T_1) \qquad (2)$$

where T_1 is the temperature of the beginning of the transition, T_2 the temperature at the end of the process, T_m the temperature where half of the reaction has taken place.

For sharp transitions the enthalpy of melting can be obtained from the van't Hoff equation; it is given in calories per mole.

From the area under the peak of the thermogram the enthalpy of melting in calories per gram is obtained. The ratio of the two calculated enthalpies is in grams per mole—by comparing this value with the monomolecular weight of the substance undergoing melting the size of the cooperative unit can be obtained.

C. Thermotropic phase transitions of lipids

1. ANHYDROUS LIPIDS

Thermotropic transitions of lipid-type molecules have been studied for many years. The complex mesomorphism of soap systems in particular received much early attention (Lawrence, 1938; Vold, 1941).

With anhydrous sodium palmitate, five phase transitions were observed between the crystalline state and the isotropic melt (Vold et al., 1941). The various regions were referred to as curd to waxy, waxy to subneat, subneat to neat, and neat to isotropic. Nordsieck et al. (1948) examined these systems using X-ray methods and showed that sharp, short spacings occur up to 110 °C and that above this temperature the short spacings are diffuse. They considered that their data showed that there was order in the long spacing direction and liquid or amorphous order laterally. The 4·8 Å halo was interpreted to indicate hydrocarbon chains in a loose hexagonal packing arrangement.

An examination of a given class of these high melting point phospholipids reveals that the melting points are mainly independent of the fatty acid residues present. For example, 1-stearoyl-2-oleoyl- and 1,2-distearoyl-L-phosphatidylcholine both have melting points of 230–231 °C. The melting points of these phospholipids are not markedly affected by the number of acyl chains present. The melting points of the lyso-derivatives are slightly higher than those of the corresponding diacyl phospholipids. It is thus quite clear that the polar head group of these phospholipids is the major factor controlling their melting points (Williams and Chapman, 1970).

In addition to the capillary melting point, phase changes occur with phospholipids at lower temperatures, e.g. when a pure phospholipid, dimyristoylphosphatidylethanolamine, containing two fully saturated chains, is heated from room temperature to the capillary melting point, a number of thermotropic phase changes occur. This was first shown by i.r. spectroscopic techniques (Byrne and Chapman, 1964), then by thermal analysis (Chapman and Collin, 1965), and has now been studied by a variety of physical techniques (Chapman et al., 1966). Optical studies show that dimyristoylphosphatidylethanolamine at room temperature is birefringent under crossed polars. On heating, three processes occur: first, some loss of birefringence at the first transition temperature ~120 °C, then a small increase at 135 °C followed by a pronounced overall loss of birefringence near the capillary melting point of 200 °C. Above 120 °C, pressure on a cover-glass with a needle causes the material to flow. When the temperature of the phospholipid reaches ~120 °C the i.r. absorption spectrum undergoes a remarkable change. Above this temperature the spectrum loses all the fine structure and detail which was present at lower temperatures, and the spectrum becomes similar to that obtained with a phospholipid dissolved in a solvent such as chloroform.

Differential thermal analysis (d.t.a.) shows that a marked endothermic transition (absorption of heat) occurs at this transition temperature. An additional heat change occurs at 135 °C and only a small heat change is involved near the capillary melting point of the lipid. This behaviour is similar to that which occurs with liquid crystals, such a p-azoxyanisole or cholesteryl acetate which form nematic and cholesteric liquid crystalline phases.

The lecithins or phosphatidylcholines have also been extensively studied (Chapman et al., 1967). Series of d.t.a. heating curves for the a_1 form of the 1,2-diacyl-L-phosphatidylcholine monohydrates have been obtained. In each case a pronounced endothermic transition is observed, corresponding to the melting of the hydrocarbon chains of the lecithins. The temperature of transition is chain-length dependent, as with the phosphatidylethanolamines. Egg yolk lecithin, which gives a much broader endotherm than do the phosphatidylcholines of a single discrete chain length, does not behave as a homogeneous phase at this transition, but behaves as a mixture of chain lengths.

The transition at higher temperatures from liquid crystal to isotropic liquid involves only a breakdown of the polar lattice since the melting points are chain length independent. Also the heat and entropy changes associated with this transition are very small and therefore there is only some fragmentation of the ordered structure and little significant change in the chain motions.

2. HYDRATED LIPIDS

Phospholipids exhibit interesting behaviour in the presence of water. As with their thermotropic mesomorphism, where they do not pass directly from the crystalline state to a liquid, the phospholipids, in general, do not pass directly from the crystalline state to a solution in the presence of water. Various hydrated phases are encountered before solution of the phospholipids in water occurs. Such behaviour is called lyotropic mesomorphism. The lyotropic phases exhibit thermotropic mesomorphism; in other words, the particular phase obtained is a function both of water content and of temperature.

The importance of the thermotropic phase transition temperature can be seen when we appreciate that when water diffuses into the lattice it does so into the polar (ionic) region only when the temperature is reached at which the hydrocarbon chains "melt". If the temperature is higher than this there is a simultaneous dissociation of the ionic lattice by the penetration of water and melting of the hydrocarbon chain region. The temperature of transition (T_c line) depends upon the nature of the hydrocarbon chains and of the polar region of the molecule, the amount of water present and on any solutes dissolved in the water. The endothermic phase transition of dimyristoyl phosphatidylcholine is shown in Fig. 2.

Once the water has penetrated into the lattice of the amphiphile and the sample is then cooled to below the T_c line, the hydrocarbon chains rearrange themselves into an orderly crystalline lattice, but the water is not necessarily expelled from the system. These phases containing crystalline paraffin chain regions are called gels; these gels may or may not be

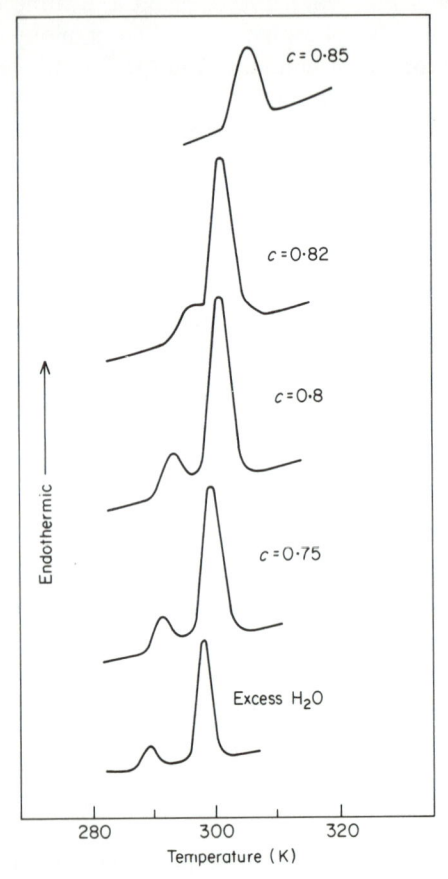

FIG. 2. The endothermic phase transition of 1,2-dimyristoyl-L-phosphatidylcholine as a function of water concentration.

metastable. If they are metastable they transform over a period of time to a suspension of microcrystals of the amphiphile in water, the coagel state. The coagel is a stable state; its structure is independent of the thermal history of the sample. When water penetrates into the polar group region the lipids separate into multilayers of two molecules thickness, i.e. many lipid bilayers are spontaneously produced as presented in Fig. 3.

The phase diagrams of the different chain length lecithin–water systems are essentially equivalent and are disposed along the temperature axis according to the melting temperature (T_c) of the hydrocarbon chains. Differential scanning calorimetric curves of a 1,2-diacyl-L-phosphatidylcholine–water system obtained over a range of water concentrations show in the range $1·0 \geqslant c \geqslant 0·8$, the temperature of the endothermic transition

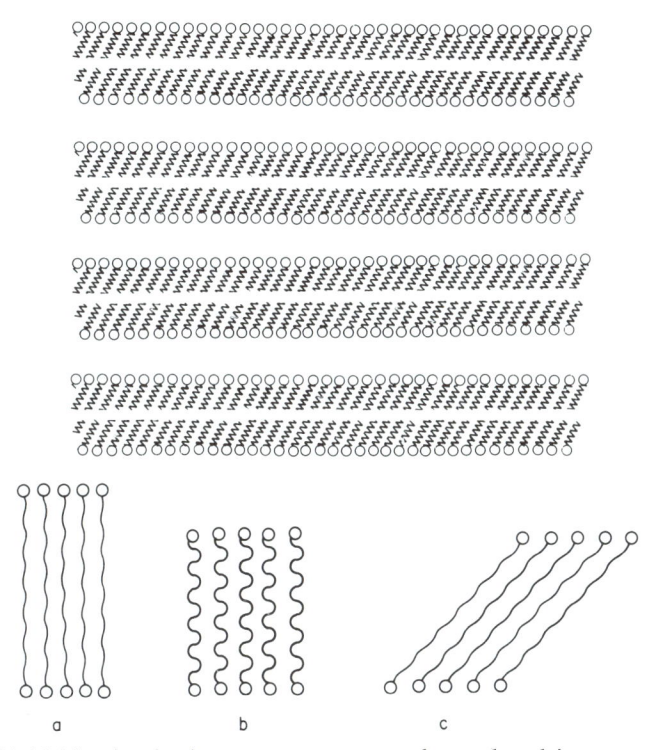

Fig. 3. Lipid bimolecular layers are spontaneously produced in water when the lipid is raised above a certain transition temperature. These fluid multibilayer structures can on cooling crystallize with the chains (a) vertical or (c) tilted to these layers.

T decreases steadily to a limiting value (T_c). The cooperativity of the transition as indicated by the narrowing of the melting range also increases with increasing water. In this concentration range, despite the fact that an appreciable amount of water is present, no transition is observed in the heating or cooling curves due to any melting of ice or freezing of the water present. When the water content is greater than 20 per cent ($c > 0·8$) the lipid endothermic transition temperature (T_c) remains constant and a peak at 0 °C, due to the melting of ice, can now be observed in the heating curve. As the concentration of water in the mixture further increases, so does the size of this peak. Quantitative studies are interpreted as showing that a proportion of the water is bound to the lecithin in a fixed ratio of 1 : 4 by weight (10 mol water per mol lecithin). This bound water is due to the formation of a hydrate structure associated with the polar group. The amount of bound water is independent of the fatty acid composition of the phospholipid, but is dependent upon the nature of the hydrophilic group.

The heat absorbed at T_c for lecithin water systems with $c < 0.7$ (Table 1) is seen to be chain length dependent. For the series of fully saturated

TABLE 1

Heats and temperatures of transition for 1,2-diacyl-L-phosphatidylcholine–water systems (Chapman et al., 1967)

1,2-Diacyl-phosphatidyl-choline	Limiting transition temperature (T_c) (°C)	Heat absorbed at T_c (kcal mol^{-1}) phospholipid	Entropy change at T_c (cal K^{-1} mol^{-1})	Temperature of "pre-transition" peak (°C)
Dibehenoyl	75	14·9	42·8	75
Distearoyl	58	10·70	32·4	56
Dipalmitoyl	41	8·65	27·6	35
Dimyristoyl	23	6·65	22·4	14
Dilauroyl	0 (under the ice peak)	—	—	—
Dioleoyl	−22	7·6	30·3	—
Egg yolk[a]	−15/−7	—	—	—

[a] The egg-yolk lecithin transition is broad in water due to the heterogeneous fatty acid composition. The lower temperature is from the heating curve, the higher from the cooling curve.

lecithins a difference of 2 kcal mol^{-1} occurs for a chain-length difference of two methylene groups. Extrapolation to zero at this transition shows that the C_7 and shorter chain-length lecithins have a negative heat of transition. A small "pre-transition" peak is observed when $c < 0.8$. The temperature interval between this peak and the main endothermic peak (T_c^*) increases (see Fig. 4) as the chain length of the lecithin becomes shorter.

Calorimetric data (Chapman et al., 1967) have been used to examine the hydrocarbon chain motion above the phase transition temperature of anhydrous lecithins (Phillips et al., 1969). The heat involved in the transition is found to be about 95 per cent of the total heat of fusion.

The total entropy per methylene group is the same for all long chain compounds in the crystalline form at their chain melting point. The entropy gain during the transition from β crystal to isotropic liquid for n-alkanes, triglycerides and fatty acids is 2·6 e.u. per CH_2 group, whereas

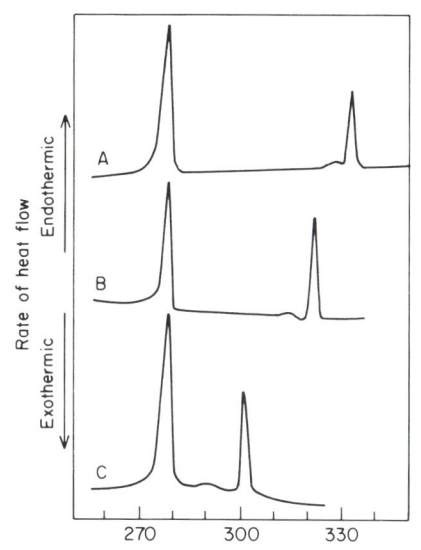

FIG. 4. DSC heating curves for some 1,2-diacylphosphatidylcholine–water systems at $c = 0.5$ for (A) the distearoyl, (B) the dipalmitoyl and (C) the dimyristoyl derivatives. The peak due to ice melting at 0 °C is also shown. (Chapman et al., 1967.)

for the crystal to liquid crystal transition for lecithins the equivalent figure is 1·1 e.u. Thus, in the liquid crystalline state the chain fluidity is about half that found in liquid n-alkanes at the transition temperature. It was suggested that this may arise by inhibition of rotation about the carbon–carbon bonds due to the presence of the neighbouring chains (Phillips et al., 1969).

3. PHASE SEPARATION

The first studies on phase separation of lipid water systems were discussed by Ladbrooke and Chapman (1969) who reported studies of binary mixtures of lecithins using calorimetry. These authors examined mixtures of distearoyl and dipalmitoyl lecithin (DSL–DPL) and also distearoyl lecithin and dimyristoyl lecithin (DSL–DML). With the DSL–DPL mixtures the phase diagram shows that a continuous series of solid solutions are formed below the T_c line. It was concluded that compound formation does not occur and that with this pair of molecules having only a small difference in chain length co-crystallization occurs. These results are presented in Fig. 5.

With the system DSL–DML monotectic behaviour was observed with limited solid solution formation. Here the difference in chain length is already too great for co-crystallization to occur so that as the system is

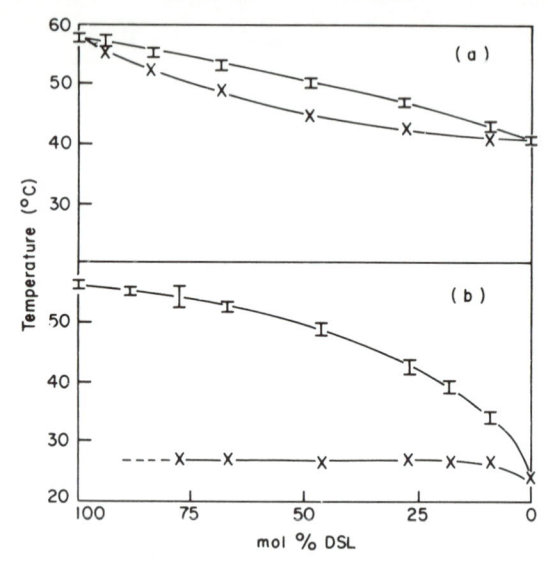

Fig. 5. Temperature composition diagrams for binary mixtures of saturated lecithins dispersed in excess (50 wt %) water. (a) 1,2-Distearoyl lecithin–1,2-dipalmitoyl lecithin–water system (DSL–DPL); (b) 1,2-distearoyl lecithin–1,2-dimyristoyl lecithin–water systems (DSL–DML). X, onset temperature from DSC heating curve; I, onset temperature from DSC cooling curve. (Ladbrooke and Chapman, 1969.)

cooled migration of lecithin molecules occurs within the bilayer to give crystalline regions corresponding to the two compounds (Ladbrooke and Chapman, 1969).

Examination of a series of fully saturated lecithins with dioleyl lecithin gave similar results with phase separation of the individual components taking place (Phillips et al., 1970). Later calorimetric studies were reported by Clowes et al. (1971) on mixed lecithin–cerebroside systems and on lecithin–phosphatidylethanolamine mixtures (reviewed by Oldfield and Chapman, 1972; Chapman et al., 1974). The lecithin–phosphatidylethanolamine systems of the same chain length give a wide melting range with some separation of the different lipid classes. The thermograms of this system are shown in Fig. 6.

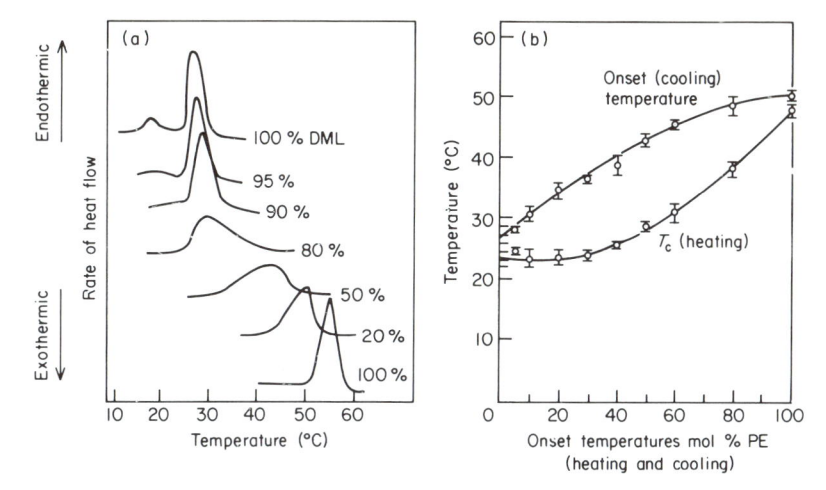

FIG. 6. (a) Differential scanning calorimetry heating curves for 1,2-dimyristoyl lecithin–1,2-dimyristoyl phosphatidylethanolamine–water mixtures. (b) Phase diagram of the 1,2-dimyristoyl lecithin–1,2-dimyristoyl phosphatidylethanolamine–water mixtures. (Chapman et al., 1974.)

4. CHOLESTEROL EFFECTS

In 1968 a paper was published (Ladbrooke et al., 1968b) describing studies on lecithin–cholesterol–water interactions by differential scanning calorimetry (DSC) and X-ray diffraction. The 1,2-dipalmitoyl-L-phosphatidylcholine–cholesterol–water system was studied as a function of both temperature and concentration of components. This particular lecithin was used because it exhibits the thermotropic phase change in the presence of water at a convenient temperature (41 °C). The addition of cholesterol to the lecithin in water lowers the transition temperature between the gel and the lamellar fluid crystalline phase, and decreases the heat absorbed at the transition. No transition at all is observed with an equimolar ratio of lecithin with cholesterol in water. This ratio corresponds to the maximum amount of cholesterol which can be introduced before cholesterol precipitation occurs.

These effects are not specific for dipalmitoyl lecithin; unsaturated lecithins and the lipid extract of human erythrocyte ghosts exhibit similar behaviour. The addition of cholesterol to the lecithin–water system in the gel phase causes a reduction in the cohesive forces between adjacent hydrocarbon chains of the lecithin; this leads to a fluidization of these chains. The ordered array of hydrocarbon chains in the gel phase is disrupted by cholesterol. The presence of water was also shown to be of prime importance in this interaction.

5. METAL-ION EFFECTS

Metal-ion interactions have been known for some years to affect the thermotropic phase transition of soap systems. The thermotropic phase transition of stearic acid occurs at 114 °C with the sodium salt and at 170 °C with the potassium salt.

Recent studies with the uranyl cation UO_2^{2+} (Chapman *et al.*, 1974) indicate that this ion causes the thermotropic phase transition temperature of lecithins to increase. Two main phase transitions were observed corresponding to the presence of complexed and uncomplexed lipid. When the titration is complete only the higher melting transition remains. The studies by Chapman *et al.* (1974) indicate that the interaction between cations and phosphatidylserine causes greater shifts of transition temperature than is observed with lecithin molecules. All the cations studied were found to shift the phase transition temperature of the phospholipids to higher values.

The precise nature of the interaction between ions and phospholipids is still open to doubt. There is some evidence that charge neutralization is the prime interaction of charged phospholipids with divalent cations (Trauble and Eibl, 1974; Verkleij *et al.*, 1974). Divalent cations were found to increase the transition temperature and the monovalent cations to fluidize the bilayer. Some authors believe that the primary effect of the

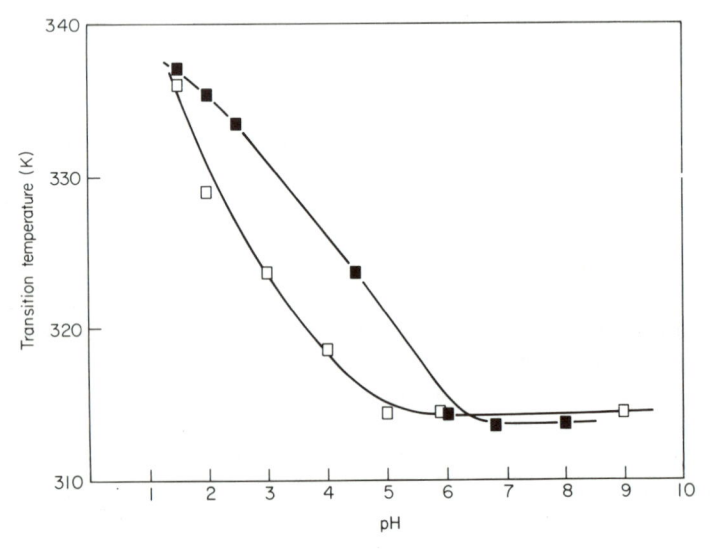

FIG. 7. Transition temperatures of dihexadecanoylphosphatidylglycerol ($C_{16}PG$) versus pH in the presence of monovalent cation. K^+ 0·24 M (□); K^+ 0·01 M (■). (Sacre *et al.*, 1978.)

cation on lecithins may be on the aqueous portion of the lipid bilayer (Gottlieb and Eanes, 1972; Ehrstrom et al., 1973; Godin and Ng, 1974). A recent study of an extensive range of salts with lecithin bilayers (Chapman et al., 1974) indicates that the anion present has a very large effect in determining the state of fluidity of the bilayer, the results obtained being best explained by a thermodynamic treatment based on relative association constants. Trauble (1971) has shown that pH can affect lipid transition temperatures, particularly lipids such as the phosphatidylethanolamines. Verkleij et al. (1974) have shown that the thermotropic phase transition of a synthetic phosphatidylglycerol is influenced by pH and Ca^{2+}. The interaction of phospholipid membranes with ions was also studied extensively by Jacobson and Papahadjopoulos (1975).

Recent calorimetric studies of phosphatidylglycerol lipids as a function of pH in the presence of various monovalent cations have been made (Sacre et al., 1978) and are presented in Fig. 7.

6. THEORETICAL TREATMENTS

A number of theoretical studies have now been made of the thermotropic phase transition of lipids and lipid–water systems. Whittington and Chapman (1965) examined the motion of chains in the hexagonal form of long-chain molecules. In later studies they also examined (1966) the way in which the end-to-end distance of chains (which are fixed at one end) varies as the distance between the chains is increased. They used two different potential functions and restricted the chains to a two-dimensional hexagonal lattice. This study emphasized the cooperative nature of the twisting and movement of the CH_2 groups of adjacent chains leading to the phase transition and also above it in the more fluid state. Rothman (1973) and Nagle (1973) have also emphasized this cooperative nature of chain movements.

Marcelja (1974) has used a molecular field approximation and examined different statistical averages over all conformations of a single chain in a field due to neighbouring molecules. This method has been extended to describe the properties of lipid monolayers and bilayers.

7. INTERACTION WITH DRUGS AND OTHER MODIFIERS

The precise biological and pharmacological actions of drug molecules are largely undetermined. There is, however, widening belief that interaction with the lipid constituents of membranes resulting in an alteration of the fluidity characteristics may be an important part of the mode of action of the drugs.

There are now available a number of studies showing that drug molecules can affect the thermotropic phase transition of lipid–water systems. Sometimes this corresponds to a removal of the transition somewhat similar to that observed with cholesterol. We can instance gramicidin A (Chapman et al., 1974) and other antibiotic molecules (Pache et al., 1972). On the other hand, sometimes the drug shifts (with increasing concentration) the lipid transition temperature to lower values, and we can instance a range of antidepressant drug molecules, e.g. desipramine (Cater et al., 1974). Hashish components: Δ^1-tetrahydrocannabinol and cannabidiol interacting with dipalmitoyl lecithin bilayers decrease both the melting temperature and the enthalpy of melting (Bach et al., 1976). At higher drug concentrations two peaks appear in the transition profile as shown in Fig. 8. The appearance of the additional peak was explained, as in the case of interaction with desipramine derivatives (Cater et al., 1974), by the phase separation due to limited solubility of the drug in the phospholipid. No difference in the modifying properties of Δ^1-tetrahydrocannabinol and cannabidiol was detected in spite of the fact that only the first is psychoactive, so it was concluded that they have no specific action on the phospholipid.

The local anaesthetic dibucaine has a strong fluidizing effect on the acidic phospholipid bilayers: dipalmitoylphosphatidyl glycerol, phosphatidylserine (Papahadjopoulos et al., 1975a). Presence of calcium ions even enhances the effect of the drug. The zwitterionic dipalmitoyl lecithin membranes are fluidized only by much higher concentrations of the drug. Recently it was reported (Frenzel et al., 1978) that cationic drugs (phenothiazine-diethazine and chloropromazine) at very low concentrations remove the pretransition peak and at higher ones lower the transition temperature of dipalmitoyl lecithin. Little or no change in the heat of transition was detected.

The effect of inducers of Friend leukaemic differentiation on the thermotropic behaviour of phospholipids was investigated (Lyman et al., 1976). It was found that these compounds stabilize the bilayers, the main temperature of melting is increased and a new transition at higher values appears.

As a model for the action of inhalation anaesthetics the interaction of cyclopropane with dipalmitoyl lecithin has been investigated (Simon et al., 1975). It was found that due to the interaction two peaks appear in the thermogram. Both of them have lower enthalpy of melting and are at lower temperatures than that of the pure lipid. If the interaction is carried out in the presence of calcium ions, the main peak is not shifted and only an additional one appears at lower temperatures. Jain et al. (1975, 1976, 1977) investigated the influence of more than a hundred compounds on

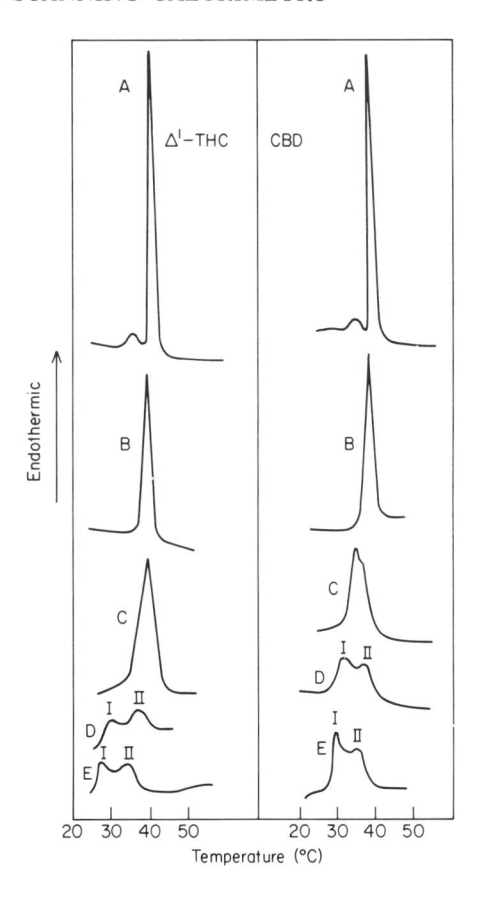

FIG. 8. The effect of Δ¹-tetrahydrocannabinol (Δ¹-THC) and cannabidiol (CBD) on the differential scanning calorimetry thermograms of dipalmitoyl phosphatidylcholine (DPL).

	Δ¹THC		CBD
(A) DPL alone		(A) DPL alone	
(B)	1 : 22 (molar ratio)	(B)	1 : 20 (molar ratio)
(C) Δ¹-THC–DPL	1 : 11	(C) CBD/DPL	1 : 10
(D)	1 : 5	(D)	1 : 5
(E)	1 : 2·4	(E)	1 : 2

(Bach et al., 1976.)

the thermotropic behaviour of dipalmitoyl lecithin. The compounds investigated were: alkanols, local anaesthetics, uncouplers, tranquillizers, inhalation anaesthetics, fatty acids, detergents, organic solvents, ionophores, spin labels and flourescent membrane probes. Different types of transition profiles were obtained and they were grouped into four classes.

1. Broadening of the peak with an increase or a decrease in the melting temperature.
2. The main peak at the normal temperature only decreases in size, but a smaller peak appears at lower temperatures.
3. Shift of the peak to lower temperatures without broadening.
4. Appearance of a new peak at higher temperatures with consequent disappearance of the original peak at higher concentrations of the modifier.

The type of profile obtained depends on the location of the compounds within the bilayer. Localization in the phosphorylcholine region will give type 4, in the glycerol backbone type 2, in C_1–C_8 methylene type 7, in C_9–C_{16} methylene type 3. The localization is determined by the presence of the polar and apolar groups in the compound and by their geometric arrangements within the molecule.

Eliasz et al. (1976) also investigated the influence of a series of alcohols, acids and quaternary ammonium salts on the thermotropic behaviour of dipalmitoyl lecithin. Various effects were detected and correlated with the structure of the compounds.

The influence of the alkyl ammonium iodides on the thermotropic phase transition of dipalmitoyl lecithin was also investigated by Frischleder and Gleichman (1977). The trimethyl ammonium iodides with an alkyl chain of C_6, C_9, C_{12} and C_{18} were investigated; they caused a shift of the melting temperature towards lower values with the maximum effect produced by C_9 and C_{12} compounds.

The anthracycline antibiotics, daunomycin and adriamycin, are used in cancer therapy. As it is suggested by different experimental findings that these compounds interact with membranes, it was of interest to study their effect on the thermotropic behaviour of phospholipid bilayers (Goldman et al., 1978). Dipalmitoyl lecithin and phosphatidyl serine were employed as there is some evidence that the drugs (most of them positively charged) have affinity for negatively charged phospholipids. The four adriamycin derivatives studied were shown to influence to a different degree the thermotropic behaviour of the dipalmitoyl lecithin. Daunomycin produces the biggest decrease in the melting temperature of the dipalmitoyl lecithin. For all the derivatives tested there is a decrease in the enthalpy of melting ranging between 25 and 15 per cent. When interacting with the negatively charged phospholipid phosphatidyl serine only daunomycin causes a significant shift in the melting temperature; the other derivatives both positively charged and uncharged did not affect the melting temperature but in all cases there was a decrease in the enthalpy of melting—the highest decrease being ~ 60 per cent. These data suggest that all these drugs fluidize the membranes; the fluidizing effect is not related directly to the charge of the compounds, but more to their structure enabling penetration into different regions of the bilayer.

8. POLYPEPTIDES AND PROTEINS EFFECTS

The influence of polypeptides and proteins on the thermotropic behaviour of phospholipids in simple model systems has been studied with an aim to provide insight into the lipid–protein interactions involved in biomembranes. Different peptides and proteins have been studied. The interaction causes a shift of the melting temperature and decreases or increases the enthalpy of melting of the phospholipids used.

Basic proteins, cytochrome C, lysozyme and polypeptide polylysine, interacting with natural phosphatidyl serine, decrease the melting temperature of the phospholipid (Chapman *et al.*, 1974). At neutral pHs phosphatidyl serine is negatively charged, so the predominant interaction is an electrostatic one between the positively charged groups of the proteins and the negatively charged groups of the lipid. This attraction leads to reorganization of the polar groups of the phospholipid with less effective packing of the lipid chains, shifting the melting temperature towards lower values. It is considered that a very limited penetration of these proteins into the bilayer is taking place.

Papahadjopoulos *et al.* (1975b) showed that cytochrome C has the same effect on the thermotropic behaviour of a synthetic negatively charged phospholipid, dipalmitoyl phosphatidyl glycerol. Other positively charged compounds, ribonuclease and polylysine, interacting with this phospholipid, caused an increase in the enthalpy of melting with or without shift of the melting temperature to higher values (Papahadjopoulos *et al.*, 1975b). It was claimed that due to electrostatic interaction there is binding of the protein to the bilayer without penetration into it, stabilizing the gel state. These proteins and polypeptides did not affect the thermotropic properties of zwitterionic lipid-dipalmitoyl lecithin. The main interaction is an electrostatic one and whether the protein fluidizes or stabilizes the bilayer depends on the additional non-polar interactions and on the possibility of the penetration of the protein molecules into the lipid bilayer.

There is some published work that suggests that the effect a protein has on a phospholipid bilayer is also a function of the fatty acids of the phospholipid, as it was found by Verkleij et al. (1974) that basic protein from myelin (A1) shifts the melting temperature of a synthetic phosphatidyl glycerol to higher values and by Papahadjopoulos et al. (1975c) that this same protein causes a decrease in the melting temperature and in the enthalpy of melting of another synthetic phosphatidyl glycerol differing in the fatty acids.

When a mixture of zwitterionic and acidic phospholipids is investigated by DSC, it is observed that the melting temperature occurs at an intermediate temperature in between that of the pure components. Addition of basic protein from human myelin induced a phase separation due to preferential binding of the protein to the acidic lipid and a shift of the peak towards the temperature of the zwitterionic lipid in the mixture (Boggs et al., 1977a, b). All the above-mentioned proteins and polypeptides are quite hydrophilic and a prerequisite for the interaction is the possibility of obtaining high enough surface concentrations on the lipid bilayer due to electrostatic attraction. With more hydrophobic peptides and proteins the requirement for electrostatic interaction is not needed and the peptides and proteins interact both with negatively charged and zwitterionic lipids.

Gramicidin A, a very hydrophobic peptide, acts as an ionophore by spanning the hydrocarbon region of the bilayer. When interacting with phospholipids at very low concentrations it removes the pretransition endotherm; at higher ones the main lipid peak becomes broadened and decreases in size linearly with the amount of the polypeptide added (Chapman et al., 1974, 1977). This behaviour was explained by interdigitation of gramicidin A molecules between lipid acyl chains preventing their crystallization.

Bee venom peptide melittin has similar effects on the melting profile of dipalmitoyl lecithin (Mollay, 1976); in its presence the onset temperature of the phospholipid is unchanged, but the area of the melting peak is reduced.

Interaction of basic polypeptide (copolymer of lysine with phenylalanine at a ratio of 1·4 : 1) with phosphatidyl serine or dipalmitoyl lecithin was investigated extensively. By employing differential scanning calorimetry it was shown that the polypeptide causes a decrease in the enthalpy of melting of the phospholipids almost without affecting the melting temperature (Bach and Miller, 1976). The decrease in enthalpy is a linear function of the amount of the polypeptide added; by extrapolation to zero enthalpy it was found that four amino acids eliminate the melting of one phospholipid molecule. At low polypeptide concentrations a limited penetration of the polypeptide into the bilayer occurs exhibiting a phase separation between the pure lipid bilayer melting with its original enthalpy and the lipid–polypeptide interaction products.

Decrease in enthalpy of melting with almost no change in the melting temperature was also detected in the case of interaction of phospholipids with serum apolipoproteins (Andrews *et al.*, 1976; Tall *et al.*, 1975, 1977) and with membrane proteins: glycophorin (MacDonald, 1975), proteolipid from human myelin (Papahodjoupolos *et al.*, 1975b, c), apoprotein from myelin (Curatolo *et al.*, 1977) and spectrin from red blood cells (Mombers *et al.*, 1977). Curatolo *et al.* (1977), by employing dimyrystoyl lecithin, observed that, in addition to a decrease in the size of the main peak, a shoulder appeared on the high temperature side of the phospholipid transition. The size of the shoulder increased with increase of the protein content with simultaneous decrease of the main peak. It was claimed that this shoulder stems from the melting of the boundary layer of dimyristoyl lecithin around the proteolipid molecule.

Mombers *et al.* (1977) studied the interaction of dimyristoyl lecithin and dimyristoyl phosphatidyl glycerol with spectrin and calcium. They found that thermotropic properties of dimyristoyl lecithin–spectrin complex were not influenced by the addition of Ca^{2+}, whereas addition of Ca^{2+} to the dimyristoyl phosphatidyl glycerol–spectrin complex caused an increase in the melting temperature and enthalpy of melting of the lipid. The increase was smaller than in the absence of the protein and specific for spectrin as albumin did not give this effect. From these data it was concluded that the lipid–protein interaction protects the lipid from the interaction with Ca^{2+}.

D. Studies of biomembranes

Biomembranes are of course quite complicated structures. In addition to containing lipids they also contain proteins. The proteins may also have attached carbohydrate groupings giving rise to glycoprotein structures.

1. ACHOLEPLASMA LAIDLAWII AND OTHER BACTERIAL AND PLANT MEMBRANES

The bacterial membranes are the simplest of all the biomembranes, containing very little or no cholesterol. In addition, their phospholipids can be manipulated by different growth conditions enabling one to correlate the physical properties of the membranes with the *in vivo* properties of the cells.

Acholeplasma laidlawii is the best candidate for such studies as it does not possess a cell wall and it is able to grow even in the presence of cholesterol and different saturated fatty acids. These bacteria were the first to be studied by differential scanning calorimetry. Steim *et al.* (1969) reported that isolated membranes and intact cells of *Mycoplasma laidlawii* (*Acholeplasma laidlawii*) exhibit an endothermic transition centred at about 40 °C. The data are presented in Fig. 9. This transition is also present in the total extracted lipids. As the value of the enthalpy of melting of the lipids in the membranes is only slightly lower than that of the total extracted lipids,

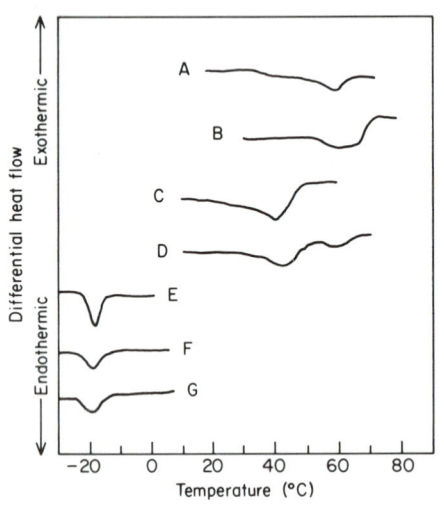

FIG. 9. Calorimeter scans of *M. laidlawii* lipids, membranes, and whole cells. A, Total membrane lipids from cells grown in tryptose with added stearate; B, membranes from stearate-supplemented tryptose; C, total membrane lipids from cells grown in unsupplemented tryptose; D, membranes from unsupplemented tryptose; E, total membrane lipids from cells grown in tryptose with added oleate; F, membranes from oleate-supplemented tryptose; G, whole cells from oleate-supplemented tryptose. The first four preparations were suspended in water; for the latter three scans, the solvent was 50 per cent ethylene glycol containing 0·15 M NaCl. (Steim *et al.*, 1969.)

Steim *et al.* (1969) concluded that the lipids in these membranes are in an extended bilayer configuration with polar groups of the lipids undergoing electrostatic interactions with the proteins. Chapman and Urbina (1971) suggested that some of the assumptions in their conclusion may not be correct. They suggested that the data were more consistent with a lipid bilayer structure *not* sandwiched by protein and also pointed out that squeezing effects due to lipid crystallization leading to high protein to lipid patches or protein aggregates had not been considered.

In an attempt to simplify the lipid composition, Baldassare et al. (1976) modified the fatty acids of *Escherichia coli* lipid mutants and studied the physical properties of the organisms and the extracted lipids. The DSC transition for the membranes and the extracted lipids for the control cells are very broad 15–25 °C in width. However when supplemented with 16 : 1 acid the transition moves to higher temperatures and becomes more asymmetric. The phase transition is not a simple two-step process as shown by changes in the values of the van't Hoff enthalpies calculated from the thermograms; probably different gel or liquid crystalline phases coexist. If the mutants were grown for four generations on cis-Δ^{11}-18:1 acid the phospholipids of the bacterial become more homogeneous and the phase transitions much sharper. The enthalpies of the transition were measured and the values of the enthalpies and the size of the cooperative units calculated. The values of the enthalpies and the size of the cooperative units increase with the decrease in the widths of the transition. The paper shows how by genetic manipulation the heterogeneity in the fatty acid chains of the bacterial membranes can be reduced leading to stronger lipid–lipid interactions as expressed by sharper melting profiles.

Fatty acid enriched *E. coli* whole cells, membranes and extracted lipids were also investigated by Jackson and Sturtevant (1977). The transitions in the whole cells are broader than those in the extracted lipids, but in both cases an asymmetry in the transition is seen. Different possibilities for the appearance of the asymmetry are discussed, but no clear cut-answer is given.

The halophilic bacterium *Halobacterium halobium*, when grown under certain conditions, produces purple-coloured membrane. This membrane is very rigid, relatively simple in its composition as it contains phospholipids, chromophore and only one protein, bacteriorhodopsin. Due to its high crystallinity it was of interest to investigate its thermotropic behaviour. The studies were undertaken independently by two groups. Jackson and Sturtevant (1978) detected no transitions in the purple membranes at temperatures below 70 °C. A small endothermic transition was seen at 80 °C and a larger one at 100 °C. The first one is reversible and probably due to a cooperative change in the crystalline structure of the membrane; the second one is irreversible, and possibly stems from denaturation of the protein.

On the other hand, Caplan et al. (1978) investigated the thermotropic behaviour of this membrane by using a different instrument. They detected a number of small reversible peaks between 10 °C and 20 °C with a bigger one at around 35 °C. At present there is no explanation for these discrepancies; it is possible that the difference comes from the different samples employed. Protein conformational changes may be occurring.

Other plant membranes have also been investigated. Recently Friedman (1977) studied the influence of fatty acid composition on the membrane lipid phase transitions of the bread-mould *Neurospora crassa*. This organism also contains ergosterol as a predominant sterol. Two mutants were used differing in the degree of saturation of their fatty acids. The extracted phospholipids (ergosterol free) from the species with higher amounts of fatty acids have higher melting temperature as predicted (the transitions are below 0 °C). These transitions are decreased by addition of ergosterol and do not appear in the intact organism. In the intact cells and in the extracted lipids another transition appears at around 30 °C. This transition is reversible and insensitive to the addition of ergosterol.

2. MAMMALIAN MEMBRANES

Mammalian membranes are even more complicated than the bacterial ones, because in addition to lipids and proteins they often contain cholesterol. It is known from model lipid–water systems that cholesterol fluidizes the lipids, removing the phase transition peaks. This is the reason why it was expected that the lipids in the membranes might be in the fluid state and not give melting peaks in the physiological range of temperatures. In the following we will show that some mammalian membranes undergo melting at particular temperatures and we shall discuss the possible implications of this phenomenon.

a. *Myelin and red blood cells*
The thermotropic study of mammalian membranes was initiated by Chapman in 1968 (Ladbrooke *et al.*, 1968a) who studied native and freeze-dried ox-brain and human myelin. It was found that no lipid phase transitions occurred in the wet membranes, whereas the dehydrated membranes showed endothermic transitions around room temperature. This behaviour was explained by dehydration causing the precipitation of cholesterol, enabling the lipids to undergo melting. Similar effects were observed with erythrocyte membranes, where only cholesterol-depleted lipids exhibited phase transitions. Recently the calorimetric studies of erythrocyte membranes were extended with an emphasis on the protein structural transitions (Jackson *et al.*, 1973; Brands *et al.*, 1977). They confirmed the previous findings of Ladbrooke *et al.* (1968a) with respect to lipids, but detected four other transitions in the region of 45–80 °C. The peaks stem from protein unfolding. The first transition is considered to arise from a transition in the spectrin complex.

b. *Microsomes, mitochondria and other membranes*

The thermotropic behaviour of rat liver microsomes and rat liver mitochondria was investigated extensively by several groups. Blazyk and Steim (1972) found reversible transitions centred around 0 °C and an irreversible peak at 65 °C in mitochondria and in microsomes. The low temperature peak stems from the gel–liquid transition of the membrane lipids and the higher temperature peak is due to protein denaturation. The lipid transitions are very broad; they extend to about 20 °C. The extracted lipids melt at lower temperatures than those in the intact membranes.

These results are different from those of Feo *et al.* (1976) who by employing DTA detected no transitions in rat liver mitochondrial membranes below 55 °C.

In recent years the development of more sensitive instruments has enabled reinvestigation of the melting properties of biomembranes. By employing a very sensitive differential scanning calorimeter it was shown that rat liver microsomal and mitochondrial membranes also melt in the region of ~ 18–40 °C (Bach *et al.*, 1977; Bach *et al.*, 1978a). The peaks of the extracted lipids are shifted to lower temperatures (10–20 °C) and their enthalpy of melting is smaller than that of the lipids in the intact membrane.

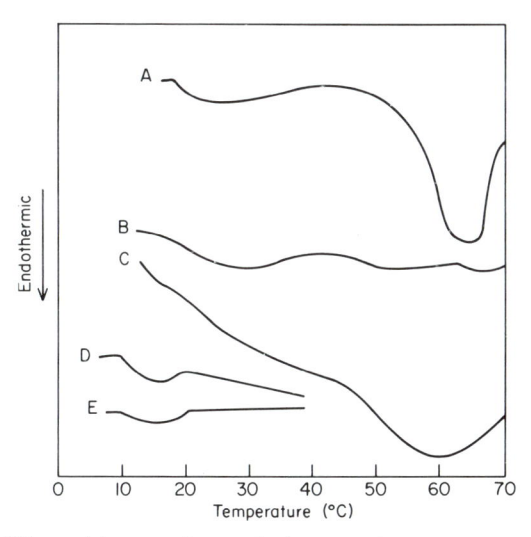

FIG. 10. The differential scanning calorimetry thermograms of microsomal membranes, extracted lipids and of microsomal membranes and extracted lipids after interaction with Δ^1-THC dispersed in water. A, wet microsomes, first scan. B, second scan of microsomes presented in A. C, wet microsomes after interaction with Δ^1-tetrahydrocannabinol, first scan. D, lipids. E, lipids with Δ^1-tetrahydrocannabinol. (Bach *et al.*, 1977.)

Incubation of the microsomal membranes or the extracted lipids with a psychoactive hashish component caused almost complete abolishment of the membrane peak and a decrease of the extracted lipids peak. The thermograms of the microsomes are presented in Fig. 10.

The melting properties of mitochondria and microsomes have also been investigated by other groups. Hackenbrock et al. (1976) examined whole mitochondria and inner and outer mitochondrial membranes. They did not run their scans to high temperatures and did not detect the transition found by Bach et al. (1978a). However, they show an interesting correlation between their calorimetric data and the freeze-fracture data (Fig. 11).

Phase transitions around room temperatures were found only in liver microsomes of rats fed a fat-free diet and not in rats fed a normal diet (Mabrey et al., 1977). According to Mabrey et al. (1977), a fat-free diet increases the degree of saturation of the membrane lipids.

By employing DSC it has been shown that other membranes also show endothermic phase transitions. Sarcoplasmic reticulum membranes exhibit lipid phase transitions at approximately 15 °C but only in the freeze-dried state. The addition of water shifts the transition to lower temperatures (Martonosi, 1974). The extracted lipids behaved similarly only their enthalpy of melting was higher than in the natural membrane.

Native chromaffin granule membranes exhibit a number of small reversible endothermic peaks, with a bigger one centred at approximately 32 °C and an irreversible peak at approximately 70 °C (Bach et al., 1978b). The enthalpy is very low, after ether extraction of cholesterol the main melting peak is shifted to higher temperatures and its enthalpy is significantly increased.

E. Interpretation of the thermal phase transitions in biomembranes

We see that phase transition and phase separation behaviour characteristic of pure lipids and lipid–polypeptide systems have also been observed in biological membranes. This has been illustrated in Fig. 11 which shows the thermogram and the electron micrographs of freeze-fracture replicas from rat liver mitochondria.

The fracture face of the membrane is divided into two separate regions, one possessing a high density of particles (believed to represent membrane proteins), the other a smooth area devoid of particles.

Similar temperature-induced protein aggregation has been reported in lipid-reconstituted Ca^{2+}-activated ATPase from sarcoplasmic reticulum (Kleemann and McConnell, 1976), and in microsomal membranes (Duppel and Dahl, 1976). Duppel and Dahl (1976) have shown that at

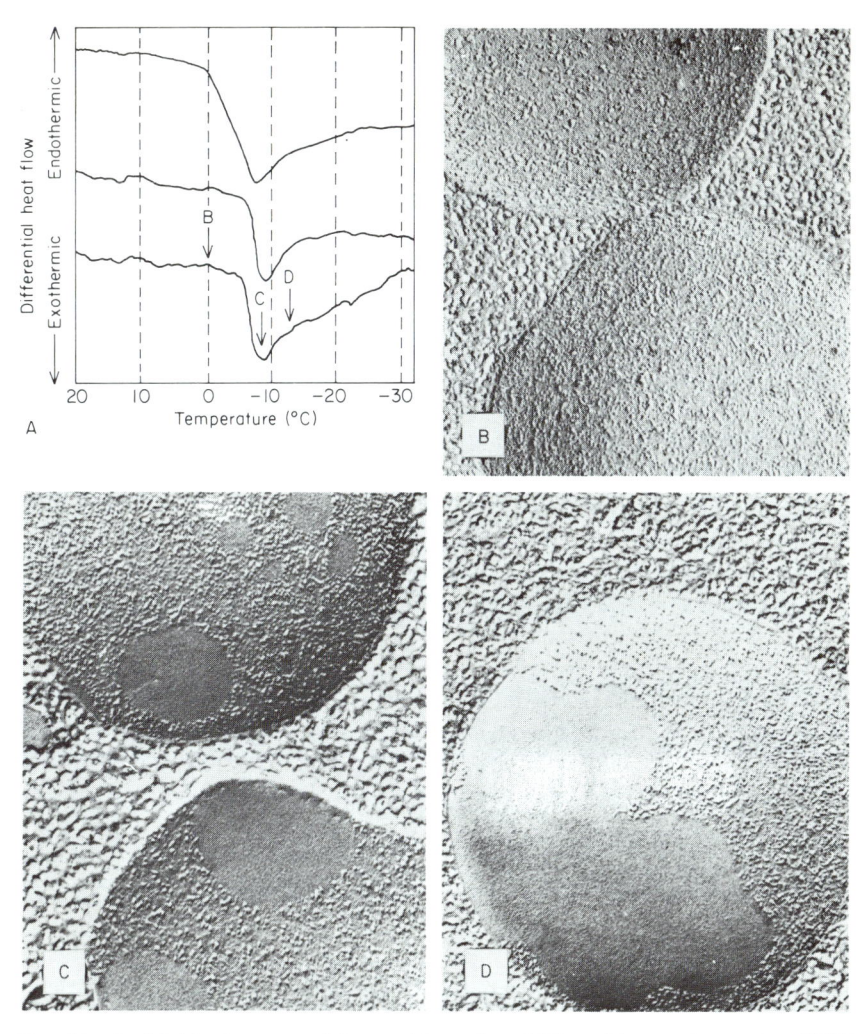

Fig. 11. A, Lipid phase transition and lateral mobility of integral proteins. DSC cooling runs of whole mitochondria (a) and purified outer membrane (b, c). Arrows identify temperatures on the curve for whole mitochondria which correspond to B–D; B, concave fracture faces of both membranes of a mitochondrion frozen from 30 °C; C, cooled to 0 °C, then frozen; D, cooled to −8 °C, then frozen. (Hackenbrock et al., 1966.)

30 °C membrane particles are randomly distributed, whereas at 4 °C aggregation of particles is induced. Addition of cholesterol to the microsomal membranes at molar ratios 1 : 1 to phospholipid prevents the aggregation of the particles. Not all membranes, however, exhibit these

effects, and it is possible that not all the proteins aggregate upon cooling of the membranes.

In a natural membrane as distinct from a reconstituted or model membrane there will be a range of lipid classes and lipid acyl chains. When such membranes are cooled slowly, the higher melting point lipids tend to crystallize out first. Depending upon the relative amount of intrinsic membrane protein, the crystallizing lipid will exclude the proteins into regions of lower melting point lipids, producing a localized increase in protein–lipid clusters (Chapman *et al.*, 1977). This has been simulated and is illustrated in Fig. 12. If the protein is easily accommodated by virtue of size within a hexagonally packed array of lipid hydrocarbon chains or alternatively is present in sufficiently low amounts so that it becomes trapped within the crystallizing lipid, the effect of lowering the temperature on protein redistribution will be negligible. This could explain some of the anomalies in which the distribution of certain proteins appears to be unaffected by equilibrium at low temperatures.

Thus cooling certain membranes to the point where the lipid crystallizes can cause the intrinsic proteins to aggregate. The formation of a crystalline array of lipids often excludes the protein from large areas of the membrane. This has important consequences for the observation of the endothermic phase transitions. *It is important to appreciate that it is the melting of these extensive crystalline lipid regions from which the protein has been excluded that gives rise to the endotherms observed by scanning calorimetry with cell membrane systems.* This observation and cautionary comment on the interpretation of calorimetric data were made some time ago by Chapman and Urbina (1971).

1. LIPID PHASE TRANSITIONS AND TRANSPORT PROCESSES

The packing arrangements of pure synthetic lecithins above and below the phase transition temperature have been shown in many experiments to influence the passive diffusion of solute across bilayers, e.g. studies of water permeation in bilayer vesicles of saturated lecithins above and below the phase transition temperature have shown a decrease in the energy of activation from 110 kJ mol^{-1} in the crystalline phase to 40 kJ mol^{-1} above the phase transition temperature, suggesting that the diffusion process itself depends on the packing arrangement of the lipid (Blok *et al.*, 1976). The permeability of phospholipid bilayers at temperatures corresponding to the phase transition temperature appears to be increased markedly compared to membranes in purely crystalline or liquid–crystalline (fluid) configuration. The apparent activation energy of 8-anilino 1-naphthalene sulphonate (ANS) transport in dimyristoyl lecithin bilayers,

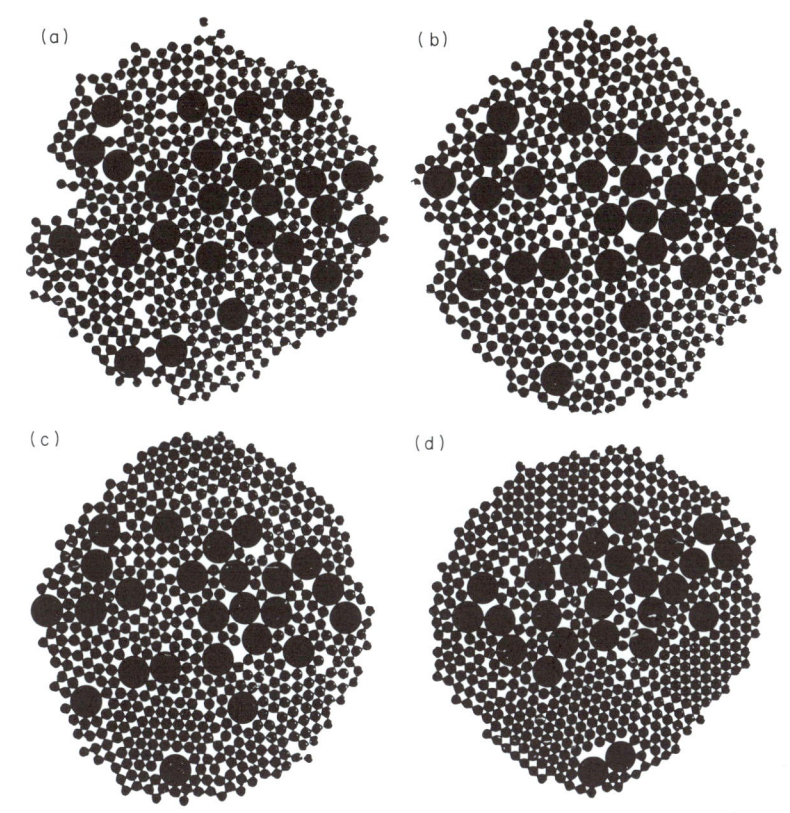

FIG. 12. A–D, Schematic diagram showing how lipid crystallization can lead to protein aggregation. A, random distribution of lipid and protein; B, more extensive areas of lipid nucleate-producing domains of crystalline lipid intersected by packing faults seen in C, the crystallization of certain areas of lipid forces protein molecules to aggregate; D, equilibrium low-temperature distribution of lipid and protein. (Chapman *et al.*, 1977b.)

for example, is large in both the crystalline and liquid–crystalline conformation but is reportedly zero at the phase transition temperature (Tsong, 1975). This enhanced permeability has been shown to depend on the size of the permeant solute as well as the length of the hydrocarbon chain of the saturated lecithin (Blok *et al.*, 1976). The usual interpretation of this phenomenon is that discontinuities originating at the boundaries of the crystal (rigid) and fluid (disordered) domains in the bilayer give rise to pores through which the solutes pass. The presence of cholesterol in the bilayer produces predictable effects on the permeability properties of the structure. Cholesterol enhances the permeability of saturated lecithin bilayers to glycerol and glycol below the phase transition temperature but depresses the permeability of the membrane towards these solutes above the lipid phase transition (Gier et al., 1969). Comparison of the permeability of bilayers of saturated lecithins containing cholesterol to various ions and glucose indicates that cholesterol affects the partition of solutes between the aqueous phase and the membranes as well as the rate of diffusion through the membrane (Papahadjopoulos et al., 1971).

The dependence of many active transport processes on membrane lipid fluidity has been demonstrated. Usually the evidence for this is derived from discontinuities in Arrhenius relationships, which in many instances appear to correspond to a phase transition of the membrane lipid. This is particularly evident in those cases in which the transport system has been reconstituted into phospholipid bilayers of defined acyl chain composition. Studies of the transport of β-galactosides and β-glucosides in fatty acid auxotrophs of *E. coli* have revealed as many as three distinct breaks in the slopes of the Arrhenius plots. The position of these discontinuities was shown to depend on the fatty acid composition of the membranes; the lower and upper breaks were observed to correlate with the onset and completion, respectively, of the melting of the hydrocarbon chains of the membrane lipids (Linden et al., 1973). The active transport of proline across cytoplasmic membrane vesicles of *E. coli* has been related to different energies of activation at temperatures above and below the lipid phase transition temperature (Esfahani et al., 1971; Shechter et al., 1974). Other examples where such behaviour has been observed include ATPase-dependent functions of *Mycoplasma* cells.

Recent studies (see Chapman et al., 1979) with reconstituted protein–lipid systems have been made using a range of physical techniques including calorimetry, nuclear magnetic resonance and laser flash photolysis. These studies have been interpreted to show that below the lipid phase transition temperature T_c the proteins are squeezed by the crystallizing lipid into patches of high protein content. This is shown diagrammatically in Fig. 12. Upon heating a reconstituted system, e.g. the sarcoplasmic reticulum Ca^{2+}ATPase in dipalmitoyl lecithin bilayers, these patches melt first. This is shown by the presence of a broad endotherm beginning at 30 °C prior to the main lipid transition at 41 °C. At this temperature the protein begins to exhibit rotational diffusion, and enzymatic activity begins. The X-ray studies also show that an appreciable melting of hexagonally packed lipid occurs at 30 °C. At the higher temperature, near 41 °C, the remaining lipid melts and the proteins become randomly distributed within the plane of the fluid lipid bilayer. In some instances however the concentration or aggregation of protein in the patches may become very high, leading to appreciable amounts of lipid which is "trapped" between the proteins. This "trapped" lipid may be relatively insensitive to temperature changes. In this case the protein rotation and enzyme activity will begin to operate normally only at the temperature where the remaining lipid melts, i.e. as the bulk of the remaining lipid becomes fluid, the lateral diffusion of this lipid will begin to separate the clusters (the reverse of the process shown in Fig. 12). The lipid will then begin to attain a more normal fluidity, and compressibility and enzyme conformational possibilities will be restored to more active forms.

F. Future studies

The future should see many more calorimetric studies of membrane systems. The existence of lipid phase transitions in biomembranes already raises a number of important questions. Why do some natural membranes contain appreciable amounts of rigid lipids at 37 °C? This question has not yet been satisfactorily answered. Other questions concern the relationship between the observations seen in the model systems where the phase transitions of the lipid can be triggered by the metal ions or by changes of pH. As yet, there is no correlation between these important changes and their biochemical and biological consequences in natural membranes.

Some biomembranes, particularly those which do not contain cholesterol, after cooling to a lower temperature, also can exhibit phase transition behaviour. This, as we have seen, is associated with the crystallization of various lipids present in the membrane exhibiting a phase separation behaviour. The freeze fracture electron microscopy studies have shown that this can lead to protein aggregation in certain instances. This may be important for various areas of cryobiology and is clearly also important in the temperature studies of enzyme action. Workers have observed Arrhenius plot discontinuities which have sometimes been associated with the phase transition temperature. Future studies should confirm whether this enzyme behaviour is related to aggregation processes and effects of trapped lipid caused by the lipid crystallization or to other non-lipid effects. Another future use to which calorimetry may be applied is to make estimates of the extent of the external amino acids as compared with the internal amino acids of membrane proteins. The external amino acids should show marked denaturation effects, whilst the internal ones should show little effect.

The phase transitions detected in the biomembranes in the physiological range of temperatures are very small; the highest sensitivities of the commercially available instruments have to be used for their detection. Development of instruments with higher sensitivity and better stability probably can lead to many more interesting discoveries in the field of biomembrane calorimetry.

References

ANDREWS, A. L., ATKINSON, D., BARRAT, M. D., FINER, E. G., HAUSER, H., HENRY, R., LESLIE, R. B., OWENS, N. L., PHILLIPS, M. C. and ROBERTSON, R. N. (1976). *Eur. J. Biochem.* **64**, 549–563.
BACH, D. and MILLER, I. R. (1976). *Biochim. biophys. Acta*, **433**, 13–19.
BACH, D., RAZ, A. and GOLDMAN, R. (1976). *Biochim. biophys. Acta*, **436**, 889–894.
BACH, D., BURSUKER, I. and GOLDMAN, R. (1977). *Biochim. biophys. Acta*, **469**, 171–179.
BACH, D., BURSUKER, I. and MILLER, I. R. (1978a). *Experientia*, **35**, 750–751.
BACH, D., ROSENHECK, K. and SCHNEIDER, A. S. (1978b). *Eur. J. Biochem*, **34** (6), 717–718.
BALDASSARE, J. J., RHINEHART, K. B. and SILBERT, D. F. (1976). *Biochemistry*, **15**, 2986–2994.
BLAZYK, J. H. and STEIM, J. M. (1972). *Biochim. biophys. Acta*, **226**, 737–741.
BLOK, M. C., VAN DEENEN, L. L. M. and DE GIER, J. (1976). *Biochim. biophys. Acta*, **433**, 1–12.
BOGGS, J. M., WOOD, D. D., MOSCARELLO, M. A. and PAPAHADJOPOULOS, D. (1977a). *Biochemistry*, **16**, 2325–2329.
BOGGS, J. M., MOSCARELLO, M. A. and PAPAHADJOPOULOS, D. (1977b). *Biochemistry*, **16**, 5420–5426.
BRANDTS, J. F., ERICKSON, L., LYSKO, K., SCHWARTZ, A. T. and TAVERNA, R. D. (1977). *Biochemistry*, **16**, 3450–3454.
BYRNE, P. and CHAPMAN, D. (1964). *Nature, Lond.* **202**, 987–998.
CAPLAN, S. R. et al. (1978). Symposium on Halophilism, Rehovot, Israel, May 1978.
CATER, B. A., CHAPMAN, D., HAWES, S. and SAVILLE, J. (1974). *Biochim. biophys. Acta*, **363**, 54–69.
CHAPMAN, D. and COLLIN, D. T. (1965). *Nature, Lond.* **206**, 189.
CHAPMAN, D. and URBINA, J. (1971). *FEBS Lett.* **12**, 169–172.
CHAPMAN, D., BYRNE, P. and SHIPLEY, G. G. (1966). *Proc. R. Soc.* **A290**, 115–142.
CHAPMAN, D., WILLIAMS, R. M. and LADBROOKE, B. D. (1967). *Chem. Phys. Lipids*, **1**, 445–475.
CHAPMAN, D., URBINA, J. and KEOUGH, K. M. (1974). *J. biol. Chem.* **249**, 2512–2521.
CHAPMAN, D., CORNELL, B. A., ELIASZ, A. W. and PERRY, A. (1977a). *J. molec. Biol.* **113**, 517–538.
CHAPMAN, D., CORNELL, B. A. and QUINN, P. J. (1977b). In "FEBS—Symposium No. 42" (G. Semenza and E. Carafoli, eds), pp. 72–85. Springer-Verlag, Berlin and Heidelberg.

CHAPMAN, D., GOMEZ-FERNANDEZ, J. C. and GONI, F. M. (1979). *FEBS Letts Review*, **98**, 211–223.
CLOWES, A., CHERRY, R. J. and CHAPMAN, D. (1971). *Biochim. biophys. Acta*, **249**, 301–307.
CURATOLO, W., SAKURA, J. D., SMELL, D. M. and SHIPLEY, G. G. (1977). *Biochemistry*, **16**, 2313–2318.
DUPPEL, W. and DAHL, G. (1976). *Biochim. biophys. Acta*, **426**, 408–417.
EHRSTROM, M., ERIKSONN, L. E. G., ISRAELACHVILLI, J. and EHRENBERG, A. (1973). *Biochem. biophys. Res. Commun.* **55**, 396–402.
ELIASZ, A. W., CHAPMAN, D. and EWING, D. F. (1976). *Biochim. biophys. Acta*, **448**, 220–230.
ESFAHANI, M., LIMBRICK, A. R., KNUTTON, S., OKA, S. and WAKIL, S. J. (1971). *Proc. natn. Acad. Sci. U.S.A.* **68**, 3180–3184.
ESSER, A. F. and LANYI, J. K. (1973). *Biochemistry*, **12**, 1933–1939.
FEO, F., COMUTO, R. A., GARCEA, R., AVOGADRO, A., VILLA, M. and CEASCO, M. (1976). *FEBS Lett.* **72**, 262–266.
FRENZEL, J., ARNOLD, K. and NUHN, P. (1978). *Biochim. biophys. Acta*, **507**, 185–197.
FRIEDMAN, J. K. (1977). *J. Mem. Biol.* **32**, 33–47.
FRISCHLEDER, H. and GLEICHMAN, S. (1977). *Studia biophysica*, **64**, 95–100.
GIER, J. DE, MANDERSLOOT, J. G. and VAN DEENEN, L. L. M. (1969). *Biochim. biophys. Acta*, **173**, 143–145.
GODIN, D. V. and NG, T. W. (1974). *Molec. Pharmacol.* **9**, 802–819.
GOLDMAN, R., FACCHINETTI, T., RAZ, A. and BACH, D. (1978). *Biochim. biophys. Acta*, **512**, 254–269.
GOTTLIEB, M. H. and EANES, E. D. (1972). *Biophys. J.* **12**, 1533–1548.
HACKENBROCK, C. R., HÖCHLI, M. and CHAN, R. M. (1976). *Biochim. biophys. Acta*, **455**, 466–484.
JACKSON, M. B. and SURTEVANT, J. M. (1977). *Biochemistry*, **252**, 4749–4751.
JACKSON, M. B. and SURTEVANT, J. M. (1978). *Biochemistry*, **17**, 911–915.
JACKSON, W., KOSTYLA, J., NORDIN, J. and BRANDTS, J. H. (1973). *Biochemistry*, **12**, 3662–3667.
JACOBSON, K. and PAPAHADJOPOULOS, D. (1975). *Biochemistry*, **14**, 152–161.
JAIN, M. K. and WU, N. (1977). *J. membrane Biol.* **34**, 157–201.
JAIN, M. K., WU, N. and WRAY, L. V. (1975). *Nature, Lond.* **255**, 494–495.
JAIN, M. K., WU, N., MORGAN, T. K., BRIGGS, M. S. and MURRAY, R. K. (1976). *Chem. Phys. Lipids*, **17**, 71–78.
KLEEMANN, W. and MCCONNELL, H. M. (1976). *Biochim. biophys. Acta*, **419**, 206–222.
LADBROOKE, B. D. and CHAPMAN, D. (1969). *Chem. Phys. Lipids*, **3**, 304–367.
LADBROOKE, B. D., JENKINSON, T. J., KAMAT, V. B. and CHAPMAN, D. (1968a). *Biochim. biophys. Acta*, **164**, 101–109.
LADBROOKE, B. D., WILLIAMS, R. M. and CHAPMAN, D. (1968b). *Biochim. biophys. Acta*, **150**, 333–340.
LAWRENCE, A. S. C. (1938). *Trans. Faraday. Soc.* **34**, 660–677.

Linden, C. D., Wright, K. L., McConnell, H. M. and Fox, C. F. (1973). *Proc. natn. Acad. Sci. U.S.A.* **70**, 2270–2275.
Lyman, G. H., Papahadjopoulos, D. and Preisler, H. D. (1976). *Biochim. biophys. Acta*, **448**, 460–473.
Mabrey, S., Powis, G., Schenkman, J. B. and Tritton, T. R. (1977). *J. biol. Chem.* **252**, 2929–2933.
MacDonald, R. I. and MacDonald, R. C. (1975). *J. biol. Chem.* **250**, 9206–9214.
Marcelja, S. (1974). *Biochim. biophys. Acta*, **367**, 165–176.
Martonosi, M. A. (1974). *FEBS Lett.* **47**, 327–329.
Mollay, C. H. (1976). *FEBS Lett.* **64**, 65–68.
Mombers, C., Van Dijck, P. W. M., Van deenen, L. L. M., De Gier, J. and Verkleij, A. J. (1977). *Biochim. biophys. Acta*, **470**, 152–160.
Nagle, J. F. (1973). *J. Chem. Phys.* **58**, 252–264.
Nordsieck, H., Roevear, F. B. and Ferguson, R. H. (1948). *J. chem. Phys.* **16**, 175–180.
Oldfield, E. and Chapman, D. (1972). *FEBS Lett.* **23**, 285–297.
Pache, W. and Chapman, D. (1972). *Biochim. biophys. Acta*, **255**, 348–357.
Pache, W., Chapman, D. and Hillaby, R. (1972). *Biochim. biophys. Acta*, **255**, 358–364.
Papahadjopoulos, D., Nir, S. and Ohki, S. (1971). *Biochim. biophys. Acta*, **266**, 561–583.
Papahadjopoulos, D., Jacobson, K., Poste, G. and Shepherd, G. (1975a). *Biochim. biophys. Acta*, **394**, 504–519.
Papahadjopoulos, D., Moscarello, M., Eylar, E. H. and Isac, T. (1975b). *Biochim. biophys. Acta*, **401**, 317–335.
Papahadjopoulos, D., Vail, W. J. and Moscarello, M. (1975c). *J. membrane Biol.* **22**, 143–164.
Phillips, M. C. and Chapman, D. (1968). *Biochim. biophys. Acta*, **163**, 301–313.
Phillips, M. C., Ladbrooke, B. D. and Chapman, D. (1970). *Biochim. biophys. Acta*, **193**, 35–44.
Phillips, M. C., Williams, R. M. and Chapman, D. (1969). *Chem. Phys. Lipids*, **3**, 234–244.
Rothman, J. (1973). *J. theor. Biol.* **38**, 1–16.
Sacre, M.-M., Hoffman, W., Turner, M., Tocanne, J.-P. and Chapman, D. (1979). *Chem. Phys. Lipids*, **69**, 69–83.
Shechter, E., Letellier, L. and Gulick-Krzywicki, P. (1974). *Eur. J. Biochem.* **49**, 61–76.
Simon, S. A., MacDonald, R. C. and Bennett, P. B. (1975). *Biochim. biophys. Res. Commun.* **67**, 988–994.
Steim, J. M., Tourtellote, M. E., Reinert, J. C., McElhaney, R. N. and Rader, R. L. (1969). *PNAS*, **63**, 104–109.
Tall, A. R., Small, D. M., Shipley, G. G. and Lees, R. S. (1975). *Proc. natn. Acad. Sci. U.S.A.* **72**, 4940–4942.
Tall, A. R., Small, D. M., Deckelbaum, R. J. and Shipley, G. G. (1977) *J. biol. Chem.* **252**, 4701–4711.
Trauble, H. (1971). *Naturwissenschaften*, **58**, 277–284.

TRAUBLE, H. and EIBL, H. (1974). *Proc. natn. Acad. Sci. U.S.A.* **71**, 214–219.
TSONG, T. Y. (1975). *Biochemistry*, **14**, 5409–5414.
VERKLEIJ, A. J., DE KRUYFF, B., VERVERGAERT, P. H. J. TH., TOCANNE, J. F. and VAN DEENEN, L. L. M. (1974). *Biochim. biophys. Acta*, **339**, 432–437.
VOLD, M. J. (1941). *J. Am. chem. Soc.* **63**, 160–168.
VOLD, M., MACOMBER, M. and VOLD, R. D. (1941). *J. Am. chem. Soc.* **63**, 168–175.
WHITTINGTON, S. G. and CHAPMAN, D. (1965). *Trans. Faraday Soc.* **61**, 2656–2660.
WHITTINGTON, S. G. and CHAPMAN, D. (1966). *Trans. Faraday Soc.* **62**, 3319–3324.
WILLIAMS, R. M. and CHAPMAN, D. (1970). *In* "Progress in the Chemistry of Fats and Other Lipids" (R. T. Holman, ed.), pp. 1–79. Pergamon Press, New York.

Interpretation of Calorimetric Thermograms and their Dynamic Corrections

STANISLAW L. RANDZIO

and

JAAK SUURKUUSK

A. Introduction

Thermal effects accompanying most known biological processes can be characterized by two parameters: *a quantity of heat* measured in energy units (joules), and its time derivative, measured in power units (watts). The quantity of heat is mostly related to the extent of a process while its time derivative, which we can call *a thermal power*, is related to the intensity of a process. The thermal power is a very important parameter in biological investigations because complex biological processes can be characterized more or less unequivocally by the power developed by them. But this can be useful only in situations where there are instruments capable of measuring the thermal power. Unfortunately most calorimeters which have found application in biological research do not measure power directly but rather a temperature or temperature difference which is recorded as a thermogram. The power developed by the investigated process is distributed and accumulated inside the calorimetric cell and exchanged with the environment. This means that the resulting temperature changes (recorded thermogram) are defined by the balance of power in the calorimeter and not by the power itself. Because the balance of power is not independent of the calorimetric system, the recorded thermogram depends not only on the process under investigation but also on the calorimeter used. As an example of this observation two thermograms describing the same process but obtained in two different calorimeters are presented in Fig. 1.

The purpose of this contribution is to analyse parameters influencing the shape of thermograms recorded by calorimetric instruments, to

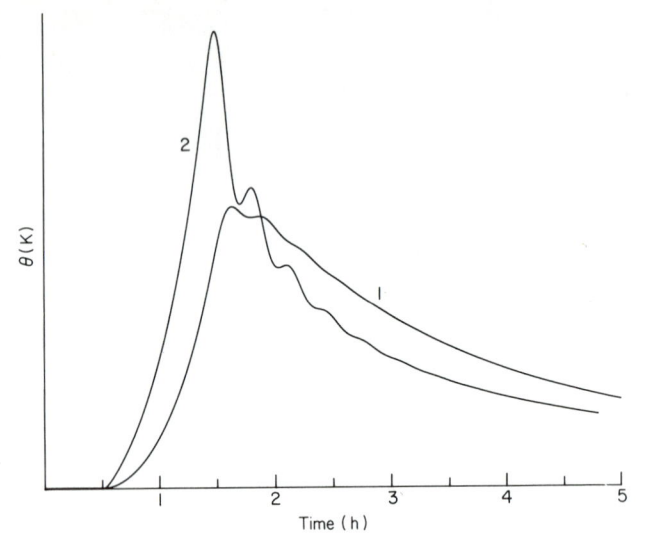

FIG. 1. Two thermograms representing the same process but obtained in two different calorimeters.

define conditions under which thermal information about the process investigated (generally presented as a thermogram) is deformed by the calorimeter, and to present different methods which can be used to correct thermograms for deformations.

B. Parameters influencing the shape of thermograms

Assume that a given biological process occurring in a calorimetric cell is generating or absorbing a thermal power, $q_x(t)$, as a function of time. Part of the total power, $q_x(t)$, is being accumulated by the calorimetric cell and part is being exchanged with the surroundings. If we denote the temperature of the calorimetric cell by $T_c(t)$, its heat capacity by C_c, and we assume that the temperature $T_c(t)$ is uniform over the whole calorimetric cell, then that part of the power $q_x(t)$ being accumulated can be expressed as $C_c[\mathrm{d}T_c(t)/\mathrm{d}t]$. That part of the power being exchanged with the surroundings is proportional to the temperature difference between the calorimetric cell and the surroundings. The temperature of the surroundings is denoted by $T_0(t)$, and the proportionality constant is the heat exchange coefficient α_{co} in W K^{-1}. Thus the balance of power can be described by the following equation:

$$C_c \frac{dT_c(t)}{dt} + \alpha_{co}[T_c(t) - T_0(t)] = q_x(t) + q_a(t) \tag{1}$$

In addition to the power $q_x(t)$ on the right-hand side of equation (1), there is a power term $q_a(t)$ which corresponds to any accidental effect accompanying the investigated process (mixing, friction, power generated in the temperature sensor, etc.). Because of these accidental effects and because of possible changes in temperature of the surroundings $T_0(t)$, the temperature of the calorimetric cell $T_c(t)$ is measured not as an absolute value but with respect to the temperature of the reference cell. Usually it is assumed that a similar accidental power $q_a(t)$ is generated also in the reference cell, for which we can write a similar power balance equation (suffix "r" stands for reference cell):

$$C_r \frac{dT_r(t)}{dt} + \alpha_{co}[T_r(t) - T_0(t)] = q_a(t) \tag{2}$$

If we define the temperature difference $\theta(t)$, which is directly measured, and its differential in the following manner:

$$\theta(t) = T_c(t) - T_r(t) \quad \text{and} \quad dT_c(t) = d\theta(t) + dT_r(t) \tag{3}$$

and introduce them into equation (1) and finally subtract equation (2) from equation (1) (differential operation of the calorimeter) we obtain the equation

$$C_c \frac{d\theta(t)}{dt} + \alpha_{co}\theta(t) = q_x(t) \tag{4}$$

It is assumed that the heat capacities and heat exchange coefficients of the calorimetric and reference cells are equal.

This equation can also be written as

$$q_x(t) = \alpha_{co}\left[\theta(t) + \tau_c \frac{d\theta(t)}{dt}\right] \tag{5}$$

often known as the Tian equation (Calvet and Prat, 1956, 1963), where τ_c is the time constant of the calorimetric cell, defined as

$$\tau_c = C_c/\alpha_{co} \tag{6}$$

This quantity has the dimension of time and characterizes the dynamic properties of the calorimetric cell because it contains at the same time information on energy accumulation (heat capacity) and energy flow (heat exchange coefficient). In general, the thermal time constant of a given body is a ratio of its heat capacity and the sum of all its heat exchange coefficients.

From equation (5) we can see that at any moment the power $q_x(t)$ is directly proportional to the sum of the actual temperature difference $\theta(t)$, and the product of the calorimetric time constant and the time derivative of $\theta(t)$. Only for the case where the time constant of the calorimetric cell is very small, or the time derivative of the temperature-difference curve is negligible, is the power $q_x(t)$ directly proportional to the recorded temperature difference, $\theta(t)$.

In the above example we have assumed that the calorimetric cell and the investigated substance are at the same temperatures. This is not always the case. Often the substance is placed in an ampoule and then introduced into the calorimetric cell. The temperature is measured, as before, somewhere on the external surface of the calorimetric cell. In this case it is not always correct to assume that the temperature of the substance $T_s(t)$ and the temperature of the calorimetric cell $T_c(t)$ are equal, because the heat exchange between them may be limited by the wall of the ampoule or by some other barrier. If we denote this heat exchange by the coefficient α_{sc} in W K^{-1}, we can write the power balance equation for the substance:

$$C_s \frac{dT_s(t)}{dt} + \alpha_{sc}[T_s(t) - T_c(t)] = q_x(t) + q_a(t) \tag{7}$$

and the power balance equation for the calorimetric cell itself:

$$C_c \frac{dT_c(t)}{dt} + \alpha_{co}[T_c(t) - T_0(t)] = \alpha_{sc}[T_s(t) - T_c(t)] \tag{8}$$

The right-hand side of equation (8), which also appears in equation (7), represents power exchanged between the substance investigated and the calorimetric cell. We can combine equations (7) and (8) into one equation by eliminating the temperature $T_s(t)$ which is not measured directly. In order to do this we differentiate equation (8) and introduce both equation (8) and its derivative into equation (7). If we introduce, as before, the concept of time constants but now defined as

$$\tau_s = \frac{C_s}{\alpha_{sc}} \quad \text{and} \quad \tau_c = \frac{C_c}{\alpha_{sc} + \alpha_{co}} \tag{9}$$

we obtain finally the following equation:

$$\tau_s \tau_c \frac{d^2 T_c(t)}{dt^2} + (\tau_s + \tau_c)\frac{dT_c(t)}{dt} + \frac{\alpha_{co}}{\alpha_{sc} + \alpha_{co}} T_c(t)$$

$$= \frac{1}{\alpha_{co} + \alpha_{sc}} q_x(t) + \frac{1}{\alpha_{co} + \alpha_{sc}} q_a(t) + \frac{\alpha_{co}}{\alpha_{sc} + \alpha_{co}} \left(T_0(t) + \tau_s \frac{dT_0(t)}{dt} \right) \quad (10)$$

We can write a similar equation for the reference cell where a reference substance is placed in conditions similar to those in the calorimetric cell and for which the accidental power $q_a(t)$ is identical. Subtracting equation (10) from the equation describing properties of the reference cell one obtains the equation

$$q_x(t) = \alpha_{co}\theta(t) + (\alpha_{co} + \alpha_{sc})\left[(\tau_s + \tau_c)\frac{d\theta(t)}{dt} + \tau_s \tau_c \frac{d^2\theta(t)}{dt^2}\right] \quad (11)$$

if the temperature difference $T_c(t) - T_r(t)$ is replaced by $\theta(t)$.

If the heat exchange between the substance and the cell is much smaller than the heat exchange between the calorimetric cell and its external environment, then

$$\alpha_{sc} \ll \alpha_{co} \quad (12)$$

and we can write equation (11) as

$$q_x(t) = \alpha_{co}\left[\theta(t) + (\tau_s + \tau_c)\frac{d\theta(t)}{dt} + \tau_s \tau_c \frac{d^2\theta(t)}{dt^2}\right] \quad (13)$$

We can see that in comparison to equation (5), describing a homogeneous calorimetric cell, in equation (13) there appears a second-order time derivative of the temperature difference and the coefficient of its first-order derivative has changed.

We could obtain an equation similar in form to equation (13) without making any assumptions concerning the heat exchange coefficient α_{sc}. We could simply introduce the following new definitions (Zielenkiewicz and Margas, 1973a, b):

$$\left| \frac{\alpha_{co} + \alpha_{sc}}{\alpha_{co}} \right| (\tau_s + \tau_c) = \tau_1 + \tau_2 \quad \text{and} \quad \frac{\alpha_{co} + \alpha_{cs}}{\alpha_{co}} \tau_s \tau_c = \tau_1 \tau_2 \quad (14)$$

and then write equation (11) in the form

$$q_x(t) = \alpha_{co}\left[\theta(t) + (\tau_1 + \tau_2)\frac{d\theta(t)}{dt} + \tau_1 \tau_2 \frac{d^2\theta(t)}{dt^2}\right] \quad (15)$$

Equation (15) has a form similar to that of equation (13) but the values of the time constants τ_1 and τ_2 are different from the values of the time constants τ_c and τ_s. When using equation (15) we refer to τ_1 and τ_2 as the first and second time constants of the calorimeter without making any reference to the substance or to the calorimetric cell.

An explanation of this statement is as follows. From the inequality (12) we assumed that the interaction of the calorimetric cell with the substance was negligible in comparison to the interaction of the calorimetric cell with the surroundings. This means that they exist separately and have non-interacting dynamic properties. In an actual case when even only two elements make a system and the elements are not separated by special barriers, then they interact with each other. This interaction is usually manifested by change of their own dynamic properties. This was very clearly shown by Webb (1964) in the case of electrical RC networks. These phenomena were thoroughly analysed by Margas et al. (1972) and by Zielenkiewicz and Margas (1973a, b, 1976) by studies of different model arrangements of calorimetric cells. In the case analysed above, equations (13) and (15) are similar but the values of the time constants will be different if the heat exchange coefficient α_{sc} has a magnitude comparable to the heat exchange coefficient α_{co}. In both cases the equations contain first- and second-order derivatives, but the shapes of the observed thermograms can be very different. As an example, two thermograms representing the same process and obtained in the same calorimeter are presented in Fig. 2. In the case of thermogram

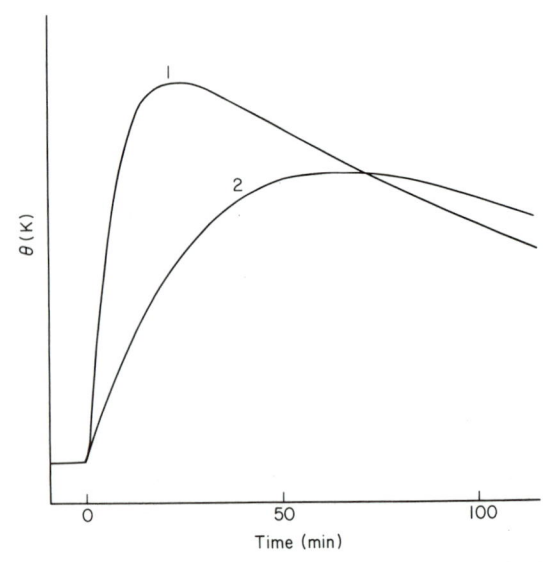

FIG. 2. Two thermograms representing the same process and obtained in the same calorimeter but with different inner heat exchange conditions.

2 the heat exchange coefficient α_{sc} was decreased by a factor of about 3 without changing significantly any other calorimetric property.

Of course one could continue the analysis and include other barriers of heat transfer and other temperature differences within the calorimetric arrangement, and consequently obtain equations with successively higher-order derivatives. On the other hand, it is possible to find a general solution to the problem, without referring to particular heat transfer barriers inside the calorimeter. The reason for this is the diffusive character of heat conduction (Camia, 1967). Even if no interfacial thermal barriers exist, significant temperature gradients may occur if heat is added locally at a sufficiently rapid rate. It has been shown by a theoretical analysis of the Tian–Calvet calorimeter that it is possible to obtain a general equation relating the power developed in the calorimeter to the recorded temperature changes, having as a starting point the Fourier equation for heat conduction (Camia, 1961; Thouvenin et al., 1967). This general relation is as follows:

$$q_x(t) = \frac{1}{\sum_i a_i \tau_i} \left[\theta(t) + \sum_i \tau_i \frac{d\theta(t)}{dt} + \sum_{i \neq j} \tau_i \tau_j \frac{d^2\theta(t)}{dt^2} \right.$$

$$\left. + \sum_{i \neq j \neq k} \tau_i, \tau_j, \ldots, \tau_k \frac{d^k\theta(t)}{dt^k} + \tau_1, \tau_2 \ldots, \tau_k \ldots, \tau_n \frac{d^n\theta(t)}{dt^n} \right] \quad (16)$$

where $i = 1, 2, \ldots, n$ and $i < j < \ldots k \ldots < n$. The derivation of equation (16) is given also by Gravelle (1975) in his review on heat-flow microcalorimetry. We can see that from a theoretical point of view a precise description of the power $q_x(t)$, generated by a process even in a homogeneous calorimetric cell, needs a series of time constants and higher-order derivatives of temperature if the process is fast. Fortunately, in practical applications one usually needs only one or two time constants at the very most. One time constant is sufficient in the case where the power generated by the process and construction of the calorimetric cell are such that the temperature is uniform at all times during the process. One can understand this statement more easily when comparing the general equation (16) with equation (5). It is possible to see that for $n = 1$ equation (16) reduces to equation (5) (with $a_i = C_c^{-1}$). In deriving equation (5) it was assumed that the temperature $T_c(t)$ was uniform. If the process is faster or the calorimetric cell arrangement has some other barriers for the transfer of heat, then the temperature is not uniform and one needs additional time constants to describe correctly the power–time curve on the basis of the recorded thermogram.

The analysis in this section has shown that the shape of the thermogram recorded in a heat-flow calorimeter for any biological process is influenced by the following parameters:

1. the number of separated elements in the calorimetric cell, their arrangement and their time constants, which depend upon their heat capacities, heat exchange coefficients and heat conductivities;
2. the degree of interaction between the elements and their mutual geometric arrangement;
3. the rate of the investigated process itself.

In the next section we describe methods applicable to the determination of the calorimetric properties which have been discussed above. We attempt to interpret the recorded thermograms, not only with the help of modelling, but primarily through the experimental data.

C. Experimental determination of static and dynamic properties of calorimeters

1. DETERMINATION OF STATIC PROPERTIES

Upon introducing a constant electrical power q_{el} into the calibration heater and achieving steady-state conditions (i.e. until $d^2\theta(t)/dt^2 = 0$ and $d\theta(t)/dt = 0$) then equations (5) and (15) are reduced to the form

$$q_{el} = \alpha_{co}\theta \quad \text{or} \quad \alpha_{co} = q_{el}/\theta \tag{17}$$

This is a steady-state solution of the differential equations (5) and (15) and is the basis for the determination of the heat exchange coefficient α_{co} which can also be called *a calibration constant of the calorimeter*. Its reciprocal is called *a static gain of the calorimeter*. It is called static because it characterizes the sensitivity of the calorimeter under steady-state conditions. Usually the temperature difference θ is measured as an electrical signal transduced by a temperature sensor, amplified by an electronic amplifier and finally recorded on an analogue strip chart recorder or digitalized and presented on punch-tape. Every component employed in this sequence has its own gain. This means that to obtain *a total static gain* k_c (in output unit W^{-1}) of the calorimetric system one should multiply the static gain $1/\alpha_{co}$ (in $K\ W^{-1}$) of the calorimeter by the static gain k_d (in $V\ K^{-1}$) of the temperature sensor, by the gain of the amplifier k_a (in $V\ V^{-1}$), by the gain of the recorder k_r (in $mm\ V^{-1}$), etc., which means that

$$k_c = \frac{1}{\alpha_{co}} k_d k_a k_r \ldots \tag{18}$$

2. DETERMINATION OF DYNAMIC PROPERTIES

a. *Time response techniques*

In order to determine dynamic parameters of the calorimeter, that is its time constants, transient state conditions must be analysed. To do this the thermal equilibrium of the calorimetric cell must be disturbed and its relaxation back to the initial state then observed. This is carried out by measuring the temperature difference $\theta(t)$ as a function of time. When carrying out these measurements it is important that there is no heat generated in the calorimeter. For this reason, as well as for theoretical reasons, it is desirable to introduce a very short but large power input into the heater (Calvet and Prat, 1956, 1963) (Fig. 3(a)). Prior to

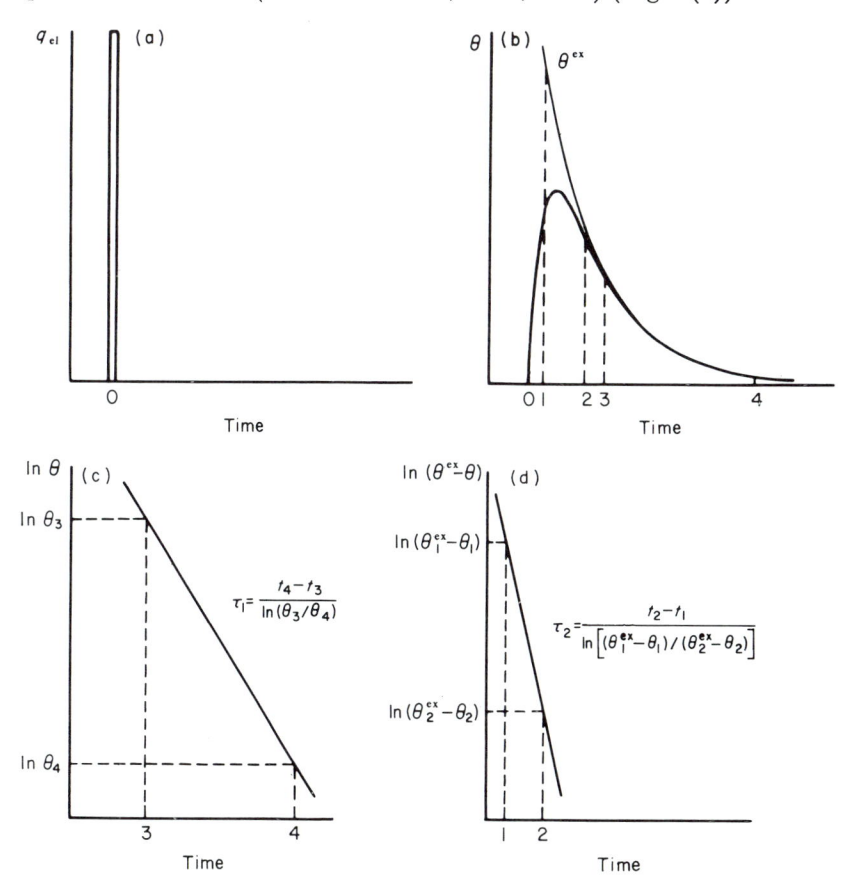

FIG. 3. Graphic presentation of the time–response procedure for determining time constants of the calorimeter.

the introduction of power the calorimeter should be in a steady state with a stable zero line. The thermogram recorded under such conditions is called *an impulse response of the calorimeter* (Fig. 3(b)). It was shown by Laville (1955) that such a thermogram can be described as a function of time by the following equation:

$$\theta(t) = A\ e^{-t/\tau_1} - B\ e^{-t/\tau_2} + \ldots \qquad (19)$$

When the second term has decreased to a negligible value then the remaining terms in equation (19) can be written in logarithmic form:

$$\ln \theta(t) = -\frac{1}{\tau_1} t + \ln A \qquad (20)$$

Thus a logarithmic plot yields a straight line whose slope is the reciprocal of the first time constant (Fig. 3(c)).

This time constant may be used to construct an extrapolated thermogram $\theta^{ex}(t)$, starting somewhere near the end of the original thermogram at an arbitrary time t_n, and then going backwards to calculate values of $\theta^{ex}(t)$ for different times $n-1$, $n-2$ etc. using the equation

$$\theta_{n-1}^{ex} = \theta_n^{ex} \exp\left(\frac{t_n - t_{n-1}}{\tau_1}\right) \qquad (21)$$

At some point significant differences between the original and extrapolated thermograms are obtained (Fig. 3(b)). These differences are the result of the influence of the other terms in equation (19), that is the higher time constants. If the actual thermogram values for given times are subtracted from the extrapolated thermogram values for the same times, and the differences are plotted (Fig. 3(d)), the value of the second time constant may be obtained from the reciprocal of the slope of the resultant straight line.

Of course, it is possible to continue this procedure to find other time constants, but in practice the determination of even the second time constant by this method is very limited in precision, because of the errors inherent in measuring small time intervals and the respective differences between the actual and extrapolated thermogram. For this reason it is sometimes better to use the frequency method described below. However, prior to this description an alternative calculation technique of the time constants from time-response curves (Zielenkiewicz and Margas, 1973) is described. This method is interesting because of the involvement of computer calculation facilities. The method essentially consists of fitting an impulse response of the calorimeter to a differential equation with proper coefficient values. Using this treatment in the case of a second-order system, it is possible to calculate directly the sum and product of the time constants of the calorimeter.

b. *Frequency response techniques*

When power of a periodical nature is introduced into a calorimetric cell, then under steady-state conditions its temperature should also change periodically. The relationship between periodical changes of power and temperature depends on the dynamic properties of the calorimeter. Determination of this dependence as a function of frequency is called a frequency analysis of the calorimeter. It has some advantages over the time analysis described above, especially since the frequency analysis is performed under steady-state conditions. This enables one to observe the response of the calorimeter for as long as necessary before changing to another frequency value. The dynamic properties of the calorimeter are determined by comparing the calorimetric responses resulting from different frequency values.

In order to perform an experimental frequency analysis of a calorimeter it is convenient to introduce a sinusoidal power signal in the calibration heater (Fig. 4). Under these conditions equation (5) takes the form

$$\tau \frac{d\theta(t)}{dt} + \theta(t) = \frac{1}{\alpha_{co}} q \sin \omega t \tag{22}$$

where q is the amplitude and ω is the angular frequency (frequency multiplied by 2π) of the power signal ($c/2$ in Fig. 4). The steady-state solution of this equation has the form

$$\theta(t) = Gq \sin(\omega t - \phi) \tag{23}$$

If this solution is introduced into equation (22) the values of G and ϕ can be found from

$$G = \frac{1}{\alpha_{co}} \frac{1}{[1+(\omega\tau_c)^2]^{\frac{1}{2}}} \tag{24}$$

and

$$\phi = \arctan \omega \tau_c \tag{25}$$

The significance of these parameters can be understood better from Fig. 4 where the upper lines represent the power input signals with two different frequency values and the lower lines represent the temperature difference signals. One can see that G is a damping factor of the amplitudes of the input power and recorded temperature difference, that is it characterizes the amount by which the recorded temperature difference is attenuated with respect to the power introduced into the calorimeter. ϕ is the phase shift between the temperature difference oscillations and the power oscillations, and it shows that the recorded temperature difference signal is delayed with respect to the power by ϕ radians or ϕ/ω seconds.

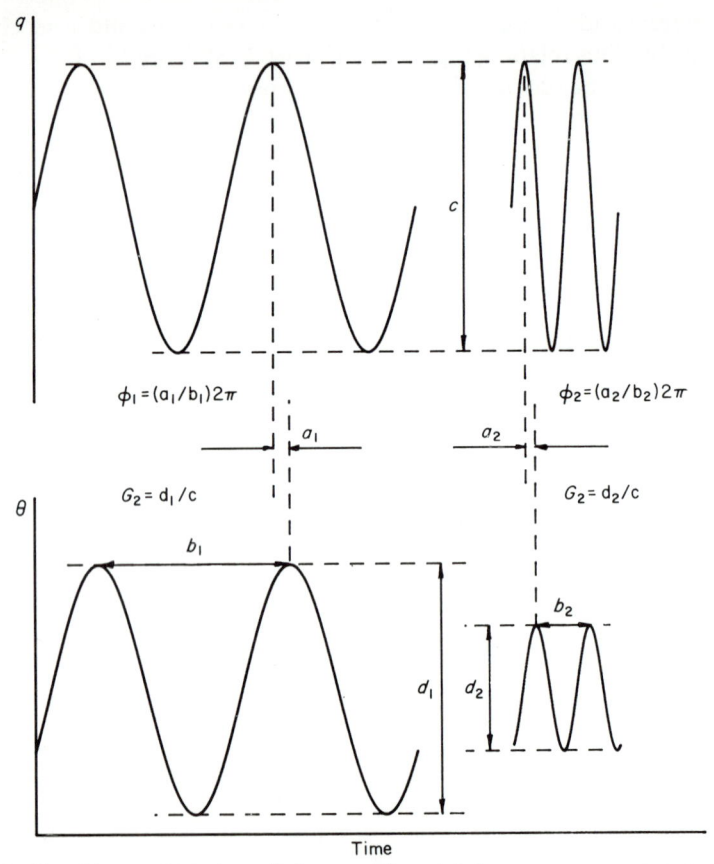

FIG. 4. Graphic presentation of the experimental frequency response procedure for determining transfer function of a calorimeter.

Both parameters, the damping factor, G, and the phase shift ϕ are related to the time constant (relations (24) and (25)) and by evaluating their values from an experiment it should be possible to determine the value of the time constant for the calorimeter. Unfortunately, it is not possible to rely on measurements made with only one frequency value. The reason for this is that one does not know *a priori* that the calorimeter is described by one time constant only. If more time constants are involved then the differential equation is of higher order and the relationships between G, ϕ and τ_i are more complicated.

It follows from the above discussion that frequency analysis gives information of a two-dimensional nature, a damping factor G and a phase shift ϕ, which can be represented by a vector in a plane. The length of the vector is made equal to G and the direction is given by the phase shift ϕ measured from the in-phase direction. The vector is shown in Fig. 5.

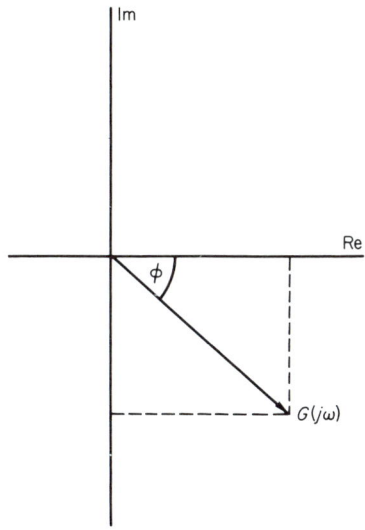

FIG. 5. Vectorial representation of the transfer function.

From the Figure it is obvious that the vector is made up of two components, the in-phase and the quadrature. If the latter component is defined as the imaginary part, we can represent the vector with a complex number

$$G(j\omega) = G(\cos \phi - j \sin \phi) \tag{26}$$

Inserting into this equation the relations for G, $\cos \phi$ and $\sin \phi$, obtained from equations (24) and (25), one obtains the following expression for the vector (Webb, 1964):

$$G(j\omega) = \frac{1}{\alpha_{co}} \frac{1}{1+j\omega\tau} \tag{27}$$

This vector defines the damping factor and the phase shift. The properties of this vector are such that by multiplying the vector by the power introduced into the calorimetric cell $q \sin \omega t$, the temperature difference $\theta(t)$ and its phase are obtained through the vector

$$\theta(j\omega) = \frac{1}{\alpha_{co}} \frac{1}{(1+j\omega\tau)} q \sin \omega t \qquad (28)$$

That is, equation (28) transforms the power into a complex temperature difference $\theta(j\omega)$. The term $(1+j\omega\tau)^{-1}$ is, for this reason, called a frequency transfer function. When rearranging equation (28) as follows:

$$\tau j\omega\theta(j\omega) + \theta(j\omega) = \frac{q}{\alpha_{co}} \sin \omega t \qquad (29)$$

it is easy to see that by replacing the time derivative of $\theta(t)$ in equation (22) with $j\omega$ and $\theta(t)$ with $\theta(j\omega)$ we obtain the same result as above. This replacement can be generalized to an n-order system where the time derivatives $(d^n/dt^n)\theta(t)$ should be replaced by the products $(j\omega)^n \theta(j\omega)$.

For practical frequency analysis of the calorimeter what is required is to determine its transfer function. For the first-order system described by equation (22) the transfer function is given by equation (27). For a calorimeter which must be described by a second-order differential equation, such as equation (15), this transfer function will have the form

$$G(j\omega) = \frac{1}{\alpha_{co}} \frac{1}{(1+j\omega\tau_1)(1+j\omega\tau_2)} \qquad (30)$$

The derivation is the same as for a calorimeter described by the first-order equation presented above.

Introducing the static properties of the electronic measuring system used (equation (18)) we obtain the following frequency transfer functions for the first- and second-order calorimetric systems:

$$G(j\omega) = \frac{k_c}{(1+j\omega\tau)} \qquad (31)$$

$$G(j\omega) = \frac{k_c}{(1+j\omega\tau_1)(1+j\omega\tau_2)} \qquad (32)$$

It is assumed that the time constants of the electronic system are negligible with respect to the time constants of the calorimeter itself.

To perform an experimental frequency analysis of a calorimeter one needs standard calorimeter calibration equipment and a voltage function generator with a low frequency range. It is not necessary to have a sinusoidal power generator. It is sufficient to supply the calibration heater with a voltage which is a sinusoidal function of time, then the

DYNAMIC CORRECTIONS 325

power will be described by the following relation:

$$q_{el}(t) = \frac{(U_0 + U \sin \omega t)^2}{R} \cong q_0 + q \sin \omega t \qquad (33)$$

where it is assumed that the amplitude of the voltage signal U is smaller by a factor of about 10 compared to the d.c. offset signal U_0. The signal from the generator should be recorded as well as the temperature signal. By measuring the maximum and minimum values of the voltage applied to the heater, the amplitude of the power changes can be calculated. From the recorded temperature changes, the amplitude of the temperature can be determined. From comparison of the time relationships of the generator signal and the temperature signal, the phase shift ϕ can be determined. The measurements should be performed for a number of different frequency values, covering as large a range as possible, especially in the middle and high-frequency range with respect to the reciprocal of the expected first-time-constant value. There are different methods of representing the experimental data. One of them, the simplest and most evident, is the Bode diagram (Bode, 1945; Randzio, 1976, 1977). An example is presented in Fig. 6.

The logarithm of angular frequency is plotted on the abscissa and the logarithm of the ratio of the temperature change and power amplitudes is plotted on the ordinate. Experimental data are represented by points. The experimental points can be combined by a polygon made up by successive asymptotes with slopes equal to 0, -1, -2, etc. (see Fig. 6). The angular frequency values at the points of intersection of these asymptotes are to a first approximation equal to the reciprocal of the respective time constants. This approximation must be made with care if the time constants are separated by much less than two orders of magnitude. The first asymptote with a slope 0 is plotted horizontally with a value equal to the logarithm of the static gain value (equation (18)). The second one must go through the largest possible number of experimental points and have the slope of -1; the third one, if it exists, must also go through the maximum number of points and have a slope of -2.

Justification for such treatment of experimental data comes from the definitions of the transfer functions, equations (31) and (32). From the logarithms of the absolute values (modulus) of the transfer functions, equations (31) and (32), we obtain

$$\log |G(j\omega)| = \log k_c + \sum_{i=1}^{n} \log \omega_i - \frac{1}{2}\sum_{i=1}^{n} \log (\omega_i^2 + \omega^2) \qquad (34)$$

where n is equal to 1 and 2 for equations (31) and (32), respectively.

We have replaced the respective time constants with specific frequency values defined by

$$\omega_i = \tau_i^{-1}, \quad i = 1, 2, \ldots, n \tag{35}$$

Equation (34) is written in such a way that it is valid for any transfer function, where n stands for the number of time constants necessary for describing the dynamic properties of a calorimeter. From equation (34) we can see that when the angular frequency is very low ($\omega \to 0$), the value of $\log |G(j\omega)|$ approaches $\log k_c$, the static gain value, independently of the order of the transfer function. This means that the experimental points always have an asymptote in the low frequency range for which the slope is 0 and which represents approximately the transfer

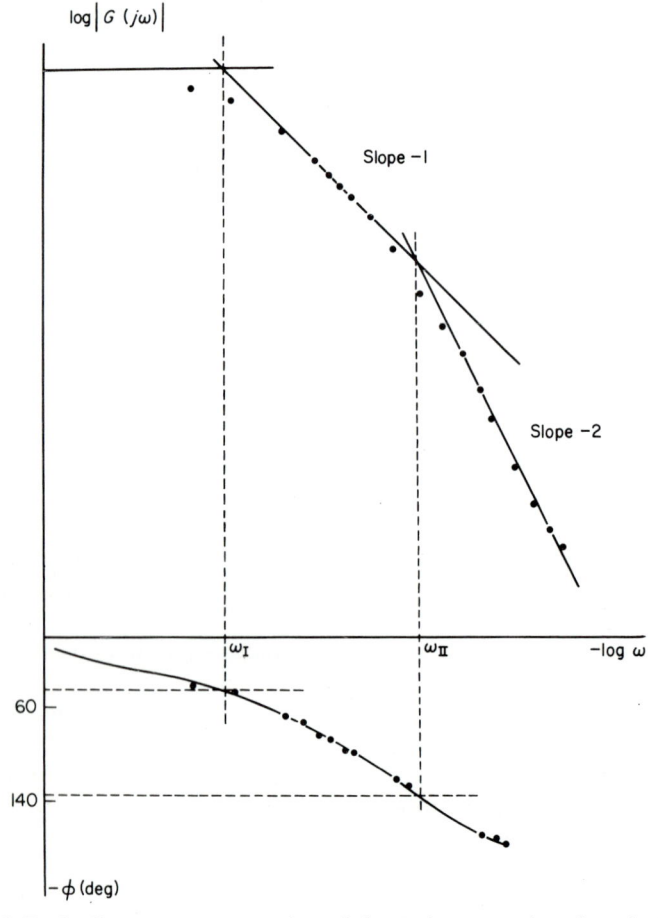

FIG. 6. Bode diagram representation of dynamic properties of a calorimeter.

function values in this frequency range:

$$\log |G(j\omega)| \cong \log k_c \qquad (36)$$

In the frequency range where the frequency $\omega > \omega_1$, for every i, equation (35) reduces to

$$\log |G(j\omega)| \cong \log k_c + \sum_{i=1}^{n} \log \omega_i - n \log \omega \qquad (37)$$

which is the equation for the asymptotes in the high-frequency range. Equation (37) shows that the slope of the last asymptote for a complete set of experimental data, obtained with a given calorimeter, gives the maximal number of time constants needed for its dynamic description.

A calorimeter which is described by the first-order transfer function (31) has only two asymptotes given by equations (36) and (37) with n equal to 1. Setting these equations equal, we observe that the asymptotes cross each other exactly at the angular frequency ω_1, which is the reciprocal value of the calorimetric time constant given in equation (35). A calorimeter which is described by the second-order transfer function (32) has three asymptotes. The low-frequency asymptote is given by equation (36). The high-frequency asymptote is given by equation (37), with n equal to 2. The middle-range asymptote, which must have the slope -1, can be obtained by the following procedure. Taking the derivative of equation (34) with respect to $\log \omega$, we obtain

$$\frac{d \log |G(j\omega)|}{d \log \omega} = \sum_{i=1}^{2} \frac{\omega^2 \omega_i^{-2}}{1 + \omega^2 \omega_i^{-2}} \qquad (38)$$

Setting this derivative equal to -1, we obtain the point with the abscissa value $(\omega_1 \omega_2)^{\frac{1}{2}}$, where the tangent of the transfer function (34) has a slope equal to -1. Substituting this value into equation (37) we obtain the value of its ordinate and the equation for the middle-frequency-range asymptote:

$$\log |G(j\omega)| \cong \log k_c - \log \left(\frac{\omega_1 + \omega_2}{\omega_1 \omega_2} \right) - \log \omega \qquad (39)$$

To find the points of intersection, which we denote by ω_I and ω_II, we must solve two sets of equations. A solution of the set of equations (36) and (39) gives the value of ω_I as

$$\omega_\mathrm{I} = \frac{1}{\omega_1^{-1} + \omega_2^{-1}} \cong \omega_1 \quad \text{if } \omega_2 \gg \omega_1 \qquad (40)$$

A solution of the set of equations (37) (with n equal to 2) and (39) gives the value of ω_{II}:

$$\omega_{II} = \omega_1 + \omega_2 \cong \omega_2 \quad \text{if } \omega_2 \gg \omega_1 \qquad (41)$$

We can see from equations (35), (40) and (41) that

$$\tau_1 + \tau_2 = \omega_I^{-1} \quad \text{and} \quad \tau_1 \tau_2 = \omega_I^{-1} \omega_{II}^{-1} \qquad (42)$$

Thus, by determining the asymptote intersection points on a Bode diagram one can determine values of coefficients of a differential equation describing the dynamic properties of a calorimeter. The first asymptote can be obtained from knowing the static gain only. A few experiments with different frequency values should then be made. If the above treatment leads to a slope of -1 only, then the calorimetric arrangement has only one time constant. It is then necessary to determine more carefully the intersection point and from its abscissa value find the time constant. If the damping of the sinusoidal signal is occurring very rapidly, it means that the temperature of the calorimetric arrangement is not uniform and there are other time constants which are necessary for a proper description of the power evolution by the investigated process. These time constants can be determined by the same procedure employed to find the first time constant. When looking at the Bode diagram of a calorimetric arrangement, one can directly characterize its properties. If the asymptote with a slope equal to -1 is long, then the calorimetric cell and the substance have the same temperature and even if in the very high-frequency range a line with a slope of -2 appears it will have only a small influence on the dynamic properties of the calorimetric response. This means that it is not necessary to determine this asymptote with high precision. If the line with a slope of -1 is short and a line with a slope of -2 appears, then it is clear that the calorimetric cell arrangement is not of uniform temperature and the influence of the second time constant on the dynamic response of the calorimeter is important, and its value should be determined carefully.

One can always verify the correctness of the amplitude frequency analysis described above by making a phase shift diagram by plotting $\log \omega$ on the abscissa as before and $-\phi$ on the ordinate. The experimental phase shift diagram should compare favourably with the values calculated from the theoretical relation (25) for the first-order system, and from the following relation for the second-order system:

$$\phi(\omega) = -\arctan \omega\tau_1 - \arctan \omega\tau_2 \qquad (43)$$

D. When is a power–time curve of a biological process deformed by a calorimeter?

It follows directly from the previous section that the temperature oscillations are much more damped when the frequency of power oscillations is higher and less damped when the frequency of power oscillations is lower. Looking at the Bode diagram one can see that in the angular frequency range $(0, 1/\tau_1)$, the first asymptote has the slope 0 and this means that there is almost no influence of the frequency on the shapes of temperature oscillations. Of course this influence theoretically always exists, as it follows from the transfer functions and especially from their modulus (equation (24)), and phase-shift definitions (equation (25)). However, in practical measurements, where one always has a given limited sensitivity and precision of measurements, there is a frequency value below which there is no observable influence. This frequency value is called *the cut-off frequency*, ω_c, but there is no accepted definition of this value. In less precise measurements one can define it as a reciprocal of the first time constant; in more precise measurements a value $\sqrt{(3)}/4\pi\tau_1$ is used (Chevillot et al., 1970). By inserting the latter value into equation (24) one can ascertain that if the frequency values are smaller than $\sqrt{(3)}/4\pi\tau_1$, the damping is less than 5 per cent. The cut-off frequency is a characteristic value of the lower limit of the calorimeter frequency characteristics. In the high-frequency range lies another value which is called *the limiting frequency* ω_0. This value is determined first of all by the noise level and the sensitivity of the instrument but also depends upon the dynamic properties of the calorimeter. When making a frequency analysis it is easy to find its value. At a certain frequency level one always finds that even if the power oscillates, the recorded temperature signal is represented by a straight line with a given constant deflection from the zero line. If the sensitivity (the static gain k_c) of the instrument is increased then it is possible to distinguish temperature oscillations if the instrument noise is not larger than these oscillations.

Until now we have been speaking only about sinusoidal changes of the power introduced into the calibration heater. Most biological processes are represented by continuous functions of time, as observed by the calorimetric or indeed any other technique. From the mathematical point of view any continuous function can be represented as a Fourier series. This means that a biological process, being a given continuous function of time, can be represented by a Fourier series where frequency is a parameter (Chevillot et al., 1970). When making this transformation it is possible to determine a *frequency bandwidth* of the process. A fast process would be described by a series containing terms with high-frequency values. A slow process would be described by a series containing only terms with low-frequency values.

In comparing dynamic properties of a calorimeter (determined by any of the methods but expressed in the language of frequency) and a frequency bandwidth of a biological process, it is possible to draw the following conclusions:

1. If the process is so slow that its frequency description does not contain terms with frequencies higher than the cut-off frequency of the calorimeter (e.g. $\sqrt{(3)}/4\pi\tau_1$), then the process is not dynamically deformed by the calorimeter. In order to obtain a power generated by the process as a function of time it is sufficient to multiply the recorded thermogram by the static gain k_c of the calorimeter.

2. If the process is so fast that its frequency description contains components with frequency values higher than the cut-off frequency of the calorimeter then the power–time curve is deformed by the calorimeter. The deformation is much larger when the calorimeter has more than one time constant. In order to obtain a normal power–time curve of the process it is necessary to make corrections to the recorded thermogram.

3. The corrections of the thermogram can be made only in the frequency range identified experimentally with the calorimeter. This means that, if the experimental determination of the calorimeter dynamic properties (with time or frequency methods) has been performed in the frequency range $0-\omega_0$, then it is possible to obtain correct power–time curves of those processes, for which the frequency description is limited to this frequency range.

An approximate answer to the question posed in this section can also be obtained from a recorded thermogram itself. From equations (5), (15) and (16), it can be seen that at every point the power generated by the process is equal to a sum of the actual recorded temperature difference and its successive time derivatives, multiplied by their respective coefficients. If the time derivative of the recorded thermogram is small in comparison to the actual $\theta(t)$ value, then one can assume that the process is not deformed by the calorimeter. The correctness of this judgement depends very much on the calorimetric time constant values.

The above considerations can also be related directly to the rate constants of a process. For a first-order reaction, which is very important from a practical point of view, because experimental conditions can often be arranged so as to obtain a first-order reaction (Becker, 1973), it follows directly from the frequency characteristics that the process under investigation is not deformed by the calorimeter if its rate constant

k_1 (in s^{-1}) is much smaller than the cut-off frequency ω_c of the calorimeter or the reciprocal of its time constant τ_1.

E. Methods of thermogram correction

The basic assumption in most of the known techniques for thermogram correction is the linearity of the calorimetric system. The calorimetric system is linear when the coefficients of the differential equation describing its properties are constant. From a practical point of view this means that the time constants of the calorimeter, $\tau_1, \tau_2, \ldots, \tau_n$, and its static gain k_c must be constant and depend neither on the power generated by the investigated process nor on the time of the measurements. Furthermore, it must be assumed that these parameters are the same both during the calibration (identification) and during the real experiment.

Among the methods of thermogram correction, one can distinguish some specific groups. The first group may be termed "differentiation techniques" because these methods are mostly based, in one way or another, on the differentiation of the measured temperature difference $\theta(t)$. The other techniques may be identified by their principal technical features: harmonic analysis, state functions, simulation and optimization and feedback techniques.

1. DIFFERENTIATION TECHNIQUES

As equations (5), (16) and (17) indicate, correction by differentiation consists mostly of differentiating the thermogram and multiplying it by the time constant, or the sum of the time constants and their products, depending on the identified calorimeter properties, and finally taking the total sum with the actual temperature difference $\theta(t)$. The shape of the curve obtained by such a procedure should correspond to the actual power evolved by the investigated process as a function of time. To obtain the power itself it is necessary to multiply this curve by the calorimeter static gain value.

The above procedure can be carried out in practice in many ways. The oldest one is the graphical correction (Calvet and Camia, 1958). More recently modern computers have been used for making the calculations. The recorded thermogram can be digitalized and then introduced at any time to a computer (off-line correction) (Randzio, 1977), or the calorimetric signal can be transmitted directly to the computer simultaneously as the process occurs (on-line correction). As an example of the differentiation technique, utilizing off-line computer calculations, corrected thermograms from Fig. 1 are shown in Figs 7 and 8. Only the first-time constant values were used in the calculations, 1330 s in the case of thermogram 1 and 170 s in the case of thermogram 2. The differentiation has been performed using simplified least-squares procedures (Savitzki and Golay, 1964).

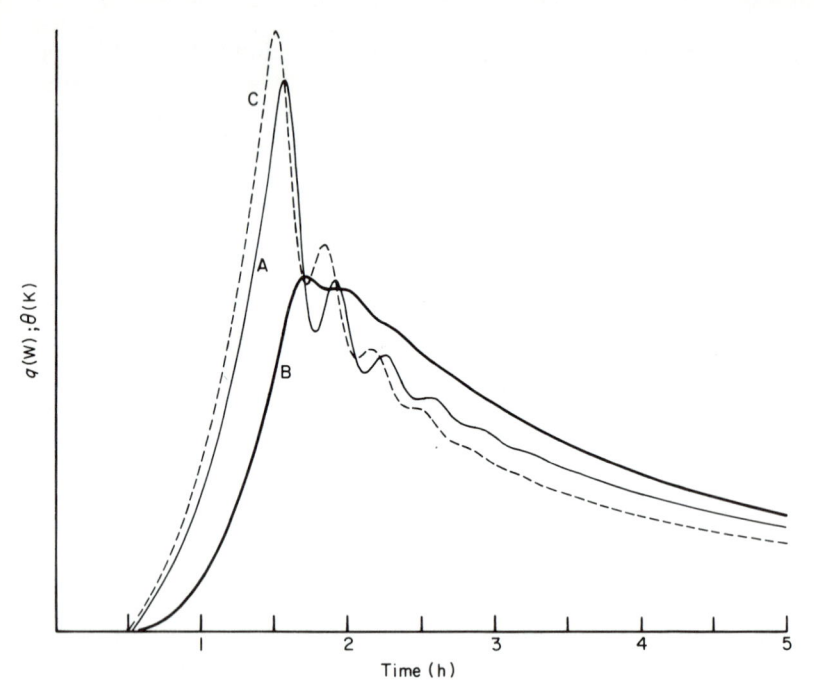

FIG. 7. Corrected thermogram from Fig. 1 (A), original thermogram 1 (B) and actual power generated in the calorimeter (C).

Another technique is a direct electronic differentiation of the calorimetric signal (Camia, 1961; Touvenin et al., 1967). The differentiation can be performed by a simple RC network or by an operational amplifier (Shukshunov, 1970; Dubes et al., 1977). The simplicity of this technique is its big advantage. But one must be careful when adjusting parameters of the electronic unit. It must be done when the system is being calibrated with a calibration heater located in exactly the same position as that in which the process will occur. When the calorimetric cell contains the calibration heater in the proper place, then the calibration with the electronic corrector is very fast and simple (Randzio and Lewandowski, 1977) because one does not need to know the time constant values.

With the techniques mentioned above there may be difficulties with "overshoots". If these appear it is more reliable only partially to correct the first time constant of the calorimeter. In this case not all the calorimetric thermal inertia is compensated for but there are no "overshoots" and the residual time constant is much smaller than the original calorimetric time constant. In conclusion, when using simple differentiation techniques of thermogram correction, one can easily and safely reduce the thermal inertia of the calorimeter by a factor of 10–20, which in a large number of biological experiments is satisfactory.

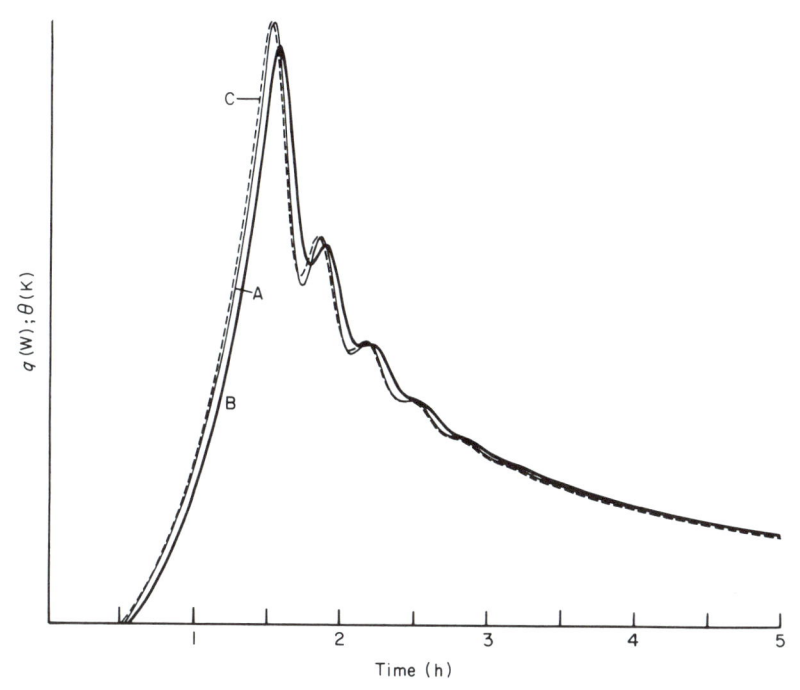

FIG. 8. Corrected thermogram from Fig. 1 (A), original thermogram 2 (B) and actual power generated in the calorimeter (C).

2. HARMONIC ANALYSIS TECHNIQUE

In the differentiation techniques the dynamic properties of a calorimeter are described by a differential equation of n order or by a set of n equations of first order. In the harmonic analysis technique these properties are described by the frequency characteristics of a calorimeter, that is, by its frequency transfer function. This frequency transfer function which has been described in published papers (Navarro et al., 1971; Rojas et al., 1971; Bokhoven and Medema, 1976; Tanaka, 1978) is calculated from the impulse response of the calorimeter. When the frequency transfer function is known, a Fourier transform is calculated for a thermogram which is to be corrected. In order to reconstruct the corrected power curve it is necessary to divide the Fourier transform of the thermogram by the frequency transfer function of the calorimeter and then to calculate the inverse Fourier transform of the resultant quotient. If the signal-to-noise ratio of the thermogram recording is sufficiently high, the final result of the correction will be satisfactory. If this is not the case, some fluctuations in the reconstructed power curve will be observed. A smoothing procedure may help to reduce these fluctuations (Bokhoven and Medema, 1976). The method needs a computer with a capacity of at least 32 K words. An example of a thermogram correction using this technique is presented in Fig. 9.

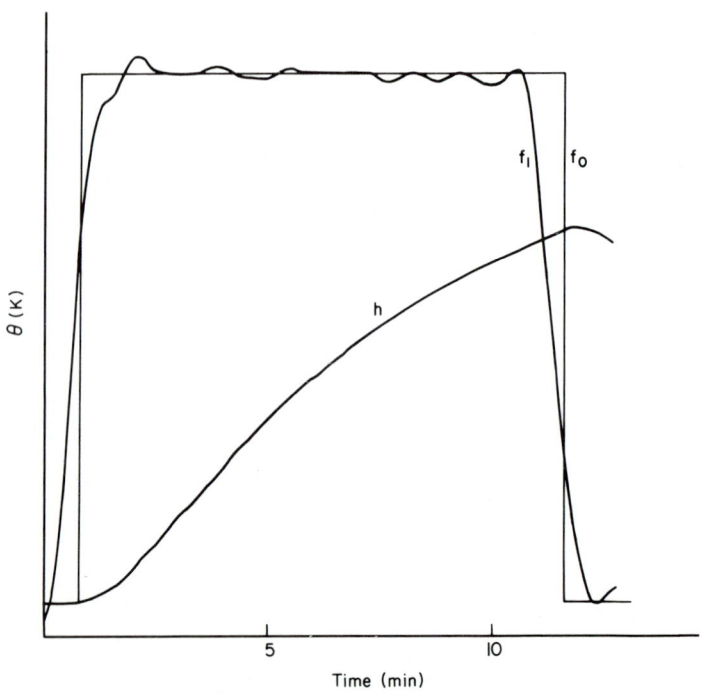

FIG. 9. An example of a thermogram correction using the harmonic analysis technique. h, Original thermogram; f_1, corrected thermogram; f_0, power introduced to the calorimeter. (Bokhoven and Madena, 1976.)

3. STATE FUNCTIONS AND NUMERICAL CORRECTOR TECHNIQUES

In the method which is described below (Brie *et al.*, 1971; Brie *et al.*, 1973a, b) the dynamic properties of a calorimeter are described by so-called state functions, a mathematical procedure used in automation theory (Schultz and Melsa, 1967). These state functions are represented by a set of two linear equations relating the input and output of a calorimeter by a state vector and state variables. The first step in utilizing this technique is to determine the state vector and state variables of the calorimeter from its transfer function, which is obtained from the calorimeter impulse response using computer numerical calculations. The second step is to calculate the unknown power $q_x(t)$ from the recorded thermogram using the state vector and state variables determined previously. The method needs only one numerical differentiation to determine the state vector, regardless of the order of the transfer function, that is, regardless of the number of time constants the transfer function contains. The precision of the method is estimated to be about 8 per cent. In order to improve this method a numerical corrector can be used. Its structure can be optimized with respect to the best reconstruction of a known input impulse (Dorf, 1964). The numerical corrector alone can also be used as an independent correction technique. In this case the method is simpler but is less precise than when it is combined with the method described first.

4. SIMULATION AND OPTIMIZATION METHODS

To some extent all the methods of correcting thermograms presented above are based on modelling. However, we can consider as a separate group the methods which rely much more than the others on computer simulation and optimization procedures. Simulation can be defined as the creation of a computer model of a physical system (e.g. calorimeter) which performs essentially in the same manner as the actual physical system.

A computer model can be created either on the basis of physical laws or as a pure mathematical task which connects input and output signals of a calorimeter. The latter model is often called "a black box". Both types of simulations have been used to correct calorimetric thermograms. Berger and Davids (1974) and Davids and Berger (1964, 1969) have developed a set of computer programs based on the simulation of heat accumulation and heat conduction. By dividing a calorimeter into small elements and then applying the law of conservation of energy and a simulation program of heat conduction (Berger and Davids, 1965) they succeeded in determining the power evolved in the calorimeter. A similar procedure for simulating a calorimeter was used by Churney *et al.* (1973) and by Höne (1978). Gutenbaum *et al.* (1973, 1976) and Romaneti and Zahra (1975) have used a purely mathematical model dependent upon a convolution integral (Lefkowitz, 1963). The results of calculations employing the convolution integral are compared with the actual thermogram and the final calculated curve is obtained by minimizing the integral of the square of the differences between the calculated thermogram and real thermogram values.

An example of a correction performed with the optimization technique as compared to the differentiation technique is given in Fig. 10. The

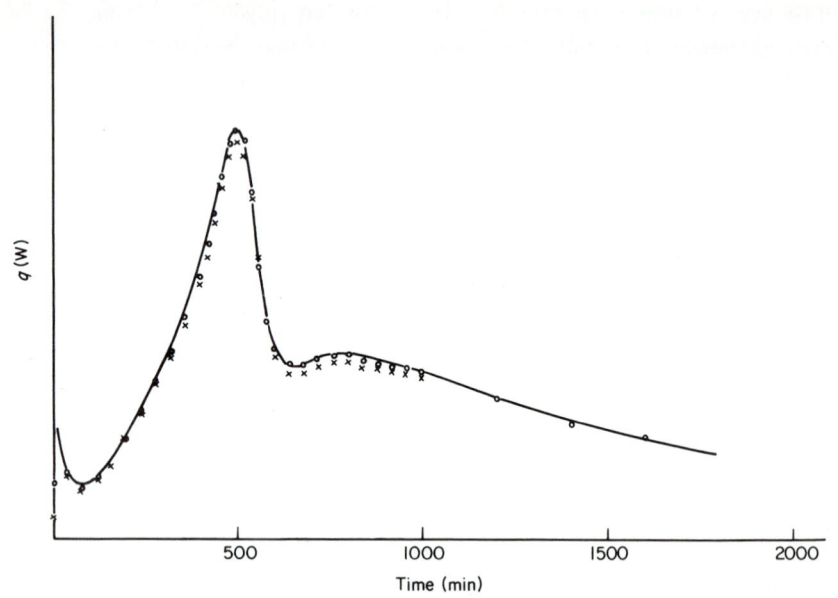

Fig. 10. Comparison between two methods of thermogram correction. The continuous line represents the actual power; closed circles represent the power calculated with the optimization technique and crosses represent the power calculated with the differentiation technique using a simple time constant. (Utzig, E., 1977.)

original thermogram (not shown in the Figure) was recorded with a calorimeter having a time constant of 220 min.

5. FEEDBACK TECHNIQUES

The last group of methods which permits determination of a thermal power by calorimetric measurements is referred to as power-compensated calorimeters. The basic difference between the methods of thermogram corrections presented above and this technique is that in the feedback technique there is a compensation of thermal effects, simultaneously with their generation, so that the temperature of the calorimetric cell remains unchanged (Weber, 1973, 1974). The recorded signal is directly proportional or equal to the power developed by the investigated process. The time constant of the calorimeter is reduced by a factor equal to the gain of the compensating feedback loop employed (Randzio and Lewandowski, 1973). The residual time constant, after feedback compensation, can be reduced further by an electronic differentiation network mounted in series (Randzio and Lewandowski, 1978) or by any of the differentiation techniques presented above. As a source of compensation power, either the Joule or Peltier effect can be used (Becker and Walisch, 1965). In the case of a Calvet calorimeter using Peltier feedback compensation, it is possible to reduce the time constant of the calorimeter by a factor of 20 (Camia et al., 1970; Coten et al., 1971). By employing a specially constructed calorimeter, Macqueron et al. (1968) obtained a final time constant value of 0·7 s. It is evident that the power compensation methods are not exactly thermogram corrections methods, because the recorded signal is not significantly influenced by the time constant of the calorimeter. However, some care must be taken as far as the configuration of the compensated calorimeter is concerned. It has been shown (Randzio and Sunner, 1977) that the above properties of a compensated calorimeter are valid only under the condition that the sample is placed inside the compensation loop. If this is not the case, the recorded signal must be corrected for the influence of the time constant of the investigated substance, even if the gain of the compensation feedback loop is very high.

F. Conclusions and remarks on reporting dynamic calorimetric data

In section E we have defined parameters influencing the shape of calorimetric thermograms, and have derived differential equations relating these parameters to temperature changes occurring in a calorimeter when thermal power is generated by a process. Two types of methods of calorimetric analyses were described: time and frequency response techniques. Of these two methods the former is better known, but the latter is more powerful and more direct, because it defines directly the frequency bandwidth of the calorimeter. From the bandwidth of the calorimeter it is possible to determine which process is damped by the calorimeter, and when one should correct recorded calorimetric thermograms for deformations. In section E we have briefly discussed presently known methods for thermogram correction. It is difficult to state categorically which of the methods is best. The answer depends very much on the experimental conditions of the calorimetric measurements and on the means available in a laboratory to make the necessary measurements and calculations. However, it should be noted that all the methods require calorimetric data of good quality, in particular the signal-to-noise ratio should be as high as possible. Under these conditions all of the methods can give satisfactory results. But even if this requirement is not satisfied, evaluation of the first time-constant can still be realized. This can lead to a reduction in the time response of the calorimeter by a factor of 10–20, which for most biological experiments can be very helpful. It is also important to notice that a calorimeter with a correction network on-line (classical electronic or computer) is a power-measuring instrument. This means that the output signal does not depend on the temperature difference but depends only on the power actually developed in the calorimeter. Thus one does not need to wait for complete equilibration after introducing the sample into the calorimeter, provided there is no initial power developed in the sample and the parameter values of the correction network have been adjusted for the dynamic properties of the calorimeter. Consequently the correction techniques can not only give much more reliable information about the dynamic process under investigation but can also considerably speed up the measurements. However, one must be careful when using correction techniques because of possible "overshoots" and non-linearities. The "overshoots" depend very much on the quality of original data and the evaluated parameters of the calorimeter. It is important when performing the identification experiments to place the calibration heater as close as possible to the actual power source. The non-linearities of the correction networks can be detected by comparing integrals of the original and corrected thermograms in the total time interval. They should always be equal.

A large number of different calorimeters used in biological investigations, together with a number of correction techniques available, may create a situation where it is difficult to compare results obtained in different laboratories. For this reason some standard methods should be accepted by all laboratories when presenting dynamic calorimetric data for publication. It is important that the dynamic parameters of the calorimeter used should always be given together with the thermogram. The dynamic parameters can be given as final values after corrections (if they were used) or separately as dynamic parameters of the calorimeter and the correction network used. For parameters describing calorimetric properties, we suggest two points of the frequency characteristics: one given by the cut-off frequency ω_c and a static gain of the calorimeter k_c, and the other given by the limiting frequency ω_0 and modulus of the transfer function near this frequency value. The value of the latter (given in the units of k_c) can be obtained by dividing the amplitude of the temperature oscillations by the amplitude of the power oscillations having a frequency value near ω_0. From these two points it is always possible to reconstruct approximately the dynamic and static properties of the calorimeter. In the case where the frequency characteristics cannot be evaluated, the time constant of the calorimeter should be given and the thermograms presented should always be accompanied by a proper description of coordinates.

ACKNOWLEDGEMENTS

This work has been supported by the Swedish Natural Science Research Council, the Swedish Board for Technical Development and by the Exchange of Scientists Program between the Polish Academy of Sciences and the Royal Swedish Academy of Sciences.

References

BECKER, F. (1973). *In* "Protides of the Biological Fluids" (H. Peeters, ed.), pp. 543–550. Pergamon Press, Oxford.
BECKER, F. and WALISCH, W. (1965). *Z. phys. Chem.* (*N. Folge*), **46**, 279–293.
BERGER, R. L. and DAVIDS, N. (1965). *Rev. scient. Instrum.* **36**, 88–93.
BERGER, R. L. and DAVIDS, N. (1974). Technical Report IV, NIH.
BODE, H. W. (1945). "Network Analysis and Feedback Amplifier Design". University Press, Princeton, New Jersey.
BOKHOVEN, J. J. G. M. and MEDEMA, J. (1976). *J. Phys.*, *E*, **9**, 123–128.
BRIE, C., GUIVARCH, M. and PETIT, J. L. (1971). *In* "Proc. 1st Int. Conf. Calorimetry and Thermodyn., Warsaw, 1969", pp. 73–90, Polish Acad. Sci. Publ., Warsaw.
BRIE, C., PETIT, J. L. and GRAVELLE, P. C. (1973a). *J. Chim. phys.* **70**, 1107–1114.
BRIE, C., PETIT, J. L. and GRAVELLE (1973b). *J. Chim. phys.* **70**, 1115–1122.
CALVET, E. and CAMIA, F. M. (1958). *J. Chim. phys.* **55**, 818–826.
CALVET, E. and PRAT, H. (1956). "Microcalorimétrie". Masson et Cie, Paris.
CALVET, E. and PRAT, H. (1963). *In* "Recent Progress in Microcalorimetry" (H. A. Skinner, ed.). Pergamon Press, Oxford.
CAMIA, F. M. (1961). *Journées Int. Transmis. Chaleur, J.F.C.E.*, 703–712.
CAMIA, F. M. (1967). *In* "Microcalorimétrie et Thermogenèse", pp. 83–93. CNRS, Paris.
CAMIA, F. M., LAFFITTE, M. and COTEN, M. (1970). *Messen-Sleuern-Regeln.* **13**, 238–240.
CERNY, S., PONEC, V. and HLADEK, L. (1970). *J. Chem. Thermodynamics*, **2**, 391–397.
CHEVILLOT, J. P., GOLDWASSER, D., HINNEN, C., KOEHLER, C. and ROUSSEAU, A. (1970). *J. Chim. phys.* **67**, 56–60.

CHURNEY, K. L., WEST, E. D. and ARMSTRONG, G. T. (1973). "A Cell Model for Isoperibol Calorimeters". Polish Acad. Sci., Inst. Phys. Chem., Warsaw.
COTEN, M., LAFFITTE, M. and CAMIA, F. M. (1971). *In* "Proc. 1st Int. Conf. Calorimetry and Thermodyn., Warsaw, 1969", pp. 67–72. Polish Acad. Sci. Publ., Warsaw.
DAVIDS, N. and BERGER, R. L. (1964). *Commun. ACM*, **7**, 547–583.
DAVIDS, N. and BERGER, R. L. (1969). *Currents in Modern Biology*, **3**, 169–179.
DORF, R. C. (1969). "Time-domain Analysis and Design of Control Systems". Addison-Wesley, Reading, Mass.
DUBES, J. P., BARRES, M. and TACHOIRE, H. (1977). *Thermochimica Acta*, **19**, 101–110.
GRAVELLE, P. C. (1975). *Adv. Catalysis*, **24**, 191–263.
GUTENBAUM, J., UTZIG, E., WISNIEWSKI, J. and ZIELENKIEWICZ, W. (1973). *In* "1st National Conf. Calorimetry, Zakopane, 1973" (in Polish). Polish Acad. Sci., Inst. Phys. Chem., Warsaw.
GUTENBAUM, J., UTZIG, E., WISNIEWSKI, J. and ZIELENKIEWICZ, W. (1976). *Bull. Acad. pol. Sci., Sér. Sci. chim.* **24**, 193–199.
HÖNE, G. W. H. (1978). *Thermochimica Acta*, **22**, 347–362.
LAVILLE, G. (1955). *C. R. Hebd. Séanc. Acad. Sci., Paris*, **240**, 1060–1063.
LEFKOWITZ, I. (1963). *In* "Temperature—Its Measurements and Control in Science and Industry" (C. M. Herzfeld, ed.), vol. III, pt 2, pp. 761–720. Reinhold, New York.
MACQUERON, J. L., GERY, A., LAURENT, M. and SINICKI, G. (1968). *C.r. Acad. Sci., Paris, Sér. B*, **266**, 1297–1298.
MARGAS, E., TABAKA, A. and ZIELENKIEWICZ, W. (1972). *Bull. Acad. pol. Sci., Sér. Sci. chim.* **20**, 239–242.
NAVARRO, J., TORRA, V. and ROJAS, E. (1971). *Anales Física* (1971), 367–374.
RANDZIO, S. (1976). *In* "Journées de Calorimétrie et d'Analyse Thermique". AFCAT, Besançon.
RANDZIO, S. (1977). *J. Phys.*, *E*, **10**, 145–148.
RANDZIO, S. and LEWANDOWSKI, M. (1973). "Fundamentals of Electronic Temperature Control in Diathermic Calorimetry" (in Polish). Polish Acad. Sci., Inst. Phys. Chem., Warsaw.
RANDZIO, S. and LEWANDOWSKI, M. (1978). *Apav. Nauk. Dyd.* **4**(6), 45–55.
RANDZIO, S. and SUNNER, S. (1977). *Analyt. Chem.*, **50**, 704–707.
ROJAS, E., TORRA, V. and NAVARRO, J. (1971). *Anales Física*, **67**, 359–366.
ROMANETTI, R. and ZAHRA, C. (1975). *Thermochimica Acta*, **12**, 343–351.
SAVITZKI, A. and GOLAY, M. J. E. (1964). *Analyt. Chem.* **36**, 1262–1639.
SCHULTZ, D. G. and MELSA, J. L. (1967). "State Functions and Linear Control Systems". McGraw-Hill, New York.
SHUKSHUNOV, V. E. (1970). "Correction Circuits in Instruments Measuring Non-stationary Temperatures" (in Russian). Energia, Moscow.
TANAKA, S. (1978). *Thermochimica Acta*, **25**, 269–275.
TOUVENIN, Y., HINNEN, C. and ROUSSEAU, A. (1967). *In* "Microcalorimétrie et Thermogenèse", pp. 65–82. CNRS, Paris.

UTZIG, E. (1976). Doctoral Thesis, Inst. of Physical Chemistry, Warsaw.
WEBB, C. R. (1964). "Automatic Control". McGraw-Hill, New York.
WEBER, H. (1973). "Isothermal Calorimetry for Thermodynamic and Kinetic Measurements", Series XII, vol. 5. European University Press, Bern.
WEBER, H. (1974). *Thermochimica Acta*, **9**, 29–36.
ZIELENKIEWICZ, W. and MARGAS, E. (1973a). *Bull. Acad. pol. Sci., Sér. Sci. chim.* **21**, 247–250.
ZIELENKIEWICZ, W. and MARGAS, E. (1973b). *Bull. Acad. pol. Sci., Sér. Sci. chim.* **21**, 251–254.
ZIELENKIEWICZ, W. and MARGAS, E. (1976). "Some Selected Problems of Thermokinetics" (in Polish). Pol. Acad. Sci., Inst. Phys. Chem., Warsaw.

Thermodynamics of Interacting Biological Systems[1]

MAURICE EFTINK

and

RODNEY BILTONEN

A. Introduction

An important and ubiquitous feature of biological macromolecules is their ability to interact with various small and large molecules with a high degree of specificity, and, by most physical chemical standards, with high affinity. Such interactions play an essential role in enzyme catalysis, antibody specificity, hormone action, and a host of other protein receptor-mediated processes. Any molecular description of such processes requires an interrelated understanding of the structural and thermodynamic details of the interactions.

The concept of structural specificity was first espoused by Fischer in his "lock-and-key" description of enzyme–substrate interactions over 80 years ago (Fischer, 1894). This model envisioned the protein as being topologically complementary to the stereochemical nature of the substrate. Over the years our understanding of the basis for specificity has evolved to include the formation of complementary hydrogen bonds, dipole–dipole interactions, salt linkages, etc., between the ligand and the interacting groups of the binding site.

---

[1] Much of the experimental work reported in this contribution was supported by grants from the National Science Foundation (PCM75–23245–A01) and from the National Institutes of Health (GM–20637–05).

The tethering of the ligand on to the macromolecule will be governed by a number of such elementary points of interaction, as well as geometric considerations. The binding loci thus define a potential energy surface for interaction with the ligand. The ligand will assume an average position on this surface that minimizes the free energy of the system with respect to the several points of interaction.

Consider as an example the schematic representation of the complex formed between 3'-cytidine monophosphate (3'-CMP) and ribonuclease A (RNAse A) depicted in Fig. 1. In this structure several elementary points of interaction between the ligand and macromolecule are suggested. For example, the phosphate group of the ligand appears to be able to interact electrostatically with a number of protein residues including His 12, His 119, and possibly Lys 41. The protein also possesses what appears to be a binding cleft for the riboside moiety. This cleft may provide a sequence of hydrogen bonding groups for interaction

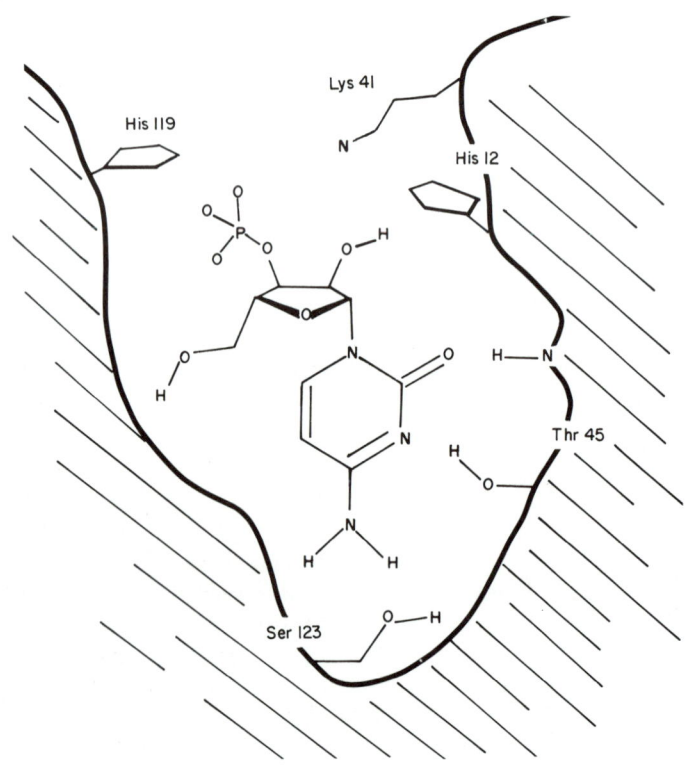

FIG. 1. The proposed structure of the ribonuclease A–3'-CMP complex. The evidence for this drawing comes from the crystallographic studies of Richards et al. (1969, 1971).

with complementary polar groups of the pyrimidine ring. The exact manner in which a ligand binds will depend on the combination of (and perhaps compromise between) such elementary interactions. In addition to these specific intermolecular forces, another very important driving force for the association process will be the liberation of water molecules upon the formation of a macromolecule–ligand complex (hydrophobic effect, release of electrostricted water, etc.).

Since the lock-and-key proposal was put forth, it has been realized that biological macromolecules, particularly proteins, are not static, rock-like structures in solution, but are dynamic, fluctuating entities (Linderstrom-Lang and Schellman, 1959; Weber, 1975). If this flexibility extends to a binding site, the potential energy surface will be less precisely defined. However, a certain amount of flexibility may be desirable to facilitate accommodation of the ligand on to the binding site (Austen *et al.*, 1975) Another possible consequence of the fluctuating nature of macromolecules is that certain small molecules (hormones, allosteric effectors, etc.) may, upon binding, be able to alter the population of accessible states of the macromolecule, and thus induce changes in its chemical and physical properties, including its affinity and reactivity towards other ligands. The interactions between enzymes and their substrates are of particular interest, since an enzyme enhances a reaction rate by bringing together reacting and catalytic groups at its active site and by having a greater relative affinity for the transition state of the reaction than for the substrate(s) (Wolfenden, 1972; Jencks, 1975).

The intent of the preceding discussion is to emphasize that many biological phenomena, such as specific binding, cooperative binding, catalysis and ligand-induced structural changes, have as a basis the interaction between a macromolecule and a ligand. The fundamental physical chemical key in understanding the molecular basis for such phenomena is to define the relationship between the structure and energetics of such interactions. For example, to understand the nature of the 3'-CMP–RNAse A complex more fully we wish to know: Which elementary interactions determine the specificity of binding? Which provide for high affinity? Does the binding of ligand affect the conformation of the protein? The answers to these questions will provide a much more complete description of the structure and energetics of the 3'-CMP–RNAse A complex, and will hopefully provide some insight as to the forces employed by this enzyme for catalysis.

The point which we would like to stress is that the relationship between structure and thermodynamics is a fundamental physical chemical key to the understanding of a variety of biological processes involving macromolecules. While the mechanisms of such processes will differ from case to case, determination and rigorous interpretation of the thermodynamics are necessary.

In this contribution means for the experimental evaluation of the principal thermodynamic quantities describing an interacting system, the apparent standard Gibbs energy change, ΔG^0, enthalpy change, ΔH^0, entropy change, ΔS^0, and heat capacity change, ΔC_p^0, will be described. Special emphasis will be focused on the use of microcalorimetric techniques. The binding of ligand to a macromolecule will often be coupled to other side-reactions and the experimentally observed thermodynamic quantities will contain contributions from these coupled reactions. Means for sorting out the various contributions will be discussed.

We will then discuss the interpretation of the apparent thermodynamic quantities in terms of the structural-mechanistic aspects of the association process. In order to do this we will briefly consider the structure-energy relationships for various types of interactions, such as hydrogen bonds, hydrophobic bonds, etc.

The relationship between thermodynamic quantities and macromolecule conformational changes and cooperative binding phenomena will also be considered. Finally, we will discuss the possible molecular interpretations of the heat capacity changes commonly observed for the binding of ligands to biological macromolecules.

B. General considerations

1. BASIC THERMODYNAMIC QUANTITIES

The equilibrium binding of a ligand, L, to a macromolecule, M, can be represented in the simplest case as

$$M + L \rightleftarrows ML \tag{1}$$

The association constant for this reaction is

$$K' = [ML]/[M][L] \tag{2}$$

This constant is usually determined by measuring the concentration of complex, ML, as a function of [L] using a variety of techniques (i.e. equilibrium dialysis, spectroscopic changes, etc.). The apparent standard Gibbs energy change is

$$\Delta G^{0\prime} = -RT \ln K' = \Delta H^{0\prime} - T\Delta S^{0\prime} \tag{3}$$

$\Delta H^{0\prime}$ can be estimated from the temperature dependence of K' using the van't Hoff relationship

$$\frac{d(\ln K')}{d(1/T)} = -\frac{\Delta H^{0\prime}}{R} \tag{4}$$

and the apparent standard entropy change is calculated from

$$\Delta S^{0\prime} = \frac{\Delta H^{0\prime} - \Delta G^{0\prime}}{T} \tag{5}$$

If data of sufficient precision can be obtained the van't Hoff plot of $\ln K'$ versus T^{-1} may not be linear. This curvature is the result of the fact that the change in heat capacity, $\Delta C_p^{0\prime}$,

$$\Delta C_p^{0\prime} = \left(\frac{d\Delta H^{0\prime}}{dT}\right)_p \tag{6}$$

is not equal to zero. The magnitudes of $\Delta C_p^{0\prime}$ can be estimated, in principle, from the degree of curvature in the van't Hoff plot.

The major disadvantage of this type of approach to the evaluation of thermodynamic quantities for binding reactions is the stress which is placed on the precision of the data. In most cases values of K' are precise within only ± 30 per cent and evaluation of $\Delta H^{0\prime}$ from the temperature dependence of K' is generally highly imprecise. It is a rare case indeed when estimates of $\Delta C_p^{0\prime}$ from the curvature of van't Hoff plots are accurate. The estimates of these quantities can be improved if care is taken to perform many experiments, but this is time-consuming and can be expensive.

These problems can be alleviated in many cases by use of calorimetric techniques. The advantage of the use of microcalorimetric techniques is that often $\Delta H^{0\prime}$ can be accurately determined from a single experiment. The imprecision of such estimates of $\Delta H^{0\prime}$ can be as low as ± 2 per cent which allows an accurate estimate of $\Delta C_p^{0\prime}$ from a few experiments. In certain cases, depending upon the magnitude of K', $\Delta G^{0\prime}$ can also be obtained from a titration experiment with a high degree of precision.

Another advantage of calorimetry is its applicability to the study of most interacting systems. This is because virtually all reactions proceed with a $\Delta H^{0\prime} \neq 0$ and thus produce a calorimetric effect. In cases where $\Delta H^{0\prime} = 0$, coupled reactions can often be used to produce or amplify a signal. However, it must be added that this aspect of calorimetry can lead to many complications since side-reactions may give rise to interfering effects. As a result, it is necessary to understand in detail all coupled reactions that may occur.

Other practical advantages of calorimetry are the large concentration range of reactants that can be employed, the fact that optically clear solutions are not required, and that the starting materials can be recovered.

2. CALORIMETRIC INSTRUMENTATION

A detailed description of the design and operation of several types of calorimeters used in biochemical studies can be found elsewhere (Wadsö, 1970; Sturtevant, 1972; Spink and Wadsö, 1976; Biltonen and Langerman, 1977; Langerman and Biltonen, 1977). The three types of instruments most commonly used for studying association reactions are the batch, titration and flow microcalorimeters. These types of calorimeters can now be obtained commercially. In a batch calorimeter, the two reactant solutions are equilibrated on either side of a split compartment cell. Mixing is then obtained by the rotation of the cell and the heat effect measured. For the flow system, the reactants are pumped into a calorimetric cell where they mix and a steady-state heat effect is observed. The flow system facilitates the rapid accumulation of data, but generally requires much more material and is less sensitive than the batch calorimeter.

Batch and flow microcalorimeters are generally based on the heat leak principle where the rate of heat flow from the calorimeter cell to the heat sink is measured. In a titration calorimeter one of the reactants is added (either stepwise or continuously) to a solution of the other reactants and the associated temperature rise is measured. In this case, the calorimeter operates in a quasi-adiabatic (isoperibol) mode and corrections for heat loss to the environment are required. In all procedures the observed signal can be converted into energy units by use of a proper calibration constant.

The specific type of calorimeter to be used is dictated by the properties of the reaction (rate, magnitude of $\Delta H^{0\prime}$, expense of material, etc.). Details of each type of calorimeter and decisions involved in the proper selection of type and experimental design are discussed in an article by Langerman and Biltonen (1977). A great deal of care must be taken in preparing the two solutions to be mixed in a calorimeter so that they are as identical in buffer components, pH and composition as possible. Heats of dilution of the reactants must be measured separately and subtracted, when necessary, from the observed heat of mixing.

3. MEASUREMENT OF THE HEAT EFFECT

For the prototype association reaction

$$M + L \rightleftarrows ML \tag{1}$$

the heat effect, W, associated with mixing the reactants will be proportional to the amount of complex formed plus the heat of dilution of components, Q_{dil}, and the heat associated with the physical mixing of solutions, Q_{mix}. The heat of reaction

$$Q = W - Q_{dil} - Q_{mix} \qquad (7)$$

where Q_{dil} and Q_{mix} have been determined in separate experiments. If the reaction goes to completion, due to the high affinity of the reactants, or is forced to completion by an excess of one reactant, the measured heat of reaction will be $Q_{max} = Q$. If the concentration of the macromolecule is limiting, as is usually the case, the quantity $Q_{max}/[M]$ is the enthalpy change per mole of macromolecule. If ligand is limiting, $Q_{max}/[L]$ is the enthalpy change per mole of ligand bound. Assuming that the macromolecule possesses a single ligand binding site, the two $\Delta H^{0\prime}$ values will be the same. This experimentally determined $\Delta H^{0\prime}$ value will in general be independent of concentration (of the limiting species) and, therefore, can be identified as an apparent standard enthalpy change.

4. DETERMINATION OF K', $\Delta G^{0\prime}$ AND $\Delta S^{0\prime}$

For systems in which K' is relatively small, a complete titration of macromolecule with ligand should be performed to determine Q_{max}. Since the heat effect, Q, for the mixing of macromolecule and ligand is proportional to the fractional completion of the reaction, K' can also be evaluated for systems as given in equation (1) by a thermal titration study.

$$Q = \frac{[ML]}{[M_0]} \Delta H^{0\prime} \qquad (8)$$

where $[M_0]$ is the total concentration of macromolecule. Expressing this equation in terms of K' and the free ligand concentration

$$Q = \frac{K'[L]\Delta H^{0\prime}}{1 + K'[L]} \qquad (9)$$

Rearranging

$$\frac{1}{Q} = \frac{1}{\Delta H^{0\prime}} + \frac{1}{K'[L]\Delta H^{0\prime}} \qquad (10)$$

where

$$[L] = [L_0] - Q[M_0]/\Delta H^{0\prime} \qquad (11)$$

$[L_0]$ being the total ligand concentration. Equation (10) is analysed by first estimating a $Q_{max}/[M_0] = \Delta H^{0\prime}$ to solve for $[L]$, then plotting $1/Q$ against $1/[L]$ to obtain new estimates of K' and $\Delta H^{0\prime}$, then iterating.

In this manner K', $\Delta G^{0'}$ and $\Delta H^{0'}$ can be obtained, with $\Delta S^{0'}$ provided by equation (3).

In Fig. 2 the calorimetric data for the binding of 3'-cytidine monophosphate to ribonuclease A are presented. These data were analysed

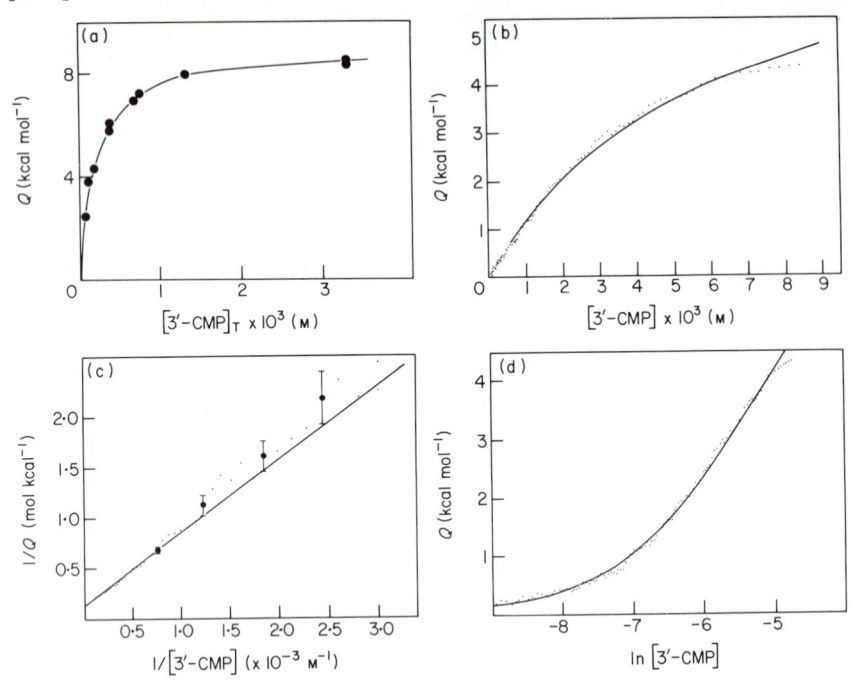

FIG. 2. (a) The binding isotherm of 3'-CMP to RNAse A as measured by batch calorimetry. [RNAse A] = 6.63×10^{-5} M; pH 5.52; ionic strength 0.5 M; 25 °C. The solid line represents a fit of the data with $\Delta H^{0'} = 9.2$ kcal mol^{-1} and $K' = 5.3 \times 10^3$ M^{-1}. (Bolen et al., 1971.) (b) The binding isotherm of 3'-CMP to RNAse A as measured by continuous dilution flow calorimetry. [RNAse A] = 2.13×10^{-4} M; pH 8.20; ionic strength 0.2 M; 25 °C. The solid line is the fit of the data to $\Delta H^{0'} = 7.7$ kcal mol^{-1} and $K = 180$ M^{-1}. (c) The reciprocal plot (equation (10) in text) of the data in (b). (d) The data in (b) plotted as Q versus ln [L]$_{free}$. The three means of plotting the data obtained by the continuous dilution experiment are shown in order to illustrate the characteristics of each type of plot. The Q versus ln [L]$_{free}$ plot is the preferred presentation. Due to the low value of K at pH 8.2 saturation was not obtained in these experiments. Since a large number of data points were obtained (along with the knowledge that the stoichiometry is 1 : 1) (Bolen et al., 1971), an analysis for values of K and ΔH was still possible.

to yield values of K', $\Delta G^{0'}$, $\Delta H^{0'}$, $\Delta S^{0'}$, as described in Bolen et al. (1971).

To facilitate the rapid, continuous generation of data, Mountcastle et al. (1976) developed an exponential concentration gradient method for use with a flow microcalorimeter. In a normal flow experiment, the two solutions are introduced at constant rates, f, into the calorimetric cell. The observed steady-state rate of heat production

$$\dot{q} = f_C \Delta H^{0\prime}[C] \tag{12}$$

where C is the limiting species and f_C is the flow rate of C. An exponential concentration gradient of C can be generated by continuously adding solvent to a constant volume vessel initially containing reactant at concentration $[C_0]$. The concentration of C flowing into the calorimetric cell as a function of time, t, will then be

$$[C] = [C_0] \exp(-f_C t/V) \tag{13}$$

where V is the volume of the vessel. The apparent heat of reaction is given by

$$\dot{q} = f_C \Delta H^{0\prime}[C_0] \exp(-f_C t/V) \tag{14}$$

Since the response time of a calorimeter is of the order of 75 s, the experimentally observed heat effect will differ from the above slightly. Mountcastle et al. (1976) have discussed means of evaluating the response time and correcting the experimental data to provide $\dot{q}$.

By providing an accurate, continuous set of data, the exponential dilution experiment allows a very precise determination of K' and $\Delta H^{0\prime}$. An example of the titration of ribonuclease A with 3′-CMP by use of a continuous dilution procedure is shown in Fig. 2(b). Good agreement between the results obtained in this manner and the standard point-by-point method was found. The advantage of this method is that a complete continuous isotherm can be obtained within approximately 1 h, compared to several hours by conventional calorimetric techniques.

C. Contributions to the thermodynamic changes associated with interacting systems

From a purely pragmatic point of view, $\Delta G^{0\prime}$ determines the equilibrium position of a chemical reaction under specified conditions. However, $\Delta G^{0\prime}$ values may be relatively insensitive to variation in the molecular details of the reaction because of a tendency for the $\Delta H^{0\prime}$ and $T\Delta S^{0\prime}$ contributions to compensate (Rajender and Lumry, 1970). In many cases which will be described, the $\Delta G^{0\prime}$ for the binding of different ligands to a macromolecule, or for the binding of a particular ligand under different conditions, may vary by only 1–2 kcal mol^{-1}, but $\Delta H^{0\prime}$ (and hence $T\Delta S^{0\prime}$) may vary by 10–20 kcal mol^{-1}. An example that clearly demonstrates this behaviour is the association of 3′-CMP to ribonuclease A. As shown in Fig. 3, the $\Delta H^{0\prime}$ for binding varies from about -6 to -15 kcal mol^{-1} with a compensating variation of $\Delta S^{0\prime}$ from 5 to 30 cal K^{-1} mol^{-1} for studies performed at ionic strengths ranging from 3·0 to 0·05 M at a constant pH of 5·5 (Flogel et al., 1972; Bolen et al., 1971). The $\Delta G^{0\prime}$ values change by only a couple of kcal mol^{-1} as a function of ionic strength.

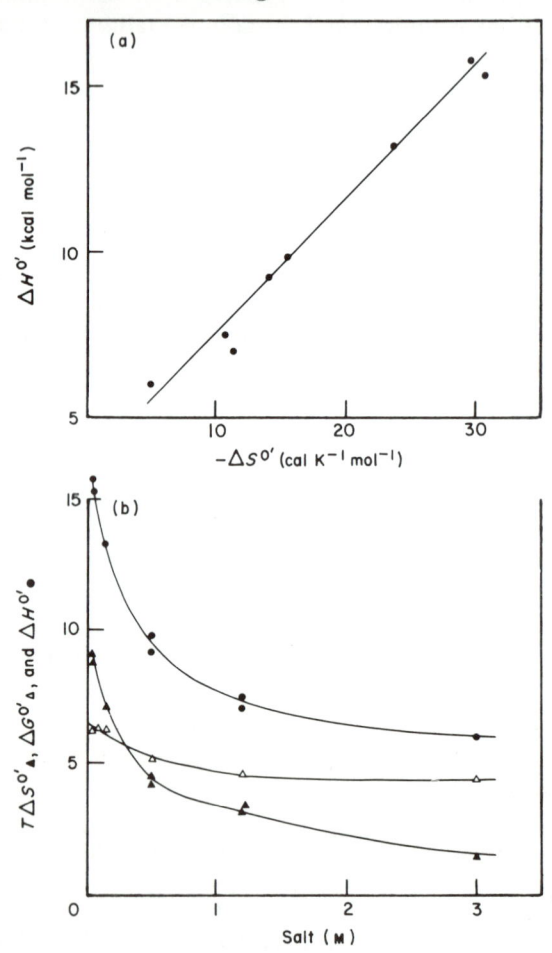

FIG. 3. (a) The apparent compensation of the enthalpy change and entropy change for the binding of 3′-CMP to RNAse A at pH 5·5, 25 °C. Various points represent different ionic strengths. (Flogel et al., 1972.) (b) The ionic strength dependence of the thermodynamic parameters for the binding of 3′-CMP to RNAse. Same data as in Fig. 2(a) plotted in a different fashion. The standard state is 1 mol l^{-1}. (Bolen et al., 1971.)

The association processes that we are considering will often be coupled to other equilibria involving either the ligand or the macromolecule (Weber, 1972). As will be discussed, the binding of ligand may perturb sensitive conformational equilibria of the macromolecule or the ligand, or may perturb equilibria involving the reactants and other molecules, including proteins, secondary ligands, protons, or water. The total intrinsic Gibbs energy available for the interaction[1] between the ligand and the macromolecule will be partitioned among these various modes. As a result, the observed Gibbs energy change will be less than the intrinsic Gibbs energy change. But this deficit will generally be small, since only delicate equilibria (ones with $\Delta G^0 \simeq 0$) will be significantly perturbed by the binding process. However, the contribution that these coupled equilibrium shifts make to the enthalpy and entropy change for binding may be much more pronounced. For this reason the evaluation of $\Delta H^{0\prime}$ and $\Delta S^{0\prime}$ will be of particular importance for the interpretation of the details of ligand association reactions.

For the 3′-CMP–ribonuclease A system discussed above, the dependence of $\Delta H^{0\prime}$ and $\Delta S^{0\prime}$ values on ionic strength can be understood in terms of a model in which there is an electrostatic interaction between the negatively charged ligand and protonated histidyl residues at the binding site. Included in the observed thermodynamic quantities is a contribution (contribution mostly to the $\Delta H^{0\prime}$ and $\Delta S^{0\prime}$ values) from the coupled uptake of protons by the histidyl residues upon ligand binding. The apparent compensation of $\Delta H^{0\prime}$ and $\Delta S^{0\prime}$ values results from the effect of ionic strength on the ionization constants of the histidyl residues and on the intrinsic electrostatic interaction between the ligand and the protein.

In this section we will discuss the ways of assessing the various contributions to the thermodynamic quantities for ligand binding; particular emphasis will be placed on enthalpy and entropy changes. The more important sources of thermodynamic changes for the binding of ligand include:

i. the change in solvation of the ligand and the macromolecule;
ii. the interaction between the ligand and the binding site through hydrogen bonds, van der Waals forces and charge–charge interactions;
iii. the release and absorption of protons by the macromolecule, ligand and buffer;

[1] The intrinsic interaction is the ideal association between some optimal state of the ligand and some optimal state of the macromolecule. The thermodynamic quantities associated with the intrinsic interaction are those that would apply if the ligand and macromolecule could somehow be locked into these optimal states before they are mixed. More discussion of the intrinsic interaction will be given in section C.5.

iv. the release or absorption of a secondary ligand upon the binding of the first;
v. a conformational change in ligand or macromolecule required for or induced by the association reaction;
vi. changes in the state of aggregation of either reactant.

Ideally one would like to relate particular molecular events to the observed changes in Gibbs energy, enthalpy and entropy. However, since they may include contributions from several of the above processes, evaluation becomes difficult. In order to sort out the various contributions to these quantities, a detailed understanding of the association process is required. As will be discussed, sources (iii), (iv) and (vi) can hopefully be accounted for, leaving the thermodynamic quantities due to the intrinsic binding (source (i) and (ii)) and conformational changes (source (v)). Separation of the latter two processes cannot be made in a strict sense, but proper experimental design can aid in drawing reasonable conclusions as to the relative importance of the two sources.

1. COUPLED SHIFTS IN EQUILIBRIA INVOLVING PROTONS

Often the binding of ligand to a macromolecule will proceed with the release or absorption of one or more protons. An example is given by the following reaction[1] in which the binding of ligand is coupled to the release of a proton from the macromolecule and to the absorption of a proton by a buffer species.

$$MH^+ + L \underset{}{\overset{K_2}{\rightleftarrows}} M + L + H^+ \underset{}{\overset{K_1}{\rightleftarrows}} ML + H^+ \quad (15a)$$

$$Buffer\ H^+ \underset{}{\overset{K_3}{\rightleftarrows}} Buffer + H^+ \quad (15b)$$

---

[1] In this contribution the symbols representing the thermodynamic parameters are an extension of the basic system of nomenclature recommended by the Interunion Commission of Biothermodynamics (1976). The pertinent features will be briefly described.
  1. All experimental values will be labelled with a superscript prime (i.e. K', $\Delta H^{0\prime}$). The prime refers to the fact that the parameter may be a function of the solvent conditions (i.e. may be a function of pH, ionic strength or buffer components). If a thermodynamic parameter is known or is expected to be a function of some variable, it should be symbolized as K' (variable) (i.e. K' (pH 6·0), or K' (0·001 M Mg^{2+})).
  2. When describing a binding process in terms of a particular scheme, each equilibrium is labelled with an equilibrium constant. There are three basic types of equilibria that may come into play for reaction in biological systems and the form of the equilibrium constants will be used as a means of their classification. The three types of equilibria and the appropriate symbols are:
      a. an association between two or more unlike species to form a non-covalent complex (ligand binding)—symbolized by an equilibrium constant having a numerical subscript, i.e. K_1, K_2, etc.;

where K_2 and K_3 are the acid dissociation constants of the macromolecule and the buffer and K_1 is the association constant of ligand to the macromolecule.

The apparent enthalpy change, $Q_{max}/[M_0]$, for such a coupled reaction is given by

$$\frac{Q_{max}}{[M_0]} = \Delta H_1^0 + \Delta N(\Delta H_3^0 - \Delta H_2^0) \quad (16a)$$

$$= \Delta H_1^{0\prime}(H^+) + \Delta N \Delta H_3^0 \quad (16b)$$

where ΔH_1^0 is the enthalpy change for the association when both reactants are in the unprotonated state, ΔH_3 and ΔH_2 are the heats of

    b. an association between like species (aggregation)—symbolized by an equilibrium constant with a numeral placed at the upper left, i.e. ^1K, ^2K, etc.;
    c. an intramolecular isomerization—symbolized by an equilibrium constant with a numeral at the lower left, i.e. $_1K$, $_2K$, etc.
  3. In reaction schemes, the first equilibrium of any of the above types will be labelled "1"; the second such equilibrium being "2", etc. For example, the first isomerization will be labelled $_1K$, the second, $_2K$, etc.
  4. Conditions of an equilibrium process can be described by the use of letters in the three positions about the symbol K. For example, the isomerization of a macromolecule that is saturated with a ligand, L, is described by $_1K_L$; in the absence of ligand, the isomerization is described by $_1K$. In this way the notation for the equilibrium constant provides a description of the process. Note that for a symbol such as $_1K_L$ or $_BK_1$ (the binding of the first ligand to state "B" of a macromolecule), the position of the numeral defines the type of process and the letter describes the conditions. The binding of a second ligand Y to a macromolecule already saturated with ligand X is symbolized as $K_{2,X}$. Protons are considered ligands.
  Letters used to symbolize bound ligands will be L, H, X, Y or Z placed as a subscript. To symbolize different isomerization states, the letters A, B, C will be placed below and to the left. To symbolize different aggregation states, the Greek letters α, β, γ etc. will be placed above and to the left, with α referring to the monomeric species, β referring to the first aggregated state (usually dimer) and γ referring to the next highest proposed aggregation state (i.e. trimer or tetramer).
  The other thermodynamic parameters are symbolized in a like manner (i.e. for $_BK_1$, the free energy, enthalpy and entropy change would be $_B\Delta G_1^0$, $_B\Delta H_1^0$ and $_B\Delta S_1^0$). The model given below exemplifies the use of this system.

$$M_A \xrightleftharpoons[]{_1K} M_B \xrightleftharpoons[]{X \quad _BK_2} M_BX$$

$$\downarrow\uparrow Y \quad _AK_1 \qquad \downarrow\uparrow Y \quad _BK_1 \qquad \downarrow\uparrow Y \quad _BK_{1,X}$$

$$M_AY \xrightleftharpoons[_1K_Y]{} M_BY \xrightleftharpoons[_BK_{2,Y}]{X} M_BXY \xrightleftharpoons[_2K_{X,Y}]{} M_CXY$$

In this scheme the thermodynamic parameters for the binding of ligand Y to the macromolecule will be a function of the concentration of X. In general, the association constant for Y will be $K_2'([X])$; a particular determination of this parameter might be symbolized as $K_2'([X]=10^{-3}\text{ M})$.

protonation of the buffer and macromolecule, respectively, ΔN is the number of protons released per mole of product formed, and $\Delta H_1^{0'}(H^+)$ ($= \Delta H_1^0 - \Delta N \Delta H_2^0$) is the pH-dependent heat of association of ligand to the macromolecule.

By determining Q_{max} for a reaction at a given pH using a series of buffers of known ΔH_3^0, one can obtain $\Delta H_1^{0'}(H^+)$ and ΔN from equation (16b). Alternatively, if ΔN is known from independent experiments, $\Delta H_1^{0'}(H^+)$ can be obtained from a single experiment. Potentiometric methods can be used to determine ΔN by measuring proton uptake or release upon the mixing of unbuffered solutions of macromolecule and ligand initially at the same pH (Flogel and Biltonen, 1975a).

In most cases ΔH_3^0 will be the heat of protonation of the external buffer used in the solution. Hinz *et al.* (1971) have tabulated such heats for a number of commonly used buffers. However, there may be cases when it is desirable not to use an external buffer or when the capacity of the external buffer is very low. In these instances the buffering due to the macromolecule or the ligand must be considered and the appropriate ΔH_3^0 will be a weighted function of the various buffering components. Flogel and Biltonen (1975a, b) have outlined a procedure for calculating ΔH_3^0 in such cases.

The parameter $\Delta H_1^{0'}(H^+)$, being pH dependent, will contain a contribution from the heat of protonation of a group on the macromolecule (or of the ligand). In order to draw any interpretations from the enthalpy change for a reaction, one must obtain a pH-independent enthalpy change related to the association of a particular state of the ligand to a particular state of the macromolecule. To exemplify how this can be done, consider the following simple scheme:

$$\begin{array}{ccc} MH^+ + L & \xrightarrow{K_2} & M + H^+ + L \\ K_{1,H} \updownarrow & & \updownarrow K_1 \\ MH^+L & \xrightarrow{K_{2,L}} & ML + H^+ \end{array} \qquad (17)$$

where K_2 and $K_{2,L}$ are the acid dissociation constants of the group on the macromolecule that is related to the binding process in the free and associated state, respectively, and $K_{1,H}$ and K_1 are the association constants of ligand to the protonated and unprotonated states of the macromolecule. The apparent association constant for ligand according to this scheme is

$$K_1'(H^+) = \frac{[ML]+[MH^+L]}{[L][MH^++M]} \quad (18a)$$

$$= \frac{K_1+[H^+]K_{1,H}/K_2}{1+[H^+]/K_2} \quad (18b)$$

and

$$\Delta G_1^{0\prime}(H^+) = \Delta G_1^0 - RT \ln \left[\frac{1+[H^+]/K_{2,L}}{1+[H^+]/K_2} \right] \quad (18c)$$

$\Delta G_1^0 = -RT \ln K_1$ is the Gibbs energy change for the association of L to unprotonated M, and $\Delta G_{1,H}^0 (= \Delta G_1^0 - RT \ln (K_2/K_{2,L}))$ is the Gibbs energy change for binding to the protonated MH^+, where electrostatic interactions may occur. The pH-dependent enthalpy change is given by

$$\Delta H_1^{0\prime}(H^+) = \Delta H_1^0 + \frac{[H^+]\Delta H_{2,L}^0}{K_{2,L}+[H^+]} - \frac{[H^+]\Delta H_2^0}{K_2+[H^+]} \quad (19)$$

and at saturating $[H^+]$,

$$\Delta H_{1,H}^0 = \Delta H_1^0 + \Delta H_{2,L}^0 - \Delta H_2^0 \quad (20)$$

where ΔH_2^0 and $\Delta H_{2,L}^0$ are the heats of protonation of the group on the macromolecule in the free and associated state, and ΔH_1^0 and $\Delta H_{1,H}^0$ are the enthalpy changes for the association to the unprotonated and protonated macromolecule, respectively. The parameters ΔG_1^0, ΔH_1^0, $\Delta G_{1,H}^0$ and $\Delta H_{1,H}^0$ are "corrected" for the coupling with proton binding, and therefore allow one to assess the contribution of other processes in determining the Gibbs energy and enthalpy change for a precisely defined reaction.

In analysing the pH dependence of an association reaction it is useful to rely on information provided by other techniques in order to obtain independent values for K_2, ΔH_2^0, $K_{2,L}$ and $\Delta H_{2,L}^0$. As mentioned above, potentiometric studies are useful in providing ΔN values at various pH values from which determination of K_2 and $K_{2,L}$ can be made. Values for the proton dissociation constants and heats of protonation can also be obtained from calorimetric titration studies of the free and liganded macromolecule (Flogel and Biltonen, 1975a). Additional information can be obtained if the titration of the critical ionizing group can be studied spectroscopically, e.g. by n.m.r.

The work of Flogel and Biltonen (Flogel and Biltonen, 1975a, b; Flogel et al., 1975) on the binding of 3'-CMP to RNAse A is a good example of the treatment of pH-dependent effects. For this interaction at least three (and possibly four) ionizing groups are perturbed during the binding process. As a result the observed Gibbs energy and enthalpy changes are quite dependent on pH (see Figs 4 and 5). The ionizing groups include the phosphate group of the ligand (pK ~6, heat of protonation ~0 kcal mol^{-1}) and two histidine residues at the binding site (pKs of 5·0 and 5·8, heats of protonation of −6·5 kcal mol^{-1} for

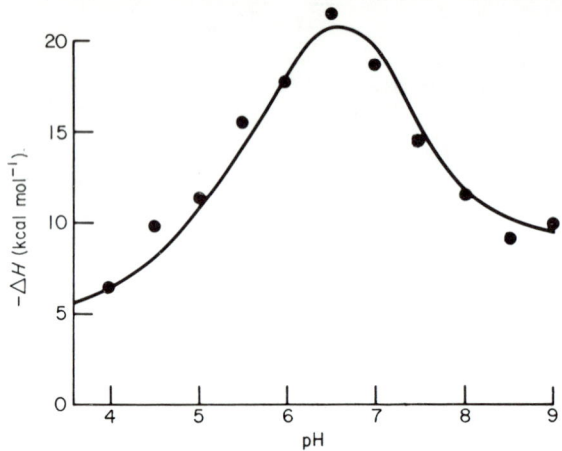

FIG. 4. The pH dependence of the apparent enthalpy change for 3'-CMP binding to RNAse A, ionic strength 0·05 M, 25 °C. The solid line was calculated assuming that three groups on the protein (1, 2, 3) must be protonated for binding with pK_1 = 5·0, pK_2 = 5·8, pK_3 = 6·7, p$K_{1, L}$ = p$K_{2, L}$ = p$K_{3, L}$ = 7·1 (where L refers to the condition of bound ligand); $\Delta H_1^0 = \Delta H_2^0 = \Delta H_{1, L}^0 = \Delta H_{2, L}^0 = -6·5$ kcal mol^{-1}, $\Delta H_3^0 = -24$ kcal mol^{-1}, $\Delta H_{3, L}^0 = -22$ kcal mol^{-1} and ΔH_4^0 (the heat of binding 3'-CMP. to the unprotonated protein) = −8·7 kcal mol^{-1}. According to the model, only the dianionic form of the ligand binds, but its heat of protonation is approximately zero and does not contribute to the apparent heat of binding (Flogel and Biltonen, 1975b.)

each). Another histidine distal from the binding site may also be perturbed, but to a lesser extent. By combining data from calorimetric, potentiometric and n.m.r. titrations of the free and associated protein, an analysis of the pH dependence of the thermodynamic quantities was made, yielding pH-independent values describing the association of the dianionic state of the ligand to the fully protonated state of the protein.

2. COUPLED SHIFTS IN EQUILIBRIA INVOLVING OTHER LIGANDS

The binding of a ligand may be coupled to side-reactions involving the same ligand or a second type of ligand. For example, coupled reactions of the type shown in equations (21a) and (21b) may take place.

$$ML_2 + L_1 \rightleftharpoons ML_1 + L_2 \qquad (21a)$$

$$RL + M \rightleftharpoons R + L + M \rightleftharpoons ML + R \qquad (21b)$$

Equation (21a) describes a situation in which the binding of one ligand results in the release (or uptake) of a second ligand. The phenomena of linked association of ligands will be discussed in greater detail in section F.2, but here we would like to point out briefly a few examples that

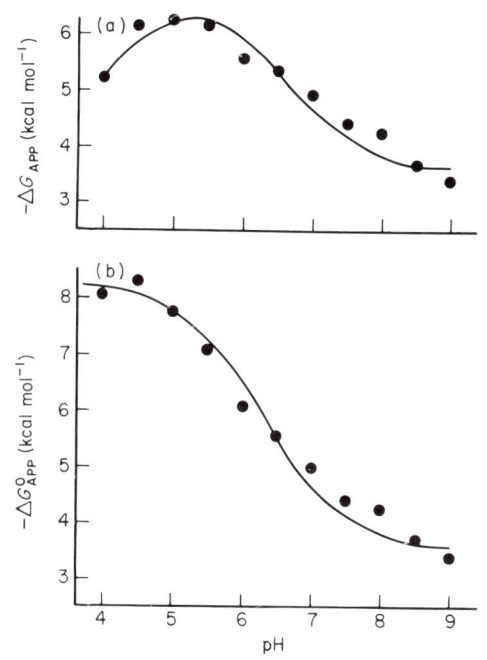

FIG. 5. (a) The apparent free energy change for the binding of 3'-CMP to RNAse A, ionic strength 0·05 M, 25 °C. (b) The free energy change corrected for the ionization of the phosphate group of 3'-CMP. The solid curves in (a) and (b) were calculated assuming the pK_i and $pK_{i, L}$ given in Fig. 4, an ionization constant of the ligand of 8×10^{-7} M, and $\Delta G_4{}^0$ (the free energy change for the binding of 3'-CMP to the unprotonated protein) = $-3\cdot4$ kcal mol^{-1}. The standard state is 1 mol l^{-1}. (Flogel and Biltonen, 1975b.)

may be of importance in designing experiments. If the binding of ligand, L_1, according to the above equation, displaces a second type of ligand, L_2, the apparent association constant for L_1 will be

$$K'_1(L_2) = \frac{K_1}{1 + K_2[L_2]} \qquad (22)$$

where K_1 and K_2 are the appropriate association constants for ligands L_1 and L_2 to the non-liganded macromolecule. If initially the macromolecule is saturated with L_2, the apparent enthalpy change for the binding of L_1 will be equal to $\Delta H_1{}^0 - \Delta H_2{}^0$. If $\Delta H_2{}^0$ is known, $\Delta H_1{}^0$

can thus be determined by measuring the heat effect for the displacement of L_2 by L_1.

Such displacement experiments may prove valuable in the following circumstances. Suppose either that the macromolecule to be studied is unstable in the absence of ligand or that macromolecule free of L_2 cannot be easily prepared. In such cases, one might study the system by simply mixing a solution of ML_2 complex with excess L_1. Of course, the problem with such experiments may lie in the independent determination of ΔH_2^0.

In other cases the binding of L_1 may be coupled to the binding of L_2. For example, nucleotides such as ATP (adenosine triphosphate) often bind to proteins as a metal–ligand complex (i.e. ATP–Mg^{2-}). In such cases the apparent enthalpy change for the binding of ATP would be dependent on the type and concentration of divalent cation present in solution.

There also may be cases in which buffer components interact with the macromolecule. If the binding of a specific ligand causes a release or uptake of these buffer components, a heat effect may result. If this is suspected to be the case one must be judicious about the choice of solution conditions. (For example, see Atha and Ackers, 1974a.)

The second type of coupled reaction described by equation (21b) involves the interaction of the ligand with another reactant, R. For example, in a study of the binding of Mg^{2+} to tRNA, Rialdi *et al.* (1972) utilized a coupled reaction of Mg^{2+} with EDTA (ethylene diamine tetraacetate). This was done because the interaction between Mg^{2+} and tRNA did not produce a significant change in enthalpy. The coupled reaction of Mg^{2+} with EDTA (a reaction having a ΔH^0 of -8 kcal mol^{-1}) allowed the workers to "amplify" the calorimetric signal. Since the association of Mg^{2+} to EDTA is much stronger than it is to tRNA, the three components were not simply mixed together in a calorimeter, but the following experimental design was employed. A tRNA sample was dialysed against a solution of known [Mg^{2+}]. The concentrations of Mg^{2+} in both the tRNA solution and the dialysate were separately determined by measuring the heat released upon mixing with excess EDTA. The difference in the heat measured for the two samples is related to the excess amount of Mg^{2+} present in the tRNA solution due to the specific binding of Mg^{2+} (and, to a lesser degree, due to Donnan effects). The results of these experiments yielded a binding isotherm (shown in Fig. 6) which subsequently was analysed in terms of two sets of independent binding sites. The association constant for four strong sites was determined to be 1×10^6 M^{-1}; for the twenty weaker sites the association constant was found to be $1 \cdot 1 \times 10^4$ M^{-1}.

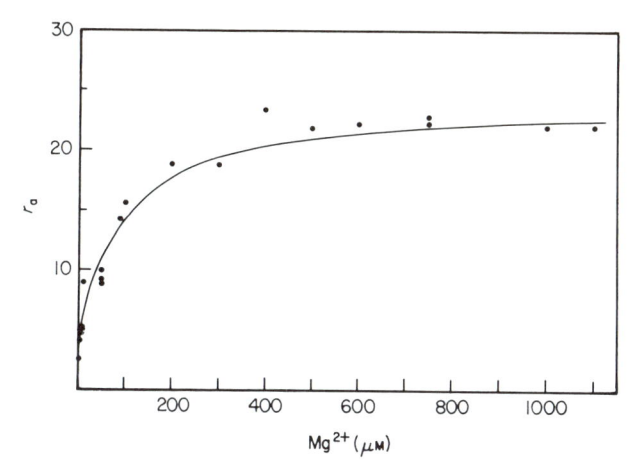

FIG. 6. Average number of Mg^{2+}/tRNA^{Phe}, r_a, as a function of free Mg^{2+} concentration at 25 °C, pH 7·2, 0·01 M salt. Solid line calculated for four strong binding sites (association constant 1×10^6 M^{-1}) and twenty weaker sites (association constant $1·1 \times 10^4$ M^{-1}). (Rialdi et al., 1972.)

3. ASSOCIATION–DISSOCIATION OF REACTANTS

In some cases it has been found that the apparent thermodynamic changes for a binding reaction is dependent upon the concentration of the macromolecule. Such a situation suggests that the binding of ligand affects the state of aggregation of the macromolecule. Consider the following scheme used by Shaio and Sturtevant (1969) to analyse the binding of ligands to chymotrypsin.

$$M + M \xrightleftharpoons{^1K} M_2 \qquad (23a)$$

$$M + L \xrightleftharpoons{^\alpha K_1} ML \qquad (23b)$$

Assume that only the monomeric form of the macromolecule is able to bind to ligand. Under such conditions the apparent association constant and the apparent enthalpy change for the binding of the ligand are given by

$$K'_1 = \frac{^\alpha K_1}{1 + (^1K[M_2]^{\frac{1}{2}})} \qquad (24a)$$

$$\Delta H_1^{0\prime} = {}^\alpha\Delta H_1^0 + \frac{[M_2]_0\,{}^1\Delta H^0}{[M_t]} \qquad (24b)$$

where $[M_t]$ is the total concentration of macromolecule and $[M_2]_0$ is the initial concentration of dimer and is given by the relationship

$$[M_2]_0 = [M_t] + {}^1K/4 \pm \frac{[[M_t]^2 + ({}^1K)^2/16 + [M_t]^1K/2 + [M_t]^2]}{2} \quad (25)$$

In order to evaluate ${}^\alpha\Delta H_1^0$ from the experimentally determined enthalpy change, one must have independent knowledge of ${}^1\Delta H^0$ and ${}^1K$. If the self-association process is isodesmic, the ${}^1\Delta H^0$ and ${}^1K$ values can be determined by measuring the heat of dilution of the macromolecule and analysis according to the following relationship (Stoesser and Gill, 1967; Shaio and Sturtevant, 1969; Atha and Ackers, 1974b):

$$Q_T = {}^1\Delta H^0 - ({}^1\Delta H^0 \, Q_T/{}^1K[M_t])^{\frac{1}{2}} \quad (26)$$

where Q_T, the molar heat of dilution, is defined as the heat change upon dilution of the macromolecule from concentration M_t to infinite dilution.

4. AN INDUCED CONFORMATIONAL CHANGE

The binding of a ligand will necessarily cause some change in the state of the macromolecule. Replacement of water molecules at a binding site with a ligand certainly affects the energetics of the groups defining the binding site. The influence of the ligand may be manifest beyond the macromolecule–ligand interface and possibly throughout the entire macromolecule. In such cases we may consider that the ligand binds preferentially to some altered state of the macromolecule, M_B, as described by equation (27) (without specifying, for the moment, the particular conformational differences between the two states of the macromolecule, M_A and M_B).

$$\begin{array}{ccc} M_A + L & \xrightleftharpoons{{}_1K} & M_B + L \\ {}_AK_1 \updownarrow & & \updownarrow {}_BK_1 \\ M_AL & \xrightleftharpoons{{}_1K_L} & M_BL \end{array} \quad (27)$$

In the above scheme, ${}_1K$ and ${}_1K_L$ are equilibrium constants describing the isomerization of either the free or the liganded macromolecule from a low affinity state (M_A and M_AL) to the altered, high affinity state (M_B and M_BL). ${}_AK_1$ and ${}_BK_1$ are the association constants of L to M_A and M_B, respectively, and ${}_BK_1 > {}_AK_1$. The route taken for the formation of M_BL will likely be $M_A + L \rightleftharpoons M_AL \rightleftharpoons M_BL$, or a concerted path, $M_A + L \rightleftharpoons M_BL$. From a thermodynamic point of view the route

$M_A + L \rightleftharpoons M_B + L \rightleftharpoons M_B L$ is equivalent and serves to distinguish more clearly between the conformational change and the intrinsic binding process. The apparent association constant for the above system is

$$K'_1 = \frac{{}_A K_1 + {}_1 K \, {}_B K_1}{1 + {}_1 K} \tag{28}$$

In the case where the $M_B L$ complex is much more stable than the $M_A L$ complex (that is, ${}_1 K \, {}_B K_1 \gg {}_A K_1$), equation (28) reduces to $K'_1 = {}_1 K \, {}_B K_1 / (1 + {}_1 K)$. If one considers the case ${}_1 K \ll 1$, then $K'_1 = {}_1 K \, {}_B K_1$ and the apparent enthalpy change, $\Delta H_1^{0\prime}$, is the sum ${}_1 \Delta H^0 + {}_B \Delta H_1^0$, where ${}_1 \Delta H^0$ is the enthalpy change for the conformational reaction and ${}_B \Delta H_1^0$ is the enthalpy change for the binding of L to M_B. Note that the apparent binding constant will always be lower than the intrinsic binding constant to the high affinity state, M_B. This is due to the expenditure of Gibbs energy required to drive the conformational change $M_A \rightleftharpoons M_B$.

From our knowledge of proteins, it seems likely that conformational changes with small changes in the Gibbs energy may readily occur upon the fixing of ligand since the globular structure of a protein in solution is determined by a large number of very weak intramolecular interactions. The minimum energy conformation of the structure will depend on a delicate balance between the enthalpy and entropy contributions of the system. The binding of a ligand to a protein will introduce added points of interaction and may serve to stabilize particular conformational states. Any difference in the enthalpy and entropy of the altered conformation from the initial one will then contribute the apparent enthalpy and entropy change for ligand binding.

It may be possible to judge whether a conformational change is coupled to a binding process from the magnitude of the $\Delta H^{0\prime}$ for the association of ligand. If $\Delta H^{0\prime}$ is very large, particularly if it is greater than can be reasonably attributed to the intrinsic interaction between ligand and macromolecule, then a contribution from a conformational change is suggested. If systematic variation of the structure of the ligand produces only a slight alteration in $\Delta H^{0\prime}$, it is likely that the measured thermodynamic values are due largely to a conformational change. Also, if the binding of one ligand affects the Gibbs energy and enthalpy change for the binding of a second ligand, a conformational change can be suspected. Independent experimental studies will, of course, be helpful in providing evidence for a conformational change.

As will be discussed in greater detail in the next section, estimates for the magnitude of the thermodynamic changes associated with various types of interactions that are believed to be involved in ligand binding have been made from studies with model systems. For example, ionic interactions are expected to be entropically driven, the enthalpy change being approximately zero. Lovrien and Sturtevant (1971) found the $\Delta H^{0'}$ for the binding of I^- to bovine serum albumin to be -18 kcal mol^{-1}. In another example, Kuriki et al. (1976) observed an enthalpy change of -49 kcal mol^{-1} for the association of Mg^{2+} to (Na^+-K^+)-ATPase. In both cases the observed $\Delta H^{0'}$ is much greater than would be expected for the electrostatic interaction of an ion with a macromolecule, suggesting the existence of an induced structural change in the protein.

If variation of the structure of the ligand does not produce a significant change in the thermodynamics of binding, it is likely that a significant conformational change in the macromolecule is occurring. This is based on the assumption that each ligand is capable of inducing the same rearrangement of the macromolecule. Shiao and Sturtevant (1969) have studied the binding of a series of inhibitors to α-chymotrypsin, including N-acetyl-D-tryptophan, and indole. The enthalpy change for binding was found to be large and negative in all cases, varying from -11 to -19 kcal mol^{-1} at pH 8. Based on the similarity of the enthalpy change for the widely varying ligands, they concluded that a conformational change of the enzyme was induced by the ligands. We must caution, however, that a variation in the intrinsic enthalpy change of 8 kcal mol^{-1} for different ligands may not be unusual. Furthermore, the above values may include coupled proton binding. The linked binding of two ligands and the evidence that it may provide for a conformational change are discussed in the section on multiple binding site macromolecules.

An indication of an induced conformational change may often be provided by other techniques. Changes in absorbance, fluorescence, circular dichroism or magnetic resonance spectra are commonly taken as evidence suggestive of conformational rearrangements. In a number of cases, a direct visualization of the adjustments of a protein structure induced by ligand binding has been provided by X-ray crystallographic techniques. Studies of the binding process by transient kinetic methods can also yield evidence for induced conformational changes. For example, such studies may reveal two relaxation times upon the mixing or perturbation of a ligand–macromolecule system, consistent with an initial diffusion-controlled binding step $(M_A + L \rightleftarrows M_A L)$ followed by an isomerization $(M_A L \rightleftarrows M_B L)$ (Eigen and de Maeyer, 1963). If a conformational change is driven by the binding of L, then the ratio of the on-and-off rate constants for the "binding" step determined by relaxation measurements will necessarily be lower than the equilibrium association constant (Eigen and de Maeyer, 1963). In cases where two such steps can be resolved, the enthalpy changes for each can be determined from the temperature dependence of the amplitude of each relaxation process.

A few words of caution must be added concerning the importance of and interpretation of changes in spectroscopic observables. While a change in an observable as a result of ligand binding is suggestive of a conformational change, it does not necessarily indicate that the conformational change is a thermodynamically significant one.[1] For example, Eisinger et al. (1970) have studied the enhancement of the fluorescence of the "Y" base of tRNA by added Mg^{2+} and have concluded that binding of the metal causes a conformational change in the tRNA. However, Rialdi et al. (1972) found the $\Delta H^{0'}$ for the binding of Mg^{2+} to tRNA to be ~ 0 kcal mol^{-1}, as expected for the electrostatic interaction between a metal and the negatively charged macromolecule. Based on this $\Delta H^{0'}$ value there does not appear to be a thermodynamically significant conformational change in the tRNA molecule. The sensitivity of the fluorescence of the "Y" base to the presence of bound Mg^{2+} may result from a very subtle local rigidification of the nucleotide chain.

As a rule it is difficult to relate a change in an observable to the amplitude or direction of a conformational rearrangement. The interpretation of changes in such spectral signals as fluorescence yield, circular dichroism or ultraviolet absorption relies on our understanding of model systems. It must be realized that macromolecules, such as proteins, may present environmental factors that cannot be adequately represented by model systems. Techniques that focus on the gross shape of macromolecules are often not sufficiently sensitive to detect ligand-induced changes (unless the macromolecule is large and asymmetric, or unless the ligand induces a change in the state of aggregation). The investigation of conformational changes by X-ray crystallography always involves the unknown factor of crystal lattice forces. Measurements designed to study the change in a dynamic property of the macromolecule induced by ligand binding (techniques such as isotope exchange, nuclear magnetic resonance relaxation measurements, nitroxide spin labelling, fluorescence

---

[1] We shall operationally define a thermodynamically significant conformational change as one having either (a) a free energy change of about 1·4 kcal mol^{-1}, or (b) an enthalpy change of ± 2 kcal mol^{-1}. The first criterion is chosen because such a ΔG value could result in a decrease in the observed association constant by a factor of ten from the intrinsic value. The second criterion is chosen because most present measurements of enthalpy changes have a precision of about ± 2 kcal mol^{-1}. If a conformational change having a smaller Gibbs energy and enthalpy changes were to occur, it would make a negligible contribution to the thermodynamic parameters for ligand binding. As experimental techniques improve and determination of free energy and enthalpy changes for ligand binding become more precise, the above criterion will, of course, need to be refined.

quenching or fluorescence depolarization) can provide information concerning changes in the motility of the macromolecule. However, a thorough understanding of the dynamic properties requires an investigation of the activation parameters associated with the measured kinetic parameter (Likhtenshtein *et al.*, 1972; Eftink and Ghiron, 1977). While changes in various observables can provide evidence that some sort of structural rearrangement is occurring, the exact details of the rearrangement are usually difficult to extract.

Besides the macromolecule, the ligand may undergo conformational transitions during the course of binding. One might consider this situation in the following manner

$$L_A \rightleftarrows L_B \qquad (29a)$$

$$M + L_B \rightleftarrows ML_B \qquad (29b)$$

in which the macromolecule binds only the ligand in conformation L_B. The observed heat effect will be (neglecting other coupled reactions) the sum of ΔH_1^0, the intrinsic heat of association of L_B to M, plus a contribution from the enthalpy change for the isomerization of L_A to L_B.

An example of this situation is given by the interaction of 3'-CMP with RNAse A. The nucleotide exists in solution as a mixture of two conformations, *anti* and *syn*, which differ in the angle between the pyrimidine ring and the ribose ring. The *anti* configuration is of lower enthalpy (the enthalpy change for $syn \rightleftarrows anti$ is $-2 \cdot 8$ kcal mol^{-1}), and at 25 °C approximately 75 per cent of the molecules exist in the *anti* form (Coulter, 1973; Lavallee and Coulter, 1973). It is believed, however, that nucleotides are bound to RNAse A in the *anti* configuration (Richard and Wycoff, 1971). The heat of binding of those nucleotides that are *anti* would be ΔH_1^0, but the heat of binding of the *syn* population would be ΔH_1^0 minus $2 \cdot 8$ kcal mol^{-1}.

To summarize briefly, the coupling of a conformational change in either the macromolecule or ligand to the binding process may be expected to occur frequently. As a result, the observed enthalpy change will include a contribution from this conformational change along with the intrinsic heat of binding. By considering the various factors discussed above, one can hopefully estimate the relative importance of a conformational change in particular binding reactions.

5. INTRINSIC BINDING

The sequestering of a ligand within a macromolecule's binding site is unfavourable from an entropic point of view. Much of the translational and overall rotational entropy of the ligand must be lost when it becomes "tethered". A complete loss of such motional freedom could result in an entropy loss of approximately 35 cal K^{-1} mol^{-1} (Kemball, 1950; Page and Jencks, 1971). Page and Jencks (1971) have noted, however, that the $\Delta S^{0\prime}$ for the formation of many weak non-covalent complexes of the type $A + B \rightleftarrows AB$ is in the range of -10 to -20 cal K^{-1} mol^{-1} (standard state, 1 molar). This includes complexes formed by intermolecular hydrogen bonds and charge transfer interactions. The $\Delta S^{0\prime}$ for such associations is believed to be more positive than -35 cal K^{-1} mol^{-1} due to the fact that various low-frequency vibrational motions and rotations are allowed within the complex. In other words, the formation of a weak non-covalent complex does not require a complete restriction of the motion of the molecules with respect to one another. Similarly, we expect that the non-covalent complexes formed between a ligand and a macromolecule will be "loose" to the extent that vibrational modes in the complex compensate, somewhat, for the restriction of translational and rotational freedom.

One can approximate the amount of entropy that may be lost due to translational restriction as the decrease in entropy upon "concentration" of the ligand from 1 molar to unit mole fraction.[1] This contribution is the cratic entropy of mixing, -8 cal K^{-1} mol^{-1}. The remaining entropy loss of -2 to -12 cal K^{-1} mol^{-1} that is commonly observed may then be attributed to the restriction of the overall rotation of the molecule. An interesting comparison can be made between this value and the entropy change of fusion for organic substances. The increase in entropy upon the melting of a substance such as methyl cyclohexane is 10·9 cal K^{-1} mol^{-1} (Glasstone, 1946). A large part of this entropy change is due to the onset of rotation in the liquid state. Cyclohexane, on the other hand, which is able to rotate freely in the solid state, shows an entropy change upon melting of only 2·3 cal K^{-1} mol^{-1}. Therefore, about 8·6 cal K^{-1} mol^{-1} of the entropy of fusion for methyl cyclohexane appears to be due to the change in overall rotational freedom of this molecule during melting.

Upon the binding of a ligand to a macromolecule it is expected that there will be a decrease in entropy of about 8 cal K^{-1} mol^{-1} due to the loss of translational freedom, plus an additional entropy loss of a similar magnitude due to the restriction of rotational freedom. Various internal

---

[1] From this point on all thermodynamic quantities will be expressed relative to a standard state of unit mole fraction. This is done to avoid the "cratic" contribution to free energy and entropy changes. To convert from the commonly used mole per litre standard state to the unitary standard state, add 8 cal mol^{-1} deg^{-1} to ΔS and approximately $-2\cdot 4$ kcal mol^{-1} to ΔG values.

rotational modes of the ligand or the macromolecule may also be lost. In order for the ligand and macromolecule to associate to any significant extent ($\Delta G^{0\prime} < 0$), other favourable interactions between the two must come into play to overcome this inherent entropic disadvantage by improving the overall enthalpy or entropy situation of the system. This intrinsic affinity between some state of ligand and some state of the macromolecule must also be sufficient to promote the various coupled reactions discussed above. For example, the energetically unfavourable transition of M_A to M_B (equation (27)) will be induced only if the affinity of ligand to M_B is sufficiently great to drive the transformation.

We define an intrinsic enthalpy change, ΔH^0, and an intrinsic entropy change, ΔS^0, as the values describing the "ideal" association of optimal states of the ligand and macromolecule. The experimentally observed enthalpy and entropy changes, $\Delta H^{0\prime}$ and $\Delta S^{0\prime}$, will contain a contribution from the intrinsic interaction plus the various coupled reactions. As discussed above, the loss of translational and rotational freedom of the ligand will make a major contribution to ΔS^0. Other entropy changes resulting from the intrinsic interaction, such as that due to the "hydrophobic effect" (to be discussed below), will also contribute to ΔS^0.

In the following section we will briefly consider a number of intermolecular interactions that are thought to be important in the binding of ligands to macromolecules.

D. Interpretation of the thermodynamic changes

While the interpretation of thermodynamic data for the binding of ligand to macromolecule must be done with caution, such data can potentially provide valuable insight as to the nature of macromolecule–ligand complexes. Spectroscopic and crystallographic studies of such complexes can provide information concerning the micro-environment of the ligand and even a direct picture of the interacting system, but they cannot provide information describing the strength of the various forces involved in the interaction. The biochemist interested in the energetics of enzyme-catalysed reactions or the factors determining the specificity of interactions between antibodies and haptens, for example, would like to be able to evaluate the relative strength and importance of hydrogen bonds, hydrophobic, electrostatic, and van der Waals forces between the ligand and binding site.

Our present understanding of the meaning of enthalpy and entropy changes for association reactions is based on model system studies and theoretical considerations. Even assuming that we fully understand the model studies, the extrapolation to macromolecular systems can be very risky. The binding of ligand to macromolecule involves changes in at least three species, the macromolecule, ligand and the solvent water. In the model systems one is primarily interested in the interaction of a solute with water and the thermodynamics of the transfer of this solute from water to some other phase (i.e. the gas phase or an organic solvent). Since ligand binding will involve the transfer of ligand from water to a binding site on the macromolecule, and may involve a change in the degree of solvent exposure of various parts of the macromolecule as well, studies of the nature of solute–water interactions are extremely valuable. These model systems studies have been important in providing a description of such forces as hydrogen bonds and the hydrophobic effect, and in describing the sources of heat capacity changes.

In this section the forces involved in binding reactions will be discussed with emphasis on the associated ΔH and ΔS values. The interpretation of the thermodynamics of conformational changes will also be discussed.

1. HYDROPHOBIC INTERACTIONS

According to the classical view of hydrophobic interactions, apolar groups in water tend to associate preferentially with themselves or other apolar groups in order to minimize their contact with the aqueous environment (Kauzmann, 1959; Tanford, 1973). This reaction is characterized by small enthalpy changes but large positive entropy changes. For example, the transfer of ethane from water to a chloroform solution is entropically driven with a ΔS of 19 cal K^{-1} mol^{-1} and a slightly unfavourable ΔH of 1·8 kcal mol^{-1} (see Table 1).

Wishnia (1969a, b) has studied the binding of small apolar molecules (alkanes) to the proteins β-lactoalbumin, ferrihaemoglobin, and serum albumin and has found the thermodynamic parameters for the association to be slightly different than those for the transfer of the same molecule to the gas phase or to a detergent micelle. Binding to the protein was still driven by a large positive $\Delta S^{0\prime}$ (although the $\Delta S^{0\prime}$ was smaller than that for the model systems), but the $\Delta H^{0\prime}$ was found to be a negative 2–3 kcal mol^{-1}. The $\Delta G^{0\prime}$ for binding to the proteins was always more negative than that for the transfer to either the gas phase or the micelle.

The binding of an apolar ligand to a protein may differ from its transfer to a non-aqueous phase in that, in the former case, the molecule must lose its translational, rotational and some internal rotational freedom as discussed above. Therefore, some of the entropic driving force of the hydrophobic effect may be lost due to the restriction of the bound

TABLE 1

Thermodynamic parameters describing various intermolecular interactive forces

Type of interaction	ΔH kcal mol^{-1}	ΔS^a cal K^{-1} mol^{-1}	ΔC_p cal K^{-1} mol^{-1}	References
Hydrophobic				
$C_2H_{6(aqueous)} \to C_2H_{6(CCl_4)}$	~0	> 0	< 0	Tanford (1973)
$C_5H_{12(aqueous)} \to C_5H_{12(SDS\ micelle)}{}^b$	1.8	19	−59	Wishnia (1969a, b)
Electrostatic	0.4	21	−103	
	~0	> 0	> 0	
$AMP^{2-} + Mg^{2+} \to AMP\text{-}Mg$	1.8	22		Belaich and Sari (1969)
$ADP^{3-} + Mg^{2+} \to ADP\text{-}Mg^-$	3.2	35		Belaich and Sari (1969)
$CH_3CO_2^- + H^+ \to CH_3CO_2H$	0.09	30	37	Edsall (1935); Edsall and Wyman (1958)
Hydrogen bond	< 0	< 0	< 0	Schellman (1955); Kreshbeck and Scheraga (1965); Gill and Noll (1972); Alvarez and Biltonen (1973)
δ-Valerolactam → dimer	−2.8^c	−15	−45	
Diketopiperazine → dimer	−2.1^c	−19		Gill and Noll (1972)

^{a} Standard state, unit mole fraction.
^{b} SDS refers to sodium dodecyl sulphate.
^{c} Values per hydrogen bond.

ligand. It would also seem that the better an apolar ligand "fits" into the binding site, the more it could be restricted, and the less favourable the net $\Delta S^{0'}$ for binding would be. However, one might expect that conformationally restricted ligands (i.e. rigid ring systems) would be able to bind very strongly to a complementary site on a macromolecule because such ligands would have little internal rotational freedom to lose (Jencks, 1975).

The fact that the $\Delta H^{0'}$ values found by Wishnia for the binding of alkanes to proteins were negative could be due to good van der Waals contacts at the ligand binding site. Another explanation proposed by Wishnia (1969a, b) is that proteins possess some regions where the interactions within the protein are suboptimal. For example, certain apolar side-chains may not be able to make good van der Waals contact with one another due to restrictions placed by nearby hydrogen-bonding groups or a protein may have a small crevice lined with apolar groups that must ordinarily be filled with one or more water molecules. Such water molecules may not be able to hydrogen bond satisfactorily and may, therefore, be relatively unstable. If an apolar ligand is able either to fill up the space between the side-chains, thus improving the van der Waals contacts, as in the first example, or to displace the unstable water molecules, as in the second example, a favourable $\Delta H^{0'}$ for binding could result.

2. ELECTROSTATIC INTERACTIONS

The electrostatic interaction between a charged ligand and a counter-charged binding locus on a macromolecule is believed to be driven primarily by an increase in entropy due to the release of hydration water. The strength of charge–charge interactions should vary inversely with the distance of approach of the two centres and the effective dielectric constant of the environment. Because of the high dielectric constant of water, charge–charge interactions are generally weak in the aqueous phase. The local dielectric constant at a binding site may be much lower than that for bulk water, but the transfer of a charge from water to a low dielectric environment is an unfavourable process and would tend to nullify any advantages for interaction at the binding site (Lumry and Biltonen, 1969).

It is well known that ionic species, particularly di- and tri-valent ions, have a very high affinity for some macromolecules. A factor that undoubtedly contributes to such strong interactions is the chelate effect. Consider a divalent cation, M^{2+}. In solution this cation will be surrounded by a coordination sphere of ligands (usually water). If, however, a molecule such as ethylenediamine tetraacetic acid (EDTA), a cyclic polyether, valinomycin, or a particular macromolecule coordinates with the M^{2+}, much less translational entropy would be lost compared to the case for the monodentate water ligands. This is, of course, because the molecules mentioned are polydentate and can present additional co-ordinating groups to the M^{2+} without an additional loss of translational entropy. The binding of Mg^{2+} to nucleotides provides a clear example of this effect (Belaich and Sari, 1969). As seen in Table 1, the association is driven by a large positive entropy change, the magnitude of which depends on the number of available coordinating centres provided by the nucleotide. The unfavourable enthalpy change for binding is probably related to distortion of the nucleotide upon formation of the complex.

Rialdi et al. (1972) found that Mg^{2+} binds to tRNA at four strong sites ($K \sim 10^6 \, M^{-1}$) and that the association is totally driven by a unitary entropy change of 36 cal K^{-1} mol^{-1}. The binding of dianionic phosphate to the low pH state of RNAse A is also characterized by a large $\Delta S^{0\prime}$ of 31 cal K^{-1} mol^{-1} with an associated $\Delta H^{0\prime}$ of only $-0\cdot1$ kcal mol^{-1}. However, there are other examples in which ion binding is enthalpically driven, such as the association of I^- to bovine serum albumin (Lovrien and Sturtevant, 1971). In the latter case it is likely that a conformational change in the protein induced by the binding of ligand dominates the observed change in enthalpy.

3. HYDROGEN BONDS AND VAN DER WAALS FORCES

The formation of specific hydrogen bonds between macromolecule and ligand is commonly thought to be of great importance in promoting a binding process. For such a hydrogen bond to form, similar hydrogen bonds between the ligand and water and the binding site and water will usually have to be broken. Taking this into consideration, the Gibbs energy and enthalpy change upon association should be small. However, again the chelate principle may play a dominant role. If a given ligand and binding site have a number of complementary hydrogen bonding groups, the entropy loss upon formation of these hydrogen bonds between macromolecule and the matching groups on the ligand may be much less than the loss of entropy required for satisfying all the hydrogen bonding groups with water.

The formation of a hydrogen bond between two molecules is thought to be enthalpy driven with a ΔH of about -2 kcal mol^{-1} and a ΔS less than zero (Schellman, 1955; Kreshbeck and Scheraga, 1965; Brandts, 1969; Gill and Noll, 1972). As mentioned above, if the transfer of ligand from water to the binding site does not result in a net formation of hydrogen bonds, but simply replaces ligand–water and macromolecule–water hydrogen bonds for those between macromolecule and ligand, then the ΔH for binding would be small. However, recent studies on model systems have led to the interesting proposal that at room temperature only about 50 per cent of all hydrogen bonds between water and solute are made at any time. If this is so, the association of ligand to macromolecule may indeed result in a net formation of hydrogen bonds and thus proceed with a favourable enthalpy change.

These studies include one by Alvarez and Biltonen (1973) on the heat capacity of thymine in water and ethanol. In both solvents the excess heat capacity of thymine was found to be large compared to apolar solutes, such as benzene. The large heat capacity can be accounted for by assuming that the polar groups of the solute are able to exist in two states, hydrogen bonded to solvent and not hydrogen bonded. At 25 °C, the heat capacity data are consistent with about half of the potential bonds between thymine and water being formed. This result is not surprising when one considers that studies of the hydration of glucose (Tait *et al.*, 1972) and various models for the structure of liquid water (Eisenberg and Kauzmann, 1969) also predict that only a fraction of the possible hydrogen bonds are made at any time.

Van der Waals interactions are often overlooked as being weak or as being unchanged during the transfer of a ligand from an aqueous environment to a binding site. However, this may not be the case. The binding sites of proteins may be densely packed (Grunwald and Price, 1964; Klapper, 1971; Richards, 1974; Jencks, 1975) thus allowing for an increased degree of van der Waals interaction with ligands, particularly when compared to the van der Waals interactions available in the less dense aqueous solution. Further, it must be realized that the binding sites may be tailored to provide a snug, three-dimensional fit for the ligand. Effective van der Waals contact between two methylene moieties would be expected to reduce the enthalpy of the system by only a few hundred calories per mole at best. However, if several such contacts can be simultaneously made, the net enthalpy change could be significant. It is also possible that dipole–dipole and dipole-induced dipole interactions between ligand and macromolecule may occur. The dipole moment of some planar molecules (e.g. nucleic acid bases) is found to be large. If there is a precise juxapositioning of such a group with respect to another dipole or polarizable group at the binding site, this could give rise to a substantial decrease in the enthalpy of the system. Consider, for example, the interaction of adenine with itself in aqueous solution through base stacking. The $\Delta H^{0\prime}$ and $\Delta S^{0\prime}$ for such an interaction are found to be $-4\cdot2$ kcal mol^{-1} and -5 cal K^{-1} mol^{-1} (Gill and Stoesser, 1966; Gill et al., 1967). Dipole–dipole interactions are a likely source of this enthalpy change.

Our understanding of the above dispersion forces, particularly as they occur in aqueous solutions, is not very sound. Jencks (1969) has provided a helpful discussion of these types of interactions.

4. CONFORMATIONAL CHANGES

Ligand-induced conformational changes may serve very important functional roles, such as promoting proper orientation of catalytic groups at an enzyme's active site (induced fit: Koshland and Neet, 1968), mediating the linked binding of ligands (ordered addition of substrates, cooperative binding and allosteric effects: Monod et al., 1965; Koshland et al., 1966; Koshland and Neet, 1968), or the storing of free energy in an enzyme–substrate complex for later release during a catalytic process (distortion, Circi effect: Jencks, 1975). Citri (1973) has extensively reviewed the evidence for ligand-induced conformational changes in proteins. For a number of macromolecule–ligand systems, a direct visualization of the induced conformational change has been obtained by X-ray diffraction and Fourier difference mapping (Blow and Steitz, 1970). Most notable among these are the studies by Perutz (1970) and Perutz and Ten Eyck (1971) on oxy- and deoxyhaemoglobin, which clearly show a change in the quaternary structure of the protein upon oxygen binding. Studies at higher resolution have also revealed that less extensive changes in the structure of the individual subunits also occur. In another example, studies with carboxypeptidase A show a very pronounced movement (14 Å) of a tyrosine residue upon the binding of ligand (Lipscomb et al., 1968). In other cases, however, only slight adjustments in the protein structure can be discerned when complexed with ligand (Wycoff et al., 1967; Richards et al., 1969; Steitz et al., 1969; Henderson et al., 1972).

While structural alterations may be directly visualized by crystallographic studies, it is still difficult to assess the energetics associated with the conformational changes. What may appear to be only a slight adjustment of atoms may be characterized by a large change in the Gibbs energy, enthalpy and entropy, and vice versa. The important questions that we wish to address are: What is the extent of the conformational change and what are the thermodynamic parameters associated with the induced conformational change?

First consider descriptions of the various types of conformational changes that are believed possible for macromolecules, proteins, in particular. Lumry and Biltonen (1969) have discussed three general types: the unfolding reaction (and its reverse, the folding reaction); the refolding reaction; and the "subtle" conformational change. It is conceivable that each of these transitions may occur during the process that we have described as $M_A \rightleftarrows M_B$.

The unfolding of a globular protein typically occurs with an enthalpy change of 50–100 kcal mol^{-1} and an entropy change of 150–300 cal K^{-1} mol^{-1} for a protein of MW $\sim$ 25 000. It is believed that such transitions are generally not involved in the specific binding of small ligands under physiological conditions. If studies are carried out under extreme conditions in which the macromolecule is totally or partially unfolded, the addition of ligand may shift the equilibrium towards the folded state. Very large enthalpy changes for binding would then be observed, but such results are probably not typical of macromolecule–ligand interactions in aqueous solution, although there is some evidence that such large thermodynamic changes may accompany ligand binding in lipid–macromolecular complexes.

In a refolding transition, the macromolecule is transformed from one folded state to a distinctly different folded state. Lumry and co-workers (Parker and Lumry, 1963; Yapel and Lumry, 1964; Biltonen and Lumry, 1969; Lumry and Biltonen, 1969; Kim and Lumry, 1971) have presented evidence for refolding transitions for the protein chymotrypsin. The enthalpy and entropy change for a refolding transition may be either large or small, and the two states may be difficult to distinguish by most solution techniques. It is likely that the only way that such conformational changes can be detected is by kinetic means, since the barrier for interconversion of the states may be large.

Despite the fact that such refolding transitions may be difficult to establish, it is our feeling that the most common type of induced conformation change will prove to be the so-called "subtle" change. Lumry and Biltonen (1969) have described the "subtle" change as an expansion ($\Delta H \simeq T\Delta S \simeq 0$) of the macromolecule. Here we will generalize this description to include any small amplitude conformational change. In solution, a given macrostate of a globular protein is believed to exist as a distribution of microstates (Lumry et al., 1966; Poland and Scheraga, 1967; Lumry and Rosenberg, 1975; Ikegami, 1977). These microstates will range from low enthalpy, low entropy, highly structured conformers to high enthalpy, high entropy, loosely structured ones. For example, these conformers may differ in terms of the number of intramolecular hydrogen bonds that are made (Lumry and Rosenberg, 1975). The ligand may bind preferentially to certain of these microstates (i.e. to either the low enthalpy or high enthalpy conformers). As a result, the binding of ligand will shift the distribution to microstates resulting in a change in the average enthalpy of the macromolecule (a change in the breadth of the distribution of microstates may also result—see section G.2). It is difficult to predict the magnitude of the enthalpy change for such "subtle" changes, but in general it would be expected to be small (10 kcal mol^{-1} or less). Further discussion of this type of conformational adjustment will be given in section G.2.b. Some type of subtle conformational change is expected to occur to a certain extent in all ligand–macromolecule interactions. It must also be realized that the water surrounding a macromolecule may be energetically coupled to this conformational rearrangement. The restructuring of this water in response to a rearrangement of the macromolecule will also contribute to the observed enthalpy change. A particular type of subtle change that deserves mention is that in which the binding of ligand results in the freezing of the rotation about certain side-chains of the macromolecule. Such a restriction might not cause a significant change in the enthalpy of the system, but could reduce the entropy by approximately 2 cal K^{-1} mol^{-1} per bond frozen.

The purpose of the above discussion is to provide a format for the description of induced conformational changes. In reality it may be difficult to classify a particular ligand-induced conformational change into any of these categories.

E. Specific examples

1. RIBONUCLEASE A

The thermodynamics of the association of nucleotides, nucleosides and orthophosphate to bovine RNAse A has been studied in detail by Flogel *et al.* (1975) and Flogel and Biltonen (1975a, b). From the results of these studies it was possible to dissect the energetic contributions due to the binding of the phosphate and riboside moieties to the protein. The thermodynamic parameters obtained for the binding of several ligands are given in Table 2.

Before discussing these values, let us first address the question of whether or not an induced conformational change in the macromolecule makes any significant contribution to the apparent thermodynamic changes. At this time there is little evidence for a thermodynamically significant conformational change induced by the binding of ligand to RNAse. X-ray crystallography reveals only a minor adjustment of the protein when 3'-CMP is bound (Wycoff *et al.*, 1967; Richards *et al.*,

TABLE 2
Thermodynamic quantities[a] for the binding of ligands to RNAse A at 25 °C,
$\mu = 0.05$

Ligand	ΔG^0 kcal mol^{-1}	ΔH^0 kcal mol^{-1}	ΔS^0 cal K^{-1} mol^{-1}
Phosphate	− 7·8	−0·1	31
Cytidine	− 5·1	−6·1	− 3
3′-CMP	−10·8	−7·1	12
2′-CMP	−12·0	−6·6	18

[a] The standard state is unit mole fraction. The quantities have been corrected for coupled ionizations and reflect the binding of the dianionic ligand to the fully protonated enzyme. The contribution from a conformational change of the protein is considered negligible. A syn⇌anti conformational change of the nucleosides and nucleotides is believed to occur upon binding, but this transition probably contributes less than 0·3 kcal mol^{-1} to the free energy change for the binding and less than 1·0 kcal mol^{-1} to the enthalpy change. The reported thermodynamic parameters are therefore considered intrinsic values.

1969). The most apparent changes involve histidine residues at the binding site. The u.v. difference spectra and the changes in the c.d. spectra observed upon the binding of nucleotides can be attributed to perturbation of the chromophoric pyrimidine ring (Irie and Sawada, 1967; Samejima et al., 1969). The ratio of the on-and-off rate constants for the binding of 3′-CMP measured by temperature jump studies is found to be roughly equal to the observed association constant[1] (Cathou and Hammes, 1964, 1965; French and Hammes, 1965). This indicates

[1] Hammes and co-workers (Cathou and Hammes, 1964, 1965; French and Hammes, 1965) have, however, argued that their temperature jump data does show a discrepancy between the ratio of the on-to-off rate constant and the equilibrium association constant, particularly at low pH. They find the kinetic value to be larger than the equilibrium value. The only way this type of discrepancy can be explained is if a certain population of one of the reactants (either L or M) exists in an "inactive binding" state. The remainder of this limiting reactant must be converted slowly to the "active" state for binding to take place.

Temperature jump studies with RNAse A alone show a relatively slow relaxation process at low pH consistent with the existence of an equilibrium between two structural forms of the enzyme. Cathou and Hammes (1965) proposed that only one of these protein forms, the least stable one at low pH, is capable of binding the ligand 3′-CMP. The major fraction of the enzyme was proposed to exist in a conformation unable to bind ligand without first undergoing the slow conversion to the "active" form. Since the $_1\Delta H^0$ they determined for the transition of the protein is 2·9 kcal mol^{-1}, the apparent $\Delta H^{0\prime}$ for ligand binding at low pH would include a contribution from this transition according to their model.

However, we believe that the basis for their model, a discrepancy between the kinetic (on/off rate constants) and equilibrium association constants, is questionable. This

[continued overleaf]

that the initially formed ML complex is of the lowest free energy. There is a small change in the tritium isotope exchange pattern for RNAse when ligand is bound (Nonnenmacher et al., 1971). However, as will be discussed in section G.2.b, this observation may be due more to a change in the breadth of the distribution function of the protein than to a change in the average enthalpy state of the protein. Also, it has been found that the heat of protonation and the dissociation constant of the conformationally sensitive His 48 of RNAse A is not significantly altered by the presence of ligand (Flogel and Biltonen, 1969a). Based on these assorted studies it can be argued that there is no important conformational change in RNAse A induced by the binding of ligand; any minor rearrangements that do occur must be thermodynamically small. Therefore, to a first approximation, the observed enthalpy and entropy changes for ligand binding (after correction for proton coupling, see below) can be attributed to the intrinsic binding of ligand.

The association of the nucleoside, cytidine, to RNAse A was found to be enthalpically driven with $\Delta H^0 = -6.1$ kcal mol^{-1} and a $\Delta S^0 = -3$ cal K^{-1} mol^{-1}. The binding of orthophosphate (as well as all nucleotides) was found to be extremely pH dependent. Figures 4 and 5 illustrate the variation of the apparent Gibbs energy and enthalpy change with pH. After treatment as described in section C.1, the association of dianionic phosphate to the fully protonated (both histidine residues at the binding site protonated) protein was found to be entropically driven with $\Delta H^0 = 0.1$ kcal mol^{-1} and $\Delta S^0 = 31$ cal K^{-1} mol^{-1}. For nucleotides, such as 3'-CMP, a pH-independent ΔH^0 of -7.1 kcal mol^{-1} and a ΔS^0 of 12 cal K^{-1} mol^{-1} was determined. Thus the binding of the nucleotide appears to be driven both by favourable enthalpic interactions with the base portion of the molecule and by the favourable entropic interaction with the phosphate group. To a first approximation it appears as if the thermodynamic quantities for the binding of nucleotide reflect the sum of the interaction between the riboside moiety and the phosphate group.

<small>comparison was made between kinetic constants determined at an ionic strength of 0·1 M and equilibrium constants determined at an ionic strength of 0·2 M. Since there is a marked dependence of the affinity of nucleotides to RNAse A on the ionic strength, particularly at low pH, it is not clear that the discrepancy between the kinetic and equilibrium association constants is real. No determination of the equilibrium association constant at an ionic strength of 0·1 M has been made to our knowledge. If we roughly interpolate from values determined at an ionic strength of 0·05 M (Flogel and Biltonen, 1975b) and 0·2 M (Anderson et al., 1968), we find that the discrepancy between the Cathou and Hammes (1964) ratio of on/off rate constants and the equilibrium value is only a factor of two, and that this discrepancy varies randomly with pH. We do not find this discrepancy to be compelling evidence to support the conformational change proposed by Hammes and co-workers, particularly when compared to the studies done with chymotrypsin to be described in the following pages.</small>

When considering the contribution of component parts of a ligand to the free energy of binding, there are a number of points that must be kept in mind. The problems associated with the additivity of free energies of binding have been discussed by Jencks (1975). First, it is desirable to discuss only unitary Gibbs energy changes (mole-fraction concentration scale) in order to eliminate the cratic contribution to the free energy and entropy. There are other salient features related to the difference in the entropy changes associated with the binding of a ligand such as A–B and with the binding of the component parts, A plus B. When A–B binds, there must be a certain loss of translational and rotational entropy of about 10–20 cal K^{-1} mol^{-1}, as discussed in section C.5. However, when the two ligands, A and B, bind simultaneously, each molecule must lose approximately the same amount of translational and rotational entropy. Therefore, the Gibbs energy for the binding of an A–B molecule will be more favourable by about 3–6 kcal mol^{-1} than the free energy for the binding of an A plus a B molecule due to this extra loss of entropy for "tethering" two molecules instead of one. One must also realize that the A–B molecule will have an additional bond, the internal rotational entropy about which may be lost upon binding. This effect would tend to make the binding of A–B somewhat less favourable. By interacting at two distinct sites on the macromolecule (a complementary "A" site and a complementary "B" site), the A–B ligand may also act to tether the macromolecule in such a way that internal entropy of the latter is reduced. The individual components, A and B, may not be able to restrict the macromolecule in such a fashion.

It is generally assumed that the enthalpy change for the interaction of the component parts should be more or less additive. However, the A–B molecule may bind in such a position that permits a full interaction between component A, for example, with its complementary site on the macromolecule, but does not allow the potential interactions involving component B to be realized.

For the interaction of 3'-CMP with RNAse A, it appears that the nucleoside portion of the ligand interacts optimally with an enthalpy contribution of $-7 \cdot 1$ kcal mol^{-1}. The phosphate group contributes a favourable entropic interaction, but the magnitude of the entropy change for 3'-CMP is not as large as that for phosphate itself. As discussed above, the explanation for this discrepancy may relate to a loss of internal rotational entropy of the ligand (about the bonds between the nucleoside and phosphate moieties) or the enzyme (the tying down of the histidines and other residues, such as Lys 41, at the binding site). Another reasonable explanation is that upon binding, the nucleoside portion realizes its full potential interaction with the protein, but that in doing so, does not allow the phosphate group to bind in its optimal position for full electrostatic interaction.

The thermodynamic information obtained by Flogel *et al.* (1975) and Flogel and Biltonen (1975a, b) provides a valuable supplement to data obtained by other techniques, such as n.m.r. and X-ray crystallography. The crystallographic structure (see Fig. 1) of the RNAse–3′-CMP complex shows two histidine residues at the binding site in a position to form ionic interactions with the phosphate group on the ligand. The increase in the pK_a of these two histidine residues upon the binding of nucleotides, as determined by n.m.r. studies (Meadows *et al.*, 1969; Markely, 1975), also indicates that an electrostatic interaction must occur and agrees excellently with the positive entropy change observed by the calorimetric studies (Flogel *et al.*, 1975). The X-ray studies also reveal several potential groups along a binding cleft that appear capable of participating in hydrogen bonds with the nucleoside moiety (Richards *et al.*, 1971). The observed enthalpy change of -6 to -7 kcal mol^{-1} can be reasonably attributed to the formation of three or four protein–ligand hydrogen bonds plus other van der Waals contacts.

The thermodynamic information also provides an interesting insight as to the source of the specificity and affinity for the binding of ligands to RNAse. The hydrogen bonds and van der Waals interactions that form between the riboside moiety and the cleft on the enzyme appear to be the prime determinants of specificity. Such interactions occur in the binding of both cytidine and the nucleotides, as evidenced by the similar binding enthalpies. However, cytidine binds very poorly (association constant of only $\sim 10^2$ M^{-1}), and the interactions involving the pyrimidine ring, while being the source of *structural specificity*, apparently are not able to provide for high affinity. The high affinity that is observed for the nucleotides must arise from the additional thermodynamic contribution from electrostatic interaction between the negatively charged phosphate group and the positively charged protein surface. This interaction is very favourable and thus provides for a degree of *thermodynamic selectivity*. By thermodynamic selectivity we mean that the protein binds 2′-CMP better than 3′-CMP, which binds better than cytidine, despite the fact that each of these ligands is specific for the binding site on the protein.

The evaluation of the energetics of the interaction between 3′-CMP and RNAse A is of particular interest because this ligand, which is actually the end-product of the hydrolysis reaction catalysed by the enzyme, is also a good candidate for a transition state analogue (Wolfenden, 1972; Flogel *et al.*, 1975; Jencks, 1975). Our understanding of the interactions that occur between the enzyme and this molecule will hopefully be helpful in providing a description of the manner in which interactions between the enzyme surface and the substrate develop during the bond-breaking process.

2. α-CHYMOTRYPSIN

Another system that has been extensively studied by calorimetric as well as other techniques is the binding of various inhibitors to α-chymotrypsin. This system is an interesting one because of the various coupled reactions that are thought to take place, including conformational changes of the protein. Using microcalorimetry, Shaio and Sturtevant (1969) and Shaio (1970) found that the apparent enthalpy change for the binding of various inhibitors to α-chymotrypsin was dependent on the concentration of the latter. This was interpreted in terms of a ligand-induced shift in an equilibrium between a dimeric and a monomeric form of the protein (with ligand binding only to the monomer). After correcting for this coupled reaction (as discussed in section C.3), the thermodynamic values listed in Table 3 were obtained for ligand binding.

It was expected that the binding of such ligands as phenol and indole would be driven by the "hydrophobic effect". However, the experimental values of $\Delta H^{0'}$ and $\Delta S^{0'}$, particularly at pH 7·8, are clearly not consistent

TABLE 3
Thermodynamic quantities for the binding of ligands to α-chymotrypsin[a, b]

Ligand (pH)	ΔG kcal mol^{-1}	ΔH kcal mol^{-1}	ΔS cal K^{-1} mol^{-1}
Calorimetric data of Shaio and Sturtevant (1969) and Shaio (1970)			
Phenol (7·8)	−5·4	−13·5	−27
Indole (7·8)	−6·7	−15·2	−29
N-Acetyl-D-tryptophan (7·8)	−5·7	−19·0	−44
Proflavin (7·8)	−8·4	−11·3	−10
Phenol (5·6)	−5·4	− 4·1	+ 5
Indole (5·6)	−6·7	− 6·4	+ 1
N-Acetyl-D-tryptophan (5·6)	−6·1	− 5·5	+ 2
Other methods			
Proflavin (8·05)[c]	−8·4	− 6·9	+ 5
(8·0)[c]	−8·3	− 7·3	+ 3·4
(7·8)[d]	−8·6	− 8·0	+ 2

[a] The standard state is unit mole fraction.
[b] Values at 25 °C.
[c] Sturgill et al. (1977).
[d] Schultz et al. (1977).

with a hydrophobically driven association. Also, the values of $\Delta H^{0\prime}$ at pH 7·8 are more or less the same in all cases, despite the drastic changes in the structure of the ligands. This fact led Shaio and Sturtevant to the conclusion that a conformational change of the protein (aside from the change in the degree of aggregation) must be coupled with the binding of ligand. The $\Delta H^{0\prime}$ values at pH 5·6 were all found to be substantially lower, suggesting that the contribution from this conformational change is less significant at this pH.

Ligand binding to α-chymotrypsin has also been studied by standard van't Hoff procedures (Schultz et al., 1977) and by a temperature perturbation technique (Sturgill et al., 1977). In these studies the enthalpy changes for binding are consistently found to be lower than the calorimetric values of Shaio and Sturtevant (1969), notably for the ligand proflavin. The reason for this discrepancy is not totally clear, but may lie in the assumption made by Shaio and Sturtevant (1969) that the binding was not dependent on the degree of aggregation of the ligand. Proflavin forms aggregates in the concentration range employed for the calorimetric studies. If one assumes that only the monomer of proflavin is capable of binding to either the monomer or the dimer of chymotrypsin, the data of Shaio and Sturtevant (1969) can be brought into line with those of the others. However, the fact that the calorimetric data can be made to comply with the other data by assuming that only the monomer of proflavin binds to all forms of the protein does not necessarily prove that this is the case. The dependence of the calorimetric binding enthalpy on protein concentration must also be taken into account. The true binding mechanism may be somewhere in between that assumed by Shaio and Sturtevant and the one suggested above.

Work by Fersht (1972) on the kinetics of proflavin binding provides independent evidence from a ligand-induced conformational change in α-chymotrypsin. The data are consistent with a model in which a fraction (~20 per cent) of the enzyme at pH 7·8 exists in an "inactive" form that is unable to bind ligand. Addition of ligand pulls the equilibrium of the protein in favour of the "active" form; the rate constant for the interconversion of these forms was found to be ~3 s^{-1}. The "active" fraction binds proflavin with a ratio of on-to-off rate constants approximately equal to the experimentally determined association constant for the ligand (Havsteen, 1967). The $_1\Delta H^0$ for the interconversion of protein forms (inactive ⇌ active) was found to be 7 kcal mol^{-1}. A certain portion of the apparent enthalpy change for the binding of proflavin must then be due to the induced shift in the protein equilibrium as discussed in section C.4. For example, at pH 7·8 approximately 1·4 kcal mol^{-1} (20 per cent × $_1\Delta H^0$) of the observed $\Delta H^{0\prime}$ of $-11\cdot3$ kcal mol^{-1} (value of

Sturtevant and Shaio, 1969) can be attributed to this shift in equilibrium. At higher pH, an ionization of the protein occurs which favours the "inactive" form. This deprotonation will affect both the ratio of active to inactive protein molecules and the enthalpy change for the interconversion as well.

After accounting for this conformational change, the enthalpy change for ligand binding is still ~ -12 kcal mol^{-1}. This value is more negative than normally expected for a "hydrophobic" interaction and suggests that either there are favourable hydrogen bonds, van der Waals or dipole–dipole interactions between the ligand and the binding site, or that there is an additional rearrangement of the protein upon binding.

F. More complex systems

Attention will now be focused on systems that are more complex than the simple type $M+L \rightleftharpoons ML$. The manner in which knowledge of enthalpy and entropy changes may add to our understanding of such systems will be stressed.

1. MULTIPLE, NON-INTERACTING SITES

If the macromolecule possesses more than one identical and independent site, the observed heat effect, $Q_{\max}$, per mole of M will be $n \Delta H^{0'}$, whereas per mole of ligand bound the heat effect will be $\Delta H^{0'}$. As pointed out by Bolen et al. (1971) the ratio of the apparent enthalpy changes determined in each manner is equal to the number of binding sites, n. Alternatively, comparison of $Q_{\max}/[M_0]$ to a van't Hoff enthalpy change can also be used to determine the stoichiometry (Cooper and Jenkins, 1972).

For the general case of j sets of distinct non-identical sites (with n sites per set), the observed heat effect will be (Steinhardt and Reynolds, 1970)

$$Q = \sum_{i=1}^{j} = \frac{n_i \Delta H_i^{0'}[L]K'_i}{1+K'_i[L]} \qquad (30)$$

where K'_i, $H_i^{0'}$ and n_i are the apparent (site) association constant, enthalpy change and number for each set of sites. Calorimetric titration curves (Q versus $[L]$) cannot, in general, be analysed for a unique fit. This is because three adjustable parameters exist for each class of binding sites. Combination of a thermal titration with some other technique that hopefully provides a linear assessment of the binding (i.e. equilibrium dialysis) is required in order to arrive at reliable estimates for K'_i, $\Delta H_i^{0'}$ and n_i (Sturtevant, 1974). It is often difficult to obtain a unique fit to data even when combining techniques (unless the K_is are widely separated, n_i is 1 and j is small), but, as pointed out by Sturtevant (1974), enthalpy titrations can provide valuable insight into such systems. A determination of $\Delta H^{0'}$ at a particular temperature can be useful in setting limits on the association constants that can describe the binding isotherms obtained at other temperatures. Also if $\Delta H_i^{0'}$ varies for each class of sites, this provides some insight as to the difference between each class. Velick *et al.* (1971) have employed calorimetric, spectroscopic and fluorometric methods to study the binding of NAD+ to yeast glyceraldehyde-3-phosphate dehydrogenase. This enzyme has four binding sites for NAD+ which under certain conditions appear to be independent and non-identical (under other conditions, cooperative binding occurs). In certain cases it was found that the calorimetric titration curves differed in shape from isotherms obtained by other methods, indicating definite variation in $\Delta H_i^{0'}$ between sites.

The interaction of AMP with phosphorylase b is noteworthy in that the thermal titration of this enzyme with ligand results in a biphasic curve, suggesting the existence of two quite distinct classes of binding sites (Ho and Wang, 1973). The $\Delta H^{0'}$ for the high affinity site was found to include a contribution from the association of phosphorylase b dimers to form monomers. The low affinity sites were thought to be due to non-specific binding.

2. INTERACTING SITES

The observation that the binding of ligands to a macromolecule is energetically linked is a common one in biochemical systems, and the functional consequences of such linkage are of vast importance (Weber, 1975). Often for multi-subsite proteins, the binding of one ligand may enhance or decrease the affinity of a second ligand. Even with single subunit macromolecules the presence of one bound ligand may have an influence on the binding of others. As will be discussed in this section, calorimetric studies are able to provide insight as to the nature of the linkage. It is possible that the linkage between the ligands is mediated by conformational changes in the macromolecule. If so, one might expect this to be apparent in the enthalpy changes associated with ligand binding.

Consider a macromolecule having binding sites for two ligands, X and Y. The equilibrium may be described as follows:

$$M + X + Y \xrightleftharpoons[K_2]{K_1} \begin{matrix} MX + Y \\ MY + X \end{matrix} \xrightleftharpoons[K_{1,Y}]{K_{2,X}} MXY \quad (31)$$

Accordingly $K_1 K_{2,X} = K_2 K_{1,Y}$. The apparent association constant for the binding of X (as a function of Y concentration) is

$$K'_1(Y) = \frac{K_1 + K_{1,Y} K_2[Y]}{1 + K_2[Y]} \quad (32)$$

When $[Y] = 0$, $K'_1(Y) = K_1$ and $\Delta H_1^{0\prime}(Y) = \Delta H_1^0$, and at saturating $[Y]$, $K'_1(Y) = K_{1,Y}$ and $\Delta H_1^{0\prime}(Y) = \Delta H_{1,Y}^0$. The ratio of $K_{1,Y}/K_1$ will reflect the influence of Y on the binding of X. The value $-RT \ln (K_{1,Y}/K_1) = \Delta G_{linkage}$ is defined as the Gibbs energy linkage between the two ligands. If $\Delta G_{linkage}$ is > 0, then Y is an antagonist to the binding of X (negative heterotropic cooperativity); if $\Delta G_{linkage} < 0$, then Y promotes the binding of X (positive heterotropic cooperativity); if $\Delta G_{linkage} = 0$, then, of course, X and Y bind independently. (Since $K_{1,Y}/K_1 = K_{2,X}/K_2$, the effect of added Y on the binding of X must be the same as the effect of added X on the binding of Y.) Weber (1975) has provided a thorough discussion of free energy linkage and cooperative binding phenomena in proteins.

The above scheme does not include information concerning the source of the coupling between X and Y. The possible molecular sources of linked binding include (Steinhardt and Reynolds, 1970):

i. the direct interaction between the two ligands at a common, overlapping binding site via hydrogen bonds, van der Waals forces, etc.;

ii. long-range electrostatic interactions between the two ligands;[1]

iii. the induction of a conformational change in the macromolecule by the first ligand producing a conformation state having an altered affinity for the second ligand.

---

[1] A distinction between (i) and (ii) is made because electrostatic interactions may be of much longer range than other interligand interactions, and do not require the two binding sites to be contiguous.

If the linkage is mediated by a conformational change in the macromolecule, the system may be described by the following scheme:

$$M_A \xrightleftharpoons{{}_1K} M_B \xrightleftharpoons[{}]{X \quad {}_BK_1} M_B X \tag{33}$$

with Y, ${}_BK_2$ on the left side and Y, ${}_BK_{2,X}$ on the right side, and $M_B Y \xrightleftharpoons[X]{{}_BK_{1,Y}} M_B X Y$ on the bottom.

Here we assume that the macromolecule can exist in two states: a low-affinity state M_A and a state M_B that has a high affinity for both ligands. According to the above scheme, the binding of one ligand shifts the macromolecular conformational equilibrium towards the high-affinity form. Actually the state of the macromolecule in the $M_B X$ and $M_B Y$ complexes need not be exactly the same for cooperative binding to be manifested. All that is required is that the first ligand shifts the macromolecule to a state having a higher affinity for the second ligand (Ikegami, 1977). (The above scheme describes only positively linked binding. Negative linkage can be treated by considering that the ligands bind to both states of the macromolecule, M_A and M_B, but that each ligand has a higher affinity for a different state.)

Assuming that the equilibrium constant for the isomerization of the macromolecule, ${}_1K$, is less than one, the apparent association constant and enthalpy change for the binding of X (at [Y]=0) would be

$$K'_1 = {}_1K_B K_1 \tag{34a}$$

$$\Delta H_1^{0\prime} = {}_1\Delta H^0 + {}_B\Delta H_1^0 \tag{34b}$$

At saturating [Y], these parameters become

$$K'_{1,Y} = {}_BK_{1,Y} \tag{35a}$$

$$\Delta H_{1,Y}^{0\prime} = {}_B\Delta H_{1,Y}^0 \tag{35b}$$

If there is no interaction between X and Y other than the coupling through the conformational change, the parameters describing the binding of X to M_B and $M_B Y$ will be the same. That is, ${}_BK_{1,Y}$ and ${}_BK_1$, and ${}_B\Delta H_{1,Y}^0$ and ${}_B\Delta H_1^0$ will be equal; likewise for the parameters describing the binding of Y to M_B and $M_B X$. If ${}_1K > 0$, the macromolecule would exist predominantly in the M_B state and no linkage would occur.

Determination of the enthalpy changes for the binding of each ligand in the absence and presence of the other ligand may provide information concerning the molecular source of the cooperative binding. An enthalpy linkage parameter can be defined as $\Delta H_{\text{linkage}} = \Delta H_{1,\, Y}{}^{0'} - \Delta H_1{}^{0'}$ (the enthalpy change for the binding of X in the presence of saturating Y minus that at [Y]=0); likewise, the entropy linkage is defined as $\Delta S_{\text{linkage}} = \Delta S_{1,\, Y}{}^{0'} - \Delta S_1{}^{0'}$. These parameters are analogous to the Gibbs energy linkage, $\Delta G_{\text{linkage}}$. A negative $\Delta H_{\text{linkage}}$ indicates that the binding of X is enthalpically more favourable when Y is bound, etc. The source of the $\Delta H_{\text{linkage}}$ (or $\Delta S_{\text{linkage}}$) value may be either the direct interaction between X and Y when bound or an induced conformational change in the macromolecule. From the magnitude of the $\Delta H_{\text{linkage}}$ (and $\Delta S_{\text{linkage}}$) it may be possible to determine the source of the coupling. If the $\Delta H_{\text{linkage}}$ and $\Delta S_{\text{linkage}}$ values are larger than can be normally attributed to the direct interaction between the ligands, then a mediating conformational change is suggested. This being the case, $-\Delta H_{\text{linkage}}$ is an approximate measure of $_1\Delta H^0$, the enthalpy change for the conformational transition (likewise, $-\Delta S_{\text{linkage}} \simeq {}_1\Delta S^0$).

$\Delta G_{\text{linkage}}$ values are often found to be in the range of -1 to -3 kcal mol^{-1} (Weber, 1975); it is likely that $\Delta H_{\text{linkage}}$ (and $\Delta S_{\text{linkage}}$) values will vary over a much wider range, thus providing greater information about the structure–function relationships of the system. There have been very few studies of the enthalpy and entropy linkage of the binding of ligands. In the following paragraphs, two examples will be presented which demonstrate the potential of such studies.

The apparent enthalpy change for the binding of Mg^{2+} to (Na$^+$, K$^+$)-ATPase at 25 °C was found to be -49 kcal mol^{-1}; the binding of phosphate to the enzyme occurred with a similar large apparent heat of -40 kcal mol^{-1} (Kuruki et al., 1976). The magnitude of these values strongly suggests that the binding of either ligand is coupled with a conformational change of the enzyme. At saturating levels of phosphate, the binding of Mg^{2+} was found to occur with an enthalpy change of only -8 kcal mol^{-1}. The $\Delta H_{\text{linkage}}$ for the binding of Mg^{2+} and phosphate is thus ~ 40 kcal mol^{-1}. This result is readily understood in terms of a scheme such as (equation (33)) that in which the coupling of the ligands is a result of the common conformational change (for which $_1\Delta H^0 \simeq -40$ kcal mol^{-1}) induced by the binding of either ligand.

This example also emphasizes that calorimetric studies of $\Delta H_{\text{linkage}}$ may reveal features that could be missed by studies of $\Delta G_{\text{linkage}}$ alone. The Gibbs energy linkage for the binding of Mg^{2+} and phosphate in the above system is close to zero at the temperature studied, and could easily have been dismissed as experimental error if only equilibrium studies were performed. Determination of binding enthalpy values clearly demonstrates the linkage. Also it allows one to make predictions about the system. For the (Na⁺, K⁺)-ATPase, one would predict that at lower temperatures, where the $M_a \rightleftharpoons M_b$ equilibrium may be shifted towards the M_b form, less linkage would be observed. Likewise, at higher temperatures, a greater degree of linkage would occur. This point will be discussed further in section G.3.

The binding of 3'-CMP and protons to RNAse A is another example of linked binding that has been studied calorimetrically. In Table 4 are

TABLE 4

Thermodynamic parameters describing the linked binding of 3'-CMP and protons to ribonuclease A[a]

	ΔG^0 kcal mol⁻¹	ΔH^0 kcal mol⁻¹	ΔS^0 cal K⁻¹ mol⁻¹
Binding of 3'-CMP			
To deprotonated RNAse A	− 5·8	−8·7	−10
To diprotonated RNAse A	−10·8	−7·1	12
	$\Delta G_{\text{linkage}} = -5\cdot0$ kcal mol⁻¹		
	$\Delta H_{\text{linkage}} = 1\cdot6$ kcal mol⁻¹		
	$\Delta S_{\text{linkage}} = 22$ cal K⁻¹ mol⁻¹		
Binding of two protons[b]			
To RNAse A	− 9·8	−6·5	11
To RNAse A-3'-CMP complex	−12·1	−6·5	19
	$\Delta G_{\text{linkage}} = -4\cdot6$ kcal mol⁻¹		
	$\Delta H_{\text{linkage}} = 0$		
	$\Delta S_{\text{linkage}} = 16$ cal K⁻¹ mol⁻¹		

[a] Standard state, unit mole fraction. Values for 25 °C, $\mu = 0\cdot05$ M. Adapted from Flogel and Biltonen (1975b) and Flogel *et al.* (1975).
[b] Values of ΔG^0, ΔH^0 and ΔS^0 are per proton. The small discrepancies that exist between the two sets of linkage parameters may be due, in part, to experimental error, but may also be due to the fact that the binding of a third proton (to His 48) is weakly coupled to the binding of 3'-CMP.

presented the Gibbs energy, enthalpy and entropy changes for the binding of the dianionic form of 3'-CMP to the high pH state of RNAse A (state in which two histidine residues, presumably His 12 and 119, are deprotonated) and to the low pH state of RNAse A (state in which His 12 and 119 are both fully protonated). Also listed in this Table are the thermodynamic parameters describing the association of two protons to His 12 and 119 of the free protein and the protein–3'-CMP complex. The values for the binding of 3'-CMP were determined by calorimetric titration studies over a wide pH range. Proton binding was studied by independent potentiometric and pH thermal titration studies.

As shown in Table 4, the Gibbs energy linkage for the binding of 3'-CMP and two protons to RNAse A is about -5 kcal mol^{-1}. From the $\Delta H_{\text{linkage}}$ and $\Delta S_{\text{linkage}}$, one can see that the linkage is almost entirely due to an entropy effect; the binding of the protons improves the ΔS for the subsequent binding of the nucleotide, and vice versa. There does not appear to be any enthalpy coupling of which to speak. From X-ray crystallographic studies it is known that His 12 and 119 are both located at the nucleotide binding site and appear able to form charge–charge interactions with the phosphate of the ligand. The fact that the linkage between the protons and the nucleotide is entropic is as expected if the two ligands electrostatically interact with each other at the binding site. In this case no conformational change of the macromolecule appears to be involved. (X-ray studies show that the position of His 119 is affected by the binding of nucleotide, but the repositioning of this side-chain must not require a very significant change in enthalpy.)

It is anticipated that future studies of the enthalpy and entropy linkage of the binding of ligands to other systems will yield valuable information concerning the molecular source of the coupling that is not provided by determination of the Gibbs energy linkage.

3. COOPERATIVE BINDING IN MULTI-SUBUNIT PROTEINS

For cooperative homotropic binding of ligands to multi-subunit assemblies to take place there must be some form of communication between the subunits. According to the familiar models for cooperative binding in such systems (Monod et al., 1965; Koshland et al., 1966), the fixing of a ligand to the first subunit induces a conformational change in that subunit. The adjoining subunits must "sense" this conformational change and respond in such a way as to increase their affinity for the ligand. The communication between the subunits must, therefore, occur at the interface between the subunits.

Consider the simple concerted model for the cooperative binding of L by the dimer M_A–M_A shown in Fig. 7. Here it is assumed that the ligand binds only to a subunit that exists in the high-affinity state M_B, with an association constant of K_1. (In this model, L binds to the high-affinity states of the macromolecule that are in equilibrium with the unreactive M_A–M_A states. The alternate route of an initial weak binding of L to a subunit of M_A–M_A, followed by a conformational change to the now lower Gibbs energy state M_B–M_BL is thermodynamically equivalent but will be neglected in the model.) For cooperative binding to be observed, the relatively unreactive M_A–M_A state must be the most stable, unliganded species. Binding of the first ligand will pull the dimer towards the M_B–$M_B L$ state. Thus the intrinsic ΔG_1^0 for ligand binding to the high-affinity state is utilized to induce a conformational change not only in the liganded subunit but in the adjoining subunit as well. This provides a second high-affinity site for the ligand. The binding of the first ligand will proceed with an observed Gibbs energy change of $\Delta G_1^0 - {}_1\Delta G^0$ (where ${}_1\Delta G^0$ is for the conformational change of M_A–M_A to M_B–M_B), but the second ligand will bind more favourably with an observed free energy change of ΔG_1^0. (Here, for simplicity, we assume that L binds to M_B–M_B and M_B–$M_B L$ with an identical Gibbs energy

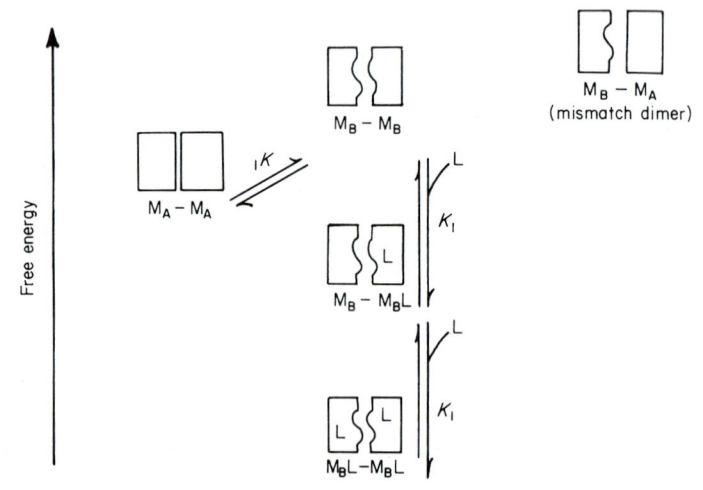

Fig. 7. Simple model for the cooperative binding of ligand, L, to a dimeric macromolecule, M_A–M_A. The model assumes that ligand binds only to a higher free energy form of the dimer, M_B–M_B. (See the text for further description of the model.) The intersubunit interactions in the M_A–M_B mismatch dimer are poor, as pictorially described above. The free energy level of the M_A–M_B dimer will be much higher than that for M_B–M_B in the concerted model (Monod et al., 1965). In the sequential model (Koshland et al., 1966), the free energy level of M_A–M_B will be between that of M_A–M_A and M_B–M_B, but closer to the latter.

change, ΔG_1^0, and enthalpy change, ΔH_1^0.) Likewise, it follows that the observed enthalpy change (and entropy change) for the binding of the first ligand will include a contribution from the conformational change, and that the values for the second ligand will be the intrinsic values, ΔH_1^0 and ΔS_1^0.

In real systems it is difficult to determine accurately the enthalpy change associated with each step for positive cooperative binding. Using microcalorimetry, Atha and Ackers (1974a) and Velick et al. (1971)

have studied the binding of O_2 to haemoglobin and NAD^+ to glyceraldehyde-3-phosphate dehydrogenase, respectively, and have been able to detect slight differences in the heat effects for each successive binding step.

The cooperativity of the sites is due to the tendency of the subunits to work as pairs (symmetry principle). That is, a mismatch dimer, M_A-M_B is relatively unstable compared to the matched M_A-M_A and M_B-M_B dimers; in the concerted model, M_A-M_B does not exist (Monod et al., 1965; Koshland et al., 1966). The ability of one subunit to control the conformation of its partner must be due to the nature and the strength of intersubunit interactions. Figure 7 depicts the changes in the macromolecular system that may occur as a result of the uptake of ligand. The binding of the first ligand to a subunit induces a conformational change that is transmitted via alterations in the interface to the other subunit.[1] The second subunit rearranges in order to optimize interactions with the first. As illustrated by the model, the mismatched dimer (M_A-M_B) would be relatively unstable due to its poorer intersubunit interactions. Although the strength of the intersubunit interactions for both the unliganded (M_A-M_A) and the liganded (M_B-M_BL or M_BL-M_BL) states may be similar, it seems more likely that they will differ. In order for the mismatched M_A-M_B dimer to be relatively unstable, the nature of the interactions at the interface must dramatically differ between the M_A-M_A and M_B-M_B states. It is also expected that the enthalpy and entropy change for the interactions at the interface will change markedly upon proceeding from the unliganded to the liganded state. In extreme cases the difference between the subunit interactions may be so large that dissociation (or association) of the macromolecule assembly may come into play (Levitzki and Schlessinger, 1974).

In cases where reactions of the monomeric species (regardless of whether it resembles the reactive or unreactive subunit) can be studied independently (i.e. at dilute solution or at proper solvent conditions), information concerning the energetics of the intersubunit interactions can be obtained. By comparison of the thermodynamic parameters for the binding of ligand to both the monomeric species (described by $^\alpha K'_1$) and the dimer (described by $^\beta K'_1$), one can obtain the difference in the

---

[1] In the above discussion, the nature of the conformational change in the individual subunits is not specified. The impression often left by such discussions and diagrams as shown in Fig. 7 is that the conformational rearrangement is a very extensive one, of the "refolding" type (section D.4). However, a "subtle" change, that is, a shift in the subunit towards a particular microstate, could also be responsible for the cooperativity, so long as the conformational change is propagated to the adjoining subunit.

intersubunit interactions between the unliganded and liganded subunits (where the intersubunit interactions are described in terms of the thermodynamics for the formation of a dimer from the respective monomeric species, i.e. $^1\Delta G^0$ and $^1\Delta G_L^0$):

$$2\overline{M} \underset{2L \ _{\alpha}K_1}{\overset{_1K}{\rightleftharpoons}} M_A - M_A \underset{2M_B L}{\overset{2L}{\underset{_1K_L}{\rightleftharpoons}}} \overset{_\beta K_1}{\rightleftharpoons} M_B L - M_B L \quad (36)$$

The difference between the Gibbs energy of binding ligand to $\overline{M}$ and M_A-M_A is equal to $^1\Delta G^0 - {}^1\Delta G_L^0$ and the difference between the enthalpy change for binding to $\overline{M}$ and M_A-M_A is $^1\Delta H^0 - {}^1\Delta H_L^0$. In the above, the monomer was represented as $\overline{M}$, and may, in fact, resemble either the reactive subunit or the unreactive subunit. If one can independently determine either $^1\Delta G_L^0$ and $^1\Delta H_L^0$ or $^1\Delta G^0$ and $^1\Delta H^0$, the energetics for each step in equation (36) can be defined.

An example in which the interactions between subunits have been studied in this fashion is given in the studies by Ackers and co-workers (1976) and Valdes and Ackers (1977) on the cooperative oxygenation of haemoglobin. Employing a combination of calorimetric, kinetic and equilibrium measurements, this group has focused attention on the energetics of the subunit interactions in haemoglobin and on the way that these interactions change upon the binding of ligand.

Haemoglobin is a tetrameric protein composed of two αβ dimers. The α and β chains of the dimer are held together by very strong intersubunit interactions. When studied independently, a dimer binds two oxygen molecules in a non-cooperative manner. Furthermore, the enthalpy of the interaction between the α and β chains in the αβ dimer does not change significantly upon oxygenation. Two dimers, however, will associate to form the tetrameric species that is capable of cooperative O_2 binding. The Gibbs energy and enthalpy change for the interdimer interaction is quite dependent on the state of oxygenation. In the deoxy form, two αβ dimers associate with a $^1\Delta G^0 = -14\cdot4$ kcal mol^{-1} and a $^1\Delta H^0 = -28\cdot9$ kcal mol^{-1}. Upon oxygenation the interdimer interaction parameters change to a $^1\Delta G_L^0 = -8\cdot05$ kcal mol^{-1} and a $^1\Delta H_L^0 = 3\cdot8$ kcal mol^{-1}. Thus the binding of O_2 must induce a conformational change in the individual subunits that leads to a large increase in the Gibbs energy and an even larger increase in the enthalpy of the interaction between the αβ dimers. Since the Gibbs energy change for the binding of O_2 by an αβ dimer is more negative than that for complete oxygenation of the tetramer, the strong interdimer interactions in the deoxy state must serve to restrict the individual chains in low O_2 affinity conformations.

As the first subunits bind O_2 and change their conformation slightly, the restraining interactions at the dimer–dimer interface must be broken, allowing further oxygenation to proceed more favourably. Also, from the calorimetric measurements it is clear that the forces responsible for maintaining the unliganded tetramer in a low-affinity state are favourable enthalpic interactions at the interface between the αβ dimers. The thermodynamic description of the action of haemoglobin provided by the work of Ackers and co-workers becomes even more meaningful when compared to the X-ray crystallographic studies of Perutz (1970) and Perutz and Ten Eyck (1971) which show large structural changes in the dimer–dimer interface upon oxygenation.

G. Heat capacity changes

The association of a ligand to a macromolecule has frequently been observed to proceed with a negative change in heat capacity, $\Delta C_p^{0'}$. Summaries of $\Delta C_p^{0'}$ values for protein–ligand association reactions have been provided by Rialdi and Biltonen (1975), Biltonen and Langerman (1977) and Sturtevant (1977), and only a few examples will be mentioned here. Suurkuusk and Wadsö (1972) measured the enthalpy change of binding biotin to avidin at various temperatures and found $\Delta H^{0'}$ (25 °C) = -22.5 kcal mol^{-1} of biotin and $\Delta C_p^{0'} = -237 \pm 12$ cal K^{-1} mol^{-1}. In a number of other cases, $\Delta C_p^{0'}$ values in the range of -50 to -300 cal K^{-1} mol^{-1} have been determined, but in a few cases, such as the binding of hexitol-1,6-diphosphate to rabbit muscle aldolase (Hinz et al., 1971), the $\Delta C_p^{0'}$ have been larger than -1000 cal K^{-1} mol^{-1}. The observed $\Delta H^{0'}$ for the binding of hexitol-1,6-diphosphate to aldolase changes from 20·5 kcal mol^{-1} at 8 °C to 2·5 kcal mol^{-1} at 25 °C. As pointed out by Hinz et al. (1971), the drastic change in $\Delta H^{0'}$ upon elevation of the temperature makes any attempt at an interpretation in terms of hydrogen bonds, hydrophobic interactions, etc., difficult. On the positive side, the $\Delta C_p^{0'}$ value itself may provide useful information concerning the ligand binding process.

The constant pressure heat capacity, defined as $(\partial H/\partial T)_p$, is a measure of the width of the distribution of enthalpy states of a system. The general trend of $\Delta C_p^{0'}$ values for ligand–protein association reactions to be negative indicates that the binding produces a system having a smaller C_p, that is, having a narrower distribution of enthalpy states. There are a number of possible molecular interpretations for the negative $\Delta C_p^{0'}$ values which will be discussed here.

The binding of ligand to macromolecule can be conceptually divided into two processes: the transfer of ligand from an aqueous environment to the binding site on the macromolecule and the alteration of the conformational properties of the macromolecule caused by the bound ligand molecule. Both of these processes may give rise to a change in heat capacity. First the ΔC_p^0 associated with the transfer of a molecule from water to the binding site of a macromolecule will be considered.

1. TRANSFER OF LIGAND FROM WATER TO BINDING SITE

The heat capacity of non-polar compounds in water is found to be much larger than that in a non-aqueous solvent. For example, the partial molar heat capacity of tetra-n-butyl ammonium bromide is 281 cal K^{-1} mol^{-1} in water, but only 108 cal K^{-1} mol^{-1} in ethanol (Sarma and Ahluwalia, 1971). It follows that the ΔC_p^0 for the transfer of a non-polar solute from water to a non-aqueous phase or to a protein surface will be negative (Wadsö, 1972). Studies with a series of compounds having the same polar functional group but differing in the number of methylene groups show that C_p in water increases by about 20 cal K^{-1} mol^{-1} for every additional methylene added (Konicek and Wadsö, 1971). The molecular explanation for the large C_p values for non-polar molecules in water is usually given in terms of the Frank and Evans (1945) "iceberg" model for the structuring of water about a hydrophobic solute. The structured water molecules are considered to be more temperature labile and are thought to "melt" away as the temperature is increased, giving rise to a large change in the enthalpy of the system with temperature.

Therefore, it is not unreasonable to expect the binding of a non-polar ligand to a macromolecule to proceed with a negative ΔC_p^0 as large as 100 cal K^{-1} mol^{-1}. However, the work of Franks et al. (1973) suggests that as the number of polar groups on a solute increases, the negative ΔC_p^0 for binding may become much lower. These workers found that the change in heat capacity for the solution (into water) of propane, propane-1-ol, propane-1,2-diol, propane-1,3-diol and glycerol were 70, 48·3, 23·3, 20·4 and 1·0 cal K^{-1} mol^{-1}, respectively. If this trend is a general one, the negative ΔC_p^0 for the binding of a ligand having a large number of polar groups would not be expected to be very large.

However, other studies have suggested that the replacement of ligand–water hydrogen bonds by macromolecule–ligand hydrogen bonds may actually be characterized by a significantly large negative ΔC_p^0. Interpretation of results on the dimerization of δ-valerolactam in water suggests a negative $\Delta C_p^{0\prime}$ of 45 cal K^{-1} mol^{-1} (Alvarez and Biltonen, 1973). In another study of the interaction of a hydrogen bonding solute with solvent, Alvarez and Biltonen (1973) found the molar heat capacity of thymine in both water and ethanol to be unusually large. These results have been interpreted in terms of an incomplete formation of hydrogen bonds between the polar solute and the protic solvents at room temperature. The large C_p would then be due to a temperature-induced change in the degree of hydrogen bonding in the system. The implication of this interpretation is that the association of a polar molecule to a macromolecule (or the dimerization of the molecule, as in the study with δ-valerolactam) may lead to a decrease in the distribution of enthalpy states of the system as reflected by a negative ΔC_p^0. A rough upper estimate of the magnitude of the negative ΔC_p^0 expected for a given ligand is 12 cal K^{-1} mol^{-1} times the number of potential hydrogen bonding groups (Alvarez and Biltonen, 1973). If one assumes that the same number of hydrogen bonding groups are present at the binding site, and that these groups also form incomplete hydrogen bonds with water, then the negative ΔC_p^0 value obtained from the above formula should be doubled. Therefore, while the presence of polar groups on a ligand may reduce the hydrophobic contribution to ΔC_p^0, incomplete ligand–water hydrogen bonding may produce an opposing change in this thermodynamic quantity.

Based on studies of the solvation of salts and the ionization of acids, it is found that the exposure of a charged group to water will also decrease the heat capacity of the system (Edsall, 1935; Edsall and Wyman, 1958). For example, the ΔC_p^0 for the ionization of acetic acid is -37 cal K^{-1} mol^{-1}. The association of a charged ligand to a macromolecule may require a significant desolvation of the charge (and also of any counter-charged group at the binding site), and thus may be characterized by a positive ΔC_p^0.

In summary, it is difficult to evaluate the contribution of the transfer of ligand from water to the binding site to the total observed $\Delta C_p^{0\prime}$. For a large hydrophobic ligand, a sizeable negative ΔC_p^0 may be associated with the transfer step. For polar or charged ligands, it is difficult to predict the magnitude and sign of the heat capacity change for transfer. The most reasonable approach is to study directly the ΔC_p^0 for the transfer of a particular ligand from water to a non-aqueous phase. Halsey and Biltonen (1975) have performed such a study with the ligand, 2,4-dinitrophenyl lysine. This ligand binds to a specific antibody with a $\Delta H^{0\prime}$ of -16 kcal mol^{-1} at 25 °C and a $\Delta C_p^{0\prime}$ of -220 cal K^{-1} mol^{-1}. The ΔC_p^0 for the transfer of this compound from water to ethanol was determined (by calorimetric heat of solution studies) to be -97 cal K^{-1} mol^{-1}. (Ethanol was chosen as a reference solvent because of its capability to form hydrogen bonds with the solute, thus allowing the assessment of the hydrophobic contribution to the ΔC_p^0.) Thus a substantial portion of the negative $\Delta C_p^{0\prime}$ for the protein–ligand interaction can be attributed to the transfer of the ligand from the aqueous environment. The remaining portion of the $\Delta C_p^{0\prime}$ must be due to other factors, such as the desolvation of a hydrophobic surface at the binding site, the incomplete formation of ligand–water hydrogen bonds, or changes in the conformational properties of either the macromolecule or the ligand induced by the binding process.

Similarly, Suurkuusk and Wadsö (1972) have employed model studies to demonstrate that most of the negative $\Delta C_p^{0\prime}$ associated with the binding to biotin to avidin may be due to the transfer of the relatively nonpolar ligand from water to a hydrophobic site on the protein.

2. INDUCED CONFORMATIONAL CHANGES

Another major source of a negative ΔC_p^0 value is a ligand-induced change in the conformational properties of the macromolecule. The three ways in which an induced structural change can lead to a significant negative ΔC_p^0 are: (i) a ligand-induced shift in the equilibrium between macromolecule macrostates; (ii) a ligand-induced shift in the distribution of microstates of the macromolecule; and (iii) ligand-induced change in macromolecule–solvent interactions.

a. *Shift in equilibrium between macrostates*

As discussed in section C.4, the binding of ligand may stabilize a state of the macromolecule that is normally not of the lowest free energy at the temperature studied. Here we will consider the induced conformational change from a low-affinity to a high-affinity form ($M_A \rightleftharpoons M_B$) as an example of a two-state refolding transition. Let us consider M_A and M_B to be two different folded states of the macromolecule. M_B may be either a higher enthalpy or a lower enthalpy conformation, depending on whether it is more or less structured than M_A. If M_B is a lower enthalpy state, then at lower temperatures this state should eventually become the most stable conformation (in the absence of alternative low enthalpy conformations). Likewise, if M_B is a higher enthalpy state, it will become more populated as temperature is increased, unless thermal denaturation occurs first.

Consider the case in which M_B is a lower enthalpy state and that the transition from M_A to M_B occurs at some temperature T_c. The sharpness of the transition will depend on the magnitude of $_1\Delta H^0$. If we assume that $_1\Delta H^0$ has a value of -20 kcal mol^{-1}, and $T_c = 286$ K, then the transition from 90 per cent M_A to 90 per cent M_B will occur within a range of about 40 K. The apparent association constant for ligand will be given by equation (25). Let us first assume that the ligand binds only to the high-affinity form, M_B ($_1K\,_BK_1 >\, _AK_1$) at all temperatures studied. In this case $K'_1 =\, _1K\,_BK_1/(1 +\, _1K)$. If T_c is much lower than the experimentally accessible temperature range, the apparent enthalpy change, $\Delta H^{0\prime}$, for the binding of ligand will be equal to $_B\Delta H_1^0 +\, _1\Delta H^0$, and will not include any temperature dependence due to the induced conformational change. (Here we neglect for the moment any differences in the heat capacity of the states M_A and M_B. As will be discussed in the following sections, the C_p of the two macrostates may not be the same. The C_p will differ if the two conformational states are dissimilar with respect to their interactions with water, their vibrational and rotational energy levels, or the number of hydrogen-bonded conformers they have available.)

If T_c is close to or within the experimental temperature range, however, a negative ΔC_p^0 may arise due to a temperature-dependent shift in the equilibrium between the two states of the macromolecule. Near

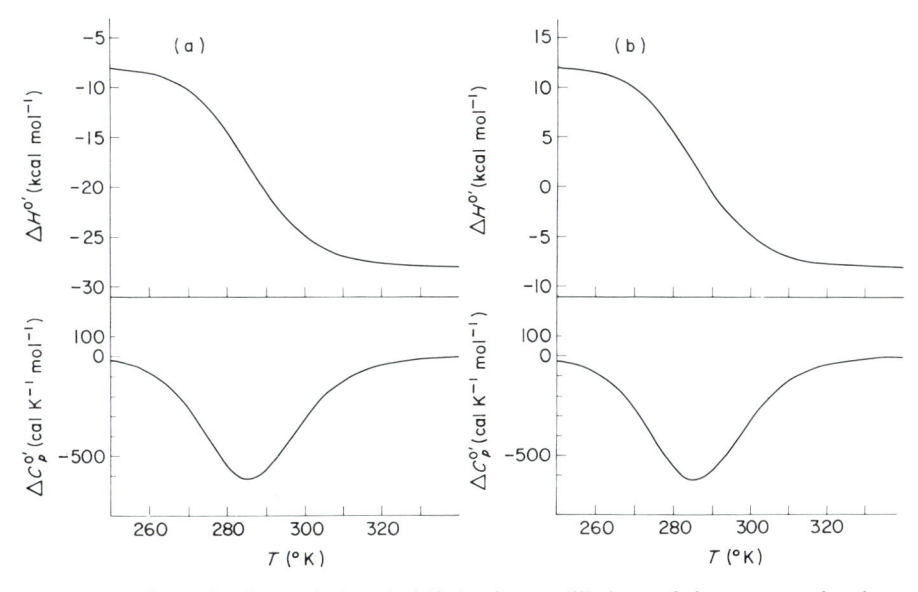

FIG. 8. Effect of a ligand-induced shift in the equilibrium of the macromolecule on the apparent $\Delta H^{0\prime}$ for binding. An equilibrium between two states of the macromolecule, $M_A \rightleftarrows M_B$ is assumed, with ligand binding only to M_B (a) $_B\Delta H_1^0 = -8$ kcal mol^{-1}, $_B\Delta S_1^0 = 0$, $_1\Delta H^0 = -20$ kcal mol^{-1}, $_1\Delta S^0 = -70$ cal K^{-1} mol^{-1}. (b) $_B\Delta H_1^0 = -8$ kcal mol^{-1}, $_B\Delta S_1^0 = 0$, $_1\Delta H^0 = 20$ kcal mol^{-1}, $_1\Delta S^0 = 70$ cal K^{-1} mol^{-1}. Bottom half of Figure shows the resulting $\Delta C_p^{0\prime}$ generated by the shift in equilibrium.

T_c, the observed enthalpy change for binding will be equal to

$$\Delta H^{0\prime} = {}_B\Delta H_1^0 + \frac{[M_A]_1 \Delta H^0}{[M_B]+[M_A]} \tag{37a}$$

$$= {}_B\Delta H_1^0 + {}_1\Delta H^0/[1+\exp(-{}_1\Delta H^0/RT + {}_1\Delta S^0/R)] \tag{37b}$$

For systems described by this model, a plot of $\Delta H^{0\prime}$ versus temperature will resemble Fig. 8(a). Note also that the $\Delta C_p^{0\prime}$ curve for this case will go through a minimum at T_c. If the high-affinity state of the macromolecule is the higher enthalpy state, a similar decrease in $\Delta H^{0\prime}$ versus temperature will result as shown in Fig. 8(b). In Fig. 9, the cases for

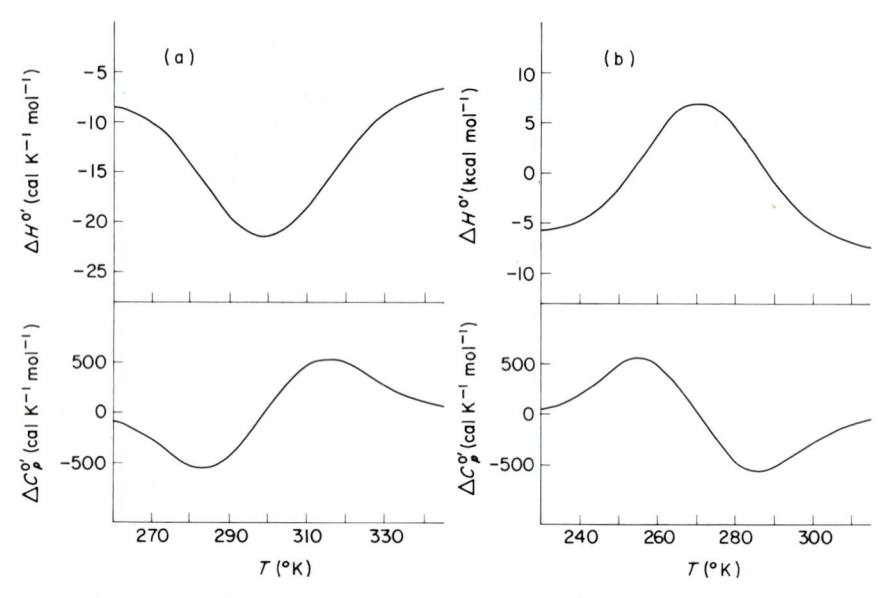

FIG. 9. Effect of a ligand-induced shift in the equilibrium of the macromolecule on the apparent $\Delta H^{0\prime}$ for binding. Same as Fig. 8, except that here it is assumed that ligand can bind to both M_A and M_B. (a) ${}_B\Delta H_1^0 = -8$ kcal mol^{-1}, ${}_B\Delta S_1^0 = 0$, ${}_A\Delta H_1^0 = -6$ kcal mol^{-1}, ${}_A\Delta S_1^0 = 0$, ${}_1\Delta H^0 = -20$ kcal mol^{-1}, ${}_1\Delta S^0 = -70$ cal K^{-1} mol^{-1}. (b) Same as (a), except that ${}_1\Delta H^0 = 20$ kcal mol^{-1}, ${}_1\Delta S^0 = 70$ cal K^{-1} mol^{-1}.

which binding to the low-affinity state M_A is significant (${}_AK_1 \simeq {}_1K \, {}_BK_1$) are also treated. The interesting feature of these curves is the increase in $\Delta H^{0\prime}$ with temperature that can occur (and the associated positive $\Delta C_p^{0\prime}$). Such a temperature dependence would be found when the most stable state of the macromolecule–ligand complex changes from M_AL at one temperature to M_BL at another temperature range. At some intermediate temperature both the complexes $M_A L$ and $M_B L$ would co-exist.

The detection of such patterns as shown in Figs 8 and 9 will obviously require the measurement of enthalpy changes at a number of temperatures. In studies to date (Barisas et al., 1971; Hinz et al., 1971; Suurkuusk and Wadsö, 1972; Niekamp et al., 1977), $\Delta H^{0'}$ values have been determined at only two or three temperatures at most.

As discussed earlier, conformational changes from one folded state to another folded state, as postulated above, have been extensively studied for biological macromolecules. For technical reasons, such rearrangements are difficult to study directly by either spectroscopic or calorimetric methods, and often only indirect evidence for such changes can be obtained from binding and kinetic studies such as in the chymotrypsin–proflavin system (Fehrst, 1972). However, structural transitions of this type have been observed for tRNA (Brandts et al., 1974; Freire and Biltonen, 1977).

If the source of the heat capacity change for a given ligand binding process is this temperature-sensitive shift in the equilibrium of the macromolecule, an estimate of the enthalpy change due to the conformational change (and the contribution of $_1\Delta H^0$ to the observed $\Delta H^{0'}$) can be made from the maximum value of the heat capacity change. The maximum $\Delta C_p^{0'}$ is related to $_1\Delta H^0$ by $(_1\Delta H^0)^2 = 4RT^2 \Delta C_p^{0'}$. For example, at 286 K in Fig. 8 one calculates that $_1\Delta H^0$ is about ± 20 kcal mol^{-1}. The observed $\Delta H^{0'}$ at 286 K will thus include a contribution of ± 10 kcal mol^{-1} (at the maximum in $\Delta C_p^{0'}$, $M_B/M_A = 1$) from the conformational change. The heat due to the intrinsic association of ligand to macromolecule is then either -8 kcal mol^{-1} or 12 kcal mol^{-1}.[1]

---

[1] Another somewhat trivial type of equilibrium shift that can be a source of a negative ΔC_p^0 involves the temperature-dependent ionization of either the ligand or macromolecule. Normally one does not consider an ionization as a conformational change, but it can give rise to a negative ΔC_p^0 and, therefore, must be considered. If the observed heat of binding is pH dependent

$$\Delta H^{0'}(H^+) = \Delta H_1^0 + \Delta N_i \Delta H_i^0 \tag{38}$$

where, as discussed in section C.1, ΔN_i is the number of protons absorbed by the reactants upon binding, and ΔH_i^0 is the heat of protonation of the group involved. The temperature dependence of $\Delta H^{0'}(H^+)$ will then be

$$\frac{d\Delta H^{0'}(H^+)}{dT} = \frac{d\Delta H_i^0}{dT} + \frac{\Delta N_i \, d\Delta H_i^0}{dT} + \frac{\Delta H_i^0 \, d\Delta N_i}{dT} \tag{39}$$

The last term, representing the contribution due to ionization, is

$$\frac{\Delta H_i^0 \, d\Delta N_i}{dT} = \frac{-(\Delta H_i^0)^2 \, K_i(T)}{RT^2[H^+](1 + K_i(T)/[H^+])^2} + \frac{(\Delta H_{i,\,L}^0)^2 \, K_{i,\,L}(T)}{RT^2[H^+](1 + K_{i,\,L}(T)/[H^+])^2} \tag{40}$$

[continued overleaf]

b. *Ligand-induced shift in the distribution of microstates*

The binding of ligand need not induce a dramatic change in the folding pattern of the macromolecule to give rise to a heat capacity change. Another possibility is that the ligand "tightens up", or decreases the heat capacity of the macromolecule by inducing a subtle conformational change. A globular macromolecule (such as a protein or a tRNA molecule) may actually exist as a distribution of microstates, each differing slightly in their enthalpy, but having the same overall folding pattern. The difference between these microstates may only be in the number of hydrogen bonds, the "structure" of associated water molecules, or the rotational or vibrational energy levels of various bonds. The distribution of microstates will broaden and shift to higher enthalpy levels as temperature is increased due to the increased population of the higher enthalpy microstates. Scanning calorimetric studies of proteins in solution have revealed a temperature-dependent increase in heat capacity of the native state of proteins prior to the thermal unfolding transition (Privalov et al., 1971). Suppose that the ligand, when bound, acts to restrict the width of the distribution of microstates. That is, the binding of ligand shifts the equilibrium distribution of the macromolecule towards the lower enthalpy microstates. A possible distribution of microstates for the macromolecule and the macromolecule–ligand complex is represented in Fig. 10. The binding of ligand may thus lead to a decrease in both the average enthalpy and the heat capacity of the macromolecule.

The reduction in the average enthalpy and C_p may be related to the more common description of the tightening or "tensing" of a macromolecule upon ligand binding. This tightening is evidenced in hydrogen isotope exchange studies of proteins and protein–ligand complexes (Hvidt and Nielson, 1966; Levitski and Schlessinger, 1974; Woodward et al., 1975). In thermodynamic terms, the binding of ligand may reduce the probability of the macromolecule existing in the higher enthalpy states that are more labile to isotope exchange. The distribution of microstates for native protein is normally skewed towards the higher enthalpy

where $K_i(T)$, the temperature-dependent dissociation constant, is $K_i(T) = \exp(\Delta H_i^0/RT - \Delta S_i^0/R)$, and $K_{i,\,L}(T)$ and $\Delta H_{i,\,L}^0$ are the temperature-dependent dissociation constant and heat of protonation of the group in the macromolecule–ligand complex. At extreme pH values where $[H^+] \ll K_i + K_{i,\,L}$ or $[H^+] \gg K_i + K_{i,\,L}$, the last term in equation (39) is small, but at intermediate pH values this term may not be negligible and may be either positive or negative. For example, assume that $\Delta H_i^0 = \Delta H_{i,\,L}^0 = 10$ kcal mol^{-1}, and $K_i = 100 K_{i,\,L} = 10^{-7}$ M. At pH 5, equation (40) equals -5.5 cal K^{-1} mol^{-1}; at pH 7, it equals -135 cal K^{-1} mol^{-1}; at pH 9 it equals 135 cal K^{-1} mol^{-1}; and at pH 11, it equals 5·5 cal K^{-1} mol^{-1}. Such values will usually be negligible, unless there are a number of ionizing groups. It is desirable, however, to use pH-independent binding enthalpies in determining the heat capacity change.

levels (Cooper, 1976). (The third moment of the distribution is positive.) If the heat capacity of the macromolecule decreases upon the binding of ligand (regardless of whether the slope of the C_p versus temperature curve remains the same, or is reduced) it follows (Cooper, 1976) that the distribution of enthalpy states will become more symmetric. This effect on the shape of the distribution of microstates further decreases the probability of the macromolecule existing in the higher enthalpy states.

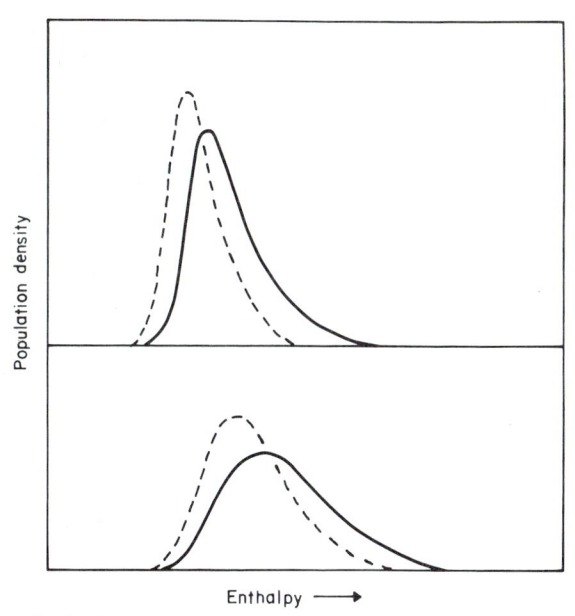

Fig. 10. Hypothetical population density of microstates of a macromolecule in the absence and presence of ligand. Top is at low temperature; the bottom is at some higher temperature.

It is difficult to predict the magnitude of $\Delta C_p{}^0$ that one would expect for a ligand-induced shift in the distribution of microstates. In order to approach this matter, let us first consider the C_p of globular macromolecules. Suurkuusk (1974), as well as others (Hutchens et al., 1969), have measured the heat capacity of solid, anhydrous proteins using a drop microcalorimeter. For chymotrypsin, for example, they obtained a value for C_p (solid) of 0·30 cal K^{-1} g^{-1}. This value is not much different from that for other organic molecules and can be attributed to the various excitable vibrational and rotational modes of the macromolecule. Kaneshia and Ikegami (1977) have calculated a value of 0·250 cal K^{-1}

g^{-1} for the C_p of an average protein by summing the contribution from the various excitable bonds. When a protein is dissolved in water, however, the measured heat capacity, C_p (solution), increases somewhat (Hutchens et al., 1969; Suurkuusk, 1974); for chymotrypsin, C_p (solution) was found to be 0·38 cal K^{-1} g^{-1}. The added heat capacity is thought to be largely due to the interactions between non-polar side-chains (and peptide bonds) and the solvent water. Another possible source of the additional heat capacity is the existence of a distribution of structural conformers, differing from each other in their degree of intramolecular hydrogen bonding and van der Waals contacts (Lumry and Rosenberg, 1975) (analogous to configurational heat capacity of liquid water; Eisenberg and Kauzmann, 1972). The concept that globular proteins in solution exist as a distribution of rapidly interconverting hydrogen-bonded conformers has recently been presented by Lumry and Rosenberg (1975) in their "mobile defect" hypothesis.

Thus, the heat capacity of a solvated globular protein is a result of contributions from excitable vibrational and rotational modes, interactions with water and the equilibrium between structural conformers. The contribution due to the structural conformers is undoubtedly small, but may be significant in that it is most likely to be affected by the binding of ligand. Let us assume a value of 0·04 cal K^{-1} g^{-1} for this contribution (by supposing that there is no equilibrium between conformers in the solid, and that at least half of the increase in C_p upon dissolution in water is due to interactions with solvent). Ikegami (1977) has arrived at a similar value for the contribution of structural conformers to the C_p of a protein from his statistical thermodynamic model for a protein. A similar estimate of this contribution can be made based on the melting of polymers (Karasz and O'Reilly, 1966; Brandts, 1969).

The decrease in heat capacity of the macromolecule that occurs upon the binding of ligand can be a result of a selecting out of a particular structural conformer in the formation of the ligand–macromolecule complex. One thus might imagine that the binding of ligand acts to tether different regions of the macromolecule and, in doing so, restricts the magnitude of oscillations between the various structural conformers. By limiting the fluctuations between conformers, the binding of ligand would, therefore, lead to a decrease in the heat capacity of the macromolecule by as much as 0·04 cal K^{-1} g^{-1} (which is 1000 cal K^{-1} g^{-1} for a macromolecule of MW 25 000).

The possibility that ligand binding reduces the heat capacity of the system by affecting the nature of the macromolecule–solvent interactions will be considered in the following section. The other possibility that the ligand somehow affects the vibrational modes of the macromolecule must also be considered. Infrared and Raman spectroscopy show that high-frequency vibrational modes are rather insensitive to their environment. It is more likely that certain low-frequency modes, such as torsional side-chain rotation and backbone bending, will be affected by ligand binding. Side-chains located at the binding site may be frozen in by direct interaction with the ligand. The freezing of rotational modes will decrease their capacity to absorb energy. For each readily excitable rotational degree of freedom that is frozen, the maximum drop in heat capacity of the macromolecule will be 2 cal K^{-1} mol^{-1}. Since there will only be a few side-chains at the binding site, the contribution from their restriction would be small. However, one might expect that rotational modes throughout the macromolecule might be attenuated by ligand binding, particularly the rotation of bulky or planar side-chains (i.e. aromatic amino acid residues). Nuclear magnetic resonance relaxation studies have demonstrated that certain side-chains in proteins have considerable rotational freedom, while others are relatively immobilized (Browne *et al.*, 1973; Hull and Sykes, 1975; Snyder *et al.*, 1975; Visscher and Gurd, 1975). The rotational freedom of such bonds is determined by the secondary bonds and steric interactions between the side-chain and the surrounding matrix of the macromolecule (Gelin and Karplus, 1975). The binding of ligand could conceivably act to restrict further the motion of these side-chains. This could be done if the structural conformers selected by the ligand are ones in which constraints are imposed upon the motion of these side-chains. Such ligand-induced restriction of the motion of protein side-chains has been recently demonstrated (Critz and Martinez-Carrion, 1977).

c. *Ligand-induced change in macromolecule–water interactions*

We have discussed the manner in which a negative heat capacity change might arise as a result of a ligand-induced shift in an equilibrium or a shift in the distribution of microstates of the macromolecule. A conformational change may also lead to a negative $\Delta C_p{}^0$ if the heat capacity of the liganded macromolecule is smaller than the heat capacity of the free macromolecule. One way in which this may occur is related to the change in vibrational and rotational heat capacity of the macromolecule. The heat capacity of the free and complexed state of the macromolecule may also differ with respect to the nature of their interactions with water. If a ligand-induced conformational change causes a change in the degree of exposure of various residues of the macromolecule, a negative $\Delta C_p{}^0$

may result. For example, if the binding of ligand results in the "burying" of a number of hydrophobic side-chains, the accompanying change in the heat capacity for ligand binding would be negative. This induced conformational change may be either of the two-state (refolding) type or merely a shift in the distribution of structural conformers (subtle change).

Sturtevant (1977) has argued that the large negative $\Delta C_p^{0\prime}$ found for the binding of NAD$^+$ to glyceraldehyde-3-phosphate dehydrogenase may be due largely to a change in macromolecule–water interactions. Buoyant density measurements (Sloan and Velick, 1973) on the enzyme have suggested that as many as 280 water molecules that normally interact with the protein are released as a result of a ligand-induced rearrangement of the macromolecule. If the bound water possesses properties like that of water in contact with hydrophobic groups, the release of this water would lead to a decrease in heat capacity.

To summarize, it can be said that there are a number of possible sources of a heat capacity change. One that can be experimentally tested is the role of the hydrophobic effect in the transfer of ligand from water to the binding site. Wishnia (1969a, b) studied the binding of pentane to β-lactoalbumin and ferrihaemoglobin and found $\Delta C_p^{0\prime}$ values of -140 and -106 cal K^{-1} mol^{-1}, respectively. These values are approximately equal to the ΔC_p^0 for the transfer of pentane from water to a detergent micelle. In this case, therefore, the source of the negative $\Delta C_p^{0\prime}$ for ligand binding can be accounted for in terms of the hydrophobic effect. There are other systems, such as the binding of benzene and naphthalene to chymotrypsin, that show a negative $\Delta C_p^{0\prime}$ of the order that is reasonable for the hydrophobic effect (Hymes et al., 1969).

The existence of changes in conformational equilibria in the macromolecule (or the ligand) may also give rise to changes in C_p. For α-chymotrypsin, there is evidence that an equilibrium between an active and an inactive protein form exists (see section D.6). Since the enthalpy change for the interconversion of these forms is small (7 kcal mol^{-1}) the maximum change in heat capacity that can be attributed to this process is less than 100 cal K^{-1} mol^{-1}. However, if the enthalpy change for a conformational transition were larger, the negative ΔC_p^0 due to a shift in the equilibrium position could be more substantial.

The contributions from the other sources discussed above are difficult to assess, but they must all be considered. Hinz et al. (1971) have noted that large negative $\Delta C_p^{0\prime}$ values have been reported for systems having very dissimilar ligands. This suggests that attention should be focused primarily on the macromolecule for an explanation of the heat capacity changes, particularly if the values are large.

3. HEAT CAPACITY CHANGES IN COOPERATIVE SYSTEMS

Consider the system $M_A-M_A \rightleftharpoons M_B-M_B$ to which a ligand preferentially binds at two sites on the M_B-M_B form (see Fig. 7). If the macromolecule exists, in the absence of ligand, in the M_B-M_B form then a normal rectangular hyperbolic binding isotherm is observed and the apparent enthalpy change for binding is the intrinsic ΔH_1^0 for binding to the form M_B-M_B. However, if the macromolecule exists predominantly in the form M_A-M_A, positive cooperative binding will be observed and the apparent enthalpy change for binding will include a contribution from the conformational change. A simple example is illustrated in Fig. 11

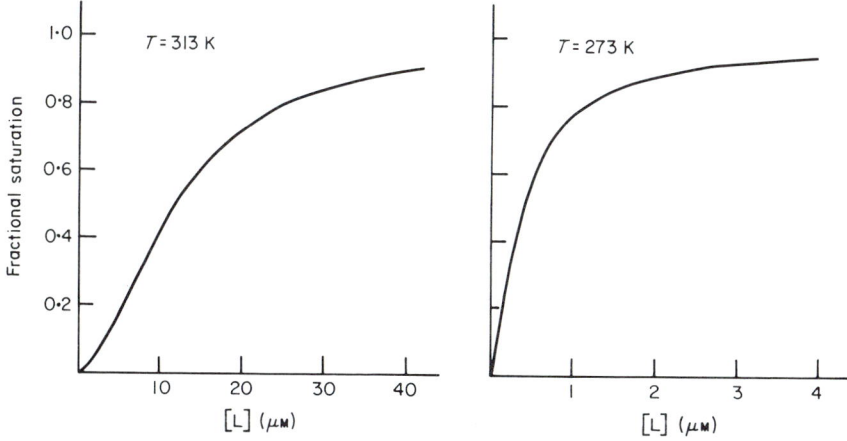

FIG. 11. The effect of temperature on the linked binding of L to a two-subsite macromolecule. The linkage is mediated by a conformational change as described by Fig. 7. $\Delta H_1^0 = -8$ kcal mol^{-1}, $\Delta S_1^0 = 0$, $_1\Delta H^0 = -20$ kcal mol^{-1}, $_1\Delta S^0 = -70$ cal K^{-1} mol^{-1}. It is assumed that L binds only to the "active" state of the dimer, M_B-M_B. The transition temperature between M_A-M_A and M_B-M_B is 286 K. The degree of saturation of the dimer is given by the relationship
$$[1/2[L]K_1 + K_1^2[L]^2]/[1 + {_1}K^{-1} + K_1[L] + K_1^2[L]^2]$$
for the case that $_1\Delta H^0$, the enthalpy change for the transition $M_A-M_A \rightleftharpoons M_B-M_B$ is equal to -20 kcal mol^{-1}. At low temperature the macromolecule will exist in the high-affinity M_B-M_B state and at high temperature in the M_A-M_A state.

For such a situation an apparent negative $\Delta C_p^{0\prime}$ will be found if the experimental temperature range includes temperatures where the Gibbs energy change for the transition $M_A-M_A \rightleftharpoons M_B-M_B$ is approximately zero. The maximal heat capacity change will be observed at the transition temperature, T_c, and will be equal to

$$\Delta C_p^{0\prime} (\text{max}) = ({_1\Delta H^0})^2/4RT^2 \tag{41}$$

In the example given the maximal $\Delta C_p^{0\prime}$ would be -555 cal K^{-1} mol^{-1}; $\Delta C_p^{0\prime}$ calculated from the difference in the apparent $\Delta H^{0\prime}$ at 0 °C and 40 °C is about -400 cal K^{-1} mol^{-1}.

Recently Niekamp et al. (1977) reported their calorimetric results on the binding of NAD$^+$ to glyceraldehyde-3-phosphate dehydrogenase. At 5 °C the binding to the four sites was essentially non-cooperative, but at 40 °C, highly cooperative. Their overall thermodynamic results are summarized in Table 5. These results are consistent with the above

TABLE 5
The thermodynamic parameters for the association of NAD$^+$ to yeast glyceraldehyde-3-phosphate dehydrogenase[a]

Temperature °C	$\Delta G^{0\prime}$ kcal mol^{-1}	$\Delta H^{0\prime}$ kcal mol^{-1}	$\Delta S^{0\prime}$ cal K^{-1} mol^{-1}	$\Delta C_p^{0\prime}$ cal K^{-1} mol^{-1}
5	-36	-27.2	30	
				-3000
25	-34	-87.6	-180	
				-3100
40	-29.5	-134	-340	

[a] pH 8·5, 0·005 M pyrophosphate, 0·002 M EDTA. The standard state is unit mole fraction. All values are for the binding of four moles of ligand per mole of macromolecule.

model assuming four sites on the high-affinity form, and that $_1\Delta H^0 \simeq -100$ kcal mol^{-1} and $T_c \simeq 10$ °C. In this situation the maximal change in heat capacity would be about -10 kcal K^{-1} mol^{-1} if the conformational change was all-or-none for the tetramer. Their experimental value of -3 kcal mol^{-1} (which, of course, may not be the maximal value) strongly suggests that conformational change may be the basis of the observed changes in the binding thermodynamics. In any case this study clearly demonstrates that accurate thermodynamic data as a function of temperature can provide valuable insight into the molecular details of binding reactions.

H. Conclusion

At the very least, the thermodynamic quantities pertaining to an interacting ligand–macromolecule system describe the temperature dependence of the affinity between the reactants. At the most, these quantities can provide very valuable insight concerning the molecular events and molecular forces that are operative within the system. Deducing the correct interpretations can be a mammoth problem. In fact, a complete definition of the relationship between thermodynamics and molecular details may be beyond our present means. However, if one can at least delineate and describe the major factors involved in an association process, useful information can be obtained.

The analysis and interpretation of thermodynamic quantities for biological systems as described in this article involve the following logical progression:

1. evaluation of the ways in which the available (intrinsic) energy of interaction is partitioned among the various equilibria that are coupled to the binding process;
2. definition of the structure–energy relationships that pertain to the intrinsic interaction between the ligand and the binding site;
3. extension of the information of points (1) and (2) to the realm of biochemistry by seeking the role that such coupled shifts in equilibria and such intrinsic binding interactions play in particular biological processes.

Different biological micromolecule–ligand systems have been designed to carry out different functions. Antibodies, for example, are designed to remove antigen from biological fluids, and thus require a very high affinity for the interaction. For such systems it would be desirable to employ very strong interactive forces between the protein and ligand. It would also seem desirable to minimize the dissipation of the intrinsic binding energy into coupled equilibria.

The situation will probably be entirely different for enzyme–substrate interactions, due to the different function of such complexes. Enzymes are designed to have a higher affinity towards the transition state of a reaction than towards the substrate(s). In order to do this, the enzyme will have only a marginal affinity towards the substrate(s); it is often postulated that there may be geometric distortion or electronic destabilization of the substrate(s) in the enzyme–substrate(s) complex. It would also be desirable for the interactions between the enzyme and the reaction product(s) to be relatively weak so that release of the latter would not be rate limiting. Coupled changes in the conformation or state of ionization of the enzyme upon the binding of substrate (induced fit: Koshland and Neet, 1968) are also likely to occur and may serve the purpose of bringing catalytic groups into the proper position and ionization state relative to the substrate(s) required for the catalysed reaction to proceed. In this manner part of the intrinsic interaction energy will be spent in driving the coupled reactions, but this will ultimately lead to a higher degree of specificity in the catalytic process.

Linked binding phenomena, which appear to be so important in the control and amplification of biological processes, may also be a manifestation of the utilization of intrinsic binding energy to drive coupled conformational changes in macromolecules.

If we have learned anything since proteins were first isolated and described, it is that their properties can and must be understood in terms of simple physical chemical principles. Modern techniques such as X-ray crystallography and n.m.r. have and will continue to provide for us detailed pictures of the complexes formed between ligands and macromolecules. In order to understand further the nature and function of these complexes we must arrive at some appreciation of the energetics involved. The thermodynamic quantities ΔH, ΔS and ΔC_p describing interacting systems can best be obtained by calorimetric studies. If carefully interpreted, such thermodynamic information will aid greatly in the development of a molecular interpretation of complex biological systems.

References

ACKERS, G. K., JOHNSON, M. L., MILLS, F. C. and IP, S. H. C. (1976). *Biochem. biophys. Res. Commun.* **69**, 135.
ALVAREZ, J. and BILTONEN, R. (1973). *Biopolymers*, **12**, 1815.
ANDERSON, D. G., HAMMES, G. G. and WALZ Jr, F. G. (1968). *Biochemistry*, **1**, 1637.
ATHA, D. H. and ACKERS, G. K. (1974a). *Biochemistry*, **13**, 3276.
ATHA, D. H. and ACKERS, G. K. (1974b). *Archs Biochem. Biophys.* **164**, 392.
AUSTIN, R. H., BEESON, K. W., EISENSTEIN, L., FRAUENFELDER, H. and GUNSALUS, I. G. (1975). *Biochemistry*, **14**, 5355.
BARISAS, B. J., STURTEVANT, J. M. and SINGER, S. J. (1971). *Biochemistry*, **10**, 2816.
BELAICH, J. P. and SARI, J. C. (1969). *Proc. natn. Acad. Sci. U.S.A.* **64**, 763.
BILTONEN, R. L. and LANGERMAN, N. (1979). *Meth. Enzym.* **61**, 287.
BILTONEN, R. L. and LUMRY, R. (1969). *J. Am. chem. Soc.* **91**, 4251.
BLOW, D. M. and STEITZ, T. A. (1970). *A. Rev. Biochem.* **39**, 63.
BOLEN, D. W., FLOGEL, M. and BILTONEN, R. L. (1971). *Biochemistry*, **10**, 4136.
BRANDTS, J. F. (1969). In "Structure and Stability of Biological Macromolecules" (S. Timasheff and G. Fasman, eds), vol. 2, p. 213. Marcel Dekker, New York.
BRANDTS, J. F., JACKSON, W. M. and TING, T. Y.-C. (1974). *Biochemistry*, **13**, 3595.
BROWNE, D. T., KENYON, G. L., PACKER, E. L., STERNLICHT, H. and WILSON, D. M. (1973). *J. Am. chem. Soc.* **95**, 1316.

CATHOU, R. E. and HAMMES, G. G. (1974). *J. Am. chem. Soc.* **86**, 3240.
CATHOU, R. E. and HAMMES, G. G. (1965). *J. Am. chem. Soc.* **87**, 4674.
CITRI, N. (1973). *Adv. Enzymol.* **37**, 397.
COOPER, A. (1976). *Proc. natn. Acad. Sci. U.S.A.* **73**, 2740.
COOPER, A. and JENKINS, F. M. (1972). In "Protides of the Biological Fluids" (H. Peeters, ed.), p. 457. Pergamon Press, Oxford.
COULTER, C. L. (1973). *J. Am. chem. Soc.* **95**, 570.
CRITZ, W. J. and MARTINEZ-CARRION, M. (1977). *Biochemistry*, **16**, 1559.
EDSALL, J. T. (1935). *J. Am. chem. Soc.* **57**, 1506.
EDSALL, J. T. and WYMAN, J. (1958). "Biophysical Chemistry". Academic Press, New York.
EFTINK, M. R. and GHIRON, C. A. (1977). *Biochemistry*, **16**, 5546–5551.
EIGEN, M. and DE MAEYER, L. (1963). In "Techniques of Organic Chemistry" (S. L. Friess, E. S. Lewis and A. Weissberger, eds), vol. III, part II, p. 895. Interscience, New York.
EISENBERG, D. and KAUZMANN, W. (1969). "The Structure and Properties of Water", p. 174. Oxford University Press, London.
EISINGER, J., FEUER, B. and YAMANE, T. (1970). *Proc. natn. Acad. Sci. U.S.A.* **65**, 638.
FEHRST, A. R. (1972). *J. molec. Biol.* **64**, 497.
FISCHER, E. (1894). *Berichte*, **27**, 2985.
FLOGEL, M. and BILTONEN, R. L. (1975a). *Biochemistry*, **14**, 2603.
FLOGEL, M. and BILTONEN, R. L. (1975b). *Biochemistry*, **14**, 2610.
FLOGEL, M., BOLEN, D. W. and BILTONEN, R. L. (1972). In "Protides of the Biological Fluids" (H. Peeters, ed.), vol. 20, p. 521. Pergamon Press, Oxford.
FLOGEL, M., ALBERT, A. and BILTONEN, R. L. (1975). *Biochemistry*, **14**, 2616.
FRANK, H. S. and EVANS, M. W. (1945). *J. Chem. Phys.* **13**, 507.
FRANKS, F., REID, D. S. and SUGGETT, A. (1973). *J. Solution Chem.* **2**, 99.
FREIRE, E. and BILTONEN, R. L. (1977). *Biopolymers*, **16**, 2641–2652.
FRENCH, T. C. and HAMMES, G. G. (1965). *J. Am. chem. Soc.* **87**, 4669.
GELIN, B. R. and KARPLUS, M. (1975). *Proc. natn. Acad. Sci. U.S.A.* **72**, 2002.
GILL, S. J. and NOLL, L. (1972). *J. Phys. Chem.* **76**, 3065.
GILL, S. J. and STOESSER, R. (1966). *J. Phys. Chem.* **71**, 564.
GILL, S. J., DOWNING, M. and SHEATS, G. F. (1967). *Biochemistry*, **6**, 272.
GLASSTONE, S. (1946). "Textbook of Physical Chemistry", p. 462. Van Nostrand, New York.
GRUNWALD, H. and PRICE, E. (1964). *J. Am. chem. Soc.* **86**, 4517.
HALSEY, J. F. and BILTONEN, R. L. (1975). *J. Solution Chem.* **4**, 275.
HAVSTEEN, B. H. (1967). *J. biol. Chem.* **242**, 769.
HENDERSON, R., WRIGHT, C. S., HESS, G. P. and BLOW, D. M. (1972). *Cold Spring Harb. Symp. quant. Biol.* **36**, 63.
HINZ, H. J., SHAIO, D. D. F. and STURTEVANT, J. M. (1971). *Biochemistry*, **10**, 1347.
HO, H. C. and WANG, H. C. (1973). *Biochemistry*, **12**, 4750.
HULL, W. E. and SYKES, B. D. (1975). *J. molec. Biol.* **98**, 121.
HUTCHENS, J., COLE, A. and STOUT, J. (1969). *J. biol. Chem.* **244**, 26.

HVIDT, A. and NIELSON, O. (1966). *Adv. Protein Chem.* **21**, 287.
HYMES, A. J., CUPPETT, C. C. and CANADY, W. J. (1969). *J. biol. Chem.* **244**, 637.
IKEGAMI, A. (1977). *Biophys. Chem.* **6**, 117.
IRIE, M. and SAWADA, F. (1967). *J. Biochem., Tokyo*, **62**, 282.
JENCKS, W. P. (1969). "Catalysis in Chemistry and Enzymology", Chapter 8. McGraw-Hill, New York.
JENCKS, W. P. (1975). *Adv. Enzymol.* **43**, 219.
KANESHIA, M. I. and IKEGAMI, A. (1977). *Biophys. Chem.* **6**, 131.
KARASZ, F. E. and O'REILLY, J. M. (1966). *Biopolymers*, **4**, 1015.
KAUZMANN, W. (1959). *Adv. Protein Chem.* **14**, 1.
KEMBALL, C. (1950). *Adv. Catalysis*, **2**, 233.
KIM, Y. and LUMRY, R. (1971). *J. Am. chem. Soc.* **93**, 1003.
KLAPPER, M. (1971). *Biochim. biophys. Acta*, **229**, 557.
KONICEK, J. and WADSÖ, I. (1971). *Acta chem. scand.* **25**, 1541.
KOSHLAND Jr, D. E. and NEET, K. E. (1968). *A. Rev. Biochem.* **37**, 359.
KOSHLAND Jr, D. E., NEMETHY, G. and FILMER, D. (1966). *Biochemistry*, **5**, 365.
KRESHBECK, G. C. and SCHERAGA, H. A. (1965). *J. Phys. Chem.* **69**, 1704.
KURIKI, Y., HALSEY, J. R., BILTONEN, R. L. and RACKER, E. (1976). *Biochemistry*, **15**, 4956.
LANGERMAN, N. and BILTONEN, R. L. (1979). *Meth. Enzym.* **61**, 261.
LAVALLEE, D. K. and COULTER, C. L. (1973). *J. Am. chem. Soc.* **95**, 576.
LEVITZKI, A. and SCHLESSINGER, J. (1974). *Biochemistry*, **13**, 5214.
LIKHTENSTEIN, G. I., GRENBENSHCHIKOU, YU. B. and AVILOVA, T. V. (1972). *Molec. Biol.* **6**, 52.
LINDERSTROM-LANG, K. U. and SCHELLMAN, J. A. (1959). *In* "The Enzymes" (P. D. Boyer, H. Lardy and K. Myrback, eds), vol. I, p. 443. Academic Press, New York.
LIPSCOMB, W. N., HARTSUCK, J. A., REEKE Jr, G. N., QUIOCHO, F. A., BETHGE, P. H., LUDWIG, M. L., STEITZ, T. A., MUIRHEAD, H. and COPPOLA, J. C. (1968). *Brookhaven Symp. Biol.* **21**, 24.
LOVRIEN, R. and STURTEVANT, J. M. (1971). *Biochemistry*, **10**, 5006.
LUMRY, R. and BILTONEN, R. L. (1969). *In* "Biological Macromolecules" (G. Fasman and S. Timasheff, eds), vol. 2, Chapter 3. Marcel Dekker, New York.
LUMRY, R. and ROSENBERG, A. (1975). *In* "L'eau: Et les Systèmes biologiques", no. 264, p. 53. Coll. Int. C.N.R.S., Paris.
LUMRY, R., BILTONEN, R. L. and BRANDTS, J. F. (1966). *Biopolymers*, **4**, 917.
MARKELY, J. L. (1975). *Biochemistry*, **14**, 3546.
MEADOWS, D. H., ROBERTS, G. C. K. and JARDETSKY, O. (1969). *J. molec. Biol.* **45**, 491.
MONOD, J., WYMAN, J. and CHANGEAUX, J. P. (1965). *J. molec. Biol.* **12**, 88.
MOUNTCASTLE, D., FREIRE, E. and BILTONEN, R. L. (1976). *Biopolymers*, **15**, 355.
NELSON, D. J., OPELLA, S. J. and JARDETSKY, O. (1976). *Biochemistry*, **15**, 5552.
NIEKAMP, C. W., STURTEVANT, J. M. and VELICK, S. F. (1977). *Biochemistry*, **16**, 436.

NONNENMACHER, G., VIALA, E., THIERY, J. M. and CALVET, P. (1971). *Eur. J. Biochem.* **21**, 393.
PAGE, M. and JENCKS, W. (1971). *Proc. natn. Acad. Sci. U.S.A.* **68**, 1678.
PARKER, H. and LUMRY, R. (1963). *J. Am. chem. Soc.* **85**, 483.
PERUTZ, M. F. (1970). *Nature, Lond.* **228**, 726.
PERUTZ, M. F. and TEN EYCK, L. F. (1971). *Cold Spring Harb. Symp. quant. Biol.* **36**, 295.
POLAND, D. and SCHERAGGA, H. A. (1967). *In* "Poly-α-amino Acids" (G. Fasman, ed.), Chapter 10. Marcel Dekker, New York.
PRIVALOV, P. L., KHECHINASHVILLI, N. N. and ATANASOV, B. P. (1971). *Biopolymers*, **10**, 1865.
RAJENDER, S. and LUMRY, R. (1970). *Biopolymers*, **9**, 1125.
"Recommendations for Measurement and Presentation of Biochemical Equilibrium Data" (1976). *Q. Rev. Biophys.* **9**, 439. (Prepared by the Inter-union Commission on Biothermodynamics.)
RIALDI, G. and BILTONEN, R. L. (1975). *In* "MTP International Review of Science. Physical Chemistry" (H. A. Skinner, ed.), Series 2, vol. 10, p. 147. Butterworths, London.
RIALDI, G., LEVY, J. and BILTONEN, R. L. (1972). *Biochemistry*, **11**, 2472.
RICHARDS, F. M. (1974). *J. molec. Biol.* **82**, 1.
RICHARDS, F. M. and WYCOFF, H. (1971). *In* "The Enzymes" (P. D. Boyer, ed.), vol. 4. Academic Press, New York and London.
RICHARDS, F. M., WYCOFF, H. W. and ALLEWELL, N. M. (1969). *In* "The Neurosciences" (F. W. Schmidt, ed.), 2nd Study Program, p. 901. MITP.
RICHARDS, F. M., WYCOFF, H. W., CARLSON, W. D., ALLEWELL, N. M., LEE, B. and MITSUI, Y. (1971). *Cold Spring Harb. Symp. quant. Biol.* **36**, 35.
SAMEJIMA, T., KITA, M., SANEYOSHI, M. and SAWADA, F. (1969). *Biochim. biophys. Acta*, **179**, 1.
SARMA, T. S. and AHLUWALIA, J. C. (1971). *Trans. Faraday Soc.* **92**, 7097.
SCHELLMAN, J. A. (1955). *C. r. Trav. Lab. Carlsberg. Sér. Chim.* **29**, 230.
SCHULTZ, R. M., KONOVESSI-PANAYOTATOS, A. and PETERS, J. R. (1977). *Biochemistry*, **16**, 2194.
SHAIO, D. D. F. (1970). *Biochemistry*, **9**, 1083.
SHAIO, D. D. F. and STURTEVANT, J. M. (1969). *Biochemistry*, **8**, 4910.
SLOAN, D. L. and VELICK, S. F. (1973). *J. biol. Chem.* **248**, 5419.
SNYDER, G. H., ROWMAN, R., KARPLUS, S. and SYKES, B. D. (1975). *Biochemistry*, **14**, 3765.
SPINK, C. and WADSÖ, I. (1976). *In* "Methods of Biochemical Analysis" (D. Glick, ed.), vol. 23, p. 1. John Wiley, New York.
STEINHARDT, J. and REYNOLDS, J. A. (1970). "Multiple Equilibria in Proteins." Academic Press, New York and London.
STEITZ, T. A., HENDERSON, R. and BLOW, D. M. (1969). *J. molec. Biol.* **46**, 337.
STOESSER, P. R. and GILL, S. J. (1967). *J. Phys. Chem.* **71**, 564.
STURGILL, T. W., JOHNSON, R. E. and BILTONEN, R. L. (1978). *Biopolymers*, **17**, 1793.

STURTEVANT, J. M. (1972). *In* "Methods in Enzymology" (C. H. W. Hirs and S. N. Timasheff, eds), vol. 26, p. 227. Academic Press, New York and London.
STURTEVANT, J. M. (1974). *A. Rev. Biophys. Bioeng.* **3**, 35.
STURTEVANT, J. M. (1977). *Proc. natn. Acad. Sci. U.S.A.* **74**, 2236.
SUURKUUSK, J. (1974). *Acta chem. scand.* **B28**, 409.
SUURKUUSK, J. and WADSÖ, I. (1972). *Eur. J. Biochem.* **28**, 438.
TAIT, M. J., SUGGETT, A., FRANKS, F., ABLETT, S. and QUICKENDEN, P. A. (1972). *J. Solution Chem.* **1**, 131.
TANFORD, C. (1973). "The Hydrophobic Effect." John Wiley, New York.
VALDES Jr, R. and ACKERS, G. K. (1977). *J. biol. Chem.* **252**, 88.
VELICK, S. F., BAGGOTT, J. P. and STURTEVANT, J. M. (1971). *Biochemistry*, **10**, 779.
VISSCHER, R. B. and GURD, F. R. N. (1975). *J. biol. Chem.* **250**, 2238.
WADSÖ, I. (1970). *Q. Rev. Biophys.* **3**, 383.
WADSÖ, I. (1972). *In* "MTP International Review of Science. Physical Chemistry" (H. A. Skinner, ed.), Series 1, vol. 10, p. 1. Butterworths, London.
WEBER, G. (1972). *Biochemistry*, **11**, 864.
WEBER, G. (1975). *Adv. Protein Chem.* **29**, 1.
WICKETT, R., IDE, G. and ROSENBERG, A. (1974). *Biochemistry*, **13**, 3273.
WISHNIA, A. (1969a). *Biochemistry*, **8**, 5064.
WISHNIA, A. (1969b). *Biochemistry*, **8**, 5070.
WOLFENDEN, R. (1972). *Acc. Chem. Res.* **5**, 10.
WOODWARD, C. K., ELLIS, L. M. and ROSENBERG, A. (1975). *J. biol. Chem.* **250**, 432.
WYCOFF, H. W., HARDMAN, K. D., ALLEWELL, N. M., INAGAMI, T., JOHNSON, L. N. and RICHARDS, F. M. (1967). *J. biol. Chem.* **242**, 3984.
YAPEL, A. and LUMRY, R. (1964). *J. Am. chem. Soc.* **86**, 187.

Heat Capacity Studies in Biology

PETER L. PRIVALOV

A. Introduction

The attraction of heat capacity studies of any object follows directly from the definition of this parameter—heat capacity is a temperature derivative of one of the basic thermodynamic functions describing the state of a macroscopic system, i.e. the enthalpy:

$$C_p = \frac{dH(T)_p}{dT} \tag{1}$$

As a derivative, heat capacity, much more precisely than enthalpy, describes the dependence of a state of a system on temperature. At the same time, by direct calorimetric measurements we determine the heat capacity directly and not the enthalpy.

In so far as the heat capacity of a system $C_p(T)$ is known in some temperature region $(T_0 - T)$, we can determine not only the enthalpy function in this region:

$$H(T) = H(T_0) + \int_{T_0}^{T} C_p \, dT \tag{2}$$

but the entropy of the system

$$S(T) = S(T_0) + \int_{T_0}^{T} \frac{C_p(T)}{T} \tag{3}$$

and hence the Gibbs energy as well:

$$G(T) = H(T) - TS(T) \tag{4}$$

From the above it is evident that the broader the temperature region in which the heat capacity is known, the more completely we can describe a system. In the limit, if we determine heat capacity down to absolute zero ($T_0 = 0$ K), having in mind that entropy at this temperature is also close to zero (or equal to zero for perfectly ordered systems) and that the enthalpy also reaches its minimum value, we can define almost (or even) the *absolute* values of thermodynamic functions.

But usually there is no need of a complete description of a system, and in many cases we can restrict studies to a certain temperature range. For example, if we are interested in the liquid state of a system, then we do not need to overstep the bounds of existence of the liquid state. This is the case met in the study of aqueous biopolymer solutions. Such solutions are the closest models for living systems. It is evident that the temperature range of 273–373 K, i.e. 0–100 °C, is the most interesting for the study of these systems. This does not mean, however, that calorimetric studies of these systems outside this range are uninteresting. As we shall see below, there are many experimental objectives at much lower temperatures which are important for clarifying our knowledge of aqueous systems.

The division of the temperature scale into definite regions for studying definite problems is practically very important. This becomes evident if we take into account that in the whole range of, say, 0–400 K, which might be interesting for biology, the thermophysical properties of many objects, and especially of solutions, change drastically. This is why precise measurements over the whole of this region by only one instrumental technique is impossible. In reality we do not need the same accuracy of measurements over all this range. The requirements for accuracy are especially rigid when studying biopolymer solutions in the temperature range of liquid-phase existence, since most physical objectives which are solved in this range by heat capacity measurements are focused on the investigation of non-interacting macromolecules in solution and their intermolecular transitions. The effects of macromolecular interactions in solution become negligible when their concentrations fall below 0·3 weight per cent. It can be shown that the apparent heat capacity of the macromolecular component in these solutions makes up only a tenth of a per cent of the total. Thus, determinations of the partial heat capacity of a macromolecule in these solutions will require an exceptionally accurate calorimetric measurement. In fact, it is crucial to carry out such measurements in the development of a new super-precise calorimetric technique—scanning microcalorimetry. The precision of instruments used for determining the relative heat capacity should be not less than 0·002 per cent and their sensitivity to small heat capacity changes with temperature of an even higher order (see section F). This extreme sensitivity and precision are not necessary when studying dry biopolymers and their non-liquid solutions below 0 °C. But here we meet another difficulty. Indeed, dry biopolymers are nothing other than powders with extremely low heat conductivity. Heat conductivities of hydrated biopolymers are somewhat higher, but they are characterized by very large internal friction which makes their relaxation very slow. As a result of both these properties the time to reach true thermal equilibrium in these systems is far too long and often places such systems outside present experimental technique.

One of the most difficult problems in biological calorimetry is the problem of the quantity of a sample available and the experimentally required amount of material. Homogeneous high-quality biological preparations are expensive and cannot always be obtained. Thus, milligram quantities which are regarded in conventional thermochemistry and thermophysics as small are too large in biochemistry and biophysics. This is why in these fields of science, and particularly in calorimetry, micromethods are so popular.

Before considering experimental methods for heat capacity studies of biological objects, it is worthwhile to review the most important results of these studies to get some idea of the basic problems and complications which arise.

B. Heat capacity of biopolymers in a broad temperature range

The heat capacities of several biopolymers were measured in the vicinity of absolute zero up to room temperatures by Hutchens and co-workers at the University of Chicago (Hutchens *et al.*, 1969) and also by Andronikashvili and co-workers at the Institute of Physics in Tbilisi, USSR (Andronikashvili *et al.*, 1974, 1975, 1976), and at temperatures above 100 K by Haly and Snaith (1968, 1971), CSIRO, Australia.

The results obtained on the heat capacities of anhydrous preparations of collagen, chymotrypsinogen, serum albumin, keratin and DNA are presented in Fig. 1. It can be seen that although the specific heat capacity of DNA differs substantially from that of proteins, the specific heat capacity of the proteins are indistinguishable from each other even for different classes of proteins. Thus we can conclude that the heat capacity function of proteins does not bear, in itself, any specific information about their structural organization (although, according to Andronikashvili *et al.*, 1976, in the case of fibrillar structures there are some peculiarities in the heat capacity function in the vicinity of absolute zero).

As for general information, in comparing the protein molar entropy with that of the constituent amino acids, Hutchens succeeded only in determining the entropy of formation of a peptide bond which he found to be 37 J K^{-1} mol^{-1}. Suurkuusk (1974) found that the heat capacities

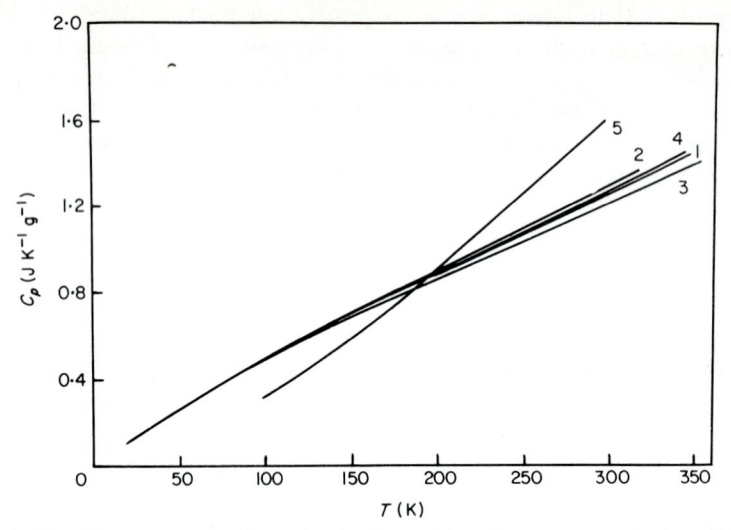

FIG. 1. Specific heat capacities of anhydrous biopolymers representing different structural classes. Globular proteins: 1, serum albumin (Hutchens et al., 1969); 2, chymotrypsinogen (Hutchens et al., 1969). Fibrillar proteins: 3, collagen (Hutchens, 1968; Andronikashvili et al., 1976); 4, keratin (Haly and Snaith, 1968). Nucleic acids: 5, DNA (Mrevlishvili, 1977, personal communication).

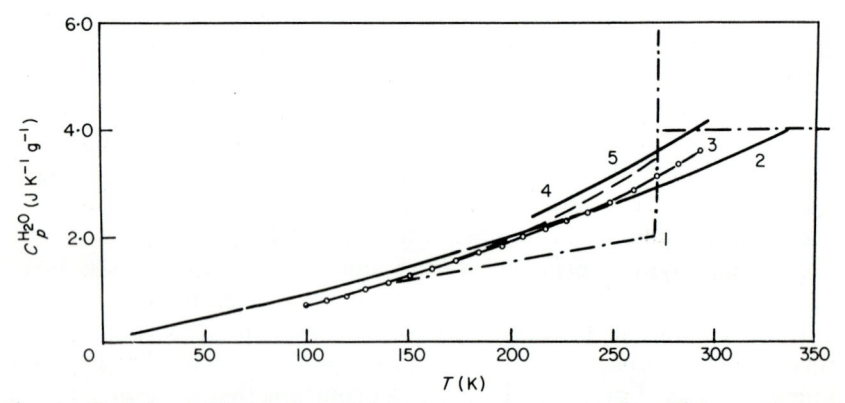

FIG. 2. Partial specific heat capacities of water in a water–biopolymer system at low water content. 1, Pure water; 2, water–collagen—13·5 per cent (Hutchens et al., 1969); 3, water–DNA—18 per cent (Mrevlishvili, 1977, personal communication); 4, water–chymotrypsinogen—10·7 per cent (Hutchens et al., 1969); 5, water–keratin—8 per cent (Haly and Snaith, 1968).

of various proteins at 298 K correlate well with the heat capacities of the constituent amino acids if the formation of peptide bonds is taken into account. The absence of any specific effects of unique polypeptide conformation is not as surprising as it seems at first sight, if we allow for the fact that proteins may lose all their specific organization whilst being dried (see, for example, Esipova and Chirgadze, 1969). This is why studies of hydrated preparations are much more justified, since it was found that the addition of a small amount of water to dry biopolymers does not significantly affect the heat capacity function. In Fig. 2 are presented values of the partial specific heat capacities of water, added to biopolymers in quantities not exceeding 15 per cent. These values were determined in the studies of Hutchens *et al.* (1969), Haly and Snaith (1968, 1971) and Mrevlishvili (1977). Although the objects studied represent absolutely different classes of biopolymers (especially DNA), we do not observe a great difference in the partial heat capacities of water. Most surprising is the lack of any heat sorption which would be associated with melting of the water in the sample. Thus, we have to conclude that in these systems there is no free water which could create a liquid phase, i.e. all the water is bound by the macromolecules. It is remarkable that the partial specific heat capacity of the bound water at temperatures below 150 K is close to the specific heat capacity of ice, and at higher temperatures it slowly approaches the specific heat capacity of liquid water, reaching it at room temperature.

At 298 K in the presence of a small amount of water there is an additivity in the heat capacities of the biopolymer and the water but with increase in the amount of water this additivity disappears. As seen from Fig. 3, the partial specific heat capacity of biopolymers increases, starting from some fixed amount of water and shifts to another level which is again independent of any further increase in the water content.

Simultaneously with the increase of the partial heat capacity of biopolymers with water content, the water–biopolymer system acquires some new anomalous properties. These anomalies appear first of all in the temperature dependence of the heat capacities in two separate temperature regions, 220–273 K and 320–373 K, and in both these regions they develop simultaneously with the increase in the water content above some critical values. In Figs 4 and 5 they are shown for the case of collagen. As is seen, with increase in water content the anomalous heat sorption increases, becomes sharper and shifts to distinct temperature limits: the low-temperature anomaly shifts up to 273 K but the high-temperature one shifts down to 310 K.

For the case of the low-temperature anomaly, there is no doubt that it is connected with the appearance in a system of unbound water,

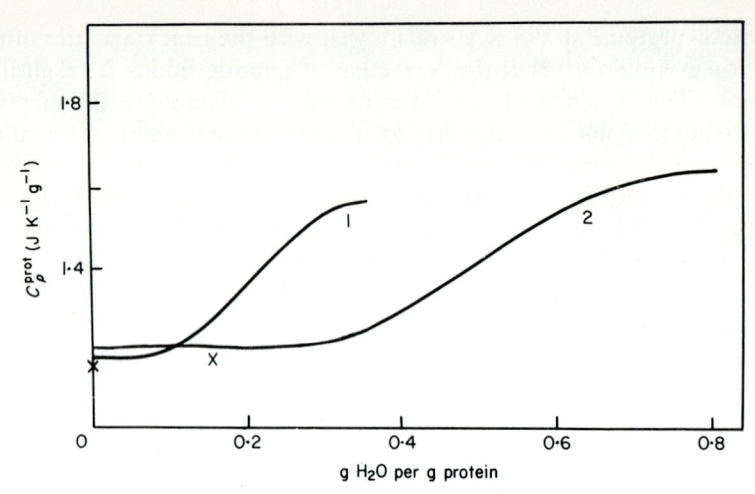

FIG. 3. Partial specific heat capacity of proteins in the presence of various amounts of water. 1, Keratin at 293 K, calculated from the data of Haly and Snaith (1968); 2, collagen at 283 K, calculated from the data of Haly and Snaith (1971); ×, collagen at 293 K, calculated from the data of Hutchens *et al.* (1969).

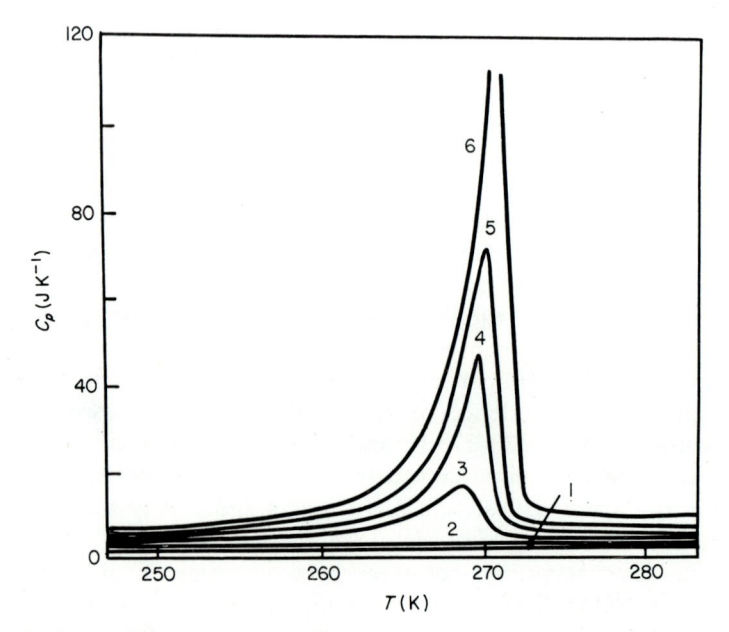

FIG. 4. Heat capacities of 1 g collagen + W g water below 300 K. Curves correspond to the following values of W: 1, 0; 2, 0·35; 3, 0·64; 4, 0·93; 5, 1·26; 6, 2·0 g of water. (Privalov and Mrevlishvili, 1967.)

i.e. the water which could freeze on cooling and melt on subsequent heating. This conclusion is supported by the known fact that the freezing–melting cycle, even when repeated many times, does not essentially influence the native properties of a biopolymer molecule in solution. As for the high temperature anomaly of the heat capacity function

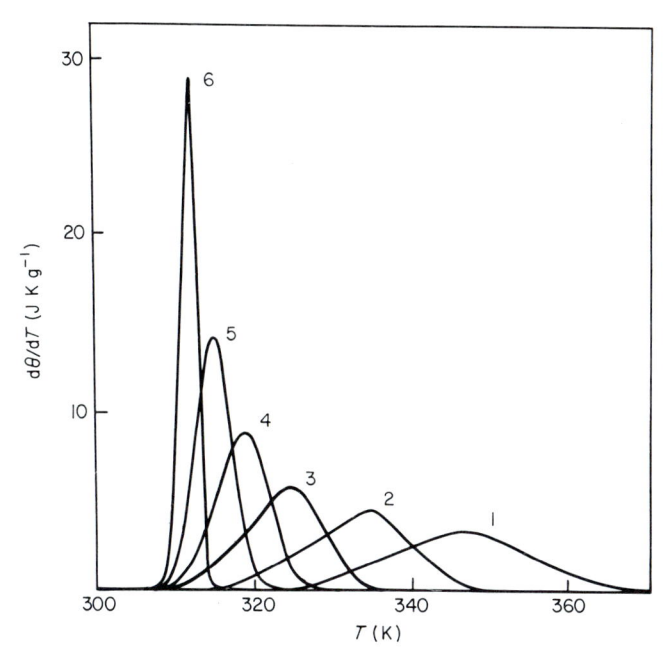

Fig. 5. Apparent partial specific heat capacities of collagen (rat skin) above 300 K in the presence of various amounts of water, calculated from the results of Andronikashvili et al. (1973). Curves correspond to the following amounts of water in g g^{-1} protein: 1, 0·43; 2, 0·71; 3, 1·45; 4, 2·70; 5, 9·5; 6, 17·0.

of a biopolymer solution, here we have quite the opposite situation: in this temperature region the macromolecule loses all its native properties, it denatures and the observed heat sorption is nothing other than the heat of denaturation. Considering denaturation as a process of disruption of the ordered native structure of biopolymers, we can say that here we have an intramolecular melting process, in contrast to the melting of the surrounding ice at lower temperatures. Since both these processes are absolutely different and are even related to the different subsystems of a complex water–biopolymer system, we shall consider them separately, the more so since they are studied by different methods and are interesting to a wide range of scientific investigations.

C. Fusion of the water phase in biological systems

Studies on the fusion of the water phase in biological systems are closely connected with the investigation of the water state in these systems. This problem became apparent after the appearance of Szent Gyorgyi's (1956, 1957) fascinating hypothesis on the extremely ordered state of water in living systems and its unique role in life processes (see also Klotz, 1962; Privalov, 1968). But the improvement of this hypothesis was not as easy as it seemed at first since the study of such complex systems was practically impossible by most physical methods available for studying the state of the water. The possibilities of calorimetry in this connection were first shown in 1966 by Privalov and co-workers (Privalov et al., 1966; Privalov and Mrevlishvili, 1966, 1967). The advantage of calorimetry in studying the state of water was based on the exceedingly large enthalpy and entropy of freezing of "free" or pure water.

To a first approximation we can make judgements about the water bound by a biopolymer by reference to the critical amount which induces an anomaly in the heat capacity function in the vicinity of 273 K. For collagen this amount is between 0·35 and 0·64 g of water per gram of dry preparation (see Fig. 4). For dry DNA preparations this is between 0·50 and 0·75 g of water per gram (Fig. 6). For a more precise evaluation of the amount of bound water it is necessary to determine accurately the water content when the heat capacity anomaly appears. But a much more practical method is to determine the deficit in the amount of heat necessary to "defrost" the given solution in comparison to an ideal one containing the same amount of dry preparation and pure water. For real systems the enthalpy and the entropy of "defrosting" are always less, and from this deficit we can estimate the amount of bound water. But here, in transforming thermodynamic parameters into structural ones, we have a difficulty. It is connected with the fact that the water, which is fusing in a system, is not in a truly "free" state—its melting temperature is significantly shifted from 0 °C. At the same time, the water which does not take part in the fusion process is also not ordinary ice—its specific partial heat capacity is larger, and even reaches the heat capacity of pure water. Since we do not know the specific enthalpy of fusion of ice which is under the influence of the macromolecule nor its dependence on temperature (i.e. the heat capacity change on fusion), we have to postulate some concrete, more or less reliable values for them in order to perform quantitative calculations. For example, we can assume that all the fusing water has the same thermodynamic parameters as pure water, i.e. its enthalpy of fusion is 333 J g⁻¹ at 273·15 K and the temperature dependence of the enthalpy or heat capacity change on fusion is 2·1 J K⁻¹ g⁻¹. But we must remember that this is only a convenient assumption which gives us the possibility to describe the very complicated phenomenon of water–macromolecular interaction in terms of the amount of bound water (Privalov, 1968). This description is very illustrative and useful, particularly in comparative studies, although in principle the presentation of a

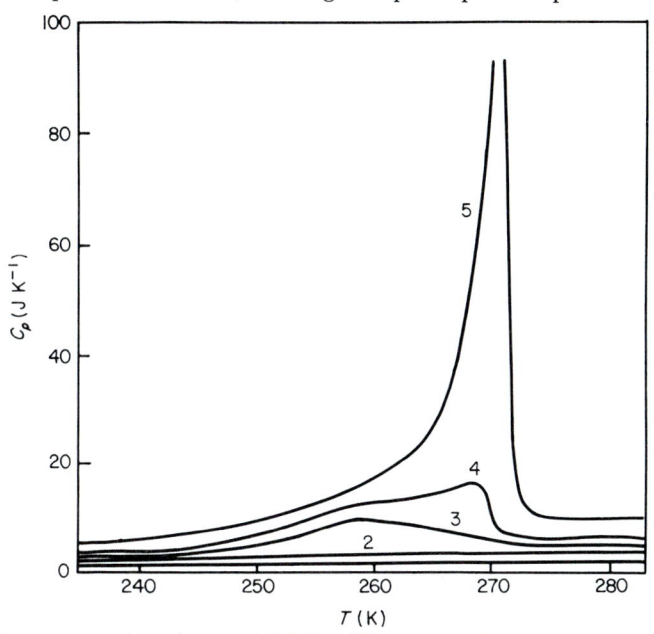

Fig. 6. Heat capacities of 1 g of DNA + W g water. Curves correspond to the following values of W: 1, 0; 2, 0·50; 3, 0·75; 4, 1·00; 5, 2·0 g water. (Privalov and Mrevlishvili, 1967.)

water–macromolecule interaction directly in thermodynamic characteristics is much more justified. It should be noted that Chatoraj and Bull (1971) attempted to avoid the difficulty mentioned above by comparing the enthalpy of defrosting the solution with the enthalpy of heating it over the same temperature interval, but in a supercooled state. The shortfall in this elegant method is that it is not easy to supercool a solution below −6 °C, as done by these authors, and even this temperature is quite insufficient as is evident from the heat capacity curves.

Analysis of the amount of bound water in biopolymer solutions of different concentrations leads to a very important conclusion. With the increase in the water content in the systems the amount of bound water rapidly reaches a certain limit which is constant and is a characteristic for a given biopolymer. Thus it seems that the water which is directly under the influence of a macromolecule (i.e. the bound water) creates a coat of definite thickness around a macromolecule. Although outside this coat the water is still under the influence of the macromolecule its state is qualitatively different. Thus, the water in the presence of a macromolecule is in itself a heterogeneous subsystem and the influence of a macromolecule on it cannot be considered as a spread homogeneous action, i.e. we cannot describe it in terms of "structural temperature". This represents one of the most remarkable differences between macromolecular aqueous solutions and aqueous solutions of small molecules which average their interaction with water because of their much greater mobility.

Table 1 presents the calorimetrically determined amounts of water bound by different biopolymers and the thickness of the corresponding

TABLE 1
Calorimetric data on hydration of biopolymers

Biopolymer	Hydration in grams of water per gram of preparation	Thickness of the hydrated coat in Ångströms	Reference
DNA	0·610	4·0	
Collagen	0·465	2·8	
Serum albumin	0·320	3·2	Privalov and Mrevlishvili (1967)
Ovalbumin	0·323	3·4	
Haemoglobin	0·324	3·8	
tRNA	0·600	—	Bakradze et al. (1973)
Keratin	0·227	—	Haly and Snaith (1968)
B-Lactoglobulin	0·290	—	Rüegg et al. (1975)

hydration coat determined from the known shape of the macromolecules. As is seen, in all cases the thickness of the hydration coat is very close to that of a monomolecular layer of water. But, as we have already mentioned, the influence of a macromolecule on water is not completely described by this monolayer. It is remarkable that this influence depends also on the state of a macromolecule. It significantly changes on denaturation of a macromolecule, as is seen in the case of DNA presented in Fig. 7 (Privalov et al., 1966). In this case coincident with a change of shape in the melting curve we have also a considerable increase in the amount of tightly bound water (see, for details, Privalov and Mrevlishvili,

1967; Privalov, 1968; Mrevlishvili and Privalov, 1969; Ruegg et al., 1975).

As noted above, the application of calorimetry in the study of the state of water in tissues seems very encouraging.

Figures 8 and 9 show the heat sorption and the apparent entropy changes observed on heating of a cooled muscle. Table 2 summarizes

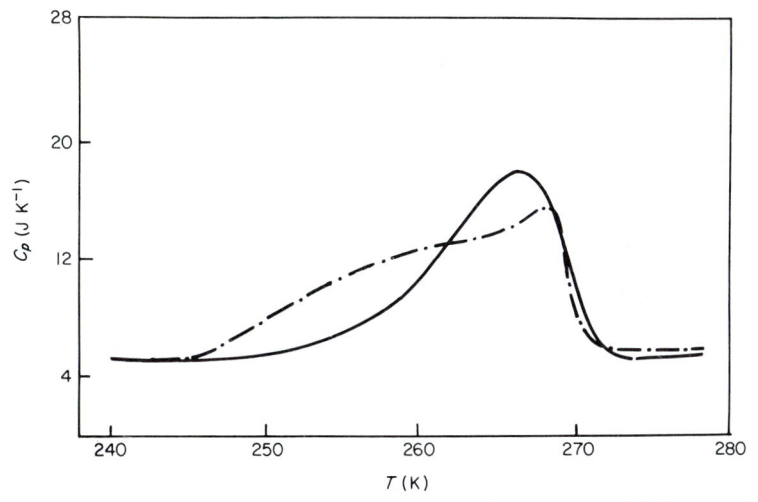

FIG. 7. Partial specific heat capacities of DNA in the native (solid line) and denatured (dotted line) state in the presence of 1 g water (Privalov et al., 1966).

some results on the determination of the amount of bound water in different tissues (see, for details, Andronikashvili et al., 1966a, b; Andronikashvili et al., 1969).

It is remarkable that the amount of bound water in tissues, calculated per gram of dry weight, is of the same order as that found for biopolymer solutions. Thus, these results do not confirm the hypothesis of the exclusive ordering of water in living tissues, but after studying solutions of biopolymers they are not unexpected. Indeed, it is evident that the boundary surface of water with the non-water component in tissues must be less than it is in biopolymer solutions, since in a tissue the macromolecules are arranged in supramolecular structures. But if the ordering of water takes place only in this macromolecular boundary surface, then this ordering should be less in tissues and the extent of its decrease should be proportional to the extent of development of supramolecular structures in the cells. Thus we can use this information to estimate the intracellular organization. In this respect the results obtained in studying the bound water content in developing silkworm

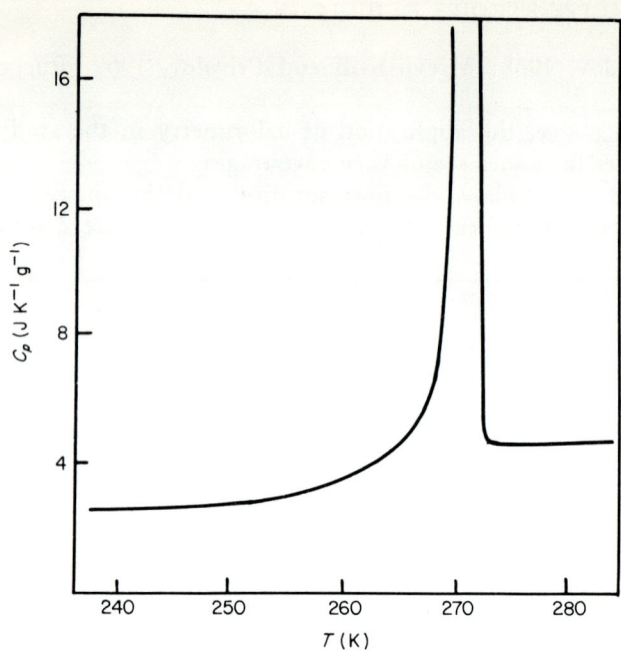

FIG. 8. The heat sorption observed on heating 1 g muscle tissue (Andronikashvili et al., 1966b).

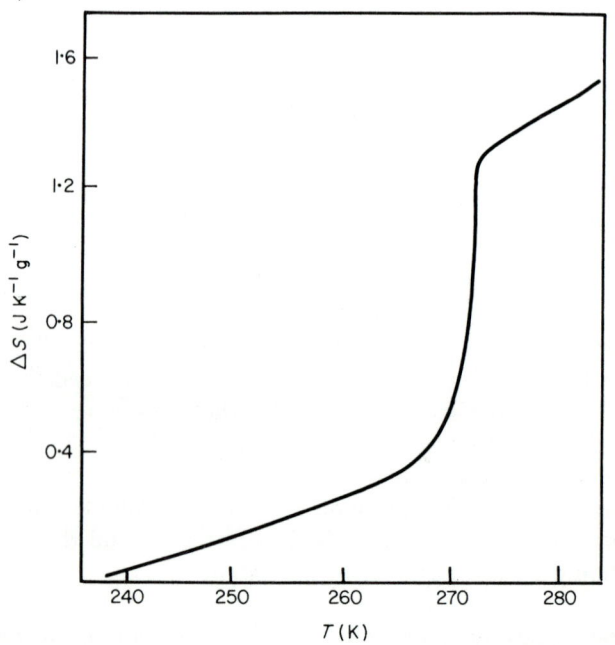

FIG. 9. The apparent entropy increase on warming cooled muscle (Andronikashvili et al., 1966b).

TABLE 2
Calorimetric data of the amount of bound water in tissues

Tissue	Amount of bound water in grams per gram of dry weight of tissue	Reference
Membranes	0·2	Williams and Chapman (1970)
Rat brain	0·15	Andronikashvili et al. (1966a, b, 1969)
Frog muscle	0·25	
Frog liver	0·40	
Silkworm eggs during hibernation	0·32	Privalov (1968)
Silkworm eggs on 8th day of incubation at 20 °C	0·18	
Cancer tissue	0·73–0·85	Andronikashvili and Mrevlishvili (1975)

eggs are very illustrative (Privalov, 1968). As seen from Table 2, the amount of the bound water in silkworm eggs decreases from, initially, 0·32 to 0·18 g per gram of dry weight on the eighth day of incubation at 20 °C. We would explain this effect by a decrease of the total boundary surface of the non-water component of the cells with water in developing organelles. Another very intriguing aspect of this phenomenon becomes clearer if we consider all the processes in the development of the eggs, and the release or disordering of the water during this process in entropy terms. It will then be evident that we can consider water as a negentropy pool for a developing living system which compensates the entropy decrease involved in this process.

In this respect the results of calorimetric studies of water in cancer tissues should also be mentioned. According to Andronikashvili and Mrevlishvili (1975), the amount of the bound water in cancer tissues is much larger than in other kinds of tissues. Is this a manifestation of the differentiation of cells or is it connected with the presence in these tissues of some specific strongly hydrated compounds? This remains to be clarified.

D. Intramolecular melting of proteins

Interest in the study of intramolecular melting or the heat of denaturation of proteins is provoked not only by the fact that this property is a specific property of biological macromolecules and particularly of proteins, but much more important is the fact that denaturation studies are the only approach to solving the problem of the stability of the unique structure of a macromolecule and the problem of its organization and maintenance in space. Calorimetry has here a special significance, since it is the only direct method of estimating the stabilities of structures in thermodynamical values and only through calorimetry can we check the adequacy of models in respect of real objects by thermodynamic criteria.

As has already been mentioned, with a change of the water content in water–macromolecular systems, there is a change not only of the heat sorption connected with the fusion of water, but also of the sorption connected with melting of the macromolecules themselves (see Fig. 5) and these changes are especially noticeable in the range of concentrations of water where the building of the hydration coat proceeds (see Monaselidze and Bakradze, 1969; Ruegg et al., 1975). It follows from this fact that the influence of the macromolecule and the water is mutual, and if the state of water changes in the presence of a macromolecule, the state of a macromolecule changes also to a no lesser extent.

In dilute solutions the conformational transition of a protein is independent of the concentration. Here the temperature interval of the transition is sharpened and has a definite breadth which does not change on further dilution. The same occurs with the midpoint temperature of the transition, T_{trs}, which is also called the melting temperature, T_m, or the temperature of denaturation, T_d, and with the partial specific enthalpy of transition ΔH_{trs}. On diluting the solutions they approach some definite limiting values which are very specific for given proteins under given conditions. For example, for rat skin collagen at pH 3·5 without salt $T_{trs} = 314$ K and $\Delta H_{trs} = 70$ J g^{-1} (see Fig. 10). Having in mind that the molecular weight of this macromolecule is $3·6 \times 10^5$ daltons, we obtain 26×10^5 J mol^{-1} for the molar denaturational enthalpy of rat skin collagen. This represents an extremely large amount of energy which is expended on the disruption of the native structure of this macromolecule.

In the native state collagen represents a hard, rod-like molecule, comprised of three polypeptide chains wound into a triple super-helix of the poly-L-proline type. As a result of denaturation, the collagen molecule breaks down into three polypeptide chains which are randomly coiled. Thus, the heat denaturation of collagen consists in the unwinding of the triple helix over a short temperature interval.

In considering the mechanism of stabilization of collagen structure it is convenient to carry out thermodynamic calculations per amino acid residue or per structural element which in the case of collagen consists of three residues. The average molecular weight of a residue of collagen is 91 daltons. Therefore the enthalpy of denaturation per residue of collagen is 6·3 kJ mol⁻¹ and thus 19 kJ mol⁻¹ per triplet. According to recent theoretical considerations, in a triplet of a collagen structure there should be no more than two hydrogen bonds (Ramachandran *et al.*,

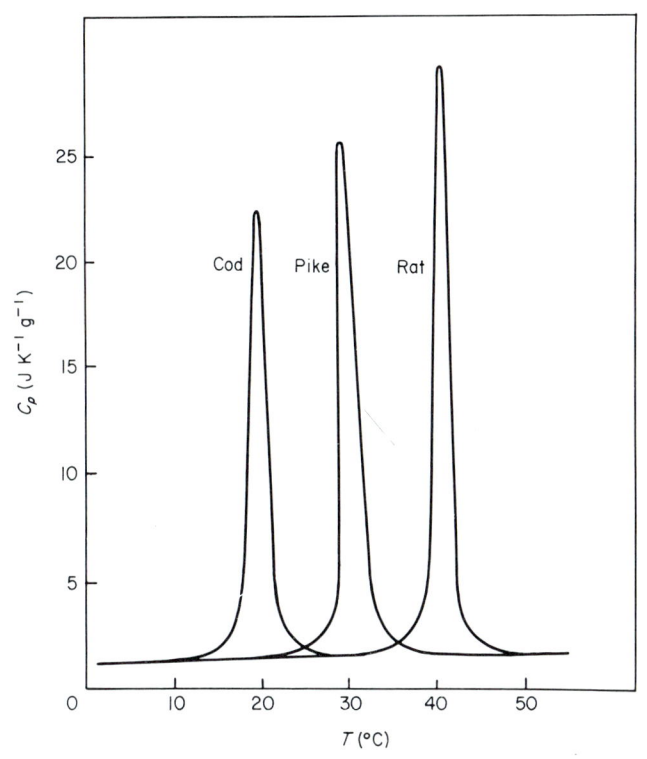

FIG. 10. Temperature dependence of partial specific heat capacities of collagens from different sources: pike skin; cod skin; rat skin.

1973). If we assume that collagen structure is stabilized only by hydrogen bonds, then the enthalpy of disruption of one hydrogen bond of collagen should be of the order of 10 kJ mol⁻¹. From the viewpoint of the water surroundings there should be a large compensation effect since hydrogen bonds will be created with water; this enthalpy change seems to be too large. But a much more serious discrepancy was discovered in another thermodynamic property of collagen. It was found that the denaturation enthalpy for collagens from different animals is different and the lower the enthalpy change the lower is the physiological temperature of the donor animal, although all the collagens have the same structure (Fig.

10). Simultaneously with the enthalpy decrease the stability of collagens decreases also in the sense that the entropy of denaturation decreases too (see Fig. 11). At the same time the enthalpy, entropy and temperature of transition are correlated with the imino acid content in the protein, although it is known that the imino acids cannot create hydrogen bonds

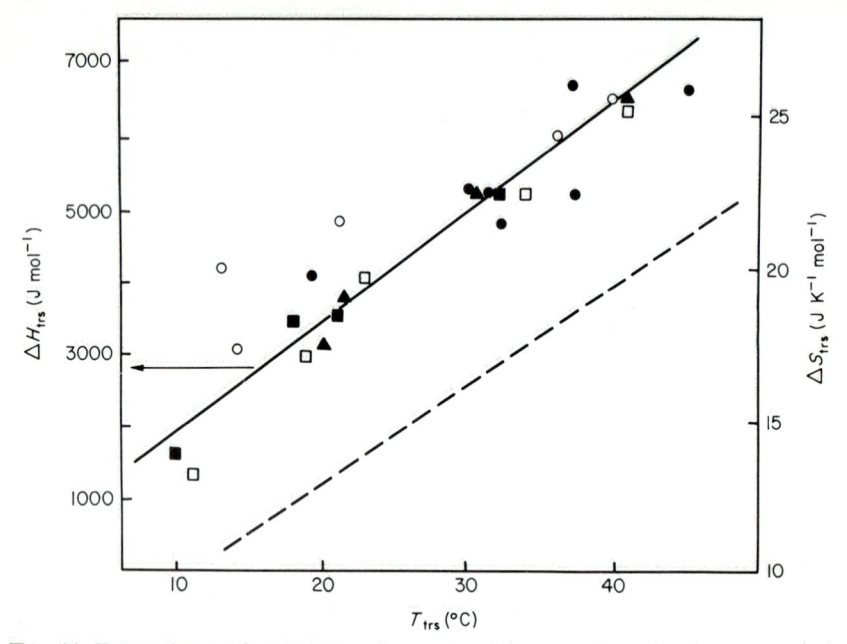

Fig. 11. Dependence of enthalpy and entropy of denaturation of collagens on their thermostability for collagens from different animals: ●, Privalov et al. (1979); ▲, Privalov and Tiktopulo (1970); ■, Burjanadze (1971); □, Monaselidze et al. (1973); ○, Menashi et al. (1976).

in the collagen helical structure. It follows that the collagen structure is stabilized by factors other than internal hydrogen bonds alone. Since the van der Waals and hydrophobic interactions could not contribute significantly in stabilizing collagen, it was concluded that a great role in the stabilization of collagen is played by the surrounding water which, mutually interacting with collagen, creates a supporting frame for it (see, for details, Privalov and Tiktopulo, 1970; Privalov et al., 1979).

Let us now consider the heat denaturation of globular proteins which represent a quite different structural class of proteins. In contrast to fibrillar collagen, they are compact macromolecules with a very complicated three-dimensional structure. Among biopolymers these proteins are, calorimetrically, the most intensively investigated (see Privalov,

1963; Beck et al., 1965; Jackson and Brandts, 1970; Tsong et al., 1970; Privalov and Khechinashvili, 1974).

Figure 12 shows the variation of the partial specific heat capacities for several globular proteins in solutions of different pH. One can see here the intensive heat sorption peaks corresponding to denaturation of

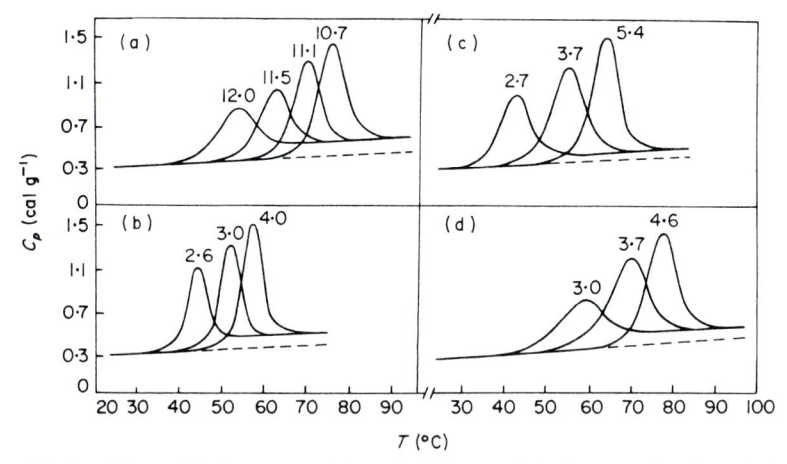

FIG. 12. Partial specific heat capacities of different globular proteins in solutions with different pH values: (a) metmyoglobin; (b) α-chymotrypsinogen; (c) ribonuclease A; (d) cytochrome c.

these macromolecules, which proceeds over different temperature regions depending on the pH. The most noticeable difference between the denaturation of the globular proteins and denaturation of the collagen is that they are much more stable and denature over a broader temperature interval. It is noticeable also that in contrast to the collagen here we have a very significant increase of the heat capacity of denaturation $\Delta_n^d C_p$. Thus the heat capacity of the denatured macromolecule C_p^d is much larger than is the heat capacity of the native one C_p^n. It can be noticed also that with the increase of thermostability the enthalpy of denaturation increases. Moreover, it was shown by Privalov and Khechinashvili (1974) that in all the cases we have the equality:

$$\frac{d\Delta H_{trs}}{dT} = \Delta_n^d C_p \qquad (5)$$

This equality means that enthalpy of denaturation is a direct function of temperature only and not of the pH which influences the stability of proteins.

The other remarkable feature of the globular protein melting curve is that the peak area correlates directly with its breadth, i.e. the greater is the denaturational enthalpy, the sharper is the transition. As a result, the height of the peak of heat sorption markedly increases with the increase in thermostability.

As is known, the sharpness of temperature-induced transitions is determined by the enthalpy of the transition. For a two-state transition this dependence is expressed by the van't Hoff equation:

$$\frac{1}{\phi(1-\phi)}\frac{d\phi}{dT}=\frac{\Delta H^{vH}}{RT^2} \quad (6)$$

where ϕ is any parameter characterizing the reaction progress. For calorimetric measurements ϕ is nothing other than the portion of heat absorbed for the given temperature $\phi = Q(T)/Q_d$, where Q_d is all the heat absorbed at transition. Thus,

$$\frac{d\phi}{dT}=\frac{1}{Q_d}\frac{dQ}{dT}=\frac{\Delta C_p}{Q_d} \quad (7)$$

i.e. the peak height normalized to peak area.

For the mid-point of transition T_{trs}, where $\phi = \frac{1}{2}$, we have

$$\Delta H^{vH}=\frac{4RT_{trs}^2}{Q_d}(\Delta C_p)_{trs} \quad (8)$$

Thus the maximal height of the heat sorption peak gives us the van't Hoff, or effective enthalpy, in contrast to the peak area, which gives the calorimetric or real enthalpy of transition. It was shown by Privalov and Khechinashvili (1974) that in the case of small compact globular proteins the van't Hoff enthalpy is very close to the calorimetric value, although it is not equal to it. Since the van't Hoff equation is applicable only to the "two-state" transitions, it follows from the coincidence of the van't Hoff and calorimetric enthalpy that the denaturation of small compact globular proteins could be considered, as a first approximation, as a two-state transition process, i.e. there are no stable intermediate states between the native and the denatured states. This is one of the most fundamental properties of compact globular proteins. It means that this macroscopic system behaves as one indivisible system or a single cooperative unit.

In the case of large proteins the situation may not be so simple. Thus, for the Bence-Jones protein with a molecular weight of 50 000 daltons, the effective or van't Hoff enthalpy is half that of the calorimetric enthalpy (Zavyalov et al., 1977). From this divergence it follows that the Bence-Jones protein should have two identical cooperative units. Since it is known that this protein has four structural domains, we can conclude that each cooperative unit consists of two domains.

Much more complicated is the case where the protein consists of cooperative regions with different stabilities. In this case we observe a very complicated melting curve of the type shown in Fig. 13 for L-meromyosin (Potekhin and Privalov, 1977). The same was found for fibrinogen by Donovan and Mihalyi (1974), and for myosin (Burjanadze et al., 1966).

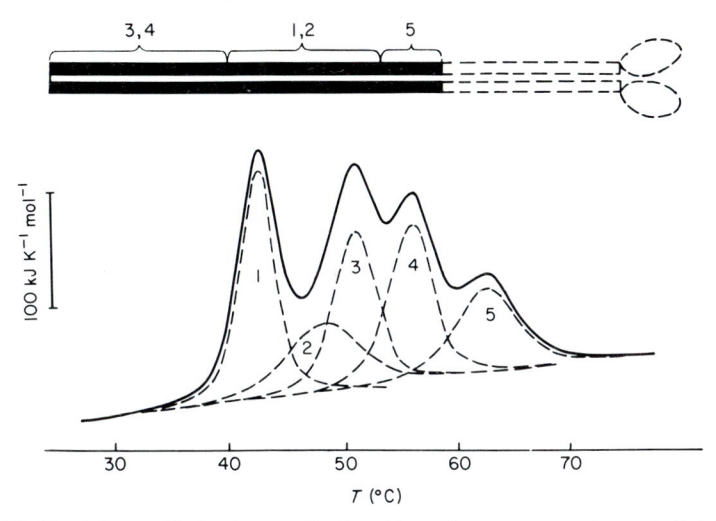

FIG. 13. Partial specific heat capacity function of L-meromyosin and distribution of corresponding cooperative regions in macromolecules (Potekhin and Privalov, 1977).

In so far as protein denaturation represents a transition between the two states—the native and the denatured—we can consider it as a phase transition. In the mid-point of the transition, T_{trs}, the Gibbs energy of both states should be equal, i.e. the Gibbs energy difference for these states should be zero:

$$\Delta_n^d G(T_{trs}) = \Delta_n^d H(T_{trs}) - T_{trs} \Delta_n^d S(T_{trs}) = 0 \tag{9}$$

From this equation for the entropy of denaturation at transition temperature we have

$$\Delta_n^d S(T_{trs}) = \frac{\Delta_n^d H(T_{trs})}{T_{trs}} \tag{10}$$

Thus calorimetrically we can obtain not only the enthalpy of denaturation, but the entropy of denaturation as well. And what is more, since we know the heat capacity change at denaturation $\Delta_n^d C_p$ we can derive the enthalpy and entropy of denaturation not only at the transition point, but over a much broader temperature region:

$$\Delta_n^d H(T) = \Delta_n^d H(T_{trs}) + \int_{T_{trs}}^{T} \Delta_n^d C_p \, dT \qquad (11)$$

$$\Delta_n^d S(T) = \Delta_n^d S(T_{trs}) + \int_{T_{trs}}^{T} \frac{\Delta_n^d C_p}{T} \, dT \qquad (12)$$

These functions for several globular proteins are presented in Fig. 14. It is remarkable that the specific enthalpies and entropies of denaturation

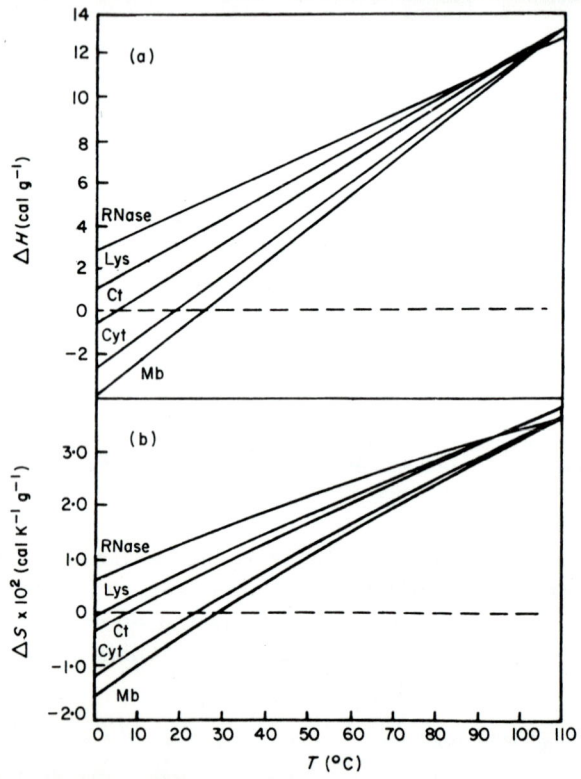

FIG. 14. Temperature dependence of enthalpies and entropies of denaturation for various globular proteins (Privalov and Khechinashvili, 1974).

of these very different proteins are not very different. Indeed, with increase of temperature this difference decreases, and it seems that their enthalpy and entropy values tend to reach some definite limits at 100 °C. This limiting value seems to be a characteristic for compact globular proteins, although the physical meaning of this limit is obscure. It might be that it reflects the fact that the native structures of all the globular proteins considered are equally compact and equally saturated by the secondary bonds maintaining it, particularly by the hydrogen bonds. The difference in contacts between the groups and their packing in the three-dimensional structures vanishes at elevated temperature. This suggestion was somewhat confirmed by the comparative structural analysis of the considered globular proteins. This revealed that the number of contacts between the non-polar groups in the native protein correlates with the slope of the enthalpy function. Thus this functional dependence seems to be directly connected with hydrophobic bonding in proteins (see, for details, Privalov and Khechinashvili, 1974).

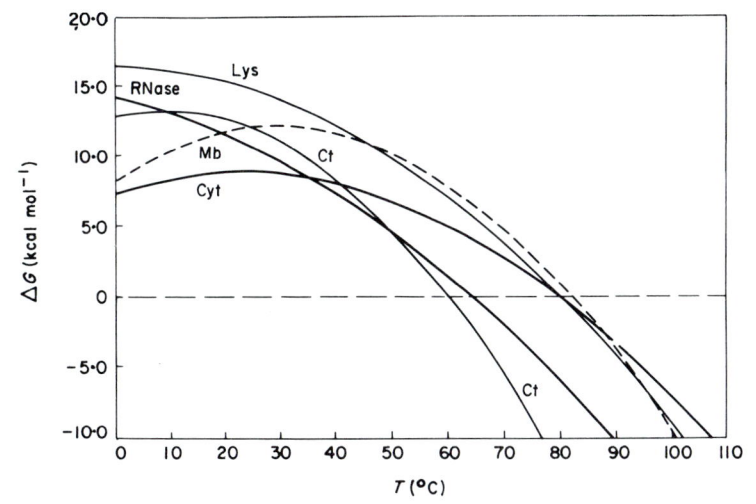

FIG. 15. Gibbs free energy of stabilization of the native structures of different globular proteins at various temperatures and at the pHs corresponding to their maximal stabilities (Privalov and Khechinashvili, 1974).

Combining the enthalpy and the entropy functions, we can obtain the Gibbs free-energy difference for the native and the denatured states directly from calorimetric measurements:

$$\Delta_n^d G(T) = \Delta_n^d H(T) - T\Delta_n^d S(T) = \Delta_n^d H(T)\left(1 - \frac{T}{T_{trs}}\right)$$
$$+ \int_{T_{trs}}^{T} \Delta_n^d C_p \, dT - T \int_{T_{trs}}^{T} \Delta_n^d C_p \, d\ln T \quad (13)$$

These functions are shown in Fig. 15 for several globular proteins in solutions at the pH corresponding to their maximal stability. But if we do the same calculations for a solution of different pH we obtain the

Gibbs energy difference as a function not only of the temperature but of the pH as well (Fig. 16) (see also Pfeil and Privalov, 1976). It is evident that this function is a direct characteristic of the stability of a native state, since it is just the work necessary to transform native protein into the denatured state. Thus we can describe it as the energy

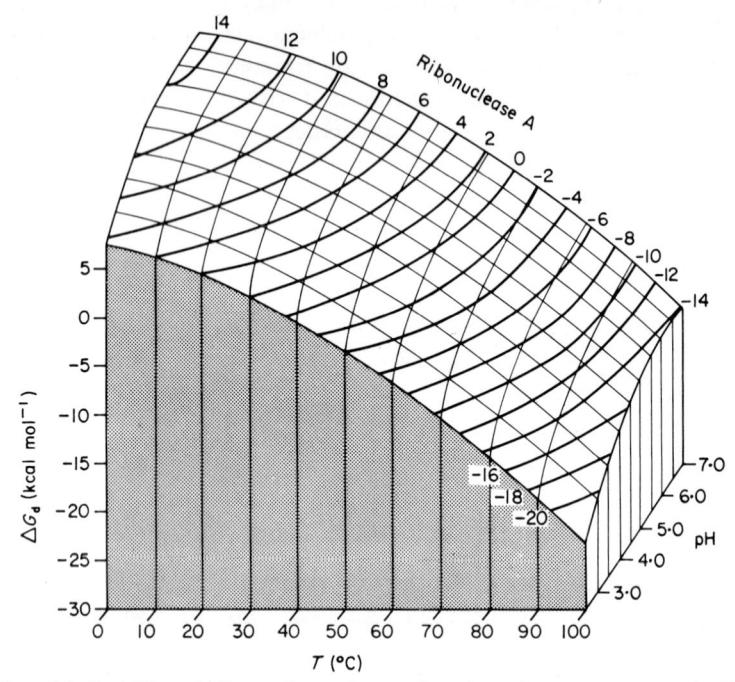

FIG. 16. Stability of ribonuclease A as a function of temperature and pH.

of stabilization of the protein structure. Three points attract attention in considering the energy of stabilization of globular proteins:

a. very different globular proteins have rather similar stabilities;
b. the stability reaches its maximal values at temperatures close to physiological;
c. even the maximal stability of globular proteins is not at all great, and what is more, calculated per link of the corresponding polypeptide chain (one residue), it is much less than the thermal energy per degree of freedom, i.e. $RT \simeq 2 \cdot 5$ kJ mol^{-1}.

Thus the protein structure is stable only because of some mechanism of cooperation between the states of the links constituting a polypeptide chain which integrates all the macroscopic system in one cooperative unit.

The mechanism of cooperativity of globular proteins is still obscure, but it is evident that this mechanism creates significant non-trivial thermodynamic properties of this macromolecule. Attempts to create artificial systems with such properties have been unsuccessful up to the present. This suggests that these properties are very specific to globular proteins and are necessary to provide some general biological requirements. We can only express some ideas on the necessity of these requirements:

a. For the effective and reliable functioning of protein, its structure must be well defined over a wide range of conditions. A system which may be damaged even by a slight action cannot be reliable. The reserve of protein stability must considerably exceed RT, ensuring a constant relative location of all its elements.

b. The system must not be too rigid. A certain mobility is necessary not only to ensure a relative displacement of the structural elements in the process of protein functioning, but also for its sufficiently easy dismantling without considerable energy expediture.

E. Melting of nucleic acids

The intramolecular melting of nucleic acids has been intensively studied by calorimetry since here a process of conformational transition plays a basic functional role and its study was necessary not only to clarify the mechanism of stabilization of the three-dimensional structure of these subjects, but also the mechanism of their functioning. From this point of view, the studies of the unfolding of the double helical structures of deoxyribonucleic and ribonucleic acids were the most important. We shall start this consideration from the simplest case, i.e. the melting of the synthetic model polyriboadenylic [poly A] and polyribouridylic [poly U] complex, since this system, in contrast to the natural one, is homogeneous and its melting process is simpler.

The melting of the poly A–poly U complex was studied calorimetrically by Danforth *et al.* (1967), and Neumann and Ackermann (1969). Figure 17 presents the calorimetric results for this double helix melting at two different concentrations. As is seen, the melting process takes place over a very narrow temperature interval with a half-width of less than 0·5 K. This circumstance allows measurement of the heat effect by scanning microcalorimetry in extremely dilute solutions down to a concentration of 0·3 mmol. But in very dilute solutions it is impossible to clarify details of the melting process, particularly the heat capacity change on melting which is very small, but distinct, as seen from the record at higher concentrations. The observed heat capacity change on melting poly A–poly U is 73 J K^{-1} mol^{-1}. From the change in heat capacity it follows that the melting enthalpy should be temperature dependent. Measuring the melting enthalpies at various temperatures

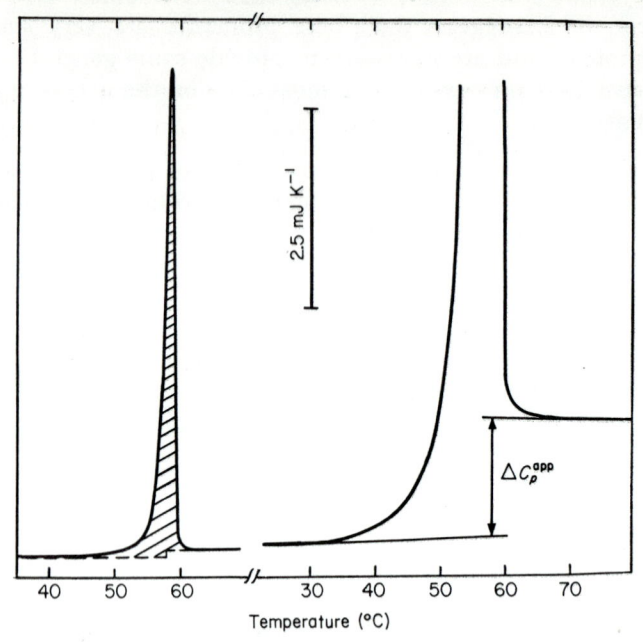

FIG. 17. Microcalorimetric recording of poly A–poly U melting at 0·3 mM concentration (left) and a fragment of recording at 5 mM in 0·1 M NaCl (right). The shaded area corresponds to the apparent melting enthalpy; arrow indicates the observed heat capacity change. (Filimonov and Privalov, 1977.)

and varying the stability of the complex by variation in the ionic strength of the solution, we find that it does indeed increase with the increase of the melting temperature (see Fig. 21) but its dependence on temperature ($[d\Delta_m H/dT] = 90$ J K^{-1} mol) greatly exceeds the observed $\Delta_m C_p^{app}$. This discrepancy cannot be due to the variation of ionic strength, since the addition of ions to nucleic acid solutions is a zero enthalpy process. The reason for this discrepancy is that after disruption of a complex the released poly A winds into a helical structure:

$$[\text{poly A–poly U}]^{hel} \rightleftarrows [\text{poly A}]^{coil} + [\text{poly U}]^{coil}$$

$$[\text{poly A}]^{hel}$$

The melting of poly A proceeds over a very broad temperature interval. From the broad heat sorption curve measured calorimetrically (see Fig. 19) we can derive corrections for the enthalpy and heat capacity change on melting of the complex (Filimonov and Privalov, 1977a).

It is evident that the correction for the enthalpy, δH, equals the area bounded by the heat sorption function of poly A from $T_m^{(A \cdot U)}$ to the end of the poly A melting. The correction for the heat capacity change δC_p at $T^{(A \cdot U)}$ equals the heat absorption intensity at this temperature. The corrected enthalpy for base interaction in the poly A–poly U complex is presented in Fig. 18 by the solid line. It can be seen that it

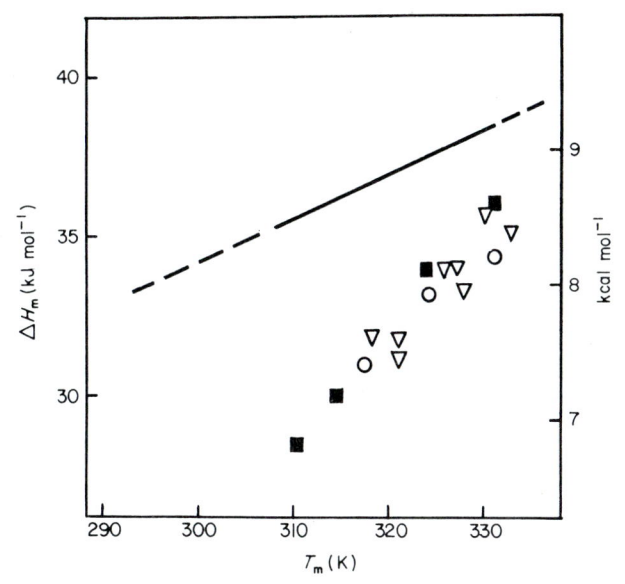

FIG. 18. Apparent molar enthalpy of poly A–poly U melting obtained by Krakauer and Sturtevant (1968) (○), Neumann and Ackermann (1969) (▽), Filimonov and Privalov (1977a) (■). The solid line corresponds to the partial molar enthalpy of base pairing.

is significantly larger than the apparent transition enthalpy at the corresponding temperature and much less dependent on the transition temperature $(dH^{(A \cdot U)}/dT) = 32$ J K^{-1} mol^{-1} whereas the corrected $\Delta C_p = 31$ J K^{-1} mol^{-1}.

Now having the corrected values of enthalpies at different ionic strengths and of the heat capacity change, we can obtain via equations (11), (12) and (13) the enthalpy, entropy and Gibbs energy functions of formation of the poly A–poly U complex from the random coiled polynucleotides, or the thermodynamic parameters for base pairing as a

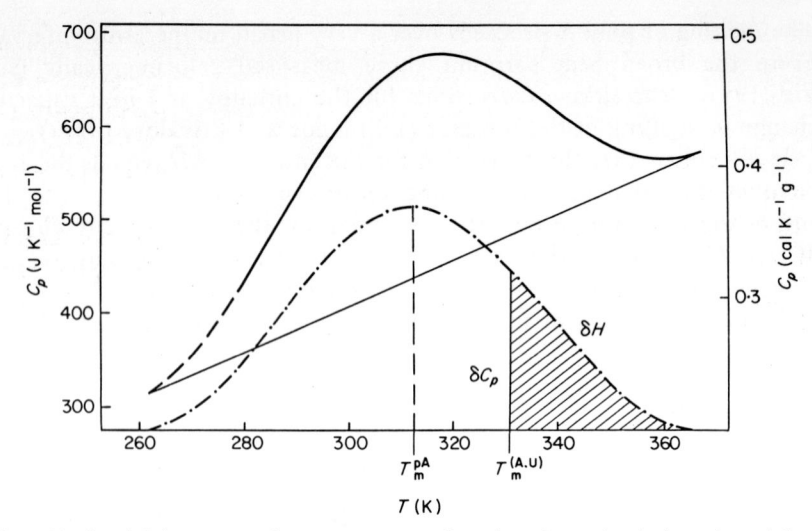

FIG. 19. Partial heat capacity temperature function for poly A (———) and heat sorption corresponding to the melting process (· – · – ·). Correction for the apparent melting enthalpy for poly A–poly U transition is hatched.

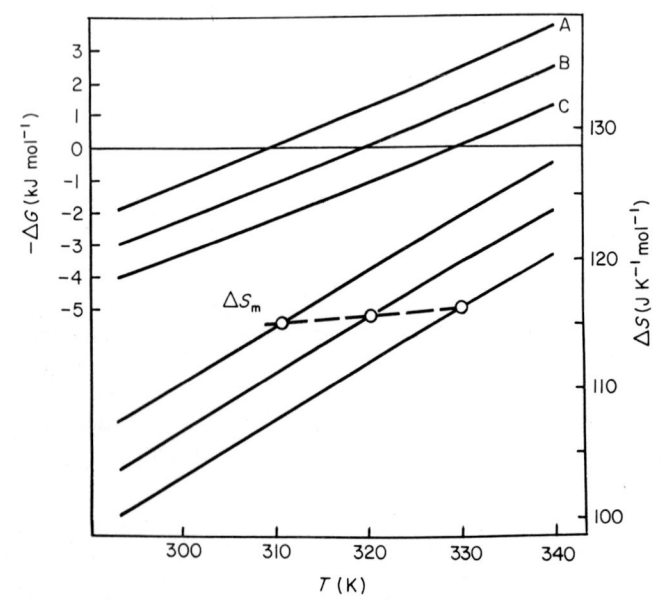

FIG. 20. Partial molar entropy and Gibbs energy of base pairing in the poly A–poly U complex in solutions with different NaCl content: A, 100 mM; B, 30 mM; C, 10 mM.

function of the temperature and the ionic strength. These are presented in Fig. 20. It can be seen that the Gibbs free energy of base interaction in the poly A–poly U complex at 298 K and 0·1 M NaCl is equal to 3·5 kJ mol^{-1} and decreases with decrease in the salt concentration. This decrease in the Gibbs free energy is entirely induced by the increase in the entropic term since the entropy of base pairing increases with decrease of ionic strength. It follows from this fact that the stabilizing action of ions is not connected with the screening of charges on double-helical rod-like molecules and decreasing electrostatic repulsive forces between the constitutive chains. This observed increase of the entropy could be explained if we recall that on the splitting of both chains the linear charge density decreases and the counterions are removed to the solution, leading to an entropy gain on mixing, proportional to the logarithm of the ionic activities in solution (see, for details, Privalov et al., 1969; Manning, 1972).

In contrast to synthetic homopolynucleotides, the melting curves of natural nucleic acids are very complicated, reflecting the great complexity of these structures which contain regions with different stabilities (see Bakradze et al., 1971, 1973; Bode et al., 1974; Brandts et al., 1974; Privalov et al., 1975; Filimonov et al., 1976; Hinz et al., 1977). The complexity of melting of the smallest RNA molecule, specific transfer RNA with a molecular weight of only 25 000 daltons, is illustrated in Fig. 21. It becomes clear that this curve represents the sum of several spread peaks and it has been shown by thermodynamic analysis that they are the sequences of quite independent consecutive transitions (see Fig. 22), corresponding to the unfolding of particular parts of this macromolecule (Privalov and Filimonov, 1978). For such a rigid and compact macromolecule as tRNA this conclusion might seem quite surprising, but it is not unexpected since in the first stage of melting tRNA disrupts its tertiary structure and this compact molecule unfolds into a topologically flat clover-leaf with independent helical segments. The binding of magnesium ion strengthens the tertiary structure and thus locks all the molecule. The action of magnesium is only of an entropic nature, since it is not accompanied by any heat effect, notwithstanding the fact that it has the largest stabilizing effect (Rialdi et al., 1972; Privalov et al., 1975; Hinz et al., 1977).

Knowing the structure of the macromolecule, we can identify the individual transitions of the melting curve with the unfolding of particular structural elements and even derive the thermodynamic parameters of base pairing. Thus it was found that the enthalpy of (A·U) base pairing is 38 kJ mol^{-1} and that of (G·C) base pairing 60 kJ mol^{-1} (Privalov and Filimonov, 1978).

If the structure of a macromolecule is not known, we can draw some general conclusions concerning its organization from the calorimetric data. For example, at present we do not know the three-dimensional structure of 5S ribosomal RNA, but judging from its melting enthalpy (see Table 3), which is similar to that of tRNA, we can conclude that this macromolecule is similarly compact. At the same time, since its

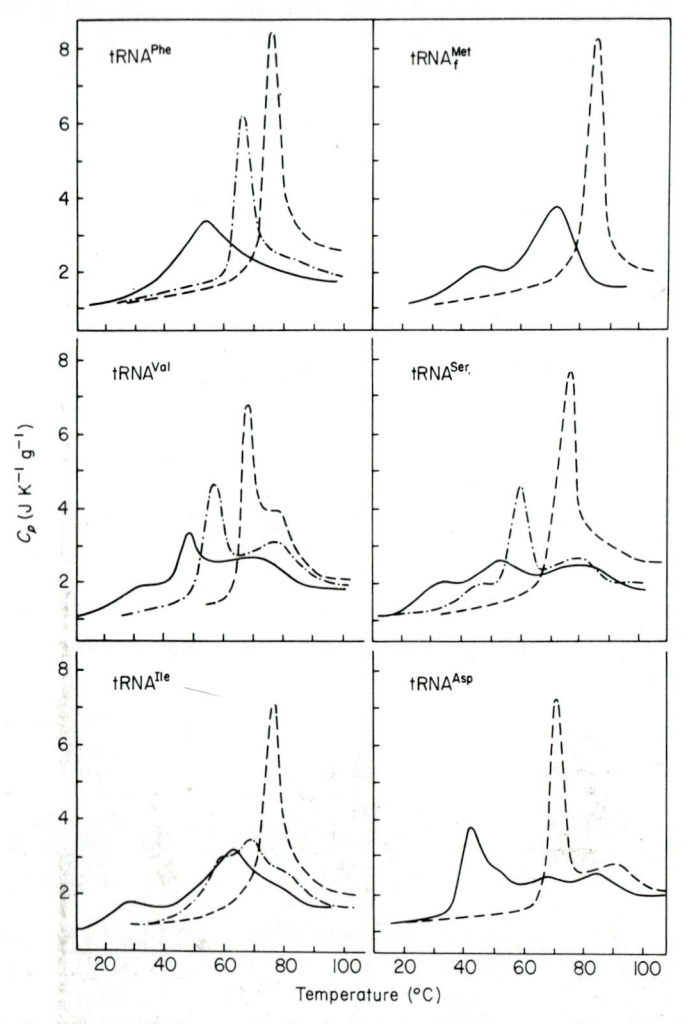

FIG. 21. Partial specific heat capacity functions of various specific transfer RNAs at different salt concentrations: ----, 1 mM MgCl$_2$; ———, 150 mM NaCl; -·-·-, 150 mM NaCl and 1 mM MgCl$_2$. (Privalov and Filimonov, 1978.)

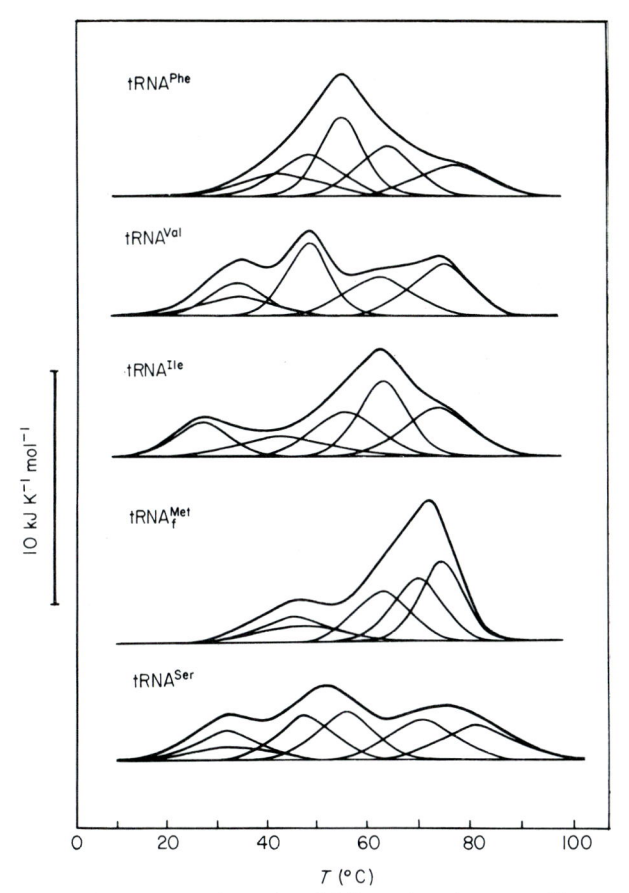

Fig. 22. Sequence of heat sorption of unfolding of various specific transfer RNAs in 150 mM NaCl solutions.

melting curve is much simpler (see Fig. 23), it should be constituted from less structural blocks.

The case of 16S ribosomal RNA, which is ten-fold larger than the 5S RNA, approaching 5×10^5 daltons, is much more complicated. Its melting curve is presented in Fig. 24. Several peaks are clearly distinguishable above the very spread heat sorption background. The sharpness of these peaks suggests that here we have a melting of several large cooperative blocks including about 20–30 base-pairs (Matveev and Privalov, 1979). This conclusion is in accord with the known sequence of this macromolecule (Ehresmann et al., 1975).

With increase in macromolecular size the possibility of obtaining any valuable structural information from the shape of melting curves

TABLE 3
Specific enthalpy and averaged temperature of melting of nucleic acids

Nucleic acid	T_m K	ΔH_m J g^{-1}	Reference
poly A	313	38 ⎫	Filimonov and Privalov (1977a)
poly A–poly U	331	60 ⎭	
poly$_d$ AT–poly$_d$ AT	346	30	Andronikashvili et al. (1974)
poly A–poly T	—	45	Scheffler and Sturtevant (1963)
tRNA^{Val}	323	52	Privalov et al. (1975)
tRNA^{Phe}	328	51	Hinz et al. (1977)
tRNA^{Ile}	333	51	Filimonov et al. (1976)
tRNA$_f^{Met}$	343	54	Filimonov et al. (1976)
tRNA^{Ser}	328	48	Privalov and Filimonov (1977b)
5S RNA	323	50 ⎫	Matveev and Privalov (1979)
16S RNA	323	50 ⎭	
DNA of phage T$_2$	354	54	Privalov et al. (1969)
Salmon DNA	—	48	Sturtevant et al. (1958)
DNA *Cl. perfrigens*	328	45 ⎫	
DNA *M. lysodeicticus*	352	50 ⎬	Klump and Ackermann (1971)
Salmon DNA	338	46 ⎭	

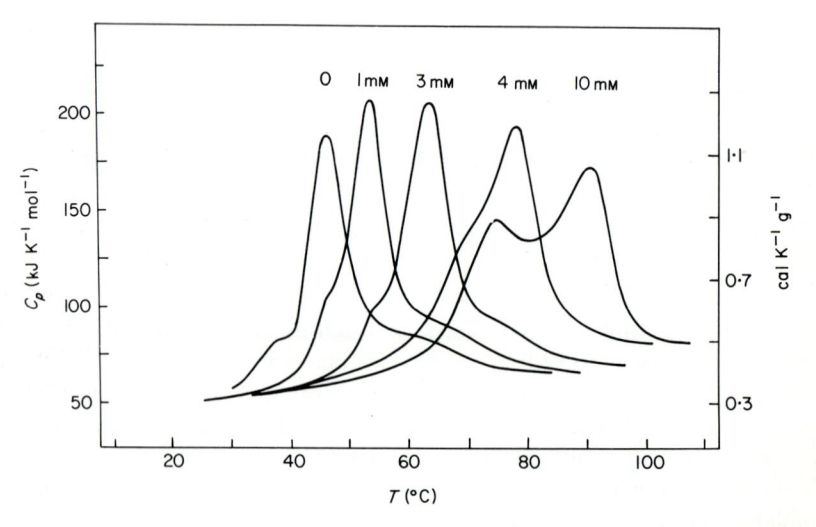

FIG. 23. Partial specific heat capacity function of ribosomal 5S RNA in solutions of MgCl$_2$ content at the concentrations shown: 0, 1, 3, 4 and 10 mM MgCl$_2$ respectively.

evidently decreases, since the shape of the melting curves becomes less expressive and its interpretation more uncertain (see Klump and Ackermann, 1971; Lyubchenko et al., 1976). But although we cannot draw any positive structural conclusions from these studies, nevertheless

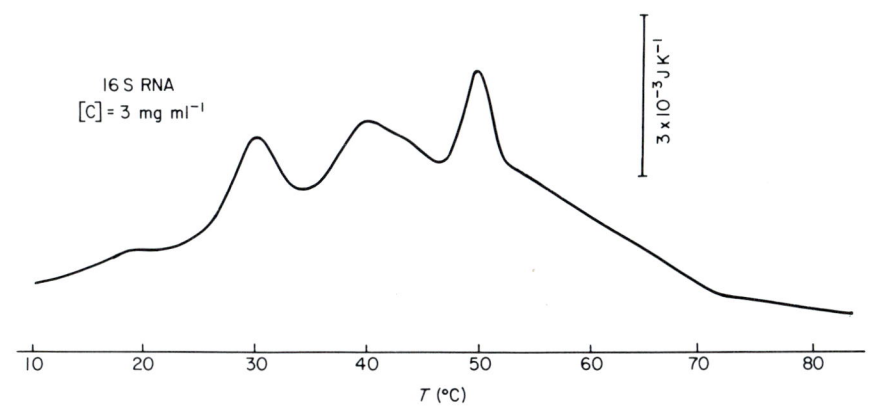

FIG. 24. Microcalorimetric recording on heating 16S ribosomal RNA in 5 mM phosphate buffer (Matveev and Privalov, 1979).

they are important for a general thermodynamic study of base interaction in the DNA double helix. For example, from comparison of the melting enthalpies of DNA with different average $(_dG \cdot _dC)$ content, values of $(_dA \cdot _dT)$ and $(_dG \cdot _dC)$ base pairing were deduced. According to Klump and Ackermann (1971) the enthalpy of $(_dA \cdot _dT)$ pairing is 32 kJ mol^{-1}, whereas that of $(_dG \cdot _dC)$ pairing is 36 kJ mol^{-1}. To compare different nucleic acid structures, it is convenient to consider their melting enthalpies in a specific basis and not the molar basis. The data obtained by different authors are presented in Table 3. The melting temperatures indicated here are approximate, since in most cases the melting of nucleic acids proceeds over too broad a temperature region and is too complicated to be averaged correctly. It is seen from this Table that the specific melting enthalpies of all RNAs are very close to 50 J g^{-1}. Thus they should all be equally saturated by the secondary bonds maintaining their structure and thus they are not as loose as was assumed before.

F. Investigations of membrane melting

At present the melting of biological membranes increasingly attracts attention of biologists and physicists. Physicists recognize that this phenomenon is one of the most vivid examples of phase transitions in two-dimensional systems. The interest of biologists is based on the belief that this phenomenon is of great biological importance. Calorimetry is one of the few physical methods by which it is possible to observe transitions in membranes without introducing additional disturbances into the systems (by labels, for example). Moreover, the observed transition can be described quantitatively in thermodynamic terms and this is important for clarifying the physical nature of this phenomenon.

Although there has already been much calorimetric work on membranes (see Williams and Chapman, 1970; Melchior *et al.*, 1970; Oldfield and Chapman, 1972; Blume and Ackermann, 1974; Chapman *et al.*, 1974), quantitative calorimetric studies of this subject were started only recently after the appearance of precise scanning microcalorimeters permitting not only an increase in the resolution of the measurements, but also a decrease in the required concentrations of phospholipids in solution (Mabrey and Sturtevant, 1976; Kantor *et al.*, 1977). Using precise instruments, it is possible to observe not only transitions of membranes, but even the denaturation of the proteins which are included in the membrane (Jackson *et al.*, 1973). The calorimetric studies of phase transitions in biological membranes are considered in detail by Chapman (p. 275).

G. Technique for calorimetric studies of biological objects

1. HEAT CAPACITY MEASUREMENTS OF ANHYDROUS AND HYDRATED BIOPOLYMERS AND TISSUES OVER A BROAD TEMPERATURE RANGE

Anhydrous and hydrated biopolymers and tissues are amorphous objects with very bad heat conductivity. For their heat capacity measurements over a broad temperature range, absolute adiabatic calorimeters of a type described by Westrum (1962) for low temperatures can be used. But one must have in mind that for biological studies it is desirable to have a calorimetric cell with as small a volume as possible and the best of adiabatic control (see Privalov and Mrevlishvili, 1966; Mrevlishvili, 1977). The requirement as to the volume follows from the shortage of homogeneous high-quality biological preparations and their high cost. As for the requirement for adiabaticity, it arises not only from the unfavourable surface–volume relation for small cells which increases the error in heat exchange, but also from an extremely slow relaxation of studied systems. In some temperature regions, particularly in those where the fusion of ice proceeds, the thermal relaxation of a water–biological system is so extended that it is practically impossible to achieve full equilibration and thus to determine the heat exchange with the surroundings. This is why it is desirable to have a calorimeter with minimal and constant heat exchange properties which do not vary from experiment to experiment. If a calorimeter has a small permanent heat-exchange parameter, it can be determined once (measuring a substance with a fast relaxation) and then used for correcting the heat exchange parameters when measuring studied objects. In such cases on measuring heat capacities there is no need to worry about the temperature drift following heating. It is sufficient to register only one temperature corresponding to some definite time equilibration.

After the appearance of the commercial scanning calorimeters of the Perkin–Elmer DSC type, such instruments have been used for studying biological objects. The undoubted advantage of these instruments is the very small expenditure on material coupled with speed of instrumental recording. The speed also is a great disadvantage in this instrument on studying biological objects, since on rapid continuous heating the distortion induced by slow relaxation becomes too large. The other shortfall of this instrument is the instability of the base line which does not permit measurement of the absolute values of heat capacities, but only of the sharp heat sorption on heating. From this point of view the scanning adiabatic calorimeters for solids of the type described by Bakradze and Monaselidze (1971) are much more useful.

2. HEAT CAPACITY MEASUREMENTS AT FIXED TEMPERATURES

For measuring absolute heat capacities of a small amount of an object at some fixed temperature, the drop calorimeter designed as a micro-modification by Suurkuusk and Wadsö (1974) can be used. Their method consists of dropping an ampoule containing the object into the isothermal calorimeter; preliminary heating of the ampoule occurs in the thermostat which is located above the calorimeter. Although this method is somewhat labour-consuming, it gives the absolute heat capacity of objects with good precision requiring only small amounts of material (100 mg) and, what is important, without respect to their aggregated state.

3. HEAT CAPACITY MEASUREMENTS OF DILUTE SOLUTIONS IN A BROAD TEMPERATURE RANGE

Heat capacity studies of diluted solutions of biopolymers require very high precision of measurements to derive the partial heat capacities of the components with good accuracy. The highest precision is achieved, at present, on scanning adiabatic differential microcalorimeters (see Privalov et al., 1964, 1975). The general outline of this instrument is given in Fig. 25. Here the measuring and reference calorimetric cells

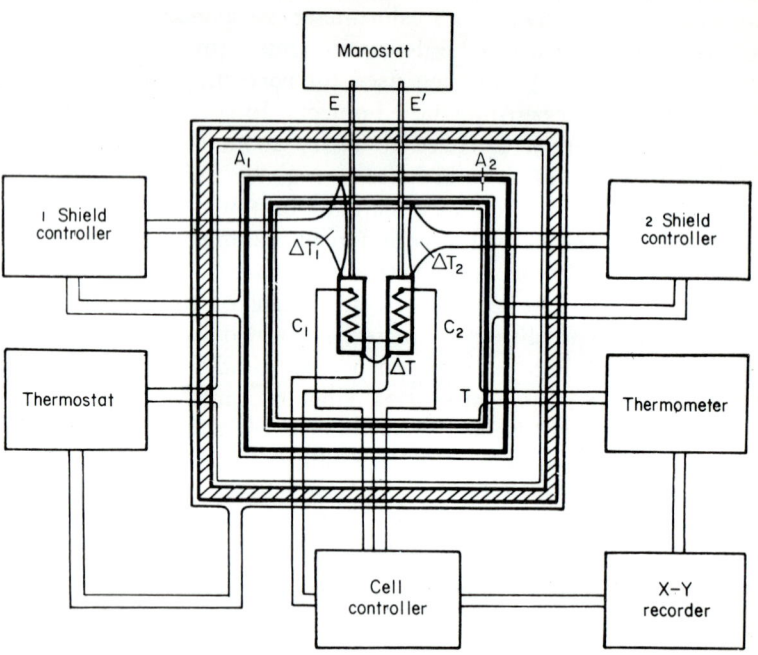

FIG. 25. Adiabatic differential scanning microcalorimeter. A_1 and A_2 are external and internal adiabatic shields with thermal sensors ΔT_1 and ΔT_2 and a thermometer T. C_1 and C_2 are measuring and reference calorimetric cells with a thermal sensor ΔT and electric heaters. E and E' are the capillary inlets.

C and C' are surrounded by a double adiabatic shield A_1 and A_2. Both calorimetric cells are heated by the same electric power and the adiabatic shields follow them in temperature, being adjusted by the shield controllers. With an increase of heat capacity in one of the cells, the cell controller monitors heating power in such a manner that no temperature difference occurs between the cells. Compensating power is recorded as a function of temperature and is nothing other than the relative heat capacity function.

One of the most serious problems in precise heat-capacity measurements is the determination of the amount of a sample. To fulfil present requirements in the precision of a heat capacity measurement (10^{-5} J K^{-1}), the accuracy of filling the cell should be of the order of 0·002 per cent. This can be achieved only by filling cells not by weight but by volume, using cells of standard volume and filling them completely. To guarantee filling and to prevent degassing during heating, a constant external pressure is applied to the liquid after filling the cell.

The precision scanning microcalorimeter of the type described is now produced by the Academy of Sciences of the USSR under the trade mark DASM–IM. Its main characteristics are the following:

Operational temperature range	270–390 K
Rate of heating	0·1–2·0 K min^{-1}
Operational volume	1 ml
Sensitivity	1×10^{-5} J K^{-1}
Precision of the relative heat capacity determination	8×10^{-5} J K^{-1} ml^{-1}
Precision of temperature recording	0·1 K

The sensitivity and precision of this instrument permit measurement not only of the heat effects of intramolecular transitions using very small amounts of material, but even permit determination over a broad temperature region of the partial heat capacities of the components in very dilute solution. This facility is most important in thermodynamic studies. This can be done by recording the deviation of the observed curve for the solution from the base line obtained on the solvent alone (see Fig. 26), i.e. by the differences of the heat capacities of the same

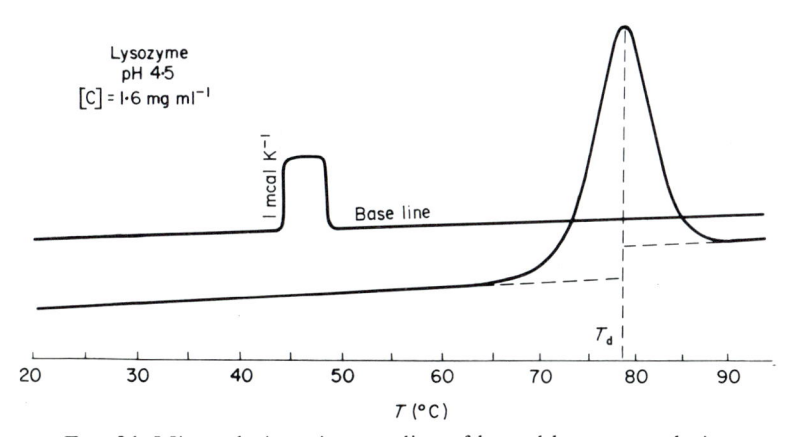

FIG. 26. Microcalorimetric recording of heated lysozyme solution.

volume of solvent and solution. It can be easily seen that at any temperature this deviation is equal to

$$\Delta C_p = C_{p,\,A} \, \Delta m_A - C_{p,\,B} \, m_B \tag{14}$$

where the indices A and B denote the solute and solvent, m_B is the mass of the studied component in the cell and Δm_A is the mass of the displaced solvent. Since

$$\Delta m_A = m_B \frac{V_B}{V_A} \tag{15}$$

where V_A and V_B are the partial specific volumes of the solvent and solute, we have

$$C_{p,\,B} = C_{p,\,A} \frac{V_B}{V_A} - \frac{\Delta C_p}{m_B} \tag{16}$$

Thus, if we know the relation of densities of the solute and solvent we can determine by the scanning calorimeter the partial heat capacity of the solute over all the experimental temperature region (see, for details, Privalov and Khechinashvili, 1974).

References

ANDRONIKASHVILI, E. L. and MREVLISHVILI, G. M. (1975). *In* "L'Eau et les Systèmes Biologiques", Colloques Internationaux du CNRS, No. 246, Roscoff, 1975, pp. 275–282.

ANDRONIKASHVILI, E. L., MREVLISHVILI, G. M. and PRIVALOV, P. L. (1966a). *In* "Biofizika Myshechnogo Sokrashcheniya" (G. M. Frank, ed.), pp. 224–228. Nauka Publishing House, Moscow.

ANDRONIKASHVILI, E. L., MREVLISHVILI, G. M. and PRIVALOV, P. L. (1966b). *Dokl. Akad. Nauk SSSR*, **171**, 1198–1200.

ANDRONIKASHVILI, E. L., MREVLISHVILI, G. M. and PRIVALOV, P. L. (1969). *In* "Water in Biological Systems" (L. P. Kayushin, ed.), pp. 63–66. Consultants Bureau, New York.

ANDRONIKASHVILI, E. L., MONASELIDZE, D. R., BAKRADZE, N. G., CHANCHALASHVILI, Z. I., MGELADZE, G. M., MIKADZE, E. L. and MREVLISHVILI, G. M. (1973). *In* "Konformatsionnie Izmeneniya Biopolymerov v Rastvorakh" (E. L. Andronikashvili, ed.), pp. 171–176. Nauka Publishing House, Moscow.

ANDRONIKASHVILI, E. L., MGELADZE, G. M., MONASELIDZE, D. R. and LEZIUS, A. G. (1974). *Biopolymers*, **13**, 1751–1756.

ANDRONIKASHVILI, E. L., MREVLISHVILI, G. M., JAPARIDZE, G. SH., SOKHADZE, V. M. and KVAVADZE, K. A. (1976). *Biopolymers*, **15**, 1991–2004.

BAKRADZE, N. G. and MONASELIDZE, D. R. (1971). *Izmeritel'naya tekhnika*, **2**, 58–60.

BAKRADZE, N. G., MONASELIDZE, D. R., BIBIKOVA, A. D. and KISSELEV, L. L. (1971). *Biochim. biophys. Acta*, **238**, 161–163.

BAKRADZE, N. G., MONASELIDZE, D. R., MREVLISHVILI, G. M., BIBIKOVA, A. D. and KISSELEV, L. L. (1973). *In* "Konformatsionniye Izmeneniya Biopolymerov v Rastvorakh" (E. L. Andronikashvili, ed.), pp. 82–86. Nauka Publishing House, Moscow.

BECK, K., GILL, S. J. and DOWNING, M. (1965). *J. Am. chem. Soc.* **87**, 901–904.
BLUME, A. and ACKERMANN, TH. (1974). *FEBS Lett.* **43**, 71–74.
BODE, D., SCHERNAU, H. and ACKERMANN, TH. (1974). *Biophys. Chem.* **1**, 214–221.
BRANDTS, J. F., JACKSON, W. M. and YAO-CHUNG TING, . (1974). *Biochemistry*, **13**, 3595–3600.
BURJANADZE, T. V. (1971). *Trudy Tbilisskogo Universiteta*, **139**, 17–23.
BURJANADZE, T. V., VEPKHVADZE, L. K., KIZIRIA, E. L., MONASELIDZE, D. R., PRIVALOV, P. L. and CHARKVIANI, G. G. (1966). *In* "Biofizika Myshechnogo Sokrashchenia" (G. M. Frank, ed.), pp. 218–223. Nauka Publishing House, Moscow.
CHAPMAN, D., URBINA, J. and KEQUGH, K. M. (1974). *J. biol. Chem.* **249**, 2512–2521.
CHATTORAJ, D. K. and BULL, H. B. (1971). *J. Col. Interf. Sci.* **35**, 220–226.
DANFORTH, R., KRAKAUER, H. and STURTEVANT, J. M. (1967). *Rev. scient. Instrum.* **38**, 484–487.
DONOVAN, J. W. and MIHALYI, E. (1973). *Proc. natn. Acad. Sci. U.S.A.* **71**, 4125–4128.
EHRESMANN, C., STIEGLER, P., FELLNER, P. and EBEL, J.-P. (1975). *Biochimie*, **57**, 711–748.
ESIPOVA, N. G. and CHIRGADZE, YU. N. (1969). *In* "Water in Biological Systems" (L. P. Kayushin, ed.), pp. 42–50. Consultants Bureau, New York.
FILIMONOV, V. V. and PRIVALOV, P. L. (1978). *J. molec. Biol.* **122**, 465–470.
FILIMONOV, V. V., PRIVALOV, P. L., HINZ, H.-J., VON DER HAAR, F. and CRAMER, F. (1976). *Eur. J. Biochem.* **70**, 25–31.
HALY, A. R. and SNAITH, J. W. (1968). *Biopolymers*, **6**, 1355–1377.
HALY, A. R. and SNAITH, J. W. (1971). *Biopolymers*, **10**, 1681–1699.
HINZ, H.-J., FILIMONOV, V. V. and PRIVALOV, P. L. (1977). *Eur. J. Biochem.* **72**, 79–86.
HUTCHENS, J. O., COLE, A. G. and STOUT, J. W. (1969). *J.biol. Chem.* **244**, 25–32.
JACKSON, W. M. and BRANDTS, J. F. (1970). *Biochemistry*, **9**, 2294–2301.
JACKSON, W. M., KOSTYLA, J., NORDIN, J. H. and BRANDTS, J. F. (1973). *Biochemistry*, **12**, 3662–3667.
KANTOR, H. L., MABREY, S., PRESTEGARD, J. H. and STURTEVANT, J. M. (1977). *Biochim. biophys. Acta*, **466**, 402–410.
KLOTZ, I. (1962). *In* "Horizons in Biochemistry" (M. Kasha and B. Pullman, eds), pp. 523–550. Academic Press, London and New York.
KLUMP, H. and ACKERMANN, TH. (1971). *Biopolymers*, **10**, 513–522.
KRAKAUER, H. and STURTEVANT, J. M. (1968). *Biopolymers*, **6**, 491–512.
LYUBCHENKO, J. L., FRANK-KAMENETSKYI, M. D., VOLOGODSKYI, A. V., LAZURKIN, YU. S. and GAUSE, G. G. (1976). *Biopolymers*, **15**, 1019–1036.
MABREY, S. and STURTEVANT, J. M. (1976). *Proc. natn. Acad. Sci. U.S.A.* **73**, 3862–3866.
MANNING, J. (1972). *Biopolymers*, **11**, 937–949.
MATVEEV, S. V. and PRIVALOV, P. L. (1979). *Dokl. Akad. Nauk SSSR*, **247**, 985–989.

MELCHIOR, D. L., MOROWITZ, N. J., STURTEVANT, J. M. and TSONG, Y. J. (1970). *Biochim. biophys. Acta*, **219**, 114–122.
MENASHI, S., FINCH, A., GARDNER, P. I. and LEDWARD, D. A. (1976). *Biochim. biophys. Acta*, **444**, 623–625.
MONASELIDZE, D. R. and BAKRADZE, N. G. (1969). *Dokl. Akad. Nauk SSSR*, **189**, 899–901.
MONASELIDZE, D. R., MAJAGALADZE, G. V., BAKRADZE, N. G., CHANCHALASHVILI, Z. I., MZHAVANADZE, A. V. and MARUDENKO, A. I. (1973). *In* "Konformatsionnie Izmeneniya Biopolymerov v Rastvorakh" (E. L. Andronikashvili, ed.), pp. 176–180. Nauka Publishing House, Moscow.
MREVLISHVILI, G. M. (1977). *Biofizika*, **22**, 180–191.
MREVLISHVILI, G. M. (1977). Personal communication.
MREVLISHVILI, G. M. and PRIVALOV, P. L. (1969). *In* "Water in Biological Systems" (L. P. Kayushin, ed.), pp. 63–66. Consultants Bureau, New York.
NEUMANN, E. and ACKERMANN, TH. (1969). *J. Phys. Chem.* **73**, 2170–2178.
OLDFIELD, E. and CHAPMAN, D. (1972). *FEBS Lett.* **23**, 285–297.
PFEIL, W. and PRIVALOV, P. L. (1976). I, II, III, *Biophys. Chem.* **4**, 23–50.
POTEKHIN, S. A. and PRIVALOV, P. L. (1979). *Biofizika*, **24**, 46–50.
PRIVALOV, P. L. (1963). *Biofizika*, **8**, 308–316.
PRIVALOV, P. L. (1968). *Biofizika*, **13**, 163–177.
PRIVALOV, P. L. and FILIMONOV, V. V. (1978). *J. molec. Biol.* **122**, 447–464.
PRIVALOV, P. L. and KHECHINASHVILI, N. N. (1974). *J. molec. Biol.* **86**, 665–684.
PRIVALOV, P. L. and MREVLISHVILI, G. M. (1966). *Biofizika*, **11**, 951–955.
PRIVALOV, P. L. and MREVLISHVILI, G. M. (1967). *Biofizika*, **12**, 22–29.
PRIVALOV, P. L. and TIKTOPULO, E. I. (1970). *Biopolymers*, **9**, 127–139.
PRIVALOV, P. L., MONASELIDZE, D. R., MREVLISHVILI, G. M. and MAGALDADZE, V. A. (1964). *Zh. Eksper. Teoret. Fiz.* (*USSR*), **74**, 2073–2079.
PRIVALOV, P. L., KAFIANI, K. A., MONASELIDZE, D. R., MREVLISHVILI, G. M. and MAGALDADZE, V. A. (1966). *In* "Nukleinovie Kisloti" (V. N. Orekhovich, ed.), pp. 15–20. Medizina Publishing House, Moscow.
PRIVALOV, P. L., PTITSYN, O. B. and BIRSTEIN, T. M. (1969). *Biopolymers*, **8**, 559–571.
PRIVALOV, P. L., PLOTNIKOV, V. V. and FILIMONOV, V. V. (1975a). *J. Chem. Thermodynamics*, **7**, 41–47.
PRIVALOV, P. L., FILIMONOV, V. V., VENKSTERN, T. V. and BAYEV, A. A. (1975b). *J. molec. Biol.* **97**, 279–288.
PRIVALOV, P. L., TIKTOPULO, E. I. and TISCHENKO, V. M. (1979). *J. molec. Biol.* **127**, 203–216.
RAMACHANDRAN, G. N., BANSEL, M. and BHATNAGAR, R. S. (1973). *Biochim. biophys. Acta*, **322**, 166–171.
RIALDI, G., LEVY, J. and BILTONEN, R. (1972). *Biochemistry*, **11**, 2472–2479.
RÜEGG, M., MOOR, U. and BLANC, B. (1975). *Biochim. biophys. Acta*, **400**, 334–342.
SCHEFFLER, I. E. and STURTEVANT, J. M. (1969). *J. molec. Biol.* **42**, 577–580.
STURTEVANT, J. M., RICE, S. and GEIDUSHEK, E. P. (1958). *Disc. Farad. Soc.* **25**, 138–149.

Suurkuusk, J. (1974). *Acta chem. scand.* **B28**, 409–417.
Suurkuusk, J. and Wadsö, I. (1974). *J. chem. Thermodynamics*, **6**, 687–679.
Szent-Gyorgyi, A. (1956). *Science*, **124**, 873–875.
Szent-Gyorgyi, A. (1957). "Bioenergetics". Academic Press, London and New York.
Tsong, T. J., Hearn, R. P., Wrathall, D. P. and Sturtevant, J. M. (1970). *Biochemistry*, **9**, 2666–2677.
Westrum, E. F. (1962). *J. chem. Educat.* **39**, 443–456.
Williams, R. M. and Chapman, D. (1970). *In* "Progress. in Chemical Fats and other Lipids", vol. II, pt 1, pp. 1–79.
Zavyalov, V. S., Troitskyi, G. V., Khechinashvili, N. N. and Privalov, P. L. (1977). *Biochim. biophys. Acta*, **492**, 102–111.

Applications of Continuous Titration Isoperibol and Isothermal Calorimetry to Biological Problems

L. D. HANSEN, T. E. JENSEN
and
D. J. EATOUGH

A. Introduction

The scope of this contribution is limited to a review of continuous titration isoperibol calorimetry and of isothermal calorimetry. Further, rather than being exhaustive, this review is intended to be selective in that only those applications are presented where these two methods show definite advantages over other calorimetric methods. The examples of such applications have been chosen to expand on this theme as much as possible while limiting the choices to those which are of real interest to biology and biochemistry.

Initial applications of thermometric titrimetry were based on the analysis of the thermogram, a plot of temperature versus time, for analytical endpoints. The development of titration isoperibol and isothermal calorimeters capable of measuring heat changes accurately has made possible the determination of additional information from the thermogram. The additional information which can be obtained includes reaction enthalpies, stoichiometries, and sometimes equilibrium constants or kinetics for the reaction(s) of interest. The accuracy, reliability and rapidity of data acquisition has been amply shown by the many chemical reactions that have been studied by the continuous titration method (Tyrrell and Beezer, 1968; Bark and Bark, 1969; Vaughn, 1973; Hansen et al., 1974;

Barthel, 1974). These have included many simple model reactions such as proton ionization (Christensen and Izatt, 1962; Christensen et al., 1967a; Christensen et al., 1970; Christensen et al., 1976), metal ligand binding (Christensen et al., 1967b; Christensen et al., 1975), and studies on surfactants (Eatough and Rehfeld, 1971; Hargraves, 1972; Kresheck and Hargraves, 1974), each of which are of interest in biochemistry. The many advantages of using a continuous titration method to study systems with several simultaneous reactions have been discussed in the literature a number of times (Christensen et al., 1967b; Christensen et al., 1972; Eatough et al., 1972a, b).

Calorimeters applicable to the measurement of thermodynamic parameters by the continuous titration technique have existed for about twenty years (Jordan et al., 1976). During this time, the components of the calorimeter (i.e. constant temperature bath, constant delivery rate buret, reaction vessel, temperature sensing circuit and data analysis procedure) have gradually been improved so that the continuous titration method now gives results comparable in accuracy to those obtained by batch calorimetry.

Continuous improvement of instrumentation has resulted in a steady increase in the accuracy achievable on both the time (buret delivery rate) and temperature-dependent axes of thermograms. Also, the miniaturization of reaction vessels now makes titration calorimetry competitive with other microanalytical techniques for the study of molecular species (Hansen et al., 1974). Since calorimetric data are linearly related to concentrations for a given equilibrium constant, sharper end-points are observed at lower concentrations than is possible using log methods such as pH and EMF (Tyrrell and Beezer, 1968; Jordan and Jesperson, 1972; Hansen et al., 1975). Calorimetric methods often can be used where spectrophotometric measurements cannot be made because of inhomogeneity of the titrate solution. Calorimetric methods are unaffected by the presence of polymers, living cells or large particles because only the net heat change is observed and this is unaffected by particle size. Other notable advantages are the speed of data collection, the quantity of data which can be accumulated, and the convenience of automated data collection.

Along with the improvements in accuracy of solution calorimeters has come a fifty-fold decrease in the volume of solution and hence of material required. Precise continuous titration calorimetric measurements on as little as 2 ml may now be done in commercially available equipment (Hansen et al., 1974).

The major advantage of continuous titration calorimetry over batch or incremental calorimetry is the increased amount of data which can be obtained for the same amount of effort and material. Recognizing that each data point in a continuous calorimetric titration run contains as much information as all of the data from a batch run, the titration method is 50–100 times more efficient in the use of material and time. Continuously variable microflow calorimeters require about the same amount of time and effort per data point but generally require about twice as much material as continuous titration calorimeters.

The isoperibol and isothermal calorimetry methods are complimentary in the advantages and disadvantages of each. Typical data presentation forms from the isoperibol and isothermal calorimeters are illustrated in Fig. 1 (Izatt et al., 1974). Isoperibol calorimeters produce a thermogram

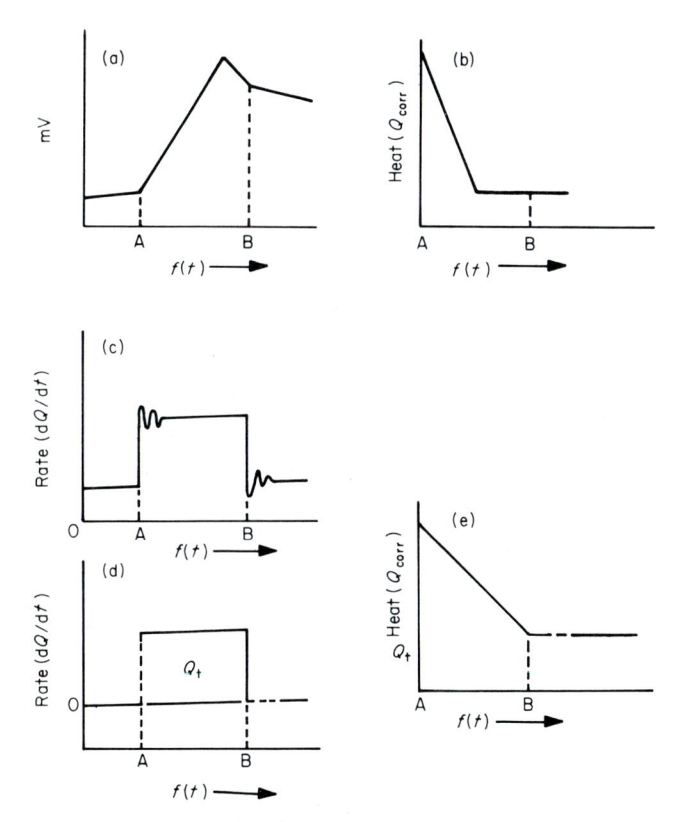

FIG. 1. Forms of calorimetric data presentation: (a) direct isoperibol data; (b) isoperibol data plotted as Q_{corr} versus time; (c) direct isothermal calorimetric data, (d) corrected isothermal data where the area Q_t is the total heat evolved during the reaction; and (e) isothermal data converted to a Q_{corr} versus time plot. In these plots $f(t)$ is the same time-dependent function, i.e. volume, moles, etc. and A and B designate the start and end of titrant addition. (Izatt et al., 1974.)

of temperature versus time (Fig. 1(a)) which can then be converted to a Q_{corr} versus time thermogram (Fig. 1(b)). In order to convert Fig. 1(a) to Fig. 1(b) corrections must be made for the heat exchanged with the surroundings during the run. These corrections for heat exchange (which require lengthy calibration experiments, especially for very small (< 10 ml) Dewar volumes) can be calculated accurately over only a limited time period. This limits the isoperibol method to experiments which are complete within 1–2 h (probably 0·5 h for very small volume (< 10 ml) Dewars). On the other hand, the isothermal calorimeter base line varies only with solution volume in the reaction vessel and the base line can be predicted accurately over long time periods. Isothermal calorimeters directly produce the time derivative of the thermogram $((dQ/dt)/\text{time})$ (Fig. 1(c)). The area under the corrected curve (enclosed area, Q_t of Fig. 1(d)) is the total change in the heat content of the system studied. A portion of the curve may be integrated to give the heat produced or absorbed during the time interval contained in that portion (Fig. 1(e)).

The dashed lines in the thermograms in Figs 1(d) and 1(e) show the major disadvantage of the isothermal compared to the isoperibol calorimetric method for experiments of short duration. The settling time of the system after a sudden change in the rate of heat production caused by the start or cessation of reactions is 3–5 min. The actual response of the instrument to a sudden change in the rate of heat production is a damped sine wave as shown in Fig. 1(c). During this time period, the instrument is not under isothermal control and the temperature of the calorimeter contents will vary from the intended control temperature. The total area under the curve and above the base line during the period of oscillation gives the heat produced (Q_t) during the time period covered by the oscillation, but the areas under increments of the curve do not give correct values for the heat produced in this portion of the curve. Thus, after any reaction begins or ends the isothermal method does not give reliable and usable information for a period of 3–5 min. End-points which occur within the first few minutes of an isothermal titration cannot be observed directly; however, the amount of heat produced during this settling period becomes small if runs are sufficiently long. Accurate data during this time period may be obtained by deconvolution of the heat rate data (Brie *et al.*, 1971; Hansen and Hart, 1977). Isothermal experiments lasting days have been run successfully. This characteristic of isothermal calorimetry is especially important for the study of intact cell systems (i.e. bacteria, algae, blood cells or tissue cells). The defined environment, long data collection periods (hours to days) and a minimum of destructive or irreversible manipulation of the cellular sample being investigated are factors that can be very significant in cellular studies.

To summarize, continuous titration isoperibol calorimeters are less expensive and simpler to operate and maintain than isothermal calorimeters. However, the data analysis and calibration procedures are more complex for isoperibol than for isothermal instruments. Recent developments in isoperibol instrumentation and operating techniques have improved this method, and currently the signal-to-noise ratios of the two methods are essentially the same for chemical systems that can be studied by both (Hansen and Hart, 1977).

From an examination of the characteristics of the two types of calorimeters discussed here, it can be concluded that isoperibol titration calorimetry can be applied to systems where the reactions are reasonably fast and can be initiated and stopped at will. On the other hand, isothermal calorimetry provides a unique method for studying metabolizing systems where often neither are the reactions fast nor can they be started or stopped by simple methods. The remainder of this contribution will be devoted to a review of selected studies which exemplify the unique capabilities of isoperibol continuous titration calorimetry and isothermal calorimetry.

B. Applications of continuous titration isoperibol calorimetry and related techniques to biomolecules

Isoperibol continuous titration calorimetry is a very useful technique for studying the multiple equilibria of proteins and other macromolecules or aggregate systems with small ions and molecules. Isoperibol batch and flow calorimetry have also been used to study the binding of small molecules and ions to proteins; but these methods are far slower, tend to miss valuable information, and are only slightly more sensitive methods. The following selection of studies has been chosen to illustrate applications where continuous titration calorimetry exhibits marked advantages over point by point (batch or flow) calorimetric methods for the determination of thermodynamic quantities.

1. PROTON IONIZATION FROM PROTEINS

The functional groups of many proteins may be reversibly titrated with acid or base. The data obtained from such studies sometimes allow identification of the specific protein functional groups from which protons ionize at different pH values and often yield information concerning the sites on proteins which complex other molecules or ions (Tanford and Epstein, 1954). Nevertheless such pH data are often difficult to interpret since sharp end-points are usually not seen in the pH titration curve and identification of the functional groups involved in proton ionization at any given pH is frequently uncertain.

Calorimetry provides a more definitive method for studies of proton ionization from proteins since end-points between pH regions where two groups have pK values close together are sharper for calorimetric than for potentiometric data (Tyrrell and Beezer, 1968; Jordan and Jesperson, 1972; Hansen et al., 1974). Also, the measured ΔH values are useful in the identification of functional groups. Extensive data exist showing that the ΔH value for proton ionization from a given group is usually less affected by substituent changes than is the corresponding pK value (Christensen and Izatt, 1962; Christensen et al., 1967a; 1970, Izatt et al., 1971).

Examples of calorimetric titration curves for the titration of human insulin with HCl and NaOH are shown in Figs 2 and 3, respectively

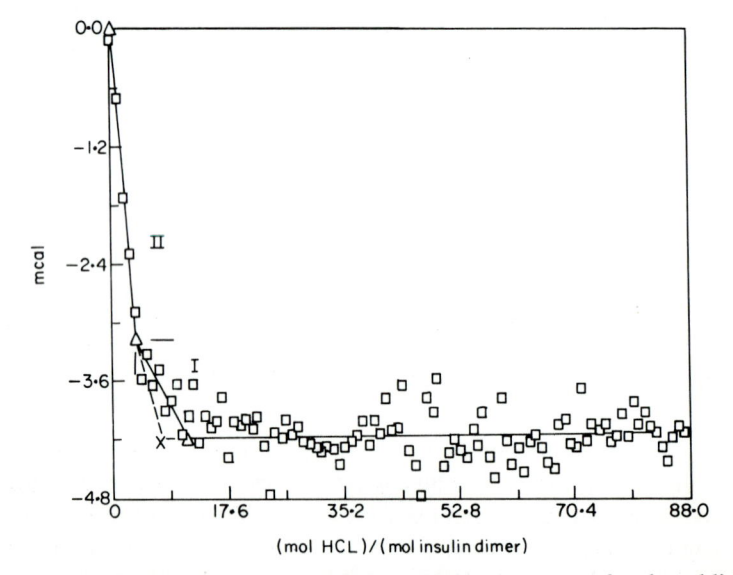

FIG. 2. Representative continuous calorimetric titration curve for the addition of 0·18 M HCl to 3·00 ml of 2·0 mg ml^{-1} insulin at 25 °C. Groups protonating in each region are: I, α-carboxyl; II, β-, γ-carboxyl.

(Jensen, 1977). The calorimetric titrations were made using a 3 ml isoperibol continuous titration calorimeter (Hansen et al., 1975). In this instrument a Dewar reaction vessel containing the titrate solution is immersed in a water bath controlled to $\pm 2 \times 10^{-4}$ °C. The titrant solution is immersed in the same bath. The stirred titrate solution is equilibrated to the bath temperature, the introduction of titrant initiated, and temperature change within the reaction vessel measured (to $\pm 25 \times 10^{-6}$ °C) as a function of the amount of titrant added. Runs were made by continuous addition of 0·2 ml of either the acid or base solution to 2·7 ml of insulin solution at 25 °C. During each run a total of 99 data points were

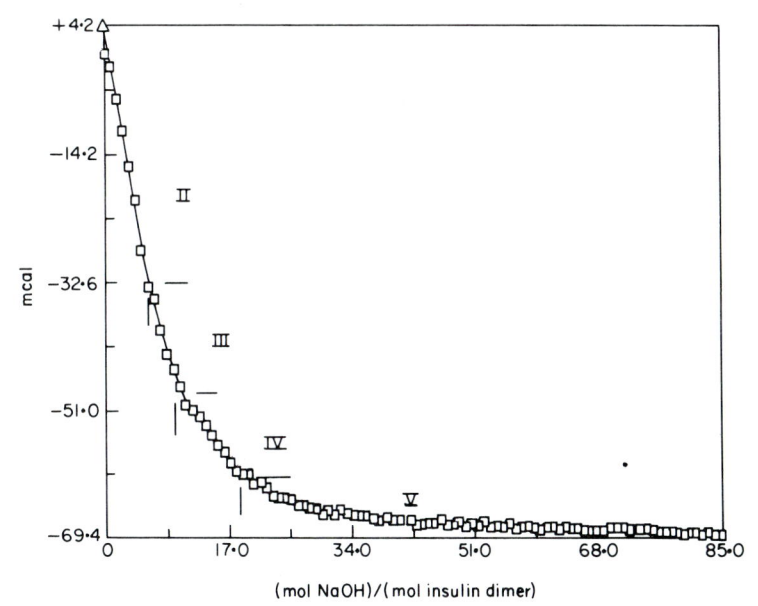

FIG. 3. Representative continuous calorimetric titration curve for the addition of 0·16 M NaOH to 3·00 ml of 2·0 mg ml^{-1} insulin at 25 °C. Groups ionizing in each region are: II, β-, γ-carboxyl; III, imidazole; IV, phenolic; V, amino and guanidinium.

recorded at 10-s intervals. The data in Figs 2 and 3 were combined with pH titration data and ΔG, ΔH and ΔS values for the proton ionization reactions of each of the functional groups in insulin were calculated.

In addition, examination of the expanded continuous titration curve of insulin with NaOH as shown in Fig. 4 showed small discontinuities in the curve which were reproducible and which are probably associated with small aggregation changes in the protein. These discontinuities would most likely have escaped notice by any method other than the continuous titration technique.

Other workers have previously done similar studies of the acid–base reactions of proteins by continuous thermometric titrimetry (Jordan and Jesperson, 1972; Marini et al., 1974). Chymotrypsinogen, chymotrypsin, haemoglobin and ovalbumin have previously been studied. All of these previous studies have used either a very rapid addition of titrant (30–45 s for a complete run) or an approximate calculation procedure to simplify the corrections for heat transfer between the reaction vessel and its surroundings. The magnitude of the effects of these procedures on the accuracy of the data is not clear, but these earlier data must remain suspect until verified by a proper procedure.

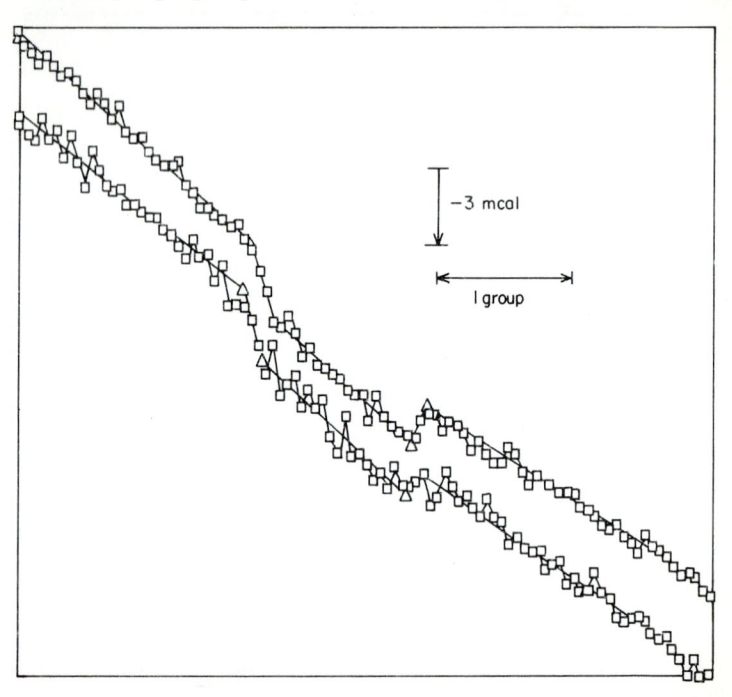

FIG. 4. Continuous calorimetric titration curve of the imidazole region of porcine insulin with NaOH at 25 °C. The reproducible exotherm at pH 5·41 and the endotherm at pH 6·11 are easily seen when 18 data points are recorded for each group titrated.

Discontinuities similar to those seen in the base titration of insulin have also been seen in the titration of bovine serum albumin with Triton non-ionic surfactants as shown in Fig. 5 (Sukow et al., 1980). The discontinuity near the critical micelle concentration (cmc) of the surfactant was not always observed and seemed to depend on both the particular surfactant and the buffer used. An example of a set of thermograms obtained where the discontinuity was absent is given in Fig. 6.

It was concluded that the only binding model which was consistent with both the calorimetric titration and equilibrium dialysis data was a model with two or three highly cooperative strong binding sites and a set of

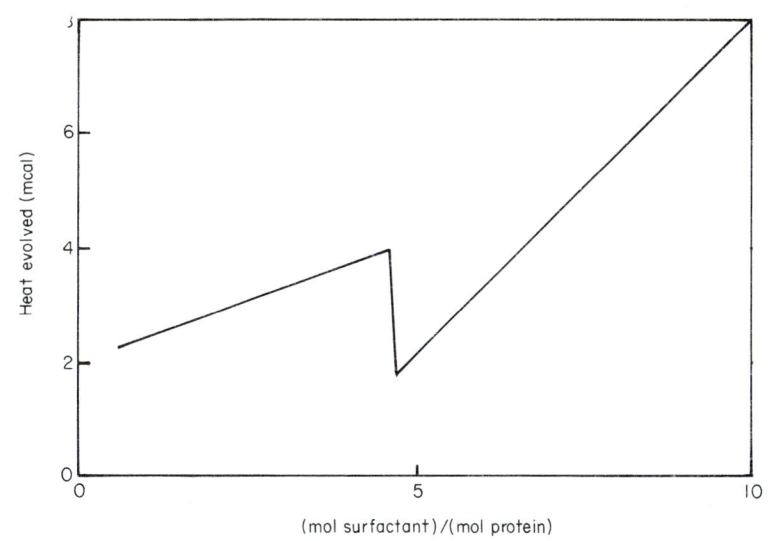

FIG. 5. Continuous calorimetric titration curve for the titration of Triton X-100 surfactant into bovine serum albumin (15 mg ml $^{-1}$) in HEPES buffer. The discontinuity which occurs at or very near the cmc of the surfactant corresponds to the addition of less than 0·1 mol of surfactant per mol of protein. The horizontal axis represents only the total solution composition and not the surfactant bound since the binding is very weak.

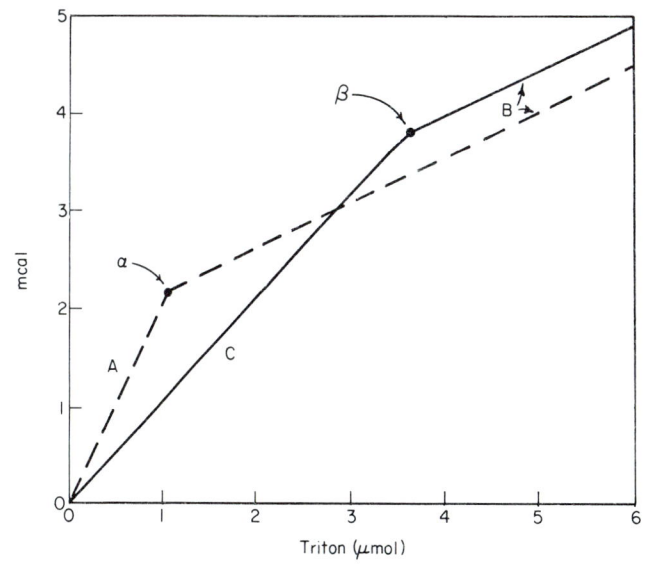

FIG. 6. Thermograms for titrations with Triton X-100. Region A corresponds to the dilution of concentrated micelles (in the titrant) to give monomers. Region B corresponds to the dilution of concentrated micelles to give micelles. Region C corresponds to the dilution of concentrated micelles to give monomers plus the reaction of monomers with BSA. Therefore, the ΔH value for reaction of monomers with BSA is equal to $\Delta H_C - \Delta H_A$ where the ΔH values are equal to the slopes of the various regions. The μmoles of Triton bound to BSA can be obtained directly from the difference in the end-points of region A and region C, α and β. α is the cmc in the absence of BSA and β is the cmc in the presence of 0·227 mM BSA.

weaker binding sites which follow a Scatchard series. The ΔG, ΔH and ΔS values were found to follow regular trends with the chain length of the surfactant molecule.

2. METAL ION BINDING TO ALBUMIN

The binding of metal ions to multiple sites on a protein has also been studied by continuous titration calorimetry as exemplified by a recent report on the binding of Ca^{2+} and Mg^{2+} to human serum albumin (Eatough et al., 1978a). This study found that the heats of binding of Ca^{2+} and Mg^{2+} to human serum albumin were approximately zero and hence could not be studied by direct titration with solutions of the metal ions. Therefore, the study was done by determining the perturbation of the proton ionization reactions of the protein by the metal ions as shown in Fig. 7.

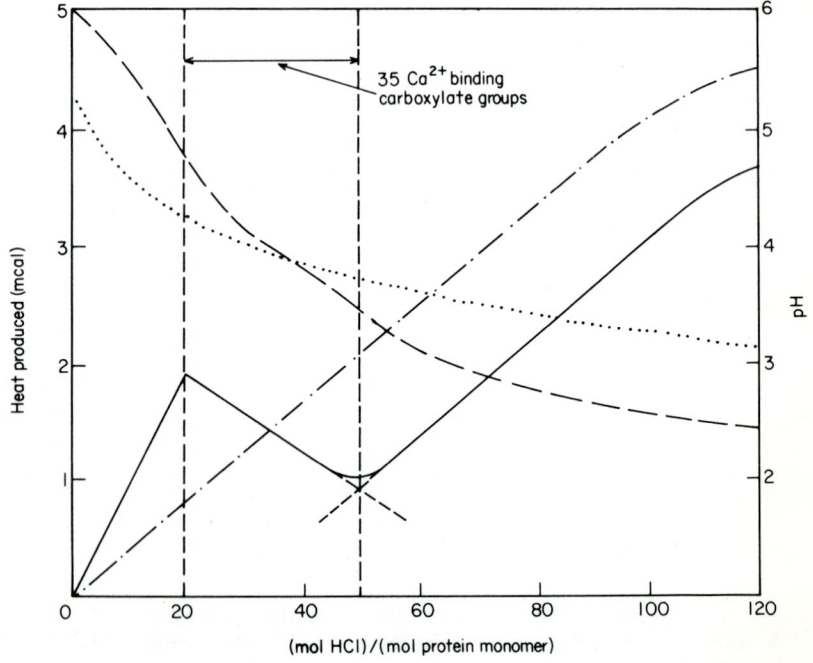

FIG. 7. Continuous calorimetric and pH titration curves for the addition of HCl to a solution of human serum albumin (4 mg ml^{-1}), —·—·— and ···· respectively, and to a solution of human serum albumin (4 mg ml^{-1}) and CaCl$_2$ (0·08 mmol ml^{-1}), ——— and ———— respectively. (Eatough et al., 1978a.)

The number of metal ions bound to the protein under a given set of conditions could be determined directly from the end-points in the titration curve as indicated in Fig. 7. The conclusions from this study were (1) Mg^{2+} and Ca^{2+} were bound to the same sites, (2) the binding sites were on carboxyl groups and neither imidazole, phenolic, nor amino groups were involved, and (3) there are about 36 carboxyl groups involved in the binding of Ca^{2+} with three groups bound to each metal ion.

3. ANTIGEN–ANTIBODY INTERACTIONS

Some successful studies of the interaction of antigens and antibodies have been carried out by continuous titration calorimetry. Jordan and Jesperson (1972) have determined the stoichiometry, free energy change and enthalpy change for the reaction of rabbit and goat antibodies (for diazotized egg albumin with egg albumin), diazotized egg albumin, p-aminobenzoic acid, o-chlorobenzoic acid and m-chlorobenzoic acid. They found that the ΔH values exhibited a strong dependence on temperature with ΔH being equal to zero at 18 °C, endothermic above 18 °C and exothermic below 18 °C.

One of the potential applications of continuous titration calorimetry to this area, namely the study and determination of the more subtle classifications of blood types, has not yet met with any success. The difficulty appears to be that the kinetics and stoichiometry of the reaction depend strongly on the rate of titrant addition as well as on small variations in the history of the samples (Rehfeld, 1978a).

4. DETERMINATION OF TOTAL PROTEIN

A thermometric titration method for total protein determination has been devised by Carr (Smith and Carr, 1973). The reaction used is the precipitation of protein by phosphotungstic acid. Precision of the results was generally better than that obtained with the currently used standard methods.

5. DETERMINATION OF BINDING SITES ON ERYTHROCYTES

The number of protein binding sites on normal and abnormal red blood cells can be determined by a direct thermometric titration with albumin or concanavalin A (Rehfeld et al., 1975). Such determinations may be of value in studying the course of pathological states of blood cells since large differences were found between red blood cells in different states of development and between normal and sickle cells.

Binding sites for sodium dodecylsulphate (NaDDS), cetyl trimethyl ammonium bromide (CTAB), lysolecithin and Triton X-100 surfactants on erythrocyte membranes or ghosts have also been studied by continuous titration calorimetry (Eatough et al., 1978b). The results suggest that NaDDS binds to the protein portion of the membrane, while CTAB and lysolecithin bind to both proteins and lipids. Both CTAB and Triton were observed to cause conformational changes in the membrane. The number of each type of binding site could readily be determined from end-points in the titration curves. In general, these agreed with results from other methods.

6. ANALYTICAL METHODS RELATED TO THERMOMETRIC TITRATION

The technique of direct injection enthalpimetry (Jordan et al., 1976) which is closely related to titration calorimetry can be a very rapid and very sensitive analytical technique (Hansen et al., 1977). This method has been applied to the determination of glucose (Goldberg et al., 1975), the activity of enzyme preparations (Carr et al., 1976), and enzymatic rate constants (Grime et al., 1977) as well as many inorganic substances (Vaughn, 1973).

Heat conduction calorimeters using a bed of immobilized enzyme in a flow cell are well characterized (Brown, 1969; Carr et al., 1976) as analytical instruments for determining the concentration of an enzyme substrate. But heat conduction calorimeters used in this way all have the disadvantage of a slow response of the instrument (Krisam and Schmidt, 1977). There is, however, a simple adaptation of the Tronac isothermal calorimeter which is being used for this purpose in industrial quality control programs (Hart, 1977) which overcomes the response time problem. By immersing a short length of tubing containing the immobilized enzyme into a few millilitres of water contained in the isothermal reaction vessel, an instrument with a total settling time of about 3 min is obtained. Except for the response time such an isothermal instrument produces data of the same type as previously described heat conduction systems.

A closely related technique is to measure the temperature rise in a flowing stream as it passes over a bed of immobilized enzyme. The temperature rise is then proportional to the substrate concentration under a given set of conditions. Such instruments can have very fast response times (Carr et al., 1976) of a few seconds, but are not as sensitive as the heat conduction or isothermal calorimetric techniques (Krisam and Schmidt, 1977).

Another thermometric method of analysis uses a pair of thermistors, one for reference and the other, being coated with immobilized enzymes, becomes a sensor for substrates of the enzyme. The temperature difference developed between the thermistors is proportional to the concentration of substrate (Cooney *et al.*, 1974).

C. Application of isothermal calorimetry to metabolizing systems

Because the reaction vessel of the isothermal calorimeter also serves as the culture chamber where the atmosphere, stirring and the medium can be controlled, this type of calorimeter is much better suited to the study of cultures of living cells than any other type of calorimeter. Isothermal calorimeters have been used to study the effects of carcinogens and mutagens on heat production by metabolizing cultures of mammalian tissues (Jensen, 1977), the effects of antibiotics and cytotoxins on heat production by Streptococci (Jensen *et al.*, 1976), the thermodynamics and kinetics of sugar transport in *Escherichia coli* (Long *et al.*, 1976, 1977), and the effects of light on photoluminescent bacteria (McIlvaine and Langerman, 1977).

1. MAMMALIAN CELL STUDIES

A recent study of normal and transformed embryonic hamster cells (Jensen, 1977) exemplifies the capabilities of isothermal calorimetry to study mammalian tissue cultures. The objective of this study was the assessment of calorimetry as an analytical method to differentiate between normal mammalian tissue cells and chemically transformed (carcinogenic) cells. The motivation behind the development of such a procedure is the problem encountered when working with human cells and attempting to prove that a cell is transformed and tumour producing. Any scheme designed to detect the transformed character of a cell requires a host. For mouse cells, this is not a problem, but for human cells, except for immunosuppressed animals, there is not a host available. Current determinations of the transformed character of human cells rely on the visual form of the cell, its growth rate, growth behaviour (CPE), a specific assay for a metabolite or enzyme, or some physical change in the cell contents (Busch, 1962). In some cases these techniques are vague or ambiguous and can be too cell specific to be generally applied to studies of all cells.

While the generality of the calorimetric measurement of total cell energetics, measured as the total heat, is its greatest recommendation, it is also its major limitation. Cellular thermogenesis measured by calorimetry is the sum of all the cell's activity and, as such, calorimetry cannot detect a small number of transformed cells amidst numerous normal neighbours. However, a homogeneous sample of transformed cells should in total behave differently from normal cells.

The rate of cellular thermogenesis as measured for normal hamster embryo cells in Delbecco's MEM media was found to decrease with increasing cellular concentration (Fig. 8). An increase of cellular concentration from 0·4 to 4·0(log(cell $_{total}$ × 10^{-7})) decreased the cellular thermogenesis more than 50 per cent. This observation reveals a critical problem that must be considered when the thermogenesis of tissue culture cells is to be studied. The normal response for an investigator would be to increase the concentration of cells used in studies where small amounts

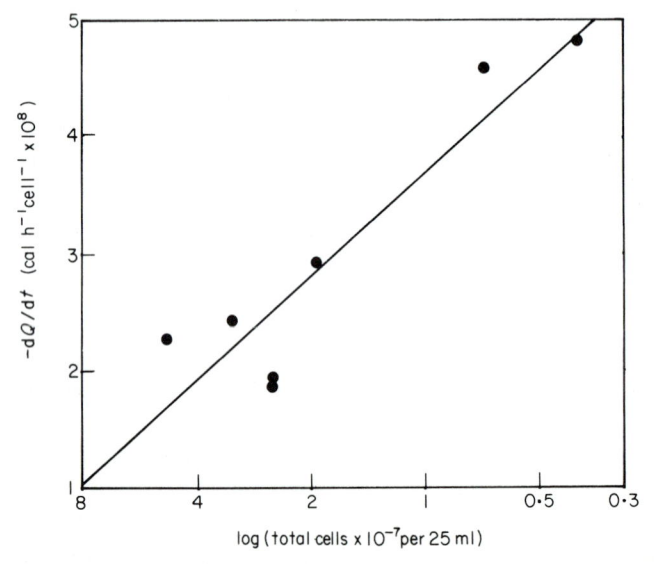

FIG. 8. Normal hamster embryo cell thermogenesis plot. The plot of the log of the total number of cells versus the heat production rate of an average cell for 1 h. The linear least-squares fit of these data is represented by the line.

of heat are expected. It is evident from Fig. 8 that it is counter-productive to attempt to increase the magnitude of $-dQ/dt$ by using higher cellular concentrations. The dependence of the thermogenesis rate on cellular concentration must, therefore, be determined for any tissue cell that is to be studied calorimetrically.

The same dependence of the rate of heat production on the log of the total cell concentration seen for normal hamster embryo cells (Fig. 8) is also evident with benzo(a)pyrene transformed cells (Fig. 9). At $\alpha = 0.05$

there is a statistical difference between the least-squares fit lines for transformed cells and for normal cells. This indicates that it may be possible to use cellular thermogenesis as an index for determining if a cell has been transformed.

The slope of the transformed cell best fit line (B in Fig. 9) is less steep compared to that of the best fit line (A in Fig. 9) for normal cells. This could be interpreted to mean that transformed cells do not exhibit the same degree of contact inhibition as do normal cells. Were the degree of inhibition the same, and the rate of heat production different, the lines would be parallel. It is known that transformed cells do not have the same degree

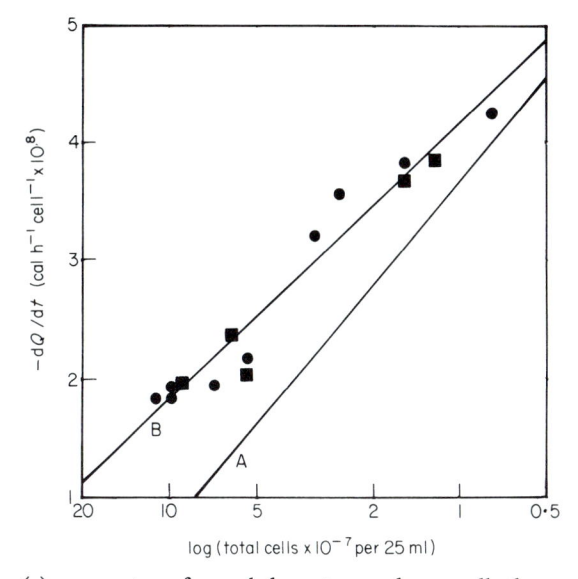

FIG. 9. Benzo(a)pyrene transformed hamster embryo cell thermogenesis plot. The plot of the log of the total number of cells versus the heat production rate of an average cell for 1 h. Line A is the best fit line of normal cells from Fig. 8, and line B is the best fit line of the data on the transformed cells.

of contact inhibition as normal cells. Therefore, the observed thermogenesis of these transformed cells is consistent with their expected cell behaviour.

Although the applicability of isothermal calorimetry to the determination of cellular transformation is still uncertain, the above study does demonstrate that mammalian tissue cultures can be studied by this method.

Part of the results of a study on human red blood cell metabolism using isothermal calorimetry is shown in Fig. 10 (Rehfeld et al., 1980). Some of the features of the thermogram could not be reproduced from one sample to another; however, the data do show the kind of data that could be obtained by this technique if all of the variables were known and could be controlled.

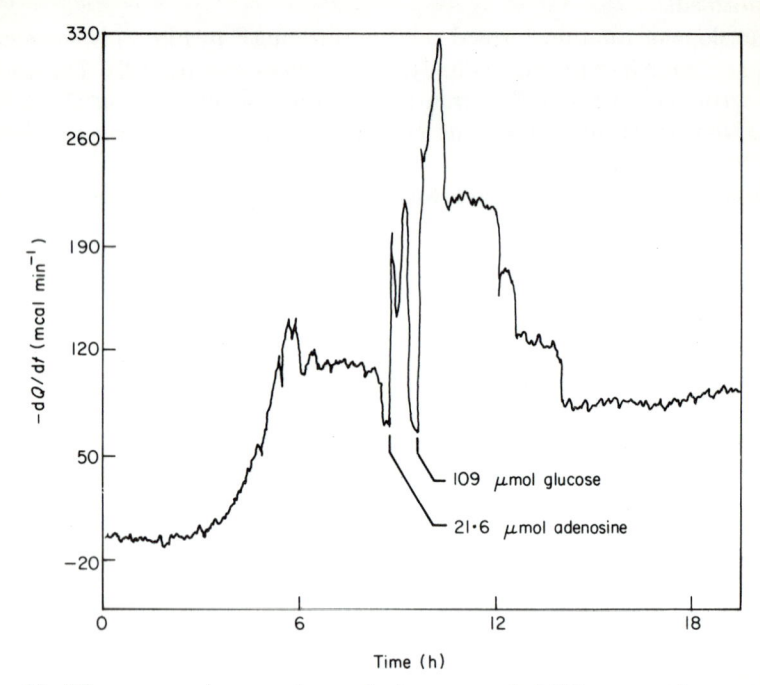

FIG. 10. Thermogenesis curve for washed, resuspended (30 per cent haematocrit) human red blood cells. At the indicated times, glucose and adenosine were added to 50 ml of the suspension. The response to the glucose and adenosine was qualitatively reproducible. (Rehfeld et al., 1978b.)

2. BACTERIAL CELL CULTURES

Three studies which exemplify the advantages of isothermal calorimeters to study bacterial cultures have been published. In two of these, the calorimeter was used to study metabolic processes occurring in the bacteria (Long et al., 1976, 1977; McIlvaine and Langerman, 1977). The other study was a study of the interaction of cytotoxic agents with bacteria with the emphasis being on developing analytical methods for the cytotoxic agents (Jensen et al., 1976).

The ability to observe directly the effect of cytotoxic agents on bacterial cells is potentially useful for clinical, pharmacological and immunological studies. Beezer et al. (1974) have shown that flow calorimetry could be used to determine the level of microbial infection in urine samples. Flow calorimetry has also been applied to bacterial antibiotic sensitivity and proposed as a clinical tool by Binford et al. (1973) but only the qualitative effect of the antibiotic (bacteriocidal, bacteriostatic, no effect) was determined for causative agents of urinary tract infections. The calorimeters (principally heat conduction flow calorimeters) which have been used in most studies of microorganisms have slow time response characteristics which make it impossible to measure short-term details of the thermogram. The faster response time of isothermal calorimeters allows the measurement of these details.

The isothermal calorimetric technique for the measurement of the effect of cytotoxic agents on bacterial cells (Jensen et al., 1976) was found to be significantly faster (<1 h compared to 18–35 h per sample) than present indirect techniques, i.e. dilution plate count, and requires only small amounts (100–250 µl) of sample. The calorimetric technique also compares favourably with newly developed agar diffusion (Sabath et al., 1971), disk-plate (Stroy, 1969) and photometric (Isenberg et al., 1971) techniques in rapidity of measurement and potentially is applicable to the

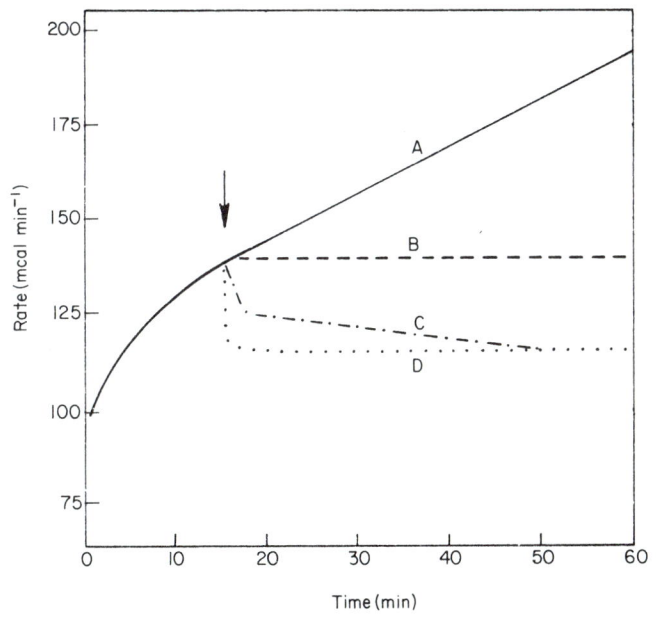

FIG. 11. Plot of rate of heat production versus time for the addition of antibiotics to S. faecalis (25 ml of 4 per cent DMSO broth). A, the standard curve (no addition). The arrow indicates the injection of: B, 100 µl of 12·5 µg Penicillin "G" per µl; C, 100 µl of 1·2 µg tetracycline HCl per µl; and D, 100 µl of 11·8 µg of tetracycline HCl per µl. (Jensen et al., 1976.)

study of any cytotoxin–cellular system. More significantly, for tetracycline the kinetic and total effects on the rate of heat evolution by *Streptococcus faecalis* were shown to be a reproducible function of the concentration of the cytotoxic agent added.

The effects of an antibiotic on thermogenesis by a test organism can be used to measure quantitatively the level of the drug present in normal human serum. The technique offers a rapid bioassay which directly measures drug activity. In a typical determination, a small volume (100–250 μl) of a solution of the cytotoxic agent in the medium or serum was added to the test growth culture (25 ml) in the isothermal calorimeter reaction vessel after the straight-line portion of the standard curve had been reached.

In Fig. 11, the results of addition of penicillin "G" (curve B) and two concentrations of tetracycline HCl (curves C and D) to the test broth are compared to the standard curve. Curves for addition of normal human serum (curve C), heated human serum (56 °C for 30 min) (curve B) and human serum plus tetracycline HCl (curve D) are shown in Fig. 12. In each case, the response of the test growth culture to the added

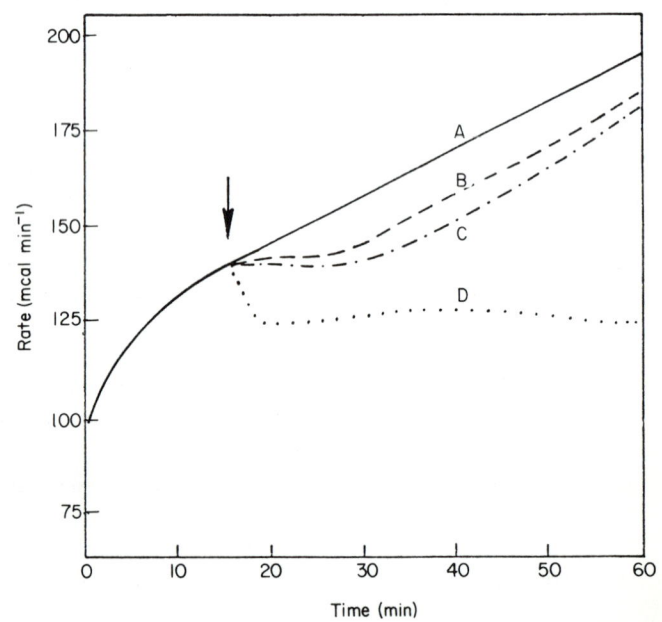

Fig. 12. Plot of rate of heat production versus time for the addition of human serum samples to *S. faecalis* (25 ml of 4 per cent DMSO broth). A, the standard curve. The arrow indicates the injection of: B, 200 μl of normal human serum; and D, 200 μl of 0·28 μg of tetracycline HCl per μl normal human serum. (Jensen et al., 1976.)

cytotoxic agent or serum sample was rapid and reproducible. The overall response of the cells to an added cytotoxic agent was different for each agent used.

When penicillin "G" was added to the test culture, the rate of heat produced became constant at about the same rate as that existing when addition occurred. This constant rate is above that of cyanide-poisoned cells. The difference could indicate that the initial effect of penicillin on cell wall synthesis does not result in significant changes in the majority of enzyme systems which affect bacterial metabolism. The ability to recognize and distinguish between such differences could provide a useful method to study the extent of the cytotoxic influence of various agents on cells by directly following the heat output response of those cells.

In the case of tetracycline HCl, the difference in antibiotic concentration of the material added results in curves which differ from each other and from the standard curve. Tetracycline at a final concentration of 47 ng μl^{-1} of 4 per cent DMSO broth produces the same eventual response as CN^-. The final base line is approached in two steps, the first a rapid reaction which is over in less than 1 min (the response time of the instrument used), and the second a slower response requiring an additional 4·8 min for the attainment of a constant rate of heat output in the bacterial system. At a lower final concentration, 4·8 ng μl^{-1} tetracycline, the end result is the same; however, the approach to the final rate of heat production is slower. Again, two distinct effects are seen, the first lasting 2·7 min and the second, 40 min. The initial response of the cells to a final concentration of 2·3 ng n^{-1} tetracycline, added as 0·2 ml of 284 ng μl^{-1} tetracycline in human serum, was similar to that seen for tetracycline in broth, with a total time of 4·8 min for the response. Any second response to tetracycline in the presence of serum was masked by the effects of the serum (see Fig. 12, curve D) along with the attainment of the non-linear portion of the standard curve.

Addition of normal serum to the test culture causes a reduction in the rate of heat production by the bacterial cells, but the reduction is only temporary. The titre of cytotoxic serum factors was proportional to the displacement of the normal serum curve from the standard curve. The return of the slopes of the curves associated with addition of human serum (B and C in Fig. 12) to a slope approximately parallel to that of the standard curve results from the continued normal metabolism of the unaffected bacterial cells.

Heated serum had less of an inhibitory effect than normal serum on the heat production of the bacterial cells. Normal serum complement would not be expected to be responsible for the difference observed between the effect of normal and heated serum since complement does not have a significant bacteriocidal effect on Gram-positive cells. Thus, the difference observed between the bacterial heat production response to normal and heated serum was probably due to non-complement heat labile serum factor(s) such as lysozyme and/or β-lysin.

When serum with tetracycline present was added to the test culture a marked response in the rate of heat production was observed. The observed change is a composite result of the action of the serum factors and antibiotic on the cells. However, the initial response to tetracycline is still clearly seen, indicating tetracycline may be quantitatively assayed in serum samples using the described calorimetric response.

It should be emphasized that even at a serum dilution factor of 1 : 126, calorimetry directly yields information about the response of bacterial cells to heat-stable and heat-labile cytotoxic agents in serum. Use of a standard viable cell plate count method at such a serum dilution would result in data that are indirect and not likely to be statistically significant. It is not yet clear what complicating factors will be present in the calorimetric assay of a system containing combinations of antibiotic drugs.

The results indicate that isothermal calorimetry should be a useful research tool to: (a) study the response of bacterial (or other) cells to cytotoxic materials; (b) determine concentrations of cytotoxic substances, i.e. serum antibiotic concentrations; (c) study cytotoxic serum factors; and (d) study other cellular processes that exhibit a change in the rate of heat produced by growing cells.

Active transport mechanisms in *E. coli* have been studied using isothermal calorimetry (Long et al., 1976, 1977). The transport of the nonmetabolizable substances, thiomethyl galactoside and α-methylglucoside, was used to study the lactose and glucose transport systems. The conclusions reached in this study were:

1. the thermal effects were largely a response to the establishment of a gradient and were not caused by the accumulation process;
2. there was no significant extrusion of protons from the cells into the medium;
3. the energy cost of maintaining a steady-state gradient is consistent with the chemiosmotic theory;
4. the initial establishment of a gradient requires more energy flow than is required by the chemiosmotic theory;
5. there is a transition in energy flow behaviour and in the biochemical processes which occur which takes place as the level of uptake increases.

The methodological design of this experiment is of interest because air was bubbled through the liquid in the reaction vessel in order to maintain a constant concentration of oxygen during the experiments. Maintenance of a constant atmosphere above or concentration of gaseous solutes in growth media is a major problem in doing cellular metabolism studies in other types of calorimeters.

Direct isothermal calorimetric determinations of the rate of heat production along with simultaneous determinations of the rate of photon emission and the number of viable cells have provided insight into the growth and mechanism of photoluminescence of *Beneckea harveyi* and *Photobacterium leiognathi*. These experiments were performed with an isothermal microcalorimeter modified with a fibre-optic light guide to allow *in situ* detection of light (McIlvaine and Langerman, 1977). The results of the observations on *B. harveyi* are shown in Fig. 13. At the point

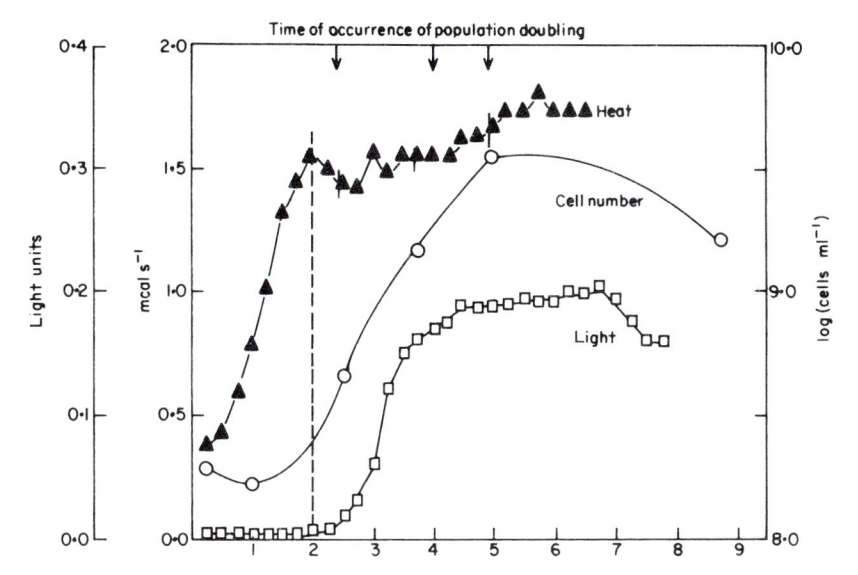

FIG. 13. Thermogenesis, light production in light units (1 light unit = $1 \cdot 44 \times 10^{12}$ photons s^{-1}), and viable cells (per millilitre) for *B. harveyi* in complete medium. The scale at the top indicates the point in time when each population doubling occurred during each experiment. (McIlvaine and Langerman, 1977.)

when luciferase begins to be expressed, as indicated by the increased light flux, the heat flux levels off (dashed vertical line). This occurs concurrently with the first doubling of the cell population.

This study, while preliminary, suggests possibilities for further studies of the effects of electromagnetic radiation on living systems that could be done by isothermal calorimetry.

References

Bark, L. S. and Bark, S. M. (1969). "Thermometric Titrimetry". Pergamon Press, Oxford.

Barthel, J. (1975). "Thermometric Titrations". John Wiley, New York.

Beezer, A. E., Bettelheim, K. A., Newell, R. D. and Stevens, J. (1974). *Sci. Tools*, **21**, 13.

Binford, J. S., Binford, L. F. and Adler, P. (1973). *Am. J. clin. Path.* **59**, 86.

Brie, C., Petit, J. L. and Gravelle, P. C. (1971). *C.r. hebd. Séanc. Acad. Sci., Paris*, **273B**, 1.

Brown, H. D. (1969). "Biochemical Microcalorimetry." Academic Press, New York and London.

Busch, H. (1962). "An Introduction to the Biochemistry of the Cancer Cell". Academic Press, New York.

Carr, P. W., Bostick, W. D., Canning Jr, L. M. and Callicott, R. H. (1976). *Am. Lab.* **8**, 45.

Christensen, J. J. and Izatt, R. M. (1962). *J. Phys. Chem.* **66**, 1030.

Christensen, J. J., Hansen, L. D. and Izatt, R. M. (1967a). *J. Am. chem. Soc.* **89**, 213.

Christensen, J. J., Hansen, L. D., Izatt, R. M. and Partridge, J. A. (1967b). Application of high precision thermometric titration calorimetry to several chemical systems. *In* "Microcalorimètre et Thermogenèse", p. 207. Pub. No. 156, Centre National de la Recherche Scientifique, Paris.

Christensen, J. J., Rytting, J. H. and Izatt, R. M. (1970). *Biochemistry*, **9**, 4907.

Christensen, J. J., Eatough, D. J., Ruckman, J. and Izatt, R. M. (1972). *Thermochim. Acta*, **3**, 203.

Christensen, J. J., Eatough, D. J. and Izatt, R. M. (1975). "Handbook of Metal Ligand Heats" (second edition). Marcel Dekker, New York.

Christensen, J. J., Hansen, L. D. and Izatt, R. M. (1976). "Handbook of Proton Ionization Heats". John Wiley, New York.

Cooney, C. L., Weaver, J. C., Tannenbaum, S. R., Faller, D. V., Shields, A. and Jahnke, M. (1974). "Enzyme Engineering" (E. K. Pye and L. B. Wingard Jr, eds), vol. 2, pp. 411–417. Plenum Press, New York.

Eatough, D. J. and Rehfeld, S. J. (1971). *Thermochim. Acta*, **2**, 443.

Eatough, D. J., Izatt, R. M. and Christensen, J. J. (1972a). *Thermochim. Acta*, **3**, 233.

Eatough, D. J., Christensen, J. J. and Izatt, R. M. (1972b). *Thermochim. Acta*, **3**, 219.

Eatough, D. J., Jensen, T. E., Loken, H., Rehfeld, S. J. and Hansen, L. D. (1978a). *Thermochim. Acta*, **25**, 289.

Eatough, D. J., Hansen, L. D. and Rehfeld, S. J. (1978b). Unpublished data. Brigham Young University, Provo, Utah.

Goldberg, R. N., Prosen, E. J., Staples, B. R., Boyd, R. N. and Armstrong, G. T. (1975). *Analyt. Biochem.* **64**, 68.

Grime, J. K., Lockhart, K. and Tan, B. (1977). *Analyt. Chim. Acta*. **91**, 243.

HANSEN, L. D. and HART, R. M. (1977). Unpublished data. Brigham Young University, Provo, Utah, and Tronac Inc., Orem, Utah.
HANSEN, L. D., IZATT, R. M., EATOUGH, D. J., JENSEN, T. E. and CHRISTENSEN, J. J. (1974). In "Analytical Calorimetry" (R. S. Porter and J. Johnson, eds), vol. 3, pp. 7–16. Plenum Press, New York.
HANSEN, L. D., IZATT, R. M. and CHRISTENSEN, J. J. Applications of thermometric titrimetry to analytical chemistry. In "New Developments in Titrimetry" (Joseph Jordan, ed.). Marcel Dekker, New York.
HANSEN, L. D., JENSEN, T. E., MAYNE, S., EATOUGH, D. J., IZATT, R. M. and CHRISTENSEN, J. J. (1975). *J. Chem. Thermodynamics*, **7**, 919.
HANSEN, L. D., RICHTER, B. E. and EATOUGH, D. J. (1977). *Analyt. Chem.* **49**, 1779.
HARGRAVES, W. A. (1972). Ph.D. Dissertation, Northern Illinois University, Dekalb, Illinois.
HART, R. M. (1977). Personal communication. Tronac Inc., Orem, Utah.
ISENBERG, H. D., REICHLER, A. and WISEMAN, D. (1971). *Appl. Microbiol.* **22**, 980.
IZATT, R. M., CHRISTENSEN, J. J. and RYTTING, J. H. (1971). *Chem. Rev.* **71**, 439.
IZATT, R. M., HANSEN, L. D., EATOUGH, D. J., JENSEN, T. E. and CHRISTENSEN, J. J. (1974). "Analytical Calorimetry" (R. S. Porter and J. Johnson, eds), vol. 3, pp. 237–248. Plenum Press, New York.
JENSEN, T. E. (1977). Ph.D. Dissertation, Brigham Young University, Provo, Utah.
JENSEN, T. E., HANSEN, L. D., EATOUGH, D. J., SAGERS, R. D., IZATT, R. M. and CHRISTENSEN, J. J. (1976). *Thermochim. Acta*, **17**, 65.
JORDAN, J. and JESPERSON, N. D. (1972). "Thermometric Methods of Analysis", p. 59. Pub. No. 201, Centre National de la Recherche Scientifique, Paris.
JORDAN, J., GRIME, J. K., WAUGH, D. H., MILLER, C. D., CULLIS, H. M. and LOHR, D. (1976). *Analyt. Chem.* **48**, 427A
KRESHECK, G. C. and HARGRAVES, W. A. (1974). *J. Coll. Inter. Sci.* **48**, 481.
KRISAM, G. and SCHMIDT, H. L. (1977). Development and properties of caloric systems for substrate determinations with immobilized enzymes. In "Application of Calorimetry in Life Sciences" (I. Lamprecht and B. Schaarschmidt, eds), chapter 1.3, pp. 39–47. Walter de Gruyter, New York.
LONG, R. A., SPROTT, G. D., LABELLE, J. L., MARTIN, W. G. and SCHNEIDER, H. (1976). *Biochem. biophys. Res. Commun.* **64**, 656.
LONG, R. A., MARTIN, W. G. and SCHNEIDER, H. (1977). "Energetics of β-Galactoside Transport in *E. coli*." National Research Council of Canada Report.
MCILVAINE, R. and LANGERMAN, N. (1977). *Biophys. J.* **17**, 17.
MARINI, M. A., MARTIN, C. J., BERGER, R. L. and FORLANI, L. (1974). In "Analytical Calorimetry" (R. S. Porter and J. Johnson, eds), vol. 3, pp. 407–424. Plenum Press, New York.
REHFELD, S. J., EATOUGH, D. J. and HANSEN, L. D. (1978a). Brigham Young University, Provo, Utah. Unpublished data.

REHFELD, S. J., EATOUGH, D. J. and HANSEN, L. D. (1975). *Biochem. biophys. Res. Commun.* **66**, 2.
REHFELD, S. J., MENTZER, W. C., SHOHET, S. B., EATOUGH, D. J. and HANSEN, L. D. (1980). *Microcalorimetry Lett.* In preparation.
SABATH, L. D., CASEY, J. I., RUCH, P. A., STUMPF, L. L. and FINLAND, M. (1971). *J. Lab. clin. Med.* **78**, 457.
SMITH, E. B. and CARR, P. W. (1973). *Analyt. Chem.* **45**, 1688.
STROY, S. A. (1969). *Appl. Microbiol.* **18**, 31.
SUKOW, W. W., SANDBERG, H. E., LEWIS, E. A., EATOUGH, D. J. and HANSEN, L. D. (1980). *Biochemistry.* In press.
TANFORD, C. and EPSTEIN, J. (1954). *J. Am. chem. Soc.* **76**, 2170.
TYRRELL, H. J. V. and BEEZER, A. E. (1968). "Thermometric Titrimetry". Chapman and Hall, London.
VAUGHN, G. A. (1973). "Thermometric and Enthalpimetric Titrimetry." Van Nostrand Reinhold, London.

Subject Index

A

Acceleration phase, 47
Acholeplasma laidlawii, 169, 221, 222, 296
Acholeplasma sp., 169
acriflavine, 73
active transport, 472
adamantane, 223
adenine, self association of, 373
  (*see also* ATP)
adhesion, bacteria, 270
  granulocytes, 270
  tissue cells, 272
adiabatic calorimeters, 248
adipose tissue, 158
Aerobacter aerogenes, 18, 26, 27, 30, 33, 34, 71
aerobic metabolism, 19, 51, 171
agitation of calorimetric medium, 257
albumin, 460, 462
alcoholic fermentation, 5
aldolase, 393
alkaloids, 240
alkanols, 225, 227
amphotericin B, 205, 208, 210, 212, 231, 235
ampicillin, 202, 203, 205, 206
ampoule drop calorimeter, 132
anabolism, 5, 48, 49
anaemias, 140
anaerobic metabolism, 19, 49, 171
anaesthetic, 225, 227, 229
anthracycline antibiotics, 229
antibiotics, 164, 187, 192, 465, 469, 470
  (*see also* under individual name of drug)
antibodies, 113, 122, 463
  −antigen reaction, 139
antigen, 113, 114, 122, 463
  −antibody reaction, 139
antifungal drugs, 205

antimycin, 122
artifacts in calorimetric measurements, 253
ash, 100, 101
assimilation, 84, 85
ATP, 49, 50, 52, 57, 69, 71, 78, 84, 116, 193
  inhibitors, 214
ATPase, 127, 128
  (Na^+, K^+), 387
avidin, 393, 396
azide, 69, 73, 77

B

bacteria, 114, 187
  characterization, 163
  identification, 188
bacteriocidal drugs, 203
bacteriostatic drugs, 203
bacteriuria, 189
baker's yeast, 179
batch calorimeter, 58, 133
batch culture, 52, 56, 78
Beneckea harveyi, 473
biomembranes, 275, 278, 295, 300, 305, 306
biopolymers, 415, 417, 420, 422
biotechnical processes, 264
blood, 115, 116, 122, 133, 269, 463, 467
  coagulation of, 135
  plasma, 133
Bode diagram, 325, 326, 328, 329
bomb calorimeter, 44, 45, 61, 98
bound dissipation function, 63
bovine serum albumen, dinitrophenylated, 121
brain, 159
brewer's yeast, 179
brown adipocytes, 130
bud, 95, 96

C

Ca^{2+}, 462
calibration constant, 119, 148
calorimetric design principles, 247
   systematic errors, 253
   test experiments, 253
candicidin, 208, 210
Candida albicans, 65, 164, 167, 177, 181, 182
   intermedia, 61
   utilis, 43, 102
carbenicillin, 202
carbohydrates, 275
carcinogen, 465
catabolism, 5, 47, 48, 49, 73, 74, 90, 102, 103
catabolite repression, 51, 83
catalytic thermometric titrimetry, 238
catecholamines, 239
cell cycle, 95, 96, 97
cellular thermogenesis, 465, 466, 470
cephaloridin, 213
caphalothin, 202
cetyl trimethyl ammonium bromide, 464
characterization, microbial, 163
chemostat, 56, 57, 92, 93, 96, 97
chloramphenicol, 34, 205
chlortetracycline, 201
chick embryo fibroblasts, 115, 116, 117, 120, 127
cholesterol, 275, 287, 301, 303
chymotrypsin, 381, 382, 383, 459
   binding of ligands to, 361, 364, 399, 404
   conformational change, 375
   heat capacity of, 401
chymotrypsinogen, 459
cinoxacin, 206
clinical specimen of urine, 188
clotrimazole, 205
coagulation, of blood, 135
colistin, 202
collagen, melting of water in, 426–428
complement, 471
computer analysis of thermogram, 182
conconavilin A, 463
conformational change, ligand induced, 362–366, 374–376, 396, 400
conservative reaction, 85
contamination, microbial, 164

continuous culture, 46, 56, 90, 92, 95, 102, 104
cooperative binding, 384, 389–393, 405
cortineurin, 66
Crabtree effect, 51
critical micelle concentration, 460
cut-off frequency, 329, 330, 331, 338
3′-cytidine monophosphate, 344
   binding to ribonuclease A, 344, 350, 352, 357, 372, 376–381, 388
cytotoxic agents, 465, 468

D

DNA, 55, 73, 214
   melting of water in, 443
   synthesis, 121, 126
damping factor, 321, 322, 323
Debaromyces hansenii, 82, 90
2-deoxy-D-glucose, 122
depots, formation of, 72
diauxic metabolism, 55, 56, 64, 67, 82, 83
differential scanning calorimetry, 219, 277
differential thermal analysis, 280
differentiation technique, 331, 332, 335, 336, 337
dilution rate, 93
2,4-dinitrophenol, 71, 85, 123, 124
   -lysine–rabbit IgG interaction, 395
dipalmitoyl lecithin, 224, 226, 233
1,2-dipalmitoyl phosphatidylcholine 221
direct calorimetry, 58
dissipation function, 63, 104
distillery yeast, 179
doxycycline, 181, 192, 203, 204, 208, 211
drugs, 289
   and *see* individual entries
Dulong formula, 98
dynamic properties, 313, 316, 318, 321, 326, 328, 330, 333, 334, 338,
dynamical calorimetry, 44, 61, 102

E

efficiency, 152, 154
electric organ, 158
electrical impedance, 191
electronic corrector, 332, 333

SUBJECT INDEX

electrostatic interactions, 371, 372
Embden–Meyerhof–Parnass pathway, 49, 51, 52, 55, 95
endogenous metabolism, 46, 48, 52, 68, 69, 71, 75, 77
endothermic phase transitions, 275
endoxan, 66
energy, balance equation, 61, 63
　metabolism, 48
　transfer chain, 51, 52
ensilage, 54
Enterobacteriaciae, 183
enthalpy change, 346, 347
　additivity of, 379
　hydrogen bonds, 372
　linkage, 387
　van der Waals, 372
entropy change, 69, 103, 346, 347
　chelate effect, 371, 372
　electrostatic interactions, 371
　hydrophobic interactions, 369
　linkage, 387
　rotational–translational, 367, 379
enumeration, 164
enzyme activity, 461
　immobilized, 464
　kinetics, 127
erythrocytes, 136, 269
erythromycin, 202, 205, 206
Escherichia coli, 25, 26, 28, 29, 37, 168, 178, 180, 182, 183, 188, 191, 192, 201, 202, 203, 204, 205, 206, 213, 214, 465, 472
Ethylene diamine tetra-acetic acid, EDTA, 119

F

fatty acids, 277
feedback technique, 331, 336
fermentation, 79, 84, 85, 87, 92, 98, 102, 103
fermentor, 52, 54, 56, 59, 61, 62, 68, 69, 84, 92, 93, 94, 95
fibrillar proteins, melting of, 431
filipin, 208, 212
flow calorimetry, 52, 56, 59, 262
flow-through vessels, 132, 263
5-fluorocytosine, 205
Fourier transform, 333, 334
frequency, analysis, 321, 323, 324, 328

frequency, bandwidth, 329, 337
　limiting, 329, 338
　response, 321, 337

G

gantrisin, 203
gas–liquid flow, 263
gas perfusion of calorimetric medium, 258
gentamycin, 202
Gibbs energy, 50, 51, 346
　additivity of, 379
　linkage, 385
globular proteins, melting of, 429
　stability of, 432, 433, 434
glucose, 7, 472
glucose oxidase, 50, 69
glyceraldehyde-3 phosphate dehydrogenase, 384, 391, 404
glycerol, heat capacity, 401
glycolysis, 134, 137
gradient-layer calorimeter, 58, 66
gramicidin A, 221
granulocytes, 269, 270
　adhesion, 270
growth and metabolism, 1

H

haemoglobin, 374, 392, 459
　binding of alkanes to, 369, 404
　binding of oxygen to, 391
　intersubunit interactions, 392
haemolytic anaemia, 141
Halobacterium halobium, 297
harmonic analysis, 331, 333, 334
hashish components, 230
heat capacity, 393, 400, 401
　of biopolymers, 415
　change on binding ligand, 347, 393–406
　of chymotrypsin, 401
　of glycerol, 394
　of hydrated biopolymers, 417
　measurements, 444, 445
　partial, 417, 447
　of propane, 394
　of tetra-*n*-butyl ammonium bromide, 394
　of thymine, 373, 395

heat of combustion, of cells, 7, 75, 78, 97, 98, 100, 101, 102
  of evaporation, 52
  of formation of bacterial cells, 6
heat-flow calorimeters, 131, 146, 248, 318
heat production, by organs and tissues, 145
  in-situ measurements, 147
  techniques, 146
heart, 154
HeLa cells, 120, 124, 130, 217, 272
hepatocytes, 130
hormone, 113
human serum albumen, 470
hydrated biopolymers, heat capacity of, 417
hydration of biopolymers, 422
  in tissue, 425
hydrogen bonds, 372, 373
hydrophobic interactions, 369–371
hyperthyreosis, 139
hypothyreosis, 139

I

ice calorimeter, 44
identification, 55, 64, 72, 82, 163
immobilized enzyme, 464
impulse response, 320, 333
indirect calorimetry, 57, 63
influenza virus, 193
inoculum history, 177
  volume, 175
insulin, 66, 458, 460
internal calorimetry, 147
ionophores, 215, 221, 225
isoperibolic calorimeters, 248

K

kanamycin, 202
kidney, 159
kinetics, 67
  enzyme, 127
  Michaelis–Menten, 23, 24, 25, 79, 83
Klebsiella aerogenes, 94, 183
Kluyveromyces fragilis, 82, 174, 175, 176, 177, 179

L

L929 mouse fibroblasts, 118, 122
LS cells, 122, 123, 124
lactic fermentation, 5

β-lactoalbumin, 369, 404
lactose, 472
lag phase, 47, 84
lecithins, 278, 281, 285
leucocytes, 132
ligand binding, heat capacity change on, 347, 393–406
limit cycle behaviour, 81, 105
liquid nitrogen stored inocula, 64, 209
liver, 159
log phase, 47
lucensomycin, 208, 209, 212
luciferase, 473
lymphocytes, 114, 115, 121, 122, 134, 269
β-lysin, 472
lysozyme, 472

M

mammalian membranes, 298
  tissue culture, 465
melting, of water in biopolymer, 417, 420
  of collagen, 426, 427, 428
  DNA, 443
  fibrillar proteins, 431
  intramolecular, 419, 425
  of membrane, 443
  polynucleotides, 435
  tRNA, 439, 440, 441, 442
  16S RNA, 443
  in tissue, 424, 425, 426
membrane, proteins, 305
  melting of water in, 443
metabolism, of growth, 77
  of maintenance, 46, 67, 68, 70, 72, 80, 84, 85, 90, 93, 94
metal dithiocarbamates, 239
metal-ion effects, 288
α-methylglucoside, 472
Mg^{2+}, 462
Michaelis–Menten kinetics, 23, 24, 25, 79, 83
microcalorimeters, 249
microsomes, 299
minocycline, 192, 203, 204, 211
mitochondria, 51, 299
mitotic division, 55
mixed cultures, 35
molasses, 54
monensin, 215

SUBJECT INDEX

morphine-like drugs, 229
multi-channel calorimeter, 65, 85
muscle, heart, 154
   smooth, 153
   striated, 148
mutagens, 465
Mycoderma yeasts, 49
Mycoplasmateles, 114, 188, 208, 216
myelin, 277, 298

N

NADH$_2$, 193
negative Pasteur effect, 51
Neisseria gonorrhoea, 192
   meningitidis, 180
nerve, 157
nigericin, 215
nonactin, 215, 221
novadral, 66
nystatin, 65, 208, 210, 212

O

oleandomycin, 205, 207
open system, 103
optimization technique, 331, 335
oscillation, 55, 74, 81, 90, 105
ouabain, 148, 157, 158
ovalbumin, 459
oxidative phosphorylation, 134
oxygen calorie equivalent (*oce*), 66
   consumption, 58
   electrode, 66
   tension, 116
oxytetracycline, 192, 201, 203, 204, 211

P

pancreatic β-cells, 130
partial heat capacity, 417, 447
Peltier heating, 148
penicillin, 201, 202, 206, 208, 470, 471
pentose shunt, 134, 137
perfusion calorimeter, 147, 156
   vessel, 260
pernicious anaemia, 140
petite yeast strains, 50
phagocytosis, 134, 139
pH, 170
   dependent thermodynamic
      quantities, 354–358
pharmacokinetics, 203
phase diagrams, 283
   separation, 285, 305

phase diagrams, shift, 322, 323, 329
   transition 275, 279, 281, 289, 290,
      300, 302, 305
phenothiazine derivatives, 234
phosphatidyl ethanolamine, 286
phosphofructokinase, 73
phosphoglycerides, 276
phospholipids, 275, 280, 281
phosphorylase B, 384
phosphotungstic acid, 463
Photobacterium leiognathi, 473
photoluminescence, 465, 473
pimaricin, 208, 210, 212
platelets, 256
ploidy, 72, 73
poisons, 66
polarography, 63
poly A, 214, 215
polyene antibiotics, 208, 231
polynucleotides, melting of, 435
   stability of, 438, 439
polypeptides, 293, 294
potassium cyanide, 50, 69, 71, 77
power compensated calorimeter, 336, 337
power–time curve, 317, 329, 330
precultures, 64
pressure, 74, 87
propane, heat capacity of, 394
protein, 293, 304, 305, 457, 463
proteolytic enzyme, 119
Proteus, 168, 172, 183, 188, 190
   morganii, 189
proton ionization, 388
Pseudomonas, 192, 202
   aeruginosa, 189, 190
pyrogallol, 50

R

radiometry, 147, 191
red blood cells, 298
reserve substances, 46
respiration, 51, 116
   repressors, 74
resting cells, 69, 70
retardation phase, 47
ribonuclease A, 344, 376–381
   binding of 3′-CMP to, 350, 352,
      357, 372, 388
rimifon, 202
RNA, 55, 121, 126

RNA, melting of 16S, 443
  melting of tRNA, 439, 440, 441, 442
RQ value, 63
rumen, 35, 36

S

Saccharomyces carlsbergensis, 81
  cerevisiae, 21, 22, 23, 43, 65, 166, 170, 173, 205, 208, 213
Salmonella, 38, 183, 201, 202
  typhi, 201, 220
saturated lecithins, 286
scanning calorimetry, 278
Schizosaccharomyces pombe, 90
second law of thermodynamics, 103
sedimentation, of cells, 60, 78, 79, 93, 266
septicaemia, 191
serum albumin, binding of I⁻ to, 364, 372
  binding of alkanes to, 369
sickle-cell disease, 141
simulation technique, 331, 335
sodium dodecylsulphate, 466
soil, 265
sorption, 117, 118
spent liquor, 54
sphingolipids, 276, 277
stability, of globular proteins, 432, 433, 434
  of polynucleotides, 438, 439
Staphylococcus albus, 188, 190
  aureus, 180, 191, 192, 206, 207, 213, 214
  aureus haemolyticus, 203, 204
state function, 331, 334, 335
static conditions, calorimetric, 257
static gain, 318, 326, 329, 330, 338
static properties, 318, 324, 328
stationary phase, 47
steady state, 45, 47, 48, 56, 66, 68, 72, 81, 103, 104, 105, 318, 320, 321
stirring of calorimetric medium, 257
Streptococcus, 169, 192, 202, 465, 470
  faecalis, 19, 20, 31, 32, 33, 37, 77, 188, 189, 190, 206, 208
  lactis, 21
streptomycin, 201, 205
structural specificity, 380
submersible calorimeter, 59

sulphadimidine, 205
surfactants, 460
survival rate, 74
synchronous cultures, 45, 55, 90, 95, 171
systematic errors, 253
Systemic Lupus Erythematosus, 139

T

Teflon vessel, 119
test experiments, 253
tetracycline, 180, 192, 203, 204, 205, 206, 208, 211, 470, 471
tetra-*n*-butyl ammonium bromide, heat capacity of, 394
thermal inertia, 333.
  power, 311, 312, 314, 317, 336, 337
thermochemistry of microbial growth, 2
thermodynamic, 102
  equilibrium, 104
  of irreversible processes, 48, 63
  selectivity, 380
thermoelements, 131
thermogram, 311, 317, 318, 320, 330, 331, 334, 339, 453
  computer analysis of, 182
  correction, 311, 330, 331, 333, 334, 335, 337, 338
thermometric titrimetry, 238
thermopile, 146, 248
thiomethyl galactoside, 472
thrombocytes, 135
thymine, heat capacity of, 373, 395
thyroid hormones, 139
time constant, 313–322, 324, 326, 327, 330, 330–333, 335, 336, 337, 339
  gap, 59
  response, 319, 320, 337, 338
tissue cells, 113, 272
  adhesion of, 272
  hydration of, 425
  melting of water in, 424, 425, 446
Torula yeasts, 50
tranquillizers, 225
transfer function, 324, 325, 326, 327, 333, 335, 338
transfer RNA, binding of Mg²⁺, to, 360, 372
  melting of water in, 439, 440–442
  structural transitions, 365, 399

SUBJECT INDEX

transfer RNA, "Y" base fluorescence, 365
transformed embryonic cells, 465
transient state, 319
transition, period, 47
  state, 105
transport processes, 302, 304
tricarboxylic acid cycle, 51, 52
Trichoderma vivide, 62
triton, 460
turbidostat, 56, 92

U

Uncouplers, 225
urinary infection, 188, 189, 191
urine, 468
  calorimetric investigation of, 187
  clinical specimen, 188

V

δ-valerolactam, 394
Van der Waals forces, 372–374
van't Hoff equation, 279
vessel design, 115
virus, 114, 192, 193
vitamin, 176

W

Warburg technique, 63

Y

Y_{ATP}, 57
Yeasts, 43, 267

Z

Zymomonas mobilis, 20, 21, 22, 23, 25, 28, 29, 30, 33, 34, 37, 38, 166